LIGHTNING

LIGHTNING

Volume 1

Physics of Lightning

Edited by

R. H. GOLDE

1977

ACADEMIC PRESS
London · New York · San Francisco

A Subsidiary of Harcourt Brace Jovanovich, Publishers

ACADEMIC PRESS INC. (LONDON) LTD
24–28 Oval Road,
London NW1

U.S. Edition published by
ACADEMIC PRESS INC.
111 Fifth Avenue,
New York, New York 10003

Library of Congress Catalog Card Number: 77–72088

ISBN: 0–12–287801–9

PRINTED IN GREAT BRITAIN BY
JOHN WRIGHT AND SONS LTD, AT THE STONEBRIDGE PRESS, BRISTOL.

Contributors to Volume 1

T. E. ALLIBONE *York Cottage, Lovel Road, Winkfield, Windsor, SL4 2ES, England*

R. B. ANDERSON *National Electrical Engineering Research Institute, P.O. Box 395, Pretoria, Republic of South Africa*

K. BERGER *Gstadstrasse 31, Zollikon CH 8702, Switzerland*

M. BROOK *Department of Physics, New Mexico Institute of Mining and Technology, Soccorro, New Mexico 87801, U.S.A.*

B. DIBNER *Burndy Library, Norwalk, Connecticut 06856, U.S.A.*

R. H. GOLDE *274 Salmon Street, London NW9 8XY, England*

R. D. HILL *P.O. Box 5484, Montecito, California 93108, U.S.A.*

J. LATHAM *Institute of Science and Technology, University of Manchester, P.O. Box 88, Manchester M60 1QD, England*

C. B. MOORE *New Mexico Institute of Mining and Technology, Soccorro, New Mexico 87801, U.S.A.*

T. OGAWA *Geophysical Institute, Kyoto University, Kyoto, Japan*

R. E. ORVILLE *State University of New York at Albany, Earth Science Building - ES214, 1400 Washington Avenue, Albany, New York 12222, U.S.A.*

E. T. PIERCE *U.S. Department of Commerce, National Severe Storms Laboratory, 1313 Halley Circle, Norman, Oklahoma 73069, U.S.A.*

S. A. PRENTICE *Department of Electrical Engineering, University of Queensland, Saint Lucia, Brisbane 4067, Australia*

H. PRINZ *Hochspannungsinstitut, Technische Universität München, Postfach 202420, Arcisstrasse 21, Munich, Germany*

S. SINGER *Athenex Research Associates, 381 South Meridith Avenue, Pasadena, California 91106, U.S.A.*

I. M. STROMBERG, *Institute of Science and Technology, University of Manchester, P.O. Box 88, Manchester M60 1QD, England*

B. VONNEGUT *State University of New York at Albany, 1400 Washington Avenue, Albany, New York 12222, U.S.A.*

Preface

"The power engineer has become aware of many investigations of lightning phenomena carried out independently by physicists and meteorologists. A survey of the relevant, very extensive, literature shows, however, that very little attempt has been made to correlate the data of various authors and of different methods of investigation or to translate the results of these investigations into terms applicable to the problem confronting the engineer", whose task it is, it was inferred, to evolve means of protection against the destructive effects of lightning strikes. Those sentences appeared in 1941 in a paper on "The Lightning Discharge" of which I was co-author. In the intervening 35 years, research workers have taken greater notice of the results obtained in other disciplines but the complexity of the many aspects of the lightning discharge makes it almost impossible for any individual to be familiar with all the accumulated results and to appreciate fully their implications.

Martin Uman's book, to which frequent reference is made, constitutes an excellent modern treatise on the physics of lightning. The chief justification for the present multi-author book is the need to supply a comprehensive survey of present knowledge of *all* major aspects of lightning and protection against its effects. In addition, I am acutely aware of the loss engendered by the death of several men who may be said to represent the first generation of lightning scientists with access to modern recording techniques: names like C. T. R. Wilson, Sir George Simpson, Sir Basil Schonland, K. B. McEachron, I. S. Stekolnikov, C. L. Fortescue, D. J. Malan spring to mind. A few of those who overlap with that generation are, however, fortunately still with us. I confidently hope that their experience will benefit the younger generation of researchers who are branching out into increasingly specialized facets of electrical discharges.

From its original concept onwards, the book was intended to present the results achieved by physicists and engineers as well as by meteorologists, historians and other research workers. In view of this wide coverage it could not be planned as a monograph. Instead, the highly distinguished contributors were requested to start with a brief survey of early work and to present a balanced critical review of present knowledge with clear indications of their own views, based, in many cases, on a lifetime of experience and thought.

Few of the chapters are entirely self-contained and the chief problem facing me as editor was to prevent duplication. The principle adopted was to allow discussion of any particular aspect to remain in that chapter to which it belonged most directly, while replacing similar statements in other chapters by cross-references. Occasional conflicting conclusions and opinions of different contributors have been retained, again with appropriate cross-references. In this context I may be permitted the customary explanation that, as editor, I am not responsible for views expressed by other contributors.

The literature on the subjects covered is so enormous that contributors were asked to limit references to any particular author to recent papers which provide a source to earlier publications. Even so, the number of references amount to several thousand.

The book is intended for serious students of lightning, be they beginners or experts in their own fields. It should prove of interest to physicists working on electrical discharge phenomena in nature and laboratory or in atmospheric electricity; to engineers concerned with the protection of electrical supply systems, telecommunication, structures, aircraft and mining; to meteorologists; to architects and to members of the medical profession.

To increase the usefulness of the book and to facilitate its reading, the bewildering jargon of terms to be found in some publications on electrical discharge phenomena in general and on the lightning discharge in particular has been unified as far as possible and the terms adopted have been clearly defined.

A book as ambitious as the present one is unlikely to be duplicated in the foreseeable future. Where then does lightning research stand? Practical problems concerning the protection against lightning can be claimed to have been largely solved. Electricity supply need no longer be interrupted every time lightning strikes an overhead line; interference with telecommunication can be reduced to a tolerable level; ordinary buildings as well as highly vulnerable installations both above and below ground can be satisfactorily protected; most of the difficulties formerly facing the aircraft manufacturer and operator have been overcome; warning systems to predict the possibility of a lightning strike have been improved and, while destructive forest fires cannot yet be prevented, the understanding of the factors involved has been notably improved.

The list of practical achievements is impressive. Most of the underlying information is due to the work by physicists, yet the chapters contributed by physicists and meteorologists clearly highlight the many questions concerning lightning, the long electric spark and the thundercloud which remain to be solved. The older research workers are happy to leave the solution of these problems in the capable hands of the younger generation.

There remain some highly controversial subjects in which serious research is still in its infancy: ball lightning, the seeding of clouds and the selective discharge of a cloud charge, say by laser beams, are typical examples. The book would have been incomplete had considerations of these topics been omitted. I regret that I was unable to find an author prepared to describe the effect of lightning on the chemical composition of the atmosphere, but reference has at least been made to the intriguing question of prebiotic synthesis.

The selection of contributors was my prerogative as editor but I was glad of an opportunity to discuss this with my old friend Dr E. T. Pierce whose assistance is greatly appreciated. Sincere thanks are due to all contributors for their compliance with my suggestions and, in many instances, for their forbearance where major changes had to be effected in the interest of the work as a whole. Permission to reproduce illustrations from published work is greatfully acknowledged in individual figure captions. In particular I am greatly indebted to Messrs Edward Arnold for their permission to reproduce Figs 2, 3(iii), 3(iv) and 4 in Chapter 16 and Figs 4, 8, 9 and 10 in Chapter 18. All these appeared in my book "Lightning Protection", published in 1973 and in the American edition, published by the Chemical Publishing Co. Inc., New York, in 1975.

My thanks are due to my son Michael for the many helpful suggestions in presentation, to my daughter Jean for assistance in proof reading and particularly to my wife who not only had to type and retype my own chapters but who had to expedite progress by typing inserts in other contributions, and who had to cope with a correspondence running to many hundred letters.

More by accident than by intent, the manuscript of this book was handed to the publisher within a few days of the bi-centenary of the founding of the American nation. One of the signatories of the Declaration of Independence was Benjamin Franklin. Franklin was an eminently practical man and, in a letter written on 20 Sept. 1761, he asks: "What signifies philosophy that does not apply to some use?" (I. B. Cohen (1941) "Benjamin Franklin's Experiments", Harvard University Press Cambridge, Mass, p. 386) and in a paper written in 1749 (*Ibid.*, p.219) he states: "Nor is it of much importance to us, to know the manner in which nature executes her laws; it is enough if we know the laws themselves." Franklin was the first to establish some of the laws governing the lightning discharge. It is my sincere hope that this book will show that knowledge on how nature executes her laws is also of some importance.

May 1977 R. H. G.

Contents

4. Point–discharge

J. LATHAM and I. M. STROMBERG

5. The Earth Flash

K. BERGER

6. The Cloud Discharge
M. BROOK and T. OGAWA

11. Thunder
R. D. HILL

12. Ball Lightning
S. SINGER

Contents of Volume 2

1. Lightning in History

H. PRINZ

Technical University, Munich, Germany

1. Introduction

Early in the history of the human race, man's realization of the order embodied in the events around him and in the forces that gave him life came from his observation of the skies and the luminous objects that had wandered in the firmament for thousands of years. Priests searched the heavens from wide open plains, from the terraces of temples and from towers, thinking about the mysteries of the universe, the essence of all being and the origins of life and death. Exorcism and offerings, whispers and shouts, rites and praises were thought to arouse the favour of the celestial deities and allow the priests to forecast the fates of the mortals.

This observation of the sky and its forces can be traced back to the oldest civilizations in the history of man as a key expression of man's will and is expressed most strongly in ritual exercises, mythological concepts and superstitious divinations.

2. Mythological Symbols and Rites

In viewing the interplay of the forces of heaven the human eye was bound to detect lightning as the most magnificent form of the fiery lightning path. Accordingly, the symbols based on the concept of fire are among the oldest pictorial representations of lightning. They emerge from the archaic high cultures of Mesopotamia, a country with more thunderstorms than Egypt, as a forked lightning emblem or a bundle of thunderbolts.

Such symbolism can be found among the earliest representations of lightning, i.e. on a roll seal from ancient Akkadian times which shows a

Fig. 1. Roll seal of Akkadian period (around 2200 B.C.) (from Porado, 1949).

goddess standing on the shoulders of a winged gryphon. Behind her, rolling on a four-wheeled car, is a weather god generating thunder with his whip. A libation is being offered to the two deities (Fig. 1).

Fig. 2. Count of Thunder Lei Kung (from Doré, 1915).

In a relief created by a Hittite artist, which was found around 900 B.C. in the outer eastern gate of the castle of Samal in northern Syria, the place later called Senjirli, the weather god Teshub holds a three-pronged thunderbolt (see Prinz *et al.*, 1965, Fig. 2).

Around 700 B.C. Greek art began to use the lightning symbols of the Middle East, attributing them to Zeus, the lightning throwing god, in a version adapted to Greek mythology. An Attic drinking vessel made around 480 B.C. depicts the heroic battle of the Olympians against the wild race of the Giants. Zeus has not yet joined the battle force, because he is just about to board his cart, swinging the flaming thunderbolts in his urge to decide the fight.

In the Chinese mythology of the opposites Yin and Yang, lightning is represented by the colourful goddess Tien Mu. She firmly holds two mirrors to direct the flashes of lightning (see Prinz *et al.*, 1965, Fig. 6). Tien Mu is among the five pre-eminent dignitaries of the "Ministry of Thunderstorms" existing in Chinese imagination, the chairman of which is Lei Tsu, the God of Thunder. His aide is Lei Kung, the drum-beating Count of Thunder (Fig. 2).

In many ethnological representations from prehistoric times lightning is depicted just as impressively as a stone falling from heaven or a stone axe hurled from the skies. Its destructive force, which becomes manifest in nature by split trees, broken rock and dead animals, is compared with the effects of a Stone Age tool. This rather imaginative concept is reflected in many languages by such words as Donnerkeil, Donnerstein, thunderstone, pierre de tonnerre, piedra de rayo, or vajra in Sanskrit.

In the Buddhist mysteries of later centuries the thunderbolt is revealed in the vajrajana, the concept of the great diamond vehicle, as the symbol of the power and omniscience of the deity. In this mysterious world, Vajrasattva with his multi-pronged vajra is a Buddha of supreme intelligence (see Prinz *et al.*, 1965, Fig. 9).

Magic powers have always been attributed to the thunderstone as Thomas Nicols (1675) in his booklet on precious stones points out:

> "Es wird gesagt, daß dieser Stein diejenigen so ihn tragen ihre Häuser für den Donner bewahren und Ruhe und Schlaff bringe und daß er helfe die Feinde überwinden und im Krieg den Sieg verleihe."

This mysticism has stayed alive to this day in many popular superstitions. Following an old custom, French peasants carry a "pierre de tonnerre" in their pockets to ward off lightning during thunderstorms and recite the little verse: "Pierre, pierre, garde moi de la tonnerre."

Similar beliefs are found in the more recent mythological concepts of some pastoral tribes in Asia. East Yakuts finding round or oblong rocks in the fields after a storm regard them as thunder axes fallen from heaven—ätin sugätä—and take them home to protect themselves against lightning. Powder scraped off these rocks is dissolved in water and used as a medicine. The Buryats living in the area of Lake Baikal are convinced that the god Esege Malan exercises his powers of thunder and lightning by throwing round stones from the skies down upon the earth.

In the minds of the African Basuto it is not a thunderstone, but a magic thunderbird called Umpundulo which flashes down to earth from a cloud in a storm to work his flapping wings.

Storm ceremonies often become manifest in rites which may seem strange to us. The Semang pygmies, a Kensiu tribe on the Malayan peninsula, believe the thundergod Pedn creates the thunder and throws lightning whenever he bears a grudge against man. In order to pacify him, they offer blood as a sacrifice. Usually one of the women of the camp concerned begins the sacrifice. She first beats her calf with the handle of a spoon made out of the shell of a coconut, then applies the tip of a knife and hits it with the handle of the spoon. The intense flow of blood gushing out is scraped into a bamboo vessel by means of the knife, mixed with water, and some of the mixture is poured into the spoon. The contents of the spoon are thrown into the sky. The ceremony is repeated several times until the thunder stops. The sacrificed blood is accepted by Pedn who boils it and puts it into the umbels of fruit trees to create new fruit.

3. Learned Opinions and Curiosities of Antiquity

The opinions of the Ancients about the origin and the effects of lightning seem strange to us. Aristotle, one of the most eminent philosophers of the time, held the view that clouds were made of telluric fumes which contracted in the cold and were ignited in a tremendous outburst.

In a work about meteorology, "Naturales quaestiones", the philosopher Lucius Annaeus Seneca around 50 B.C. makes a distinction between the wonderfully fast penetrating lightning, "genus quod terrebrat", which strikes the interior without damaging the outside, the smashing lightning, "genus quod dissipat", which is accompanied by violent storms and thunder and finally the igniting or blackening lightning, "genus quod urit". The wonderfully fast penetrating lightning flashes are repeatedly reported to have consumed the wine in earthen casks without damaging them. In the instructional poem "De Rerum Natura" by the Roman poet Titus Lucretius Carus (1534) these incidents are described as follows:

> "Curat item, ut vasis integris Vina repente
> Diffugiant, quia nimirum facile omnia circum
> Conlaxat, rareque facit lateramina vasis,
> Adueniens calor eius ut insinuatur in ipsum:
> et Mobiliter soluens differt primordia Vini."

Later, these miracles were rather believed to be the pranks of some slaves who were as cunning as they were addicted to drinking.

According to a widely held belief, sleeping persons were protected from the stroke of lightning. Plutarch in his "Discourses at Table" explains this view by stating that the human body was loose and without any spirit of life while asleep and therefore did not offer any resistance to lightning, letting it pass through easily and quickly. Also laurel was said to protect against the stroke of lightning, which is why the Roman emperor Tiberius was always careful to wear a wreath of it when a storm was impending and the emperor Augustus in these situations used to don a sealskin.

Great importance was attached to the interpretation of lightning at that time. The science of lightning of the Etruscan era, reported in the "Libri fulgurales", subdivided the sky into 16 regions. Distinctions were made, for instance, between the colour of lightning and the direction in which it flashed. Lightning coming from the east was always regarded as a good omen for any action about to be taken. In times of important decisions the observation of lightning was of great importance. In ancient Rome this was the duty of the augurs who learned to turn it into a highly efficient political instrument. The Roman senator Calpurnius Bibulus is said to have profoundly disturbed Caesar's administration of the government through his celestial observations on comitial days (days when the Roman Assembly met).

It is perhaps due to the strong influence of early ritual and mythological ideas that many ancient opinions were accepted by the scholars in the monasteries without any modification. Hence, it is hardly surprising to find artistic interpretations of the Old Testament in Scheuchzer's "Physica Sacra" (1731), in which the effects of a fiery stroke of lightning appear as something man cannot avoid, almost as a sort of fate (Fig. 3). Two years later, Zedler (1733) in his "Universal Lexikon" writes about lightning:

> "So viel Mühe als sich die Naturkündiger in Untersuchung dieser Sache gegeben haben, so wenig hat man doch von derselben Gewißheit."

This and the desire to serve the truth explain the learned urge to find out about the secrets of man's true origin and, at the same time, imagine and invent those means of protection which could guard him against the dangers of a flash of lightning.

4. Sparks and Lightning

Proving that the stroke of lightning and the stroke of thunder were identical with sparking electricity became possible at a time in which the most sophisticated experiments were conducted with glass tubes, glass spheres and glass

Fig. 3. "Igne peccantes igne necantur." Leviticus x, 2 (from Scheuchzer, 1731).

cylinders and the transmission of electric force through communication lines made of hemp and the intensity of its presence was discovered by the sparking and lighting of the matter causing it. No less important was the knowledge of the existence of resinous electricity with its action—an action different from that of the vitreous electricity previously known.

One of the first attempts in this regard is linked with the name of Francis Hauksbee (1706/7), one-time curator of the Royal Society in London, who in 1707 tried to investigate in more detail the luminous effects of rubbed glass cylinders. He used a device which consisted of an open outer cylinder with a closed inner cylinder, both of which could be brought into rapid rotation by a drive wheel actuated by two rollers. He made a strange observation: when he laid his bare dry hands upon the surface of the outer cylinder, luminous phenomena resembling lightning occurred in the inner cylinder which had previously been evacuated. He writes:

> "... but approaching my hand near the surface of the outward glass to produce flashes of light like lightning in the inward one."

One year later his compatriot William Wall (1708) conjectured from his experiments with rubbed amber that the use of a larger piece of that material would necessarily give rise to stronger cracking and firing which might be compared with thunder and lightning, and he remarked:

> "Now I make no question, but upon using a longer and larger piece of amber, both the cracklings and light would be much greater ..., and it seems, in some degree, to represent thunder and lightning."

Finally Stephen Gray (1735/6) of London concluded from his important experiments with electricity that the electric fire must be of the same nature as that of the thundering flash of lightning. In his notes he says:

> "... and consequently to increase the force of this electrick fire, which, by several of these experiments seems to be of the same nature with that of thunder and lightning."

These were unmistakably the first beginnings of some tentative guesses in this field of science.

Eleven years later, a learned treatise was published in Leipzig by Johann Heinrich Winkler (1746), a professor of classical philology working at the university of that city, who investigated the then much admired effects of electricity intensified by means of Leyden jars. In the tenth Main Piece of his treatise (Prinz, 1973, Fig. 3) he deals with the question of whether the stroke and the sparks of this intensified electricity might be regarded as a kind of thunder and lightning. The treatise, which comprised 27 pages, is based on the assumption that the electric sparks to be observed in intensified electricity would easily give rise to such questions because hearing and seeing thunder and lightning indicated the same phenomena that could be observed from electric sparks. Moreover, he said, lightning could travel through the air just as quickly as the parts of an electric spark were able to turn into their respective bodies, and both appeared to be equally bright and pure. This made it rather understandable, argued Winkler, that a thunderclap was stronger only because lightning contained a larger number of constituents producing the sound than did an electric spark.

In the further discussion of his treatise the professor mentions that the flashes of lightning, as those of an electric spark, travelled through the air on a serpentine course and that lightning was sometimes followed by many additional small strokes, as in an electric cascade. In addition, some thunderclaps penetrated soft and loose bodies without injuring them, but they damaged the solid bodies enclosed by the former. Thus, the flash could melt metals enclosed in leather without burning the latter, in the same way as this was done by electric sparks. In this way, thunder and lightning could be regarded as the effects of electric matter in the atmosphere.

In this analogy Winkler finally arrives at the following surprising summary:

> "§ 155. Es scheint demnach, daß die electrischen Funken, welche durch Kunst erwecket werden, der Materie, und dem Wesen, und der Erzeugung nach, mit den Blitzen und Donnerstrahlen von einerley Art sind, und ihr Unterschied nur in der Stärcke und Schwäche ihrer Wirkungen bestehe."

What could be added to this theory other than the proposal to prove by way of an experiment this agreement between the two phenomena?

5. Lightning Experiments

Following a suggestion advanced by Benjamin Franklin (see Chapter 2) the experimental proof of the identity between the electric spark and the natural lightning discharge was for the first time brilliantly provided by Thomas François Dalibard (1773). In his mémoire presented to the Académie Royale des Sciences in Paris he wrote as follows:

> "En suivant la route que M. Franklin nous a tracée, j'ai obtenu une satisfaction complète. Voici les préparatifs, le procédé et le succès."

and the report concluded with the words:

> "L'idée qu'en a eu M. Franklin cesse d'être une conjecture, la voilà devenue une réalité!"

Louis Guilleaume Le Monnier (1752), a botanist in the Royal Gardens of Louis XVI and later his physician, continued the same experiments in a garden of St Germain in June and July of the same year. He designed a set-up consisting of a wooden pole 9 m high with a glass tube on top of it as a substitute for the usual pitchcake, and on top of that a metal tube 1·8 m long to collect the electricity. An iron wire insulated by a silk cord led from the metal tube to the pavilion in the garden. When a storm arose on 7 June 1752, everyone in the garden felt the sharp pangs of the sparks—initially to their great amusement but gradually with such violence that the ladies attending the ceremony implored the scientist to stop the experiment.

Irrespective of the electricity from the thunderstorm, which made itself felt more and more intensely, Le Monnier ordered for his experiment, which had ended on 26 June, that poles resting on wine bottles be laid beside the wooden pole so as to be successively connected with the iron wire. In this way the sparking effects could be greatly intensified in the fashion of the Leyden experiment.

Le Monnier's courage is demonstrated by the fact that once during a storm he himself stood on a pitchcake and held a wooden pole more than 5 m high which was covered with iron-wire windings. As a consequence he became so electrified that long, crackling sparks could be drawn from his hands and his face.

Further development was characterized more and more by the desire to observe the electricity of thunderstorms in cabinets specially designed and equipped for this purpose.

Some of the first proposals came from the Abbé Nollet, a noted scientist of his time and former professor of physics at the Collège de Navarre in Paris. In the course of 1753 he installed a "cabinet de campagne" in his villa in such a way that an insulated collecting rod could be connected with the cabinet proper by means of a linkage; in this way it was possible to observe the electricity tapped comfortably and at any time (Fig. 4). A small chime indicated what was happening, even during the night

> "sans être obligé de sortir de son lit ni d'ouvrir les yeux!"

A few years later, Johann Friedrich Hartmann (1764), the royal electoral registrar at Hanover, described his storm house (Prinz, 1973, Fig. 9) the erection of which required a place far away from the city and from all other buildings. It consisted of a light structure of wooden boards which offered a good view from the outside, and a metal pole of about 10 m penetrating through the roof which, for purposes of insulation, had been set on asphalt. In order to collect as much electricity as possible, Hartmann made sure that the pole was equipped with a star consisting of many points. He also believed

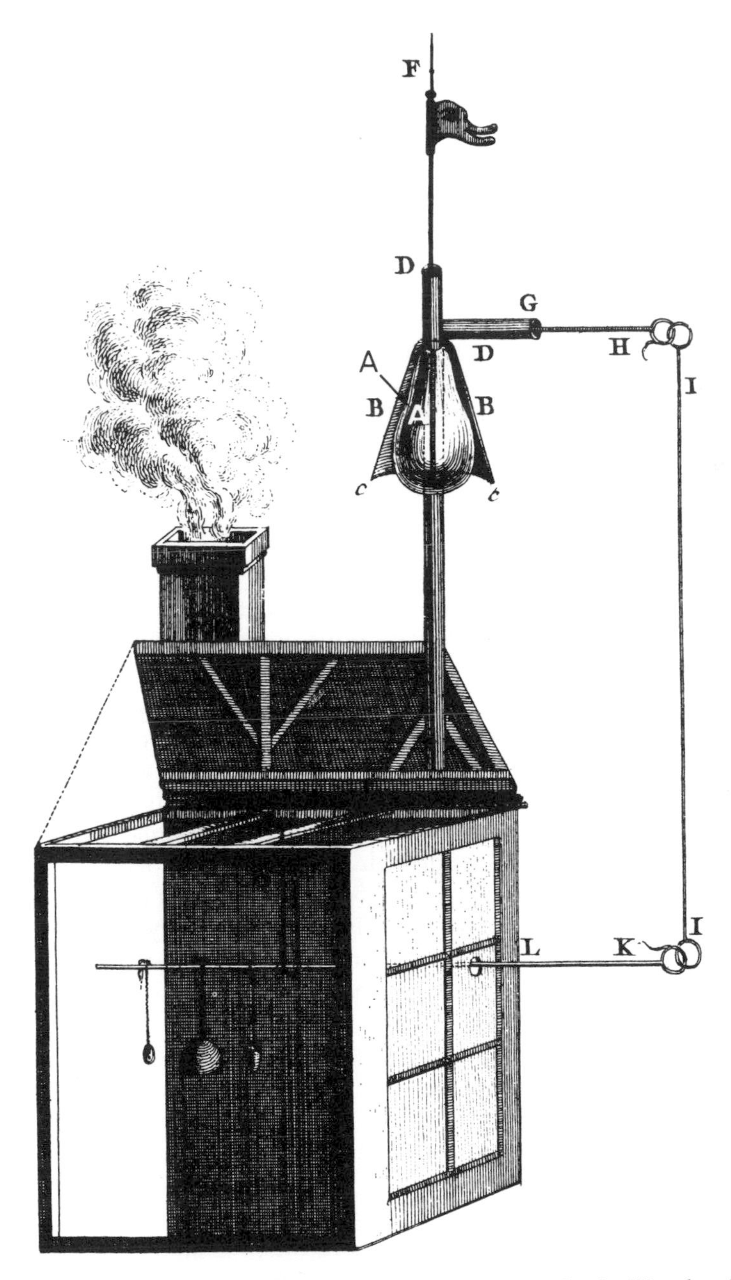

Fig. 4. Nollet's "cabinet de campagne" (from Cotte, 1774). A. Wood pole with glass vessel. B. Rain shield. F. Iron tip.

that the rain water should be collected in a separate vessel insulated with asphalt in order not to lose any electricity.

6. Kites, Balloons, Mortars and Rockets

Franklin's famous kite experiment is described in Chapter 2.5. It attracted enormous interest and, in 1753, was repeated by de Romas (1776), a lawyer working at the Regional Court in the Gascogne. He found, however, that the cord of the kite was a bad conductor for the electricity captured. Consequently, he used cords which were wrapped in violin wire over their entire length of 240 m, and during a thunderstorm on 7 June 1753 he succeeded in drawing sparks up to 20 cm long from the cord of the kite. In subsequent experiments, for which he used longer cords, he was able to observe even more electricity in the form of luminous sparks which attained surprising lengths, up to 3 m.

Eventually, because of the trouble that had sometimes been caused by electric kites, some other devices were invented to tap the electricity of clouds in thunderstorms and in this way to prove its existence. Air balloons, or air balls as they were then called, proved particularly useful in this respect. They had aroused much interest after the first successful flights of the Montgolfier brothers in aerostatic machines. One of the papers published in those years (Bertholon, 1787) said that if such an air ball was used to conduct electricity from thunderstorms then a metal thread had to be attached to the air ball, the other end of which was to be retained by a silk thread to prevent the electric fluid from dissipating right into the ground. In addition, it was advisable to equip the air ball with one or more spikes of iron and to connect these with the metal thread, the way this was customarily done with kites, in order to allow the thread to accumulate as much electricity as possible. The strength of the electricity was measured by stimulating the muscles of animals—for instance by connecting a horse to the metal thread of the air ball and at the same time making sure that its hind legs had good contact with the ground. A title engraving (Fig. 5) published at that time left no doubt about the reaction of the horse to this most cruel procedure. Sheep could also be used for this purpose, but they apparently reacted less violently to the electricity they received.

Mortars and rockets were no less useful in transmitting electricity even from more remote cloud regions. The eminent Giambatista Beccaria, former professor of experimental physics and meteorology at the University of Torino and in many ways connected with research into the electricity generated by thunderstorms, did experiments with rockets as early as in

Fig. 5. Air balls, mortars and rockets (from Landriani, 1784).

October 1753, and described them in a very convincing report (Beccaria, 1758):

> "I razzi . . . mi parvero accomodatissimi a conseguire il mio intento. Pensai d'infiggere sul capo di essi un filo di ferro acuto, che potesse trarre o spandere il fuoco elettrico da lontano."

Three out of the six rockets fired on October 8 helped to prove the existence of electricity; on October 10 it was only one out of three. Although it was hard to believe, Beccaria's idea of using rockets became important later on in the concept of triggering lightning flashes (see Chapters 5.4.4(c) and 21.3).

7. Protection against Lightning

In his assessment of the lightning rod in "Poor Richard's Almanac" of 1753 (see Chapter 2) Franklin had spoken of the protection of "Habitations and other Buildings". Although he devoted considerable thought to the action of the lightning rod, one question to which he does not appear to have paid any attention is the area which such a rod was likely to protect.

The first person to have raised this question which continues to exercise the minds of research workers (see Chapter 17.3) was Barbier de Tinan (1779), a member of the Dijon Academy, who had translated a book about the use of metal rods by Giuseppe Toaldo (1774), former professor of astronomy and meteorology at the University of Padua. He believed:

> "On peut aisément juger combien il est impossible de fixer cette distance. Elle dépend d'une infinité de circonstances variables",

and argued that the size of the clouds and the electricity stored in them, their distance and, finally, the direction in which they moved had to be included in the numerous variables.

The paratonnerre was regarded very favourably also in many other respects, and it even caused the Parisian *haute couture* to create a "chapeau-paratonnerre des dames" (Prinz *et al.*, 1965, Fig. 21):

> "pour défendre du feu du ciel les précieuses têtes des jolies femmes".

There even was an invention called "parapluie paratonnerre" of which the German poet Jean Paul (1809) has his military chaplain Schmelze say satirically:

> "Mit diesem Paradonner in der Hand will ich mich wochenlang ohne die geringste Gefahr unter dem blauen Himmel herumtreiben".

In 1822, Davy described a walking stick paratonnerre:

> "Man richtet einen Spazierstock so vor, daß an den beiden Enden desselben ein Draht sich herausziehen läßt, dessen eines Ende man in die Erde steckt und das andere 8 bis 9 Fuß über die Erde emporragen läßt. Nachdem der Stock gehörig befestigt wurde, legt man sich in Entfernung von ein paar Klaftern von demselben auf die Erde und ist so gegen den Blitz geschützt."

There is one other curiosity that should be mentioned. Seventy years after Franklin's publication, Lapostolle (1820), a pharmacist at Amiens, had an idea for a lightning and hail diverter made of straw cords attached to wooden poles:

> "Rien d'aussi évident en effet, que la propriété que possède la paille de conduire la foudre partout où l'on voudra."

He felt, however, that his invention undoubtedly would meet with much opposition before it was properly recognized by the scientists. In this assumption he proved to be correct.

Several booklets were published in those years which explained what to do when thunderstorms approached. Inside rooms they recommended you to touch nothing made of metal, such as stoves, door locks, golden wallpaper and, above all, to put away keys, watches and money. Extensive tobacco smoking was to be stopped; people were advised to move into the centre of the room and, if the house was not equipped with lightning rods, to enter a bed suspended from silk strings.

Special care should be taken when in open fields, as the secretary to the Duke, Ludwig Christian Lichtenberg (1774) of Gotha, a brother of the famous professor of the University of Göttingen, remarked in his booklet. In particular he advised against such major objects as trees, horses, loaded carts and the like and against stepping under a single tree because it could safely be said that a man approaching the tree might direct the lightning into that tree. Also elevated places should be avoided as much as possible because here man was likely to attract the matter of lightning just like a metal rod. Furthermore, it was recommended to lie flat on the ground if the thundercloud was rather close in order to avoid too frequent approaches of its matter. The booklet also recommended avoiding places in which there was a violent draught because lightning follows such currents of air. Finally, the booklet cautioned against the general fear of thunderstorms which, it said, is so pronounced that it renders its victims incapable of taking any decision so that there is no chance of even finding the nearest means of protection.

Forty years later, the fifth edition of a booklet by the beneficiary Joseph Kraus (1814) was published, in which the general public was instructed, in a popular way, in the possibilities of protecting one's life against lightning strikes. The safest way of protecting against a flash of lightning inside a house, it said, is by erecting a lightning rod, for this should allow the matter of a thundercloud to flow down the conductor into the ground without even touching a beam of the house (Fig. 6(a)). Those who are in an open field during a storm, it continued, and would like to stay alive, should not seek shelter under a tree because there lightning might strike them sooner than anywhere else (Fig. 6(b)). Likewise, one should never seek shelter from the rain in a sheaf of corn, because this involves the same danger. The reader is reminded that it is better to get soaked by rain than be killed by a stroke of lightning, for jackets and shirts will dry, but a man killed by lightning will never come alive again. For this reason, it concluded, it is better to set up two sheaves some thirty steps from each other and sit down halfway between them, because lightning coming from a thundercloud over the field would then most probably strike one of the two sheaves (Fig. 6(c)). Also riders on horseback were warned, especially if they ride fast, that they are in one of the most dangerous situations during a thunderstorm, which they can escape only by dismounting and slowly walking beside the horse until the clouds have passed (Fig. 6(d)).

8. Simulating Lightning

It is remarkable that the first tentative attempts at simulating lightning phenomena can be traced back to the time when lightning was recognized as an electrical phenomenon. Thus, the registrar Hartmann in a treatise published in 1759 described an experiment capable of demonstrating impressively the zigzag shape of a lightning flash (Fig. 7) as follows:

> "Und zwar besteht derselbe aus 40 dreipfündigen eisernen Kanonenkugeln, welche auf eben so viele gläserne Gefässe, woraus man Brantewein und andere starke Getränke zu trinken pflegt, ruhen . . . und zwar solchergestalt, daß sie nach einer Reihe von 5 Kugeln allemal einen gewissen Winkel verstelleten, wodurch ich also die Gestalt eines Zick Zacks erhielt."

When this ball machine was connected with a Leyden jar, the sparks simultaneously breaking forth from between the balls produced a shape which was completely identical with the shape of lightning. However, Hartmann added, the experiment should be conducted in weather favourable to electricity in order to be successful.

In 1773 the Edinburgh physician James Lind described his lightning house (Fig. 8) which, built in the discharge circuit of a Leyden jar, clearly demonstrated the explosive effects of a stroke of lightning when the metal connection

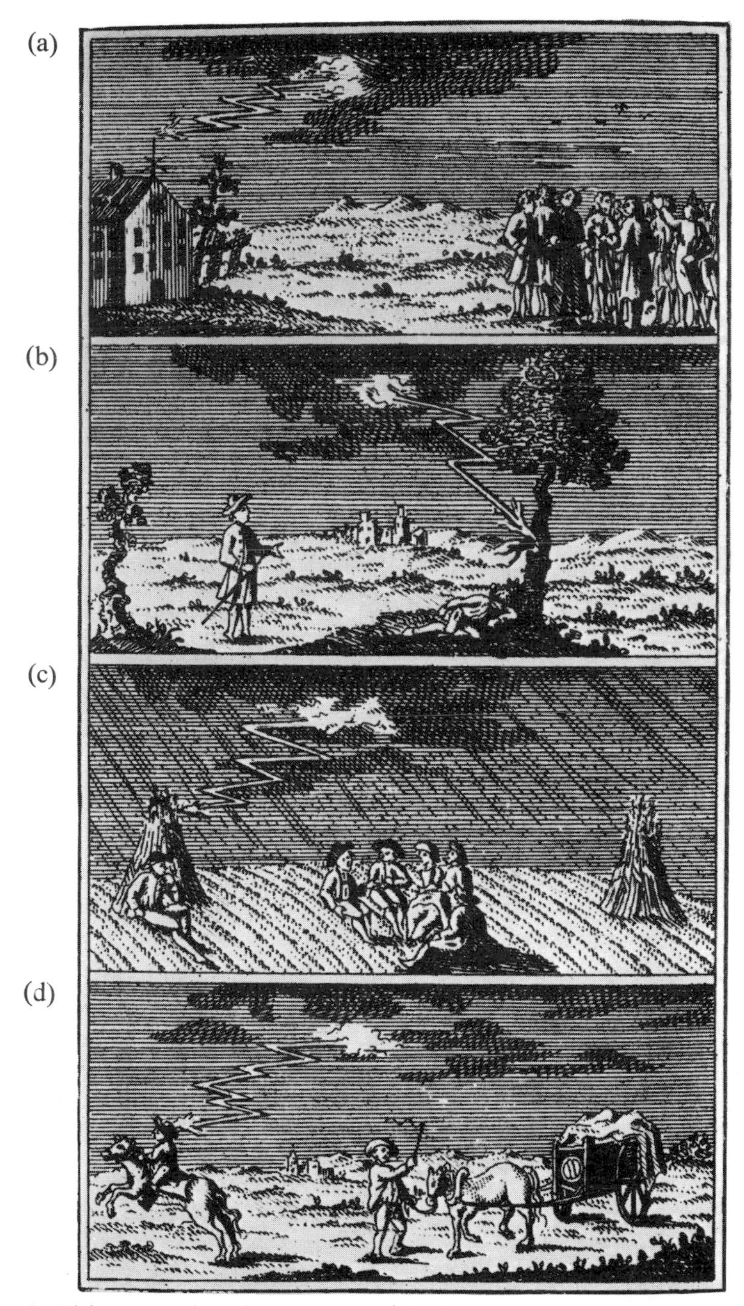

Fig. 6. Title engraving from Kraus (1814). "Catechism of Thunderstorms."

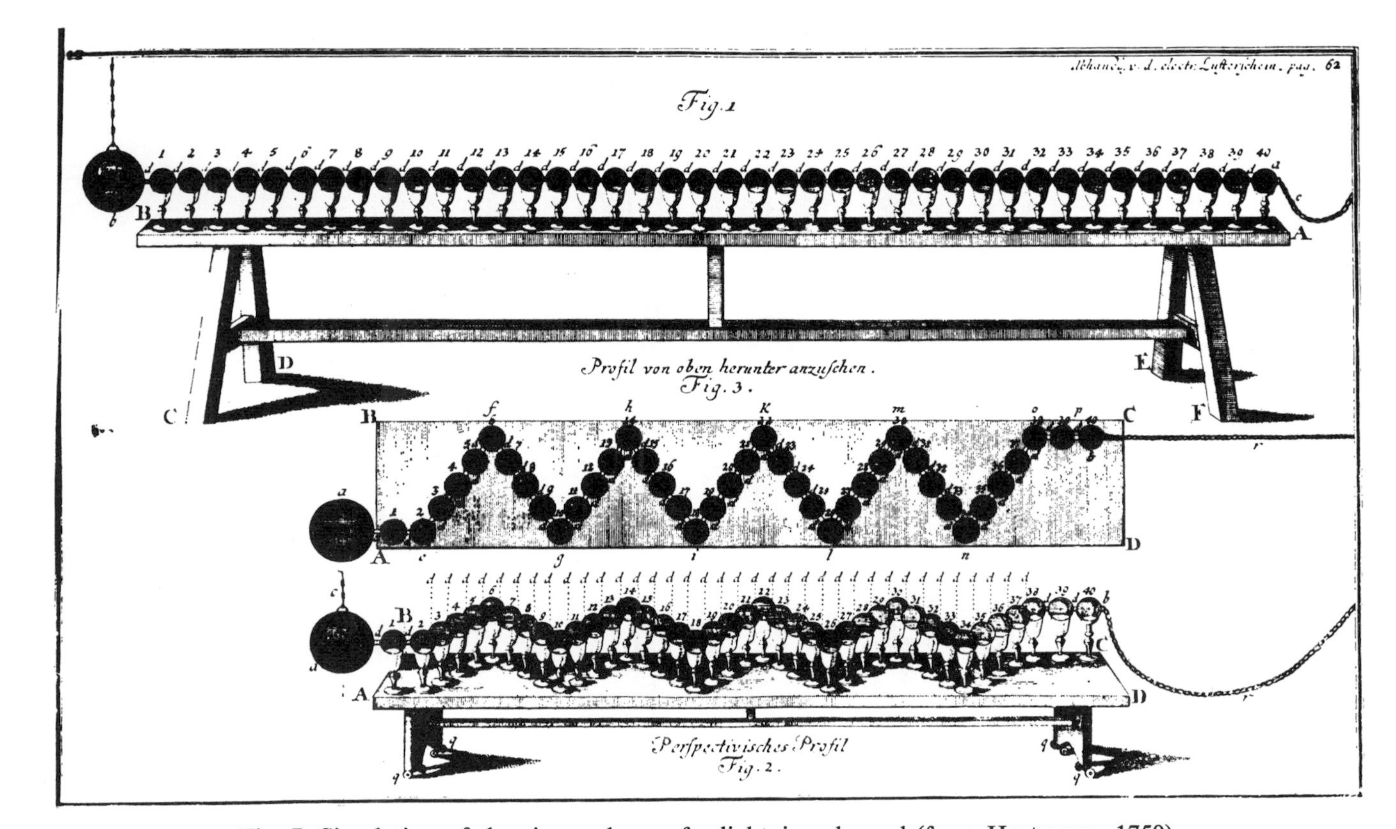

Fig. 7. Simulation of the zigzag shape of a lightning channel (from Hartmann, 1759).

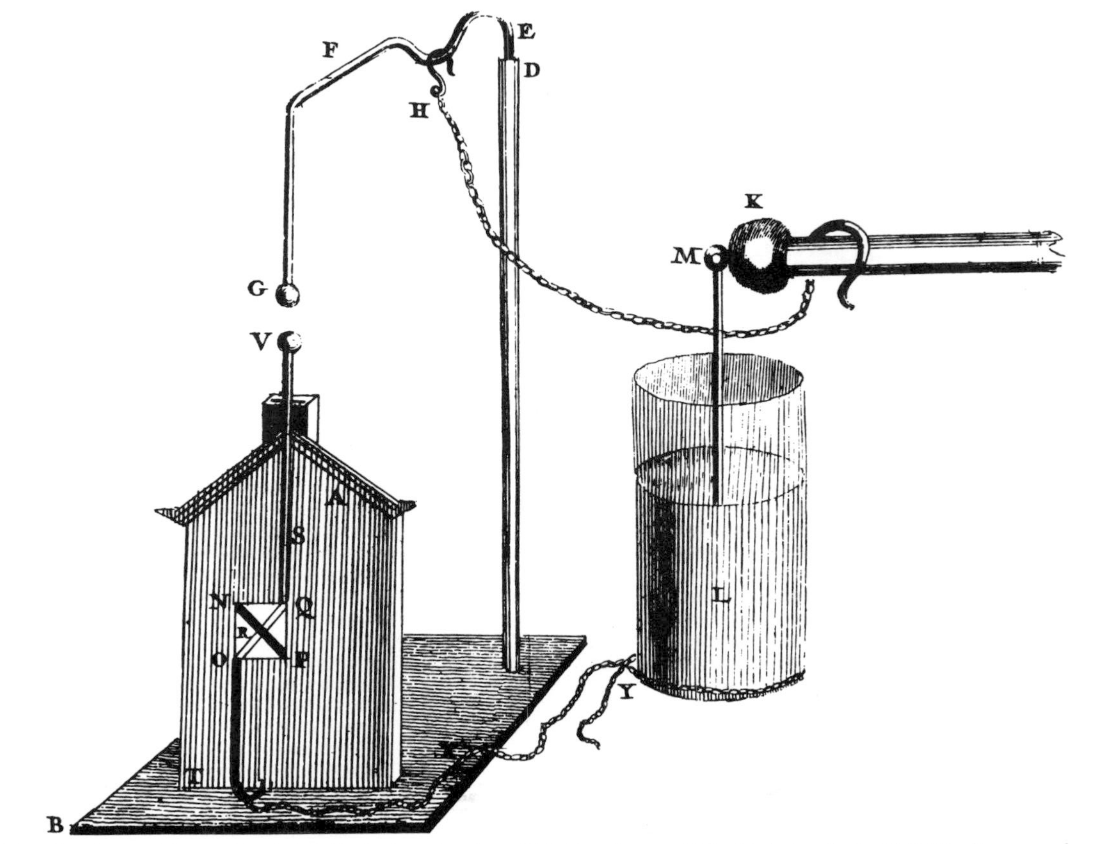

Fig. 8. Lightning house (from Lind, 1773). L. Leyden jar. GV. Spark gap. NP. Metal switch. NOPQ. Square piece of wood.

is replaced by a square piece of wood. If the connection is not interrupted, the discharge, via the spark gap, can take place without damaging the house:

> "Ce qui prouve manifestement l'utilité des conducteurs métalliques, pour préserver les maisons d'être endommagées par le tonnerre."

9. Epilogue

Inseparably linked to the history of electricity, the old desire of many scientists and sages to learn more about the electric fluid contained in the clouds, with its outward manifestations such as St Elmo's fire, sheet lightning and the fiery flash of lightning, has led to a better understanding of the cause of such phenomena and helped to replace the fatalism of former mythologies by the determinability of a physically oriented way of thinking. However, this rational approach should not conceal the fact that these phenomena will always be experienced as horrifying manifestations of an almighty nature filling the human mind with fear and awe. Beethoven has expressed these feelings in the fourth movement of his Pastoral symphony. Let us therefore not only investigate lightning on its fiery path, but also regard its majestic beauty as a phenomenon of nature which has been omnipresent since times unknown.

References

Beccaria, G. B. (1758). "Elettricismo Atmosferica." 2nd ed. Colle Ameno, Bologna.

Bertholon, P., Abbé de St Lazare (1787). "De l'Électricité des Météores." Croullebois, Paris.

Cotte, P. (1774). "Traité de la Météorologie." Imprimerie Royale, Paris.

Dalibard, T. F. (1773). Extrait d'un mémoire, lu à l'Académie Royale des Sciences le 13. mai 1752, *in* "Œuvres de M. Franklin", Vol. 1. Quillau, Paris.

Davy (1822). Davy's tragbarer Wetterableiter im Spazierstock. *Dingler's Polytech. J.* **9**, 133.

De Romas, J. (1776). "Mémoire sur les Moyens de se Garantir de la Foudre dans les Maisons." Bergeret, Bordeaux.

Doré, H. (1915). "Recherches sur les Superstitions en Chine", Vol. X. Mission Catholique à l'Orphenilat de T'Ou-Sé-Wé, Shang-Hai.

Gray, S. (1735/6). Experiments and observations upon the light that is produced by communicating electrical attraction to animal or in animate bodies, together with some of its most surprising effects. *Phil. Trans. R. Soc.* **39**, 16–24.

Hartmann, J. F. (1759). "Abhandlung von der Verwandschaft und Ähnlichkeit der electrischen Kraft mit den erschrecklichen Luft-Erscheinungen." Hannover.

Hartmann, J. F. (1764). "Anmerkungen über die nötige Achtsamkeit bey Erforschung von Gewitter-Electricität, nebst Beschreibung eines Electricitätszeigers." Helwing, Hannover.

Hauksbee, F. (1706/7). An account of an experiment touching the production of light by the effluvia of one glass falling on another in motion. *Phil. Trans. R. Soc.* **25**, 2413–2415.

Kraus, J. (1814). "Gewitterkatechismus, oder Unterricht über Blitz und Donner, und wie man bey einem Gewitter sein Leben gegen den Blitz schützen und die vom Blitz getroffenen Menschen retten kann." 5th ed. Doll, Augsburg.

Landriani, M. (1784). "Dell'utilità dei conduttori elettrici." Marelli, Milano.

Lapostolle (1820). "Traité des Parafoudres et des Paragrêles en cordes de Paille." Imprimerie de Caron-Vitet Amiens, Weimar.

Le Monnier, L. G. (1752). Observations sur l'électricité de l'air. *Mém. Acad. r. Sci.* 233–243.

Lichtenberg, L. C. (1774). "Verhaltens-Regeln bey nahen Donnerwettern." Ettinger, Gotha.

Lind, J. (1773). Maison d'épreuve du tonnerre. *J. Phys.* **2**, 443–445.

Lucretius Carus, T. (1534). "De Rerum Natura." Gryphium, Lugduni.

Nicols, T. (1675). "Edelgestein-Büchlein oder Beschreibung der Edelgesteine." Nauman und Wolff, Hamburg.

Nollet, J. A., Abbé (1753). "Lettres sur l'Électricité." Guérin et Delatour, Paris.

Paul, J. (1809). "Des Feldpredigers Schmelzle Reise nach Flätz mit fortgehenden Noten." Cotta'sche Buchhandlung, Tübingen.

Porada, E. (1949). "Corpus of Ancient Near Eastern Seals", Pl. XXXIV, No. 220, Vol. I. The Collection of the Pierpont Morgan Library, Washington.

Prinz, H. *et al.* (1965). "Feuer, Blitz und Funke." Bruckmann, München.

Prinz, H. (1973). Fulminantes über Wolkenelektrizität. *Bull. Schweiz. elektrotech. Ver.* **64**, 1–15.

Scheuchzer, J. J. (1731). "Physica Sacra", Vol. 1, Wagner, Augsburg und Ulm.

Tinan, B. de (1779). Considérations sur les conducteurs en général, *in* Toaldo, J. "Mémoires sur les Conducteurs pour préserver les Édifices de Foudre." Strasbourg.

Toaldo, G. (1774). "Dell'Uso de Conduttori Metallici a Preservazione degli Edifizi contro de Fulmine." Venezia.

Wall, W. (1708). Experiments of the luminous qualities of amber, diamonds and gum lac. *Phil. Trans. R. Soc.* **26**, 69–76.

Winkler, J. H. (1746). "Die Stärke der elektrischen Kraft des Wassers in gläsernen Gefäßen, welche durch den Musschenbroek'schen Versuch bekannt geworden." Breitkopf, Leipzig.

Zedler, J. H. (1733). "Universal Lexikon", Vol. 4, Zedler, Halle und Leipzig.

2. Benjamin Franklin

B. DIBNER

Burndy Library, Norwalk, Connecticut, U.S.A.

1. Introduction

No figure so truly personifies the universal advances of the 1700s—the Enlightenment—as does the colonial American, Ben Franklin. To him is owed the concept and the proof that lightning was an electrical phenomenon, the proposal and demonstration that the duality of electrical charges was really a single fluid, and the sagacity to help lead his people to the formation of a new nation. This runaway apprentice who, in his ascendency to participation in councils of state and royal courts, steadily referred to himself as a printer, was held by his peers to be a Newton of his time.

Although Franklin actively engaged in science for only six or seven years, he was drawn to it throughout his whole life as an intellectual challenge rather than for any possible practical purpose or material gain. It was only after he was 40 years old that he devoted a major interest to any science, and that chosen was—electricity. His concentration on that field, especially during 1747 to 1749, resulted in his formulation of the fundamentals of electrical science. During his packed life of 84 years he helped found a nation, established the lines of prime communications in colonial America, directed the advance of the new electrical science and negotiated the acquisition of instruments of science and the weapons of war for his people.

2. Points and Clouds

By chance, electrical science began neatly in the year 1600 with the publication in London of William Gilbert's "De Magnete". This quarto initiated electrostatic (triboelectric) literature which flourished in Britain and on the Continent and seeded activity among the American colonists. The second phase of electrical development occurred when Alessandro Volta (1800) announced his invention of the voltaic pile for chemically generating an electric current. During this bracketed period of exactly two centuries the understanding of electrical behaviour was patiently studied and knowledge so gained was spread by correspondence, demonstration and publication (see Chapter 1).

Fig. 1. Engraved portrait of Franklin at the age of 72 by Tardieu after Duplessis.

To the layman, Franklin (Fig. 1) is remembered as postmaster, patriot, ambassador and sage; to the historian of science he is the originator of the one-fluid theory of electrical force, the demonstrator of the identity of friction-electricity and lightning, and as the inventor of the lightning rod. Born in Boston in 1706, fifteenth in a family of seventeen children, his father a candler, he learned early to fend for himself. He arrived in Philadelphia in 1723, began work as a printer and retired early in order to devote time to science studies. In his famous "Autobiography", Franklin first referred to his electrical interests in a one-sentence statement relating his purchase of

> "all Dr Spence's* (Cohen, 1943) apparatus, who had come from England to lecture here; and I proceeded in my Electrical Experiments with great Alacrity; . . .".

Later,

> "In 1746, being at Boston, I met there with a Dr Spence, who was lately arrived from Scotland, and showed me some electric Experiments. They were imperfectly performed as he was not very expert; but being on a Subject quite new to me, they equally surprised and pleased me."

The account of Franklin's expansion into electrical experimentation, formulation of conclusions and correspondence with European colleagues fills four pages of his memoirs, and culminates in the account of success at Marly and the honours subsequently accorded him (Franklin, 1964).

This introduction to electricity was supplemented by a gift in late 1746 from Peter Collinson of London of an electrical glass-rubbing tube sent to Franklin for use by the Philadelphia Library Company, which Franklin acknowledged. It inspired several members of the Company to make electrical experiments that "totally engrossed" Franklin. Collinson, a fellow Quaker, acted as unpaid London agent of the Library Company, placed informative letters from Franklin for publication in the *Philosophical Transactions of the Royal Society*, sponsored his book "Experiments and Observations on Electricity" for publication in London in 1751 and welcomed the American to London in 1757.

By being so far removed from the European centres of experimentation and discussion of electrical events, Franklin was able to view his own observations with a freshness not encumbered by the earlier notions of others. He therefore regarded an electrically undisturbed body as being under neutral charge or as in a state of electrical equilibrium. When, as in a Leyden jar, the device was electrified, he found the jar's outside to be charged "positive" and the inside "negative", thus providing a "plus" and "minus" divergence from its neutral, uncharged state.

> "Dr Franklin had discovered . . . that the electric matter was not created but collected by friction, from the neighbouring non electric bodies",

wrote Priestley (1767, p. 161; see also Cohen, 1952a). This rational and simple notion of a quantitative gain being equalled by a quantitative loss became established as the first exact electrical law: that of the conservation of

* The lecturer was Archibald Spencer (not Adam S.), a Scottish male nurse who was an itinerant lecturer on electricity in America. Franklin met Spencer in Boston in 1743 and next year in Philadelphia where he lectured on "Experimental Philosophy", with Franklin acting as his business agent.

charge. By substituting a flat glass pane coated on each side by metal foil for the jar-shaped condenser, Franklin was able to determine which electrical properties depended on shape. These observations and conclusions were communicated to Collinson in 1747 and constitute Franklin's greatest contribution to electrical theory. Called by his contemporaries the "Newton of Electricity", Franklin had read Newton's "Opticks" and emulated him by resorting to experiment to determine the validity of a theory. Joseph Priestley (1767, p. 160) wrote:

> "Dr Franklin's principles bid fair to be handed down to posterity, as equally expressive of the true principles of electricity, as the Newtonian philosophy is of the true system of nature in general."

Franklin's observation that a pointed conductor more readily drew off an electric charge than a blunt-ended conductor led him to experiment with the factor of shape in the effectiveness of a Leyden jar. He discovered from the flattened condenser shape that it was just as effective as the jar and, further, he joined a number of such flattened condensers into an "electrical-battery" of compounded capacity. Then, thought Franklin, might the earth and the charged clouds behave like his flat glass-plate condenser, a huge planetary Leyden jar? His emphasis on the efficacy of pointed conductors is expressed in his book (1751, p. 10):

> "The first is the wonderful effect of pointed bodies, both in *drawing off* and *throwing off* the electrical fire."

The realization by Franklin that points of bodies possessed special electrical properties led him to a series of experiments that guided him toward their capabilities not appreciated by other experimenting electricians. His inquiries soon led to probing larger and more powerful charges and ultimately to lightning and the lightning rod. Priestley wrote (1767, p. 172):

> "But before I relate any of Dr Franklin's experiments concerning lightning, I must take notice of what he observed concerning the power of pointed bodies, by means of which he was enabled to carry his great designs into execution. He was properly the first who observed the entire and wonderful effect of pointed bodies, both in drawing and throwing off the electric fire."

John Freke in England (1746), J. A. Nollet in France (1743) and J. H. Winkler in Germany (1746), among others, published works identifying lightning and electricity, each outlining some similar patterns of action and listing elements of similarity. None proposed experimental procedures to prove their assumptions. This Franklin did.

Six years after Franklin had listed in his journal (now lost) twelve proven similarities between lightning and electricity, he was asked by Dr John

Lining of Charleston, South Carolina, how he first came upon the idea of similarity between them. Franklin quoted in extract form:

> "7 Nov. 1749. Electrical fluid agrees with lightning in these particulars: 1. Giving light. 2. Colour of the light. 3. Crooked direction. 4. Swift motion. 5. Being conducted by metals. 6. Crack or noise in exploding. 7. Subsisting in water or ice. 8. Rending bodies it passes through. 9. Destroying animals. 10. Melting metals. 11. Firing inflammable substances. 12. Sulphureous smell—The electric fluid is attracted by points—We do not know whether this property is in lightning —But since they agree in all the particulars wherein we can already compare them, is it not probable they agree likewise in this? Let the experiment be made" (see Franklin, 1774, pp. 47, 50, 331).

The difference between Franklin and any predecessor lies in the significant "Let the experiment be made".

This quantitative comparison of man-made electricity and lightning is expanded by him in a paragraph of his book:

> "When the gun-barrel, (in electrical experiments) has but little electrical fire in it, you must approach it very near with your knuckle before you can draw a spark. Give it more fire, and it will give a spark at a greater distance. Two gun-barrels united, and as highly electrified, will give a spark at a still greater distance. But if two gun-barrels will strike at two inches distance and make a loud snap, to what a great distance may 10,000 acres of electrified cloud strike and give its fire, and how loud must be that crack?"

Prior to the publication of Franklin's book in London in April 1751, a notice appeared in the May 1750 issue of the *Gentleman's Magazine*, prepared by Cave, the editor, and later the publisher of Franklin's book:

> "There is something however in the experiments of points, sending off, or drawing on, the electrical fire, which has not been fully explained, and which I intend to supply in my next. For the doctrine of *points* is very curious, and the effects of them truly wonderful; and, from what I have observed on experiments, I am of opinion, that houses, ships, and even towns and churches may be effectually secured from the stroke of lightning by their means; for if, instead of the round balls of wood or metal, which are commonly placed on the tops of the weather-cocks, vanes or spindles of churches, spires, or masts, there should be put a rod of iron 8 or 10 feet in length, sharpen'd gradually to a point like a needle, and gilt to prevent rusting, or divided into a number of points, which would be better—the electrical fire, would, I think, be drawn out of a cloud silently, before it could come near enough to strike; only a light would be seen at this point, like the sailors corpusante."*

* *Corpusante*, or *comazant*, is a corruption of "corpo santo", the luminescence observed at the end of ships' yard arms and masts during a low mist and charged atmosphere. These were explained by Franklin as electrical discharges from a ship's "points"; they were also known as "St Elmo's fire".

Franklin's early efforts at probing the presence of electricity in the clouds took on two approaches—the first, as represented by the "sentry box" experiment, was to attract a sufficient charge down a pointed metal rod resting on a resin cake to permit cautious bleeding of a small charge, just to prove the presence of electricity on the rod, and therefore overhead in the cloud. This was an ungrounded form of rod. The other form was the same pointed rod, but one having its lower end grounded in moist soil, to provide a direct and ample path for the cloud's charge to be drawn and dissipated into the earth's vast neutral capacity.

3. "Experiments and Observations"

The letters that Franklin had been sending to Collinson in London contained such novel and provocative experimental results and such a clear explanation of complex electrical phenomena, especially in using the recently discovered Leyden jar, that his letters were read at meetings of the Royal Society, and correspondence flourished with European savants engaged in electrical experimenting. Edward Cave perceived readership potential in Franklin's letters. With the sponsorship of Collinson, and Dr John Fothergill as editor and writer of the unsigned preface, Franklin's "Experiments and Observations on Electricity" was published in London in 1751.

This was Franklin's first published work in book form; although it is sometimes referred to as a pamphlet, it consists of 90 pages with one plate of ten illustrations. The work repeats Franklin's conclusion that lightning was an electrical phenomenon, including the description of an experiment to prove this. The lightning rod is also described on p. 62 which contains the noteworthy statement:

> "I say, if these things are so, may not the knowledge of this power of points be of use to mankind, in preserving houses, churches, ships etc. from the stroke of lightning, by directing us to fix on the highest parts of those edifices, upright rods of iron made sharp as a needle, and gilt to prevent rusting, and from the foot of those rods a wire down the outside of the building into the ground, or down round one of the shrouds of a ship, and down her side till it reaches the water?"

This and the note in the *Gentleman's Magazine* of May 1750 seem to be Franklin's first references to the lightning rod, predating the announcement in "Poor Richard's Almanack" (see Section 6).

The success of the 1751 book prompted the publisher, Cave, to issue two supplementary pamphlets, and all three parts were subsequently reissued, thereby further spreading the awareness of Franklin's views and efforts. An anonymous contemporary English reviewer of the fourth (1769) edition assessed the work as

> "... the experiments and observations of Dr Franklin constitute the principia of electricity, and form the basis of a system equally simple and profound" (see Cohen, 1941, p. 125).

The fifth edition, issued during his lifetime, appeared in 1774; this contained corrections of errors in the earlier editions, made personally while on a visit to London during the preparation of the fourth edition. Added also was more current information on the use of lightning rods and D'Alibard's account of the Marly experiment in French (not present in the earlier editions), as read at the Académie on 13 May 1752. It is preceded in the text by a two-page summary of what occurred at Marly, in English, sent on 20 May by the Abbé Mazeas from St Germain to Stephen Hales in London (Franklin, 1774, pp. 106–107). At the time of infancy of the new science of electricity, the popularity of Franklin's book is evident by its reissues in his lifetime, in five editions in English (1751, 1754, 1760, 1769 and 1774), three editions in French, one in German and one in Italian, totalling ten editions in four languages. In France, a copy of Franklin's book reached the Count de Buffon, an eminent naturalist, who urged D'Alibard to translate it.

The euphoria felt among the early experimenters during those revealing years was expressed by Franklin at the end of a long letter he sent to Collinson in 1748. In its 28 paragraphs he detailed experiments with Leyden jars, magic pictures and spinning wheels, actuated into vertical and horizontal rotation by the effusion of an electrostatic charge from tangential tips. Franklin signed off with:

> "Chagrined a little that we have been hitherto able to produce nothing in this way of use to mankind; and the hot weather coming on, when electrical experiments are not so agreeable, it is proposed to put an end to them for this season, somewhat humorously, in a party of pleasure, on the banks of Skuylkil (the river that washes one side of Philadelphia). Spirits, at the same time, are to be fired by a spark sent from side to side through the river, without any other conductor than the water; an experiment which we some time since performed, to the amazement of many. A turkey is to be killed for our dinner, by the *electrical shock* and roasted by the *electrical jack*, before a fire kindled by the *electrified bottle*:* when the healths of all the famous electricians in *England*, *Holland*, *France*, and *Germany* are to be drank in *electrified bumpers*, under the discharge of guns from the *electrical battery*" (Franklin, 1774, pp. 21–38).

* The "bottle" is meant to describe the Leyden jar (means of storing electrostatic charges) observed by von Kleist of Kamin in late 1745 and independently discovered and described by van Musschenbroek of Leyden in January 1746. (See Ronalds, 1880, p. 268; Benjamin, 1895, pp. 512, 517; Mottelay, 1922, p. 173).

Franklin detailed the construction of the electric wheel:

> "This is called an electrical jack; and if a large fowl were spitted on the upright shaft, it would be carried round before a fire with a motion fit for roasting."

The final footnote described:

> "An *electrified bumper* is a small thin tumbler, nearly filled with wine, and electrified as the bottle. This when brought to the lips gives a shock, if the party be close shaved, and does not breathe on the liquor."

The electrical battery is described as consisting of a condenser of eleven large sash-glass vertical panes armed with lead plates on each side. In describing the various electrical devices Franklin is meticulous in crediting novelty to earlier inventors; three such—Bevis, Smeaton, Kinnersley—are mentioned on one page (p. 29).

4. The Sentry-box and Marly Experiments

In Philadelphia in 1749, Franklin had performed some experiments to prove his hypothesis that lightning was an electrical discharge between clouds, or between clouds and earth. He related his thoughts and convictions in a letter sent to Peter Collinson in London on 29 July 1750 describing what became known as the "Sentry-box Experiment". Franklin (1774, pp. 65–66) stated his purpose and procedure clearly:

> "To determine the question, whether the clouds that contain lightning are electrified or not, I would propose an experiment to be tried where it may be done conveniently. On the top of some high tower or steeple, place a kind of sentry-box big enough to contain a man and an electrical stand. From the middle of the stand let an iron rod rise and pass bending out of the door, and then upright 20 or 30 feet, pointed very sharp at the end. If the electrical stand be kept clean and dry, a man standing on it when such clouds are passing low, might be electrified and afford sparks, the rod drawing fire to him from a cloud. If any danger to the man should be apprehended (though I think there would be none) let him stand on the floor of his box, and now and then bring near to the rod the loop of a wire that has one end fastened to the leads, he holding it by a wax handle; so the sparks, if the rod is electrified, will strike from the rod to the wire, and not affect him."

The iron rod conductor standing upright 20 or 30 ft "on the top of some high tower or steeple" the lower end of which terminated at the middle of "an electrical stand" seemingly provided Franklin with a sufficiently electrified source of sparks by "The rod drawing fire to him from a cloud". Caution was recommended in determining the magnitude of the sparks by interposing a loop of wire and a wax handle between the electrified rod and the investigator who, as an additional precaution, had stepped down from the stand.

The sentry-box experiment appeared in London in 1751 in Franklin's "Experiments".* The Abbé Nollet, the foremost French electrical experimenter of the time, read it, criticized the translation and remained sceptical of some of the Franklin claims. Greater enthusiasm was shown by the French King (Louis XV), who requested that the experiments be repeated and thereupon applauded Franklin and his sponsor, Collinson, for their novel contributions.

As the message and meaning of Franklin's book of "Experiments" of 1751 slowly permeated the laboratories and lecture halls of Europe, an alert few in France were prompted to confirm what the obscure American proposed—the electrical nature of lightning. The test came in May 1752 and was reported by the Abbé Mazeas (1752). The event was obscure but its influence—profound:

> "The Philadelphian experiments, that Mr Collinson, a Member of the Royal Society, was so kind as to communicate to the public, having been universally admired in France, the King desired to see them performed. Wherefore the Duke D'Ayen offered his Majesty his country house at St Germain, where M. de Lor, master of Experimental Philosophy, should put those of Philadelphia in execution. His Majesty saw them with great satisfaction, and greatly applauded Messieurs Franklin and Collinson. These applauses of his Majesty having excited in Messieurs de Buffon, D'Alibard and de Lor, a desire of verifying the conjectures of Mr Franklin, upon the analogy of thunder and electricity, they prepared themselves for making the experiment. M. D'Alibard chose, for this purpose, a garden situated at Marly, where he placed upon an electrical body a pointed bar of iron, of 40 feet high (Fig. 2). On the tenth of May, 20 minutes past two in the afternoon, a stormy cloud having passed over the place where the bar stood, those that were appointed to observe it drew near and attracted from it sparks of fire, perceiving the same kind of commotions as in the common electrical Experiments. M. de Lor, sensible of the good success of this experiment, resolved to repeat it at his house in the Estrapade at Paris. He raised a bar of iron 99 feet high, placed upon a cake of resin, two feet square, and three inches thick. On the 18th day of May, between four and five in the afternoon, a stormy cloud having passed over the bar, where it remained half an hour, he drew sparks from the bar, like those from the gun barrel when, in the electrical experiments, the globe is only rubbed by the cushion, and they produced the same noise, the same fire, and the same crackling. They drew the strongest sparks at the distance of

* A manuscript copy of the sentry-box account was discovered by Professor I. B. Cohen among a collection of Franklin's complete electrical papers, not holograph, but with Franklin's corrections (see Cohen, 1952b, pp. 331–366).

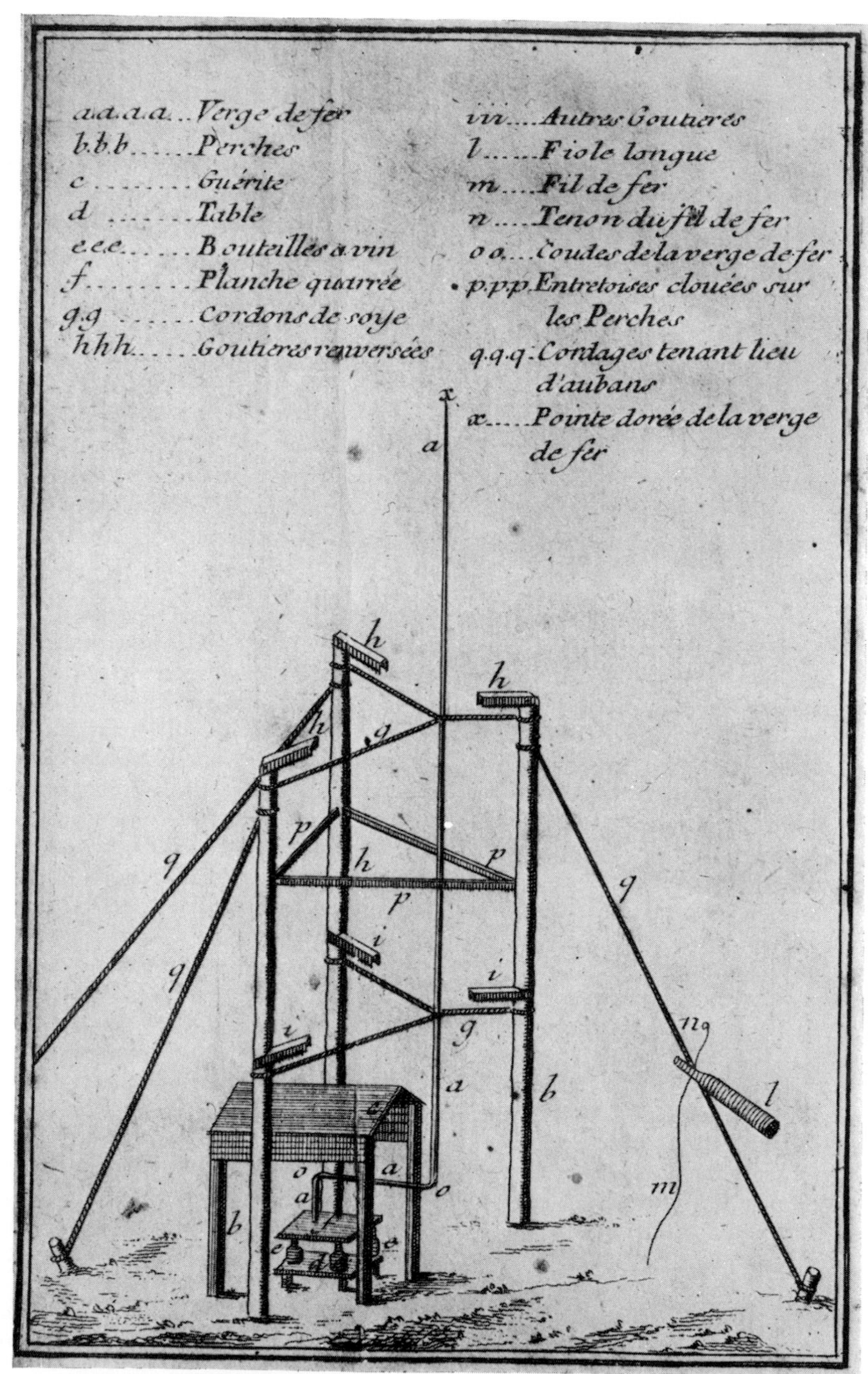

Fig. 2. Experimenters who had read the French translation of Franklin's book on electricity erected a 40-ft lightning rod at Marly, near Paris. On 10 May 1752 storm clouds gathered, and sparks were drawn by two attendants.

nine lines, while the rain, mingled with a little hail, fell from the cloud, without either thunder or lightning; this cloud being, according to all appearance, only the consequence of a storm, which happened elsewhere" (Franklin, 1774, pp. 106–107).*

The French experiment described D'Alibard erecting the 40-ft rod, pointed at the top and insulated from a sentry-box set in a garden at Marly-la-Ville, seven or eight miles from Paris. On 10 May 1752, while D'Alibard was absent and a retired dragoon named Coiffier attended the apparatus, a storm cloud passed over the area. It was observed by Coiffier and the village priest Roulet that sparks of fire appeared, similar to the effects of common electrostatic experiments. Word of this was sent to D'Alibard who, on 13 May, presented a report of the Marly occurrence and a discourse on Franklin's theories and experiments before the Académie Royale des Sciences in Paris.† Five days later the experiment was repeated by de Lor. The experiment was again successful and was repeated several times by Lemonnier and Romas in France, by Mylius and Ludolf in Germany and in England by Canton, by Wilson and by Bevis.‡ Efforts to repeat the Marly experiment in England were negated by the damp summer of 1752, as reported by William Watson.

Nollet (1753), incredulous that so rational a theory might have come out of barbaric America, reacted to the intrusion by publishing a volume of nine letters, six of which were intended to destroy Franklin's claims and postulates. Franklin, unbeholden and confident, was not for becoming involved in debate, preferring to let experimental confirmation take its course. The advice he gave to Jan Ingenhousz regarding controversy and polemics in matters of science is worthy of repetition:

"I hope you will omit the polemic piece in your French edition and take no public notice of the improper behaviour of your friend; but go on with your excellent experiments, produce facts, improve science, and do good to mankind. Reputation will follow, and the little injustices of contemporary labourers will be forgotten; my example may encourage you, or else I should not mention it. You know, that when my papers were first published, the Abbé Nollet, then high in reputation, attacked them in a book of letters. An answer was expected from me, but I made none to that book nor to any other. They are now all neglected, and the truth seems to be established. You can always employ your time better than in polemics." (Cohen, 1952c, p. 364).

* The letter by Mazéas was read on 28 May 1752 at the Royal Society and was published in the same year. The report is followed by four additional reports by Mazéas on related experiments, two by Nollet, one by Mylius of Berlin and the letter of Franklin to Collinson, on p. 565, "concerning an electric kite".

† Contained in the D'Alibard translation of Franklin's book, 2nd ed., Paris 1756, Vol. II, pp. 99–133.

‡ Reports by Canton, by Wilson and by Bevis appeared in *Phil. Trans. R. Soc.*, **48** (1753, 1754), pp. 350, 780, 347.

In contrast to Nollet's antagonism, Franklin was widely acclaimed in Italy by such admirers as Francesco Algarotti who wrote:

> "Behold from English America there come to us not only tobacco and indigo, but also philosophical systems. From Philadelphia a Quaker has sent us the most beautiful observations, the most beautiful reasonings in the world on electricity; and all our European electricians must doff their hats to this American." (Pace, 1958, p. 17).

As a result of the Marly experiment, Franklin was acclaimed in Europe as a modern Prometheus for having shown how to draw electric fire from the heavens.

5. The Kite

Franklin wrote to Dalibard in 1768 complimenting him on being

> "the first of Mankind, that had the Courage to attempt drawing Lightning from the Clouds to be subjected to your experiments" (Cohen, 1952c, p. 338.)

This was a reference to the Marly experiment and was penned sixteen years after the event and after many electrical developments. Franklin referred again to the success of the "Experiments, made by Messers Dalibard and Delor at Marly, for drawing lightning from the Clouds" in his "Autobiography" (p. 244) written on his return from France in 1788:

> "I will not swell this Narrative with an Account of that capital Experiment, nor of the infinite Pleasure I received in the Success of a similar one I made soon after with a Kite at Philadelphia, as both are to be found in the Histories of Electricity."

Franklin seemingly left no report of the kite experiment other than the oral account given to Priestley.

Priestley was the first historian of electrical science whose "History and Present State of Electricity" appeared in London in 1767 on Franklin's advice, and was translated into French and German, giving the new science a much wider readership, and also spreading the account of Franklin's experiments of 1752. Priestley's published report containing "particulars which I have from the best authority"—evidently Franklin himself—is fuller than Franklin's stating:

> "The Doctor, after having published his method of verifying his hypothesis concerning the sameness of electricity with the matter of lightning, was waiting for the erection of a spire in Philadelphia [the tall Christ Church steeple was still being built] to carry his views into execution; not imagining that a pointed rod,

of a moderate height, could answer the purpose; when it occurred to him, that, by means of a common kite, he could have a readier and better access to the regions of thunder than by any spire whatever [Fig. 3]. Preparing, therefore, a large silk handkerchief, and two cross-sticks, of a proper length, on which to extend it; he took the opportunity of the first approaching thunderstorm to take a walk into a field, in which there was a shed convenient for his purpose. But dreading the ridicule which too commonly attends unsuccessful attempts in science, he communicated his intended experiment to nobody but his son, who assisted him in raising the kite. The kite being raised, a considerable time elapsed before there was any appearance of its being electrified. One very promising cloud had passed over it without any effect; when, at length, just as he was beginning to despair of his contrivance, he observed some loose threads of the hempen string to stand erect and to avoid one another, just as if they had been suspended on a common conductor. Struck with this promising appearance, he immediately presented his knuckle to the key, and (let the reader judge of the exquisite pleasure he must have felt at that moment) the discovery was complete. He perceived a very evident electric spark. Others succeeded, even before the string was wet, so as to put the matter past all dispute, and when the rain had wet the

Fig. 3. The kite experiment, as shown on an old banknote engraving, was witnessed only by his son, actually then 21 years old. With this Franklin confirmed the electrical nature of lightning.

string, he collected electric fire very copiously. This happened in June 1752, a month after the electricians in France had verified the same theory, but before he heard of anything they had done" (Priestley, 1767, pp. 180–181).*

Franklin's own description is in the form of a set of instructions and not a narrative with details of time and place of one of the classic experiments in history. It appears as a letter to Collinson from Philadelphia dated 19 October 1752 and his Letter XI pp. 117–118 in his "Experiments", 1774. The letter ends:

> "At this key the phial may be charged; and from electric fire thus obtained, spirits may be kindled, and all the other electric experiments be performed, which are usually done by the help of a rubbed glass globe or tube, and thereby the sameness of the electric matter with that of lightning completely demonstrated."

The printed version stops here, but Cohen (1952c, p. 333) traced the original to the archives of the Royal Society which has the following additional sentences, often omitted in reprints of the letter:

> "I was pleased to hear of the Success of My experiments in France, and that they there begin to Erect points on their buildings. We had before placed them upon our Academy & State House Spires."

This clearly indicates priority in the erection of "points"—lightning rods—on the Philadelphia structures.

What Franklin provided with his kite and string was a conductor suspended diagonally in space connecting a sharp metal wire at the upper tip of the kite, down a moist string to a terminal key at its bottom. Also attached to the string at its lower end was an insulating silk ribbon, the lower end of which was grasped by Franklin. The charge drawn from the storm clouds spread down the string, but was not drained until Franklin's knuckle was brought close to the suspended key. Franklin's grounded body provided the path of discharge from the key to ground.

The involved chronology still permits the conclusion that the kite effort was an extension of the sentry-box experiment. Both were proposed by Franklin, both involved drawing charges from storm clouds to points of test and observation, both involved risk on the part of the observer, Franklin not setting himself up above being the one to take the risk. All this prompted Priestley (1767, p. 172) to write:

> "Moreover, though Dr Franklin's directions were first begun to be put in execution in France, he himself compleated the demonstration of his own

* Franklin's description of a kite experiment appeared in the *Philosophical Transactions* dated 1 October 1752; the later date often used by biographers is that of the issue of the *Pennsylvania Gazette* in which the account also appeared.

problem, before he heard of what had been done elsewhere and he extended his experiments so far as actually to imitate all the known effects of lightning by electricity, and to perform every electrical experiment by lightning."

Imprecise as the date of the kite experiment might be, it was reported by Priestley as having been made in June 1752, framing it therefore as a month later than the Marly test. News of the latter brought Franklin the

"infinite Pleasure I receiv'd in the Success of a similar one I made soon after with a kite in Philadelphia" (Franklin, 1964, p. 244).

Franklin chose to publish a description of his electrical kite experiment in the 19 October 1752 issue of the *Pennsylvania Gazette*, and in the same issue there appeared the notice of the forthcoming "Poor Richard's Almanack" for 1753 containing his account "How to Secure Houses, etc. from Lightning". Although the use of lightning rods had been recommended in the London 1751 edition of Franklin's book on electricity, there seemed to be some reticence to describe the kite experiment. This may have been based on possible fear of ridicule in having performed so important an event in the presence of, and with the knowledge of, only his young son, or possibly in the expectation of repeating the demonstration at some future, more propitious, time.

6. The Lightning Rod

Franklin having by means of the sentry-box experiment determined that this device was effective in drawing down an electrical discharge, there remained the larger question as to "whether the clouds that contain lightning are electrified or not . . .". Lightning rods seem to provide the answer in depriving the clouds of their electric fire and the damage to life and to structure. Assuming that sparks from laboratory apparatus differed from lightning bolts only in the magnitude of discharge, it required commensurately larger conductors to drain a lightning stroke than a laboratory charge.

In the sentry-box experiment, Franklin indicated a rod the lower end of which ended at the insulated stand. The purpose of box, rod and stand was to demonstrate the co-action of rod and charged cloud whereby the passing cloud energized the rod with manifest presence of charges upon it. On the other hand, the purpose of the lightning rod was to draw off, quietly and continually, the potentially destructive bolt of lightning by draining this potential for harm into the equalizing ground. Franklin (1774, p. 65) described this intent to

"draw the electrical fire silently out of a cloud before it came nigh enough to strike, and thereby secure us from that most sudden and terrible mischief".

The sentry-box rod if grounded would deprive the investigator of any electrical presence, since any charge would drain to ground. Franklin did consider the risk of exposure in the sentry-box experiment and therefore provided a grounded wire having an insulated wax handle that would permit the spark to strike from rod to wire without including the investigator in the circuit.

From Paris in 1767 Franklin (1774, pp. 498–500) wrote,

> "If the communication be through the air without any conductor, a bright light is seen between the bodies, and a sound is heard. In our small experiments we call this light and sound the electric spark and snap; but in the great operations of nature, the light is what we call *lightning*, and the sound (produced at the same time, though generally arriving later at our ears than the light does to our eyes) is, with its echoes, called *thunder*".

Later in this letter,

> "The clouds have often more of this fluid in proportion than the earth; in which case as soon as they come near enough (that is, within the striking distance) or meet with a conductor, the fluid quits them and strikes into the earth. A cloud fully charged with this fluid, if so high as to be beyond the striking distance from the earth, passes quietly without making noise or giving light; unless it meets with other clouds that have less.. . . An iron rod being placed on the outside of a building, from the highest part continued down into the moist earth, in any direction, strait or crooked, following the form of the roof or other parts of the building, will receive the lightning at its upper end, attracting it so as to prevent it's striking any other part; and affording it a good conveyance into the earth, will prevent it's damaging any part of the building."

Lightning had been regarded as a divine expression, a manifestation against which there could be no possible protection, except prayer and the ringing of church bells. Such bells cast in mediaeval times often bore the legend *Fulgura frango* ("I break up the lightning"). With the passage of time, however, it was realized that bell-ringing during a storm was a very hazardous remedy, especially for the ringer on the ropes because so many were killed by the very strokes they attempted to disperse. A book by Fischer published in Munich in 1784, as quoted by Schonland (1964, p. 9), records the sombre results of 33 years of lightning strokes on 386 church steeples and the killing of 103 bell-ringers.

The modest announcement in "Poor Richard's Almanack" for 1753, which Franklin published, contains the concise description of one of the great benefactions to mankind—the lightning rod. Thus:

> "How to secure Houses etc. from LIGHTNING
>
> It has pleased God in his Goodness to Mankind, at length to discover to them the Means of securing their Habitations and other Buildings from Mischief by Thunder and Lightning. The Method is this: Provide a small Iron Rod (It may be

made of the Rod-iron used by the Nailers) but of such a Length, that one End being three or four Feet in the moist Ground, the other may be six or eight Feet above the highest Part of the Building. To the upper End of the Rod fasten a Foot of Brass Wire the Size of a common Knitting-needle, sharpened to a fine Point; the Rod may be secured to the House by a few small Staples. If the House or Barn be long, there may be a Rod and Point at each End, and a middling Wire along the Ridge from one to the other. A House thus furnished will not be damaged by Lightning, it being attracted by the Points, and passing thro the Metal into the Ground without hurting any Thing. Vessels also, having a sharp pointed rod fix'd on the Top of their Masts, with a Wire from the Foot of the Rod reaching down, round one of the Shrouds, to the Water, will not be hurt by Lightning."

There was no rush to follow Franklin's simple and clear specifications; caution and doubt, especially in England, prevailed. In France, the Marly experiment was organized, with dramatic though controversial results.

To those who held that lightning rods were to be avoided as just being means of inviting trouble by "attracting" lightning that might otherwise not be present, Franklin (1774, p. 169) repeated the dual purpose of rods:

"I have mentioned it in several of my letters, and except once, always in the alternative, viz. that pointed rods erected on buildings, and communicating with the moist earth, would either prevent a stroke, or, if not prevented, would conduct it, so as that the building should suffer no damage. Yet whenever my opinion is examined in Europe, nothing is considered but the probability of those rods preventing a stroke of explosion, which is only a part of the use I proposed for them; and the other part, their conducting a stroke, which they may happen not to prevent, seems to be totally forgotten, though of equal importance and advantage."

Among the houses and other buildings in Philadelphia protected by rods, after those erected at the Academy and the State House, was Franklin's own house in which a rod was installed in September 1752. The rod of iron was fixed to the top of the chimney and extended about 9 ft above it. To the bottom of the rod was attached a wire ("the thickness of a goosequill") that passed through a covered glass tube set in the roof and then down the well of the staircase, with its lower end connected to the iron spear of a pump. On the staircase the wire was interrupted, the ends separated about 6 in. with a little bell attached to each end. Between the bells hung a small brass ball suspended by a silk thread. When clouds which were electrically charged passed over, the ball played between and struck the bells (Fig. 4). More than just that, narrated Franklin:

"After having frequently drawn sparks and charged bottles from the bell of the upper wire, I was one night awaked by loud cracks on the staircase. Starting up

Fig. 4. A lightning rod of special design was erected by Franklin in his home in September 1752 intended for continued study of lightning. As shown in this engraving by Chamberlain in 1762, both upper and lower rods terminated in a bell; a small ball suspended between them vibrated when electrically energized. From the right bell hung a two-ball electroscope to indicate a charged cloud overhead.

and opening the door, I perceived that the brass ball, instead of vibrating as usual between the bells, was repelled and kept at a distance from both; while the fire passed, sometimes in very large, quick cracks from bell to bell, and sometimes in a continued, dense, white stream, seemingly as large as my finger, whereby the whole staircase was inlightened as with sunshine, so that one might see to pick up a pin" (Smyth, 1907).

The gap of less than 6 inches between the bells was assumed by Franklin as being a manageable distance to be traversed by a lightning flash moving from the upper rod to the lower, based on his knowledge of such paths on unprotected, struck buildings. Franklin (see Cohen, 1952c, p. 352) wrote a letter from London in 1758 to his wife suggesting that if the ringing of the bells frightened her, she tie a piece of wire from one bell to the other,

> "and that will conduct the lightning without ringing or snapping, but silently".

Franklin's exposure to electric shocks were many (see Chapter 16.1), for that was the way stronger electric charges were identified and measured, weaker charges being determined by gauging with an electrometer. In a letter dated 18 March 1755 to John Lining, Franklin wrote:

> "You suppose it a dangerous experiment, but I had once suffered the same myself, receiving by accident, an equal stroke through my head, that struck me down, without hurting me . . ." (see Pepper, 1970, p. 26).

Early experience with the rod clearly showed a disregard in keeping the top of the pointed conductor sufficiently high (at the recommended 7–8 ft above the highest part of the structure) and its bottom point sufficiently deep in moist earth. Franklin advised:

> "The lower end of the rod should enter the earth so deep as to come at the moist part, perhaps two or three feet; and if bent when under the surface so as to go in a horizontal line six or eight feet from the wall, and then bent again downwards three or four feet, it will prevent damage to any of the stones of the foundation."

This note is contained in Letter LIX of Franklin's "Experiments", 1774, a letter that consists of 24 paragraphs, each touching on one summary item under the title "Of Lightning and the Method (now used in America) of securing Buildings and Persons from its mischievous Effects". The letter was sent from Paris in September 1767.

The earliest successful action of the rod was during a storm in Philadelphia when the home of the merchant Mr William West of Water Street was saved by the Franklin rod. Franklin's own home survived a flash in 1787. The first English installation of the rod was made in 1760 to protect the then wooden Eddystone lighthouse. In France the Academy of Sciences at Dijon had a rod installed by Professor de Moricau; in Germany it was in 1768 on the home of Dr Reimarus of Hamburg, the first of 226 buildings in that city protected by rods by 1780.

After reviewing thoroughly the contemporary and recent literature on the subject of Franklin and the introduction of the lightning rods as a security against unmitigated lightning, Cohen (1952c, p. 353) concludes:

> "I believe that the lightning rods erected in Philadelphia in 1752 were the first grounded lightning rods to be erected anywhere in the world for the purpose of protecting buildings from the lightning discharge."

He then lists the nearly dozen early reports of lightning rod attempts and shows them to be not of the grounded, protective-rod type but of the insulated, investigative variation.

In his Letter XII, sent to Collinson in September 1753, Franklin outlined a series of experiments to determine if storm clouds were, as he supposed, always charged negatively. Using the newly installed rod and bells in his Philadelphia home, he proposed comparing the constantly positive charge from his electrical machine to the unknown charge on the lightning rod; he then could decide the nature of the cloud charge. After many of his own trials and those of his friend Ebenezer Kinnersley, he concluded (Franklin, 1774, pp. 122–125):

> "That the clouds of a thunder-gust are most commonly in a negative state of electricity, but sometimes in a positive state"

and:

> "So that, for the most part, in thunderstrokes, it is the earth that strikes into the clouds, and not the clouds that strike into the earth."

Further,

> "Those who are versed in electric experiments, will easily conceive, that the effects and appearances must be nearly the same in either case; the same explosion, and the same flash between one cloud and another, and between the clouds and mountains, etc., the same rending of trees, walls, etc. which the electric fluid meets with in its passage, and the same fatal shock to animal bodies; and that pointed rods fix'd on buildings, or masts of ships, and communicating with the earth or sea, must be of the same service in restoring the equilibrium silently between the earth and clouds, or in conducting a flash or stroke, if one should be, so as to save harmless the house or vessel: For points have equal power to throw off, as to draw on the electric fire, and rods will conduct up as well as down."

It was the realization that points on a conductor were capable of drawing off electric charges from a body at considerable distances that led Franklin to the generalized idea of the lightning rod and the discharging of thunder from (tonitruous) clouds. Also, that it was essential that the rod be well grounded. Franklin believed that the pointed rods would quietly drain the charge and thus prevent the sudden bolt. In a letter to M. D'Alibard, dated 29 June 1755, he described an incident of a bolt of lightning striking a church steeple in Newburyport, Massachusetts, which Franklin had examined. The structure consisted of a tower 70 ft tall at the top of which a bell was hung. Above this rose a spire also 70 ft high. From the bell a thin wire was led down the structure for a hammer-ringing device. The lightning struck and demolished the spire, followed the wire's path, vaporizing it, yet causing little damage otherwise, but beyond its termination it did great damage to the building and stone foundation. Franklin (1774, pp. 170–172) summarized six conclusions from his observations crediting the wire with sparing that section of the structure through which it had passed.

7. The Lightning Rod in Europe

In a letter written from England in 1772, Franklin reported the relative progress made in America and on the Continent in adopting lightning rods. In America, with its higher incidence of lightning storms, there was a greater use of the device to protect private houses, churches, public buildings, gunpowder magazines and country homes. In England, such important edifices as St Paul's, St James' Church, the Queen's Palace, Blenheim Palace and a number of country seats were protected. He also reported on the unharmed condition of the ship *Endeavour* equipped with a lightning chain, in contrast to a Dutch man-of-war at anchor nearby in Batavia which "was almost demolished by the Lightning" (see Cohen, 1941, p. 133).

In Germany a physician, Johann Reimarus of Hamburg, supplemented his many writings on the performance of lightning rods by being the first to install a lightning rod on the spire of the St Jakobi Church. In Italy the earliest application is credited to Giovanni Battista Torrè who mounted the device in Florence on 23 June 1752 (Pace, 1958, p. 20).

In France some opposition to rods came from the clergy which continued its faith in the ringing of church bells to scatter the lightning's force. The Abbé Nollet (1753) opposed rods for personal reasons. Having also opposed Franklin on the interpretation of how a Leyden jar functioned, he carried his opposition over to the Marly experiment, and on to the installation of lightning rods. A storm that moved upon the coast of Brittany on the night of 14 April 1718 caused great damage when lightning struck 24 churches, completely destroying one. Of the four bell-ringers attending the ropes in the ritual of ringing the bells to dissipate the lightning, two were killed (see Cohen, 1952b).

Interest in lightning rods in Italy was sharpened by the violent destruction of a powder magazine at Brescia in 1772, and this brought the problem of magazine protection to Whitehall. The Royal Society became involved and a committee was appointed to study the problem and make recommendations. The committee of eminent electricians, including Franklin, met and rendered its report (see Chapter 17.1.2.1), proposing pointed rods, grounded, to provide the required protection. One of the five members of the committee, Benjamin Wilson, took exception to the decision for pointed rods, on the ground that these would unnecessarily "invite" strokes, whereas rounded ends would not. A small tempest was stirred but Franklin would engage in no polemics, and other partisans stepped into the fray of "points" versus "blunts". With the coming of the American Revolution the protagonists took on political colour and George III ordered blunts on the rods protecting the royal palace. The contest reached the Crown and the presidency of the Royal Society, but Franklin remained aloof, knowing wisely that such controversies wear themselves out, and that right usually prevails.

To provide houses with lightning rods seemed prudent since there was little to lose and much to gain if the rod really did avoid lightning damage. However, to protect a church spire with a rod seemed a sacrilegious act and an expression of no confidence in the mercy of God in striking a sacred structure. So the authorities of Venice decided to store a hundred tons of gunpowder in the vaults of a church as the safest possible location. In 1767 the church was struck by lightning, a thunderous explosion occurred in which three thousand people were killed and a large section of the city was destroyed. The Campanile of San Marco in Venice was struck and severely damaged or completely destroyed nine times between 1388 and 1762. In 1766 a Franklin rod was installed and it since has suffered no further damage from lightning. The cathedral tower of Sienna with a similar history was protected in 1777. The Council of the East India Company ordered the lightning rods at Fort Malaga in Sumatra to be removed as a potential danger; in 1782 lightning struck and ignited 400 barrels of gunpowder with catastrophic results. In a similar incident as late as 1856, in a church on the Island of Rhodes, four thousand people were killed.

Firmest support for Franklin's electrical contributions came from Giambattista Beccaria, professor of experimental physics at the University of Turin whose book (1753) is devoted to atmospheric electrical phenomena and strongly supports Franklin's position on Leyden jar behaviour; it was translated into English in London in 1776. Beccaria was elected a Fellow of the Royal Society a year before Franklin, albeit the Italian's fame rested on the support he gave Franklin's work. The adoption of the lightning rod in Italy was very rapid, with powder magazines in Tuscany duly protected, and installations that spread from Livorno to Arezzo in the 1760s. Venice followed with numerous installations, with the Cathedral of San Marco protected in 1777. By direct sanction of Pope Pius VI, St Peter's in Rome was secured by Franklin's device as well as the papal villa on the Quirinale (see Pace, 1958, p. 26). Among the many champions of the Franklin rod was Giuseppe Toaldo, professor of astronomy at Padua, who wrote a series of didactic pamphlets (started in 1770) on applying rods for lightning protection, in 1776 he described the triumph of structure protection—the 340-ft Campanile of San Marco in Venice (Toaldo, 1778).

A dramatic confrontation between the Franklin partisans and those who, through inertia, superstition or ignorance, rejected the rod, occurred in the spring of 1777 in the ancient city of Sienna in Tuscany. There the noble Marquis Alessandro Chigi, chamberlain to the Grand Duke, published a dissertation (1777) in which the electrical nature of lightning was denied and rods were considered ineffective. The progressive Grand Duke had ordered rods to be placed on the cathedral and on the Gothic tower of the city hall facing the Piazza, where the famous Palio is run. Controversy ran high, with Chigi one locus of faith and, in opposition, university professors strongly holding for the rod, now installed. On the afternoon of 18 April clouds began to form, distant thunder was heard, and the Siennese began moving to their Piazza with all eyes focused on the lightning rod tip.

> "At about five o'clock—lightning struck. A ball of fire, accompanied by sparks, smoke, and an odor of sulphur ran the full length of the tower and disappeared into the ground leaving the tower unharmed."

Domenico Bartolini, professor of physics at the university, examined the rod and the tower and issued a triumphant report that was widely read throughout Italy. Later, Bartolini referred to Beccaria as "the Italian Franklin", and a fellow professor published an account of the populace

> "who had their eyes turned toward an extremely high tower to observe the action of the conductor recently placed upon it, and who, without a long wait, had the fortune to admire, to the glory of Philosophy, the intelligence and genius of the immortal Mr Franklin . . ." (see Pace, 1958, p. 29).

Acknowledging the receipt of a copy of Marsiglio Landriani's book of 1784 from its author, Franklin replied to the author's request for information regarding the experience in America with lightning rods by observing that since his return to Philadelphia from service abroad, the use of rods had greatly multiplied and that many houses, including his own, had escaped damage during his absence, adding

> "So that at length the Invention has been of some use to the Inventor . . ." (see Pace, 1958, p. 41).

Franklin, like Michael Faraday, Joseph Henry and Galileo Ferraris, refused to patent his inventions, stating

> "That, as we enjoy great advantages from the inventions of others, we should be glad of an opportunity to serve others by any invention of ours; and this we should do freely and generously" (see Cohen, 1953, p. 45).

The move of electrical investigation from the laboratory and lecture hall to the scale of lightning and mass destruction also brought on the first martyrdom of the still arcane science. In St Petersburg, Georg Wilhelm Richmann (1711–1753) was professor of physics at the university and author of a score of papers in Latin on meteorological and physical subjects published by the Imperial Academy.

Richmann also described his electrical experiments in the *St Petersburg News*. Using a 6 ft long iron rod held by a glass cylinder as an antenna, he attached an iron wire and brought it through insulation into a room. From the free end of the wire an iron ruler was suspended, and a silken thread was attached to the upper end of the ruler, the thread's free end passing over a graduated arc. When the wire became charged electrically, the thread moved away from the ruler, "running after the finger". During a thunderstorm on 18 July 1753 sparks could be drawn from the metal ruler. Lomonosov, Richmann's closest friend and best known Russian scientist of the time, shared in these electrical experiments. On 26 July 1753, Richmann observed an approaching thunderstorm and hurried to his home accompanied by the Academy engraver Sokolov. Richmann approached

> "a foot away from the iron rod, looked at the electric indicator again; just then a palish blue ball of fire, as big as a fist, came out of the rod without any contact whatsoever. It went right to the forehead of the professor, who in that instant fell back without uttering a sound . . ." (see Fig. 5, also Menshutkin, 1952, pp. 85–89, and Chapter 16.1).

In his full years, Franklin summed up what he then considered his contribution to electrical science in a letter to Dr Byles of Boston written on 1 Jan. 1788:

> "It gives me much pleasure to understand, that my points [lightning rods] have been of service in the protection of you and yours. I wish for your sake that electricity had really proved, what at first it was supposed to be, a cure for the palsy."

In a later letter to another Bostonian, Franklin wrote (1833):

> "I have been long impressed with . . . the conveniences of common living, and the invention and acquisition of new and useful utensils and instruments; so that I have sometimes almost wished it had been my destiny to be born two or three

Tod des Physikers Richmann.

Fig. 5. The first martyr to the new electrical science was the Russian physicist G. W. Richmann when attempting to repeat Franklin's sentry-box experiment. On 6 Aug. 1752 lightning struck the ungrounded apparatus in his St Petersburg study, instantly killing him.

centuries hence. For invention and improvement are prolific and beget more of their kind. The present progress is rapid. Many of great importance, now unthought of, will before that period be produced; and then I might not only enjoy their advantages, but have my curiosity gratified in knowing what they are to be."

We, whose destiny it has been to fulfil the accomplishments that Franklin dreamed of, can now review that long and broad phalanx of electrical inventions that are traced back to those that he begot. Franklin died in Philadelphia in 1790 with his nation at peace and his work in electrical science about to commence its epochal expansion.

References

Beccaria, G. (1753). "Dell'Elettricismo Artificiale e Naturale." F. A. Campana, Torino.

Benjamin, P. (1895). "Intellectual Rise in Electricity." Longmans, Green and Co, London.

Chigi, A. (1777). "Del'Elettricità Terrestre-Atmosferica." L.e B. Bindi, Sienna.

Cohen, I. B. (1941). "Benjamin Franklin's Experiments." Harvard University Press, Cambridge, Mass.

Cohen, I. B. (1943). Benjamin Franklin and the mysterious "Dr Spence". *J. Franklin Inst.* **235**, 1–25.

Cohen, I. B. (1952a). Benjamin Franklin: An experimental Newtonian scientist. *Am. Acad. Arts Sci.*

Cohen, I. B. (1952b). Prejudice against lightning rods. *J. Franklin Inst.* **253**, 393–440.

Cohen, I. B. (1952c). The two hundredth anniversary of Benjamin Franklin's two lightning experiments and the introduction of the lightning rod. *Proc. Am. phil. Soc.* **96**, 331–366.

Cohen, I. B. (1953). "Benjamin Franklin." Bobbs-Merrill Co, Indianapolis.

Fischer, J. N. (1784). "Beweis dass das Glockenläuten bey Gewitter schädlich sey." Munich.

Franklin, B. (1751, 1774). "Experiments and Observations on Electricity made at Philadelphia." E. Cave, London (1751, 1774) and I. B. Cohen edition (1941); French translation, 2nd ed., Paris (1756).

Franklin, B. (1833). "Collection of Familiar Letters." Charles Bowen, Boston.

Franklin, B. (1964). "The Autobiography of Benjamin Franklin." Yale University Press, New Haven.

Freke, J. (1746). "An Essay to Shew the Cause of Electricity", p. 29. London.

Landriani, M. (1784). "Del'Utilità dei Conduttori Elettrici." Marelli, Milano.

Mazéas, G. (1752). Letters to Stephen Hales concerning the success of the late experiments in France. *Phil. Trans. R. Soc.* **47**, 534–552.

Menshutkin, B. N. (1952). "Russia's Lomonosov." Princeton University Press.

Mottelay, P. F. (1922). "Bibliographical History of Electricity and Magnetism." Charles Griffin and Co., London.

Nollet, J. A. (1743). "Leçons de Physique Expérimentale", Vol. IV, p. 314, Paris.

Nollet, J. A. (1753). "Lettres sur l'Électricité." H.-L. Guerin and L.-F. Delatour, Paris.
Pace, A. (1958). "Benjamin Franklin and Italy." American Philosophical Society, Philadelphia.
Pepper, Wm (1970). "The Medical Side of Benjamin Franklin." Argosy–Antiquarian, Ltd, New York.
Philosophical Transactions of the Royal Society, London (1752, 1753, 1754).
Priestley, J. (1767). "History of the Present State of Electricity." Dodsley, Johnson, Davenport and Cadell, London.
Ronalds, F. (1880). "Catalogue of Books and Papers Relating to Electricity, etc." E. and F. N. Spon, London.
Schonland, B. F. J. (1964). "The Flight of Thunderbolts." Clarendon Press, Oxford.
Smyth, A. H. (1907). "The Writings of Benjamin Franklin," pp. 421–422 (10 vols). Macmillan, New York.
Toaldo, G. (1772). "Della Maniera di Preservare gli Edificj dal Fulmine." Venezia (and Parma).
Toaldo, G. (1778). "Dei Conduttori per Preservare gli Edificj da'Fulmini", pp. 63–76. G. Storti, Venezia.
Volta, A. (1800). On the electricity excited by the mere contact of conducting substances of different kinds. *Phil. Trans R. Soc.* 403–431.
Winkler, J. H. (1746). "Die Staerke der electrischen Kraft", p. 137. Leipzig.

3. The Thundercloud

C. B. MOORE and B. VONNEGUT

New Mexico Institute of Mining and Technology, Socorro, New Mexico and State University of New York at Albany, New York, U.S.A.

1. Introduction

For a storm to be classified as a thundercloud or thunderstorm, meteorologists require that thunder must be heard. Since this peculiar and intense acoustic disturbance has no other natural source, the presence of lightning is necessarily required. These storms, invariably composed of strongly convective cumulonimbus clouds, are usually accompanied by strong wind gusts and rain, or sometimes hail or snow. Thunderclouds are usually a consequence of atmospheric instability and develop as the warm, moist air near the earth rises and replaces the denser air aloft. This overturn often results in the condensation of atmospheric water vapour forming a visible cloud of water droplets. When this occurs, the heat associated with the phase changes of water acts to speed the overturn: release of the heat of vaporization by condensing water vapour enhances the updraughts, while cooling, caused by evaporation of condensed water, can help drive the downdraughts which replace some of the ascending subcloud air.

In effect, thunderclouds are large atmospheric heat engines with water vapour as the primary heat-transfer agent. The output of these engines is the mechanical work of the vertical and horizontal winds produced by the storm, electrical work on free charge resulting in lightning discharges and an outflow of condensate in the form of rain and hail from the bottom of the cloud and of small ice crystals from the top of the cloud. In addition, thunderclouds increase the local stability of the atmosphere, transport horizontal momentum vertically, and are believed to maintain the atmosphere's electrical potential relative to the earth. During daylight, the snow crystals that are blown away from the top of the cloud in the form of cirrus by the high level winds, reflect much of the incoming solar radiation back to space thus reducing solar heating of the earth's surface and suppressing the development of other thunderclouds.

The processes that operate in a thundercloud to produce these actions are varied, complex and poorly understood. Since thunderclouds are vast, turbulent and hazardous, their interiors have been inaccessible and good information is unavailable on the conditions within their boundaries. This chapter catalogues some of what is known and discusses some of the speculations that have been offered about unsolved problems.

2. Thunderstorm Occurrence

Estimates made on the basis of climatological observations, satellite data and inferences from electrical observations indicate that, at any time, about 1,000 thunderstorms are continuously in progress over the surface of the earth. Although the primary activity occurs in the lower latitudes, thunderclouds are sometimes observed in the polar regions. The global distribution of thunderstorms corresponds to what might be expected from their convective origin. The greatest frequency is to be found where and when vertical convective activity is at a maximum and much of this is controlled by radiation processes: solar heating warms the surface of the earth each day with a thermal input of about 1 kW m^{-2} of perpendicular surface while the upper troposphere is cooled continually by the outward thermal radiation from water molecules and aerosol particles. As the earth rotates beneath the sun, new thunderstorms form in the subsolar area so that a wave of thunderstorm development moves westward each day.

Thunderstorms occur commonly over warm sea coasts when breezes from the surrounding ocean are induced to flow inland after sunrise as the land surface is warmed by solar radiation. Similarly, because mountains are heated before the valleys, they often aid the onset of convection in unstable air. In many parts of the world, diurnal thunderstorms occur over mountainous terrain.

In addition to air-mass convection (and sometimes superimposed upon it) is the intense convective activity that occurs when cold air meets warmer, moister air, slides under and lifts it: vigorous thunderstorms occur along an active cold front and in squall lines in the warm air ahead. While most thunderstorms develop around midday in the spring and summer months (Chapters 14.2.3.3 and 15.3) when the potential for convection is usually the greatest and adequate water vapour is available, they have been observed at all times throughout the year in temperate latitudes, as a result of frontal activity. Further, thunderstorms frequently develop over the North Atlantic Ocean during the winter when cold Arctic air flows over the warmer Gulf Stream.

Water vapour concentrations in excess of 7 g of water kg^{-1} of dry air are required generally for warm-season thunderstorm formation although lightning has been reported in winter clouds over the unfrozen Great Lakes when the water vapour mixing ratio could not have exceeded 4 g kg^{-1} of air.

The global thunderstorm activity now occurring has presumably been taking place from times early in the development of the earth's atmosphere. The evidence for this inference is obtained from geological information: when lightning strikes dry sand, the resultant high temperature melts and vitrifies the silica, forming tubes known as fulgurites, similar to the contemporary formations which are found in our beaches and deserts. Ancient fulgurites believed to be 250 million years old have been detected in geological deposits (Harland and Hacker, 1966). Speculations have been advanced by a number of scientists that lightning has played a significant role in the modification of the early atmosphere to its present state and in the origin of life on the planet (Chapter 4.5).

Lightning rarely occurs in cumuli with depths smaller than about 3 km but it has been observed in volcanic eruption clouds with dimensions of less than 500 m. Normal lightning and thunder occur frequently in large clouds. The greatest activity occurs in the largest convective systems which approach 20 km in depth. The temperature of the air and the phase of cloud water at cloud-base and cloud-top levels do not seem to be critical for the development of cloud electrification. The main requirements are that sufficient water vapour be present to power the cloud and that the atmosphere be unstable to vertical motions. Lightning in clouds everywhere warmer than 0 °C has been reported (Foster, 1950; Moore *et al.*, 1960; Pietrowski, 1960; Hiser, 1973), conversely, electrification has been observed in clouds everywhere colder than 0 °C (Colson, 1960; Fuquay, 1962) and in clouds of ice crystals.

Active thundercloud systems range in horizontal extent from about 3 km in diameter to dimensions greater than 50 km. Along cold fronts merged thunderclouds may occur in lines extending for hundreds of kilometres.

The life of an "air-mass" thundercloud over the New Mexican mountains is typically about 2 hours. Smaller, electrically active clouds in the subtropics have been observed with durations of less than 30 minutes. On the other hand, large, frontal-cloud disturbances have persisted for more than 48 hours and moved more than 2,000 km.

3. Development of Thunderclouds

3.1 Convection

The lower atmosphere becomes unstable and can be overturned with a release of mechanical energy whenever the vertical lapse of air temperature becomes more negative than $\partial T/\partial z = -9{\cdot}8\ ^{\circ}\mathrm{C\ km^{-1}}$, the adiabatic lapse of temperature with altitude.

When the earth is heated by sunlight, much of the energy goes into warming the air in contact with its surface. At midday, the normal solar input is sufficient to heat a surface layer of air 1 m in thickness by about $1\ ^{\circ}\mathrm{C\ s^{-1}}$. As a result of this heating from below, the density of the lower air is decreased and the atmosphere can become unstable for vertical motions. During the warm seasons of the year, solar heating often causes local overturns of the lower atmosphere with the formation of updrafts of warm air surrounded by compensating downdrafts of colder air from aloft. When this occurs, atmospheric convection often becomes organized into polygonal cells similar to those observed in heated fluids by Thomson (1881) and Bénard (1901). In this mode of overturn, air ascends in the interiors of the cells and descends at the common boundaries between the cells. The height of the cell depends on the vertical distribution of temperature in the atmosphere. One effect of this convection is the mixing of the lower atmosphere making it homogeneous in its properties.

Atmospheric pressure decreases with altitude so that rising air expands; this effect causes the temperature of rising air to decrease as it exchanges its internal heat energy for geopotential energy of height. If the ascent of a parcel of air continues, its cooling eventually causes some of the water vapour to condense onto dust particles in the air forming small liquid droplets scattered through the volume thus creating a "cloud". The cloud droplets grow rapidly thereafter by further condensation, soon reaching diameters of about 10 μm and concentrations of a hundred or more droplets per cubic centimetre. Liquid water concentrations amounting to one or two grams of liquid water per cubic metre of air are frequently developed. The thermal energy given up as the water molecules pass from vapour to liquid prevents the rising air from cooling as rapidly with increasing height as does the surrounding atmosphere; this makes an updraft more buoyant so that often it accelerates upward condensing more water and causing additional unstable air from below to ascend.

More dense, dry air surrounding the cloud and displaced by the updraft is caused to subside as air is removed from below. The shearing stresses between the ascending and the descending air parcels cause local turbulence in which dry air from outside the cloud can be mixed with saturated cloudy air containing condensed liquid water. When this occurs, the liquid water evaporates into the dry air at the expense of the internal energy of the adjacent air (and water vapour) thus cooling it and promoting the downdraughts.

On occasion, a few adjacent convection cells grow much larger than their neighbours and concentrate the atmospheric overturn in a small region of the unstable layer. A convergent flow of warm moist air develops in the subcloud region so that the localized updraughts are supplied with air from a wide area beneath the cloud system.

As a result of the cellular nature of convection, cumuli usually contain several regions with simultaneous updraughts and adjacent downdraughts in differing degrees of development. The dimensions of the regions with updraughts vary widely as they depend on the thickness and density of the subcloud layers, on the atmospheric stability and viscosity, on wind speed and on other properties. Typical updraught diameters ranging from 300 to 2,000 m have been reported (Jones, 1952; Vonnegut *et al.*, 1962). As may be expected from continuity considerations, the widths of the annular regions with downdraughts are generally less than those of the central regions with updraughts but adequate data are not available to describe them.

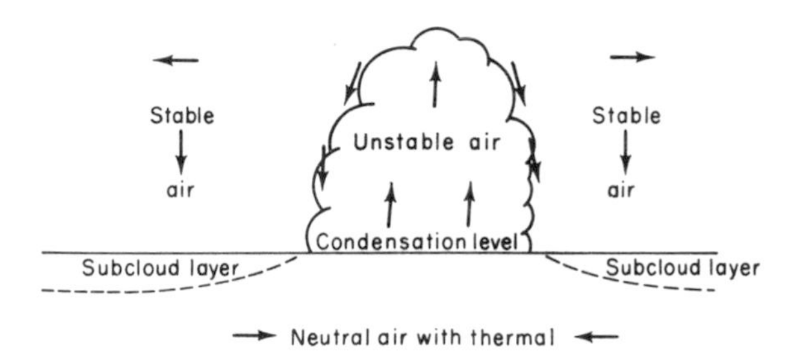

Fig. 1. The motions of air in and around a cumulus growing into stable air according to Scorer (1958). The cloud develops as a result of thermal updraughts in air beneath the condensation level.

The dynamics of turbulent convection are difficult to describe analytically but the motions of a buoyant cloud element are often seen as similar to that of a bubble rising as a spherical vortex through a less active fluid. Scorer's (1958) view of these motions is shown in Fig. 1. Levine (1959) and Malkus (1960) give an equation of motion for a cloud bubble as

$$\frac{\mathrm{d}W}{\mathrm{d}t} = g\frac{(\theta_v - \theta_0)}{\theta_0} - \tfrac{3}{8}(\tfrac{3}{2}K + C_d)\frac{W^2}{a},$$

where

W is the rate of rise of the bubble,

g is the gravitational force per unit mass,

θ_v is the cloud element's virtual potential temperature, which is described below,

θ_0 is the environmental virtual potential temperature,

K is the fraction of mass exchanged in a half cycle of vortex motion,

C_d is the form drag coefficient,

a is the radius of the cloud bubble.

The potential temperature of the air, θ, is defined as the Kelvin temperature it would have if compressed adiabatically to a pressure of 10^5 N m^{-2}, i.e.

$$\theta = T_0\left(\frac{10^5\ \mathrm{N\ m^{-2}}}{P_0}\right)^{R/C_p},$$

where C_p/R is the heat capacity of the air in universal gas constant units and is approximately equal to 3·5. The use of the virtual temperature allows for the changes in air density caused by the presence of water vapour. The virtual temperature of a parcel of air containing water vapour is defined as the temperature that an equivalent parcel of dry air would have at the same pressure if the densities of the two parcels were identical.

In Levine's derivation, the upward velocity near the centre of the bubble is about 2·5 W, while the maximum downdraught speed around its periphery is about $-0{\cdot}5$ W. Similar motions have been shown by Saunders (1962) and others.

Other students of convection have suggested that the upward motions of buoyant air may be similar to that of a rising jet or "plume" with environmental air being mixed into the ascending air from its sides by the local shear. As a result, a vertically rising plume of buoyant air often increases its diameter with its ascent. Quasi-steady-state, jet-like updrafts are observed frequently over volcanoes and sometimes found in severe storms, but measurements with instrumented aircraft of the vertical motions in small active cumuli often tend to support the bubble model of cloud convection. Malkus (1954) and Telford and Warner (1962) report updraughts of 4 to 6 m s^{-1} surrounded by downdraughts of 2 to 4 m s^{-1}. Similar determinations in mature thunderclouds by Vul'fson (1957) and by Steiner and Rhyne (1962) indicate some updraughts with velocities greater than 60 m s^{-1} and downdraughts of 35 m s^{-1}.

In the absence of precipitation, these downdraughts seem to be limited to the vicinity of the updraughts; they usually do not extend down to the earth's surface or even far into the subcloud air. There is some evidence indicating that downdraught air flows outward from the base of a thundercloud above the unstable subcloud layer.

The vertical penetration of a convective parcel of rising air is limited by atmospheric stability, by its dilution as a result of mixing in the surrounding air and by frictional forces. These limitations are greatest for small convective parcels. Accordingly, cumuli often grow in steps with the residue from one "burst" of convection protecting the next one; the active life of a single convective parcel is often of the order of 2 to 5 minutes.

If convection continues, more vigorous turrets often develop and rise with vertical speeds greater than 10 m s^{-1}. Often the rapid rising of a turret can be observed as it penetrates a stratum of moist air and lifts it causing local adiabatic cooling and a *pileus*, or mantle cloud, draped around the turret. The term *cumulus congestus* is applied usually to a convective cloud with strong vertical development.

When the temperature of the air in a cloud decreases below 0 °C, some of the condensed liquid water may freeze and release its heat of fusion which can increase the cloud's buoyancy. This does not occur early in the life of natural clouds: pure water in small quantities has a low probability of freezing spontaneously until its temperature is lowered to −40 °C, the Schaefer temperature. As a result, appreciable quantities of liquid water can be found in many convective clouds at temperatures far below 0 °C. Snow crystals may form in these clouds by sublimation of water vapour onto crystalline aerosol particles but ice-forming nuclei are frequently absent and snow usually does not form until the cloud tops reach high altitudes where the atmosphere is very cold. Precipitation is first detected as a light drizzle of small raindrops in many developing clouds when they are in the cumulus congestus state. As the precipitation particles grow and fall, they cause local downdraughts and interact with the convective overturn in several ways that are discussed later.

Vigorous cumuli with strong updraughts often continue to grow in volume and height until they encounter a thermally stable layer in the atmosphere that serves as a ceiling on their further vertical development. The stable stratosphere frequently limits the final height of most large cloud systems; however, over the southwestern United States lower stable layers often check the ascent of marginal thunderclouds so that many of them go no higher than 7 or 9 km above sea level, with bases about 4 km above sea level.

When a rising cloud parcel encounters a stable layer, its vertical motion is usually deflected horizontally. The cloud top often loses its cumuliform appearance and exhibits a toroidal or a flattened top as shown in Fig. 2. Vertical striations appear around the periphery of the cloud top; penetrations of large clouds by an aircraft at this time indicate that the particles in the visible cloud top are now largely snow crystals. Often strong horizontal

Fig. 2.

winds occur at cloud top altitudes; these blow the flattened layer of ice crystal cloud downwind, forming a *Cb incus*, or *anvil* cloud.

If the thundercloud continues to develop, very active turrets in the centre are often observed to rise for a short time above the equilibrium cloud top level, penetrating some distance into the overlying stable air and then falling back. The overshoot of the equilibrium level is a result of the vertical velocities acquired by the rising cloud parcel during its ascent through the troposphere. A photograph taken from Apollo 9 over central America in Fig. 3 shows the view from above an active thundercloud with a turret penetrating into the stratosphere.

The penetration, P, of a cloud parcel into an isothermal stratosphere has been related to the requisite updraught speed W' by equating the parcel's vertical kinetic energy to the work required to lift an adiabatically cooling parcel against the buoyancy forces (Vonnegut and Moore, 1958; Saunders, 1962) as

$$P \approx W' \Big/ \sqrt{\left(\frac{g}{\theta}\frac{\partial \theta}{\partial z}\right)},$$

where g is the gravitational force per unit mass and $\partial\theta/\theta\,\partial z$ is the fractional change of stable-layer potential temperature with height. For an isothermal stratosphere at 218 K, $P \approx 49W'$ s.

Stratospheric penetrations of up to 6 km have been reported over vigorous electric storms. From these, thundercloud updraught velocities of about 100 m s^{-1} have been inferred (Roach, 1967).

Temperature deficits in the top surfaces of thunderclouds in excess of 5 or 10 °C (Roach, 1967) relative to the surrounding air have been measured by the use of infrared sensors mounted on aircraft flying over penetrating turrets. A turret at its apogee is therefore negatively buoyant and it thereafter rebounds downward.

The duration τ of a cloud top excursion into stable air above its equilibrium level is approximately given by the half period of the Brunt–Väisälä oscillation as

$$\tau \approx \pi\{T/[g(\Gamma_0 - \Gamma_1)]\}^{\frac{1}{2}} \quad \text{(Emmons \textit{et al.}, 1950)},$$

where T is the air temperature in the vicinity of the cloud top, Γ_0 is the adiabatic lapse of air temperature with altitude ($-9{\cdot}8$ °C km^{-1}) and Γ_1 is the

---

Fig. 2. A thundercloud developing over Langmuir Laboratory in the mountains of central New Mexico. The cloud base was about 4 km above sea level while the cloud top was about 6 km higher. The 0 °C isotherm within the cloud was about 5 km above sea level. The cloud top was composed of ice crystals while the droplets in the cloud base were liquid water.

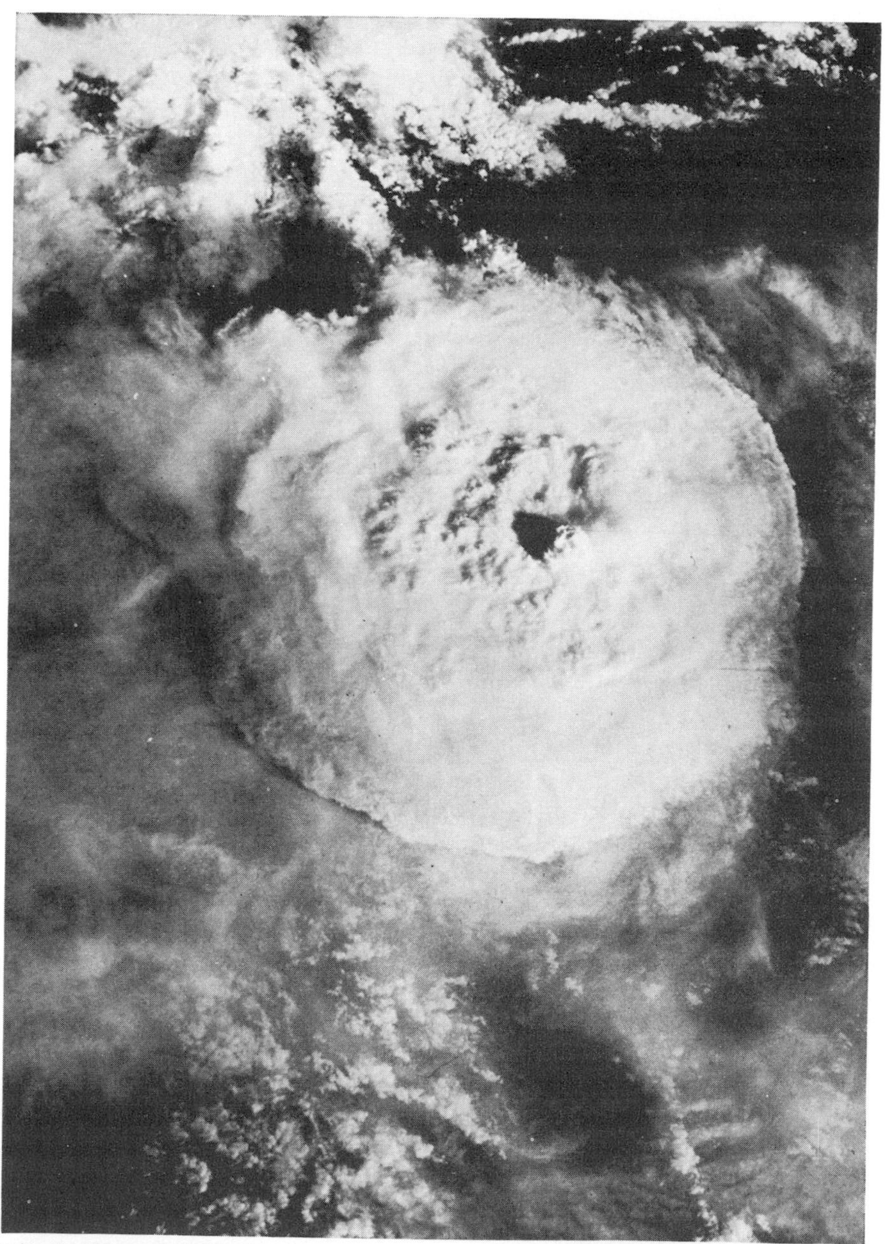

Fig. 3. View of a thundercloud over Honduras from space. This photograph was taken from Apollo 9 at an altitude of 190 km. The dark spot in the centre of the anvil cloud is the shadow of a turret that penetrated more than a kilometre above the equilibrium cloud top level. (Photograph courtesy of NASA.)

environmental temperature lapse. For penetrations into an isothermal stratosphere at $T = 218$ K, τ is of the order of 150 s and the locus of the excursion plotted against time follows the upper branch of a sinusoid.

The motions of air on the underside of an anvil cloud often seem to be convergent with streaks of falling snow at appreciable angles to the vertical and usually directed backward toward the parent cloud. Although little is known about the circulation around an anvil, these motions appear to be a consequence of the overturn process.

Often accompanying the growth of a mature thundercloud is an appreciable convergent inflow of air in the subcloud layer as it is drawn in by the storm to continue the overturn process. Byers and Braham (1949) report convergences of the order of 4 h^{-1} beneath active thunderclouds. Mass flows of subcloud air into the base of some single large thunderstorm systems in excess of 10^5 kg s^{-1} have been estimated by Newton (1968).

A large downdraught penetrating to the earth's surface often develops as intense precipitation drags air downward. Cooling by evaporation of cloud and some of the precipitation into the descending air accelerates its downward motion and causes an outflow of cold stable air when the downdraught reaches the earth. These downdraughts often shut off the supply of warm unstable air in the subcloud layer to the updraughts thus terminating the local overturn. Portions of a cumulus are negatively buoyant; when the updraughts cease, the cloud subsides and much of the condensed water evaporates.

Byers and Braham (1949) reported that the precipitation-driven outflow from a thundercloud can act as a local cold front and initiate new convective cells on the periphery of the older, now dissipating, cloud. Severe thunderstorms can also interact with strong external winds and become migratory, atmospheric-overturning disturbances operating effectively in a quasi-steady-state mode. Discussion of these complicated storms is outside the scope of this survey; the reader is referred to the meteorological literature for further information (Newton and Newton, 1959; Browning and Ludlam, 1962; Raymond, 1975).

Updraughts in clouds on occasion become so vigorous that angular momentum in the air supplying the updraught becomes concentrated. The resulting circulation may limit the further horizontal influx of air into the updraught although air may continue to flow into the updraught from below. This causes the circulation to propagate downward as air flowing into the updraught continues to leave its angular momentum behind. The resulting vortex is a region of low pressure around which air circulates in balance, with its centripetal acceleration supplied by the pressure-gradient force. A vortex extending downward beneath the base of a thundercloud often produces a characteristic funnel-shaped cloud known as a *tornado*. Its surface is often defined by the dew-point isotherm of the ambient air when it is expanded adiabatically into the low pressure region around the vortex.

The strongest winds in our atmosphere are found in tornadoes, where wind speeds in excess of 130 m s^{-1} have been inferred (Fujita, 1975). In the same regions concentrations of angular momentum as great as 7,500 m^2 s^{-1} have been deduced (Lewis and Perkins, 1953). The minimum pressures in tornadoes are not known but some observers have reported values of about 0·8 that of the surrounding atmosphere (Flora, 1953).

The means by which the low pressure is maintained have also not been established; although meteorologists generally agree that heat releases into the atmosphere can produce low pressures by increasing the temperature of the air and thus decreasing its density. Fire-storm vortices of tornadic violence have been reported over large conflagrations produced by burning cities, forests and oil tank farms (Hissong, 1926; Graham, 1952; Ebert, 1963). These observations suggest that large updraughts caused by local heating can become organized and that they can concentrate angular momentum to produce strong whirlwinds.

Vonnegut (1960) has concluded that temperature contrasts in the atmosphere are insufficient to account for severe tornadoes and suggested that some of them may be powered electrically: repeated lightning and other discharges through the low pressure vortex may heat the air and intensify the updraught. While luminosities and other electrical phenomena have been repeatedly observed in tornado vortices (Flora, 1953; Montgomery, 1956; Vonnegut and Weyer, 1966; Vaughan and Vonnegut, 1976), a full discussion of them is beyond the scope of this survey. A recent article on tornadoes by Davies-Jones and Kessler (1974) provides a more extensive examination and some alternative views on atmospheric vortices.

A curious feature of some tornado-producing thunderstorms is the "vault", which can be detected by vertically scanning radars. It is a weakly echoing, vertically elongated volume that extends from cloud base to altitudes as great as 13 km above sea level (Browning, 1965). It contains little precipitation but usually the most intense precipitation echoes in the storm are found beside and above it. The vault is usually located above tornadoes and beneath a sustained cloud dome that penetrates significantly into the lower stratosphere. It appears to contain sustained updraughts with velocities of the order of 50 m s^{-1} at the tropopause (Browning and Donaldson, 1963). The absence of strong precipitation echoes in an active updraught that extends through the depth of the troposphere is quite surprising; speculations as to the cause of this feature include centrifugal evacuation of the particles within a rotating column of air, evaporation of the particles by heating or possibly the short time available for their growth in a high velocity updraught.

Although significant progress is being made both with improved observations and with modelling of convection processes, the preceding should illustrate that the motions of air in thunderclouds are complex and that our present understanding is inadequate.

3.2 Precipitation

Much of the precipitation that falls from thunderclouds occurs in characteristic, intense, abrupt bursts of rain and sometimes of hail. Long duration, gentle rains fall most usually at the end of the storm during its dissipation. A typical plot of rain intensity beneath a thundercloud is shown in Fig. 4.

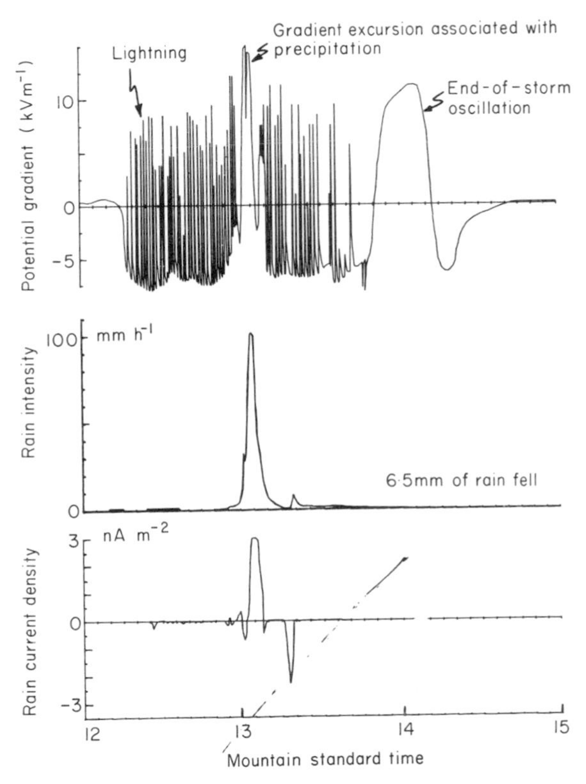

Fig. 4. Plots of the precipitation intensity and the electrical potential gradient beneath an isolated, horizontally stationary thundercloud over Mt Withington, New Mexico. The rain intensity and the potential gradient records are typical of those obtained beneath the centre of a thundercloud where bursts of rain, gradient excursions associated with precipitation and lightning, and the end of storm oscillation in the potential gradient are regularly observed.

Precipitation is first detected in a developing cumulus congestus, of depth 2 or 3 km, by aircraft penetrations or by a sensitive, ground-based radar when some of the cloud droplets have grown by coalescence and condensation to diameters of about 200 μm. Although snow can form early in clouds penetrating to levels above that of the 0 °C isotherm, most of the observed precipitation develops as liquid water drops in growing cumuli even though the air temperature may be appreciably lower than -10 °C. The first echoes provided to a radar by developing precipitation are often located near the top of an active turret and frequently appear near the time that the first cloud electrification can be detected by airborne instruments.

The initial development of precipitation and of cloud electrification are often closely linked although the cause-and-effect relation is not clear. Intense precipitation can develop without subsequent electrification but appreciable cloud electrification is invariably accompanied with detectable and often significant rainfall.

Precipitation echoes within a few kilometres of a 3-cm radar can be detected at reflectivity, Z, levels as low as 10^{-3} mm^6 m^{-3}, indicating precipitation intensities aloft of less than 0·02 mm h^{-1}. Immediately after detection the echoes often intensify at initial rates exceeding 10 dB min^{-1} but as the echo reflectivity increases, its rate of intensification decreases. Two or three minutes after initial detection the echo often achieves Z values of about 1,000 mm^6 m^{-3} indicating precipitation intensities in the cloud of 3 mm h^{-1} or so. Within 5 or 6 minutes, scattered raindrops 2 or 3 mm in diameter arrive at the earth with intensities of 1 or 2 mm h^{-1}.

If the cloud continues to increase in size, so do the precipitation particles within it. A sequence often noted is one in which the first lightning flashes, the precipitation echo overhead abruptly intensifies by 20 dB or more, and an intense burst of rain or hail arrives at the earth a few minutes after the discharge. The subsequent precipitation intensity frequently exceeds 75 mm h^{-1} for 30 to 60 s and then decreases approximately exponentially to 2 or 5 mm h^{-1} within 5 or 6 minutes after the lightning. If another discharge occurs nearby, the sequence may be repeated.

About 75% of the summertime precipitation that has fallen on our mountaintop laboratory during the past 13 years has done so within 5 minutes of a nearby lightning discharge. The close association between lightning and subsequently developing precipitation is not limited to New Mexico; the gush of rain arriving after nearby lightning has been reported since antiquity (Lucretius, 58 B.C.). For example, Robert Hooke (1664) wrote:

> "If it rained when it thundered, it immediately after the clap poured down much faster, much as if a gale of wind had suddenly shook a tree, all of whose leaves are full with drops of water."

Most of the modern guesses about this phenomenon have attributed its cause to the levitation of charged raindrops by a strong electric field in the cloud which, when lightning occurs, is reduced allowing the precipitation particles to fall (Schonland, 1950; Ziv and Levin, 1974). Schonland's guess was made without any direct information on the development of precipitation within clouds. A number of radar observations have been reported subsequently in which intense precipitation was not even present in the cloud before the first discharge; but after the discharge it developed abruptly in the same regions from which the flashes originated (Moore *et al.*, 1962, 1964; Wilk, 1963; Battan and Theiss, 1970; Hiser, 1973; Few, 1974). Schonland's explanation, therefore, does not fit the observed sequence of events.

Another speculation as to the origin of the phenomenon is that the air motions caused by the acoustic wave of thunder through the cloud suddenly increase the frequency of collisions between cloud droplets thus enhancing their rate of coalescence into raindrops. Whilst this fits the observed time sequence better than the levitation explanation, it limits the time for growth to the duration of intense thunder and this is a severe constraint.

The authors have proposed yet another guess in which the final rearrangement of charge deposited along lightning channels in an electrified cloud enhances the subsequent collisions between cloud particles (Vonnegut and Moore, 1960). The process is shown schematically in Fig. 5 and is discussed in detail by Moore *et al.* (1964). The evidence gathered in the past decade suggests that this process may often be operative in thunderclouds but there are many cases in which intense precipitation may also be formed without the intervention of lightning. The initial intensifications and arrival of this precipitation, however, are different and less abrupt than the usual gushes of rain and hail after lightning.

A different explanation of precipitation growth in thunderstorms has precipitation particles falling downward against a vigorous updraught so that they act as a suspended "filter" extracting the rising cloud droplets by colliding and coalescing with them. This process continues until the weight of the accumulated water overbalances the updraft buoyancy and the precipitation is dumped from the cloud. The importance of this process relative to the rain gush that follows lightning has yet to be established for we do not know the details of all the processes by which thunderstorm precipitation may be formed.

The growth of large cloud droplets by collision and coalescence with more slowly falling, smaller ones has long been advanced, e.g. Le Grand (1680). The process cannot be as simple as suggested, for Rayleigh (1879) and others have shown that in the atmosphere small water droplets colliding in the absence of electric forces usually rebound as a result of an air film between

It is suggested that a lightning stroke terminating in a cloud is finite in extent and that it does not subdivide into very small sparks that remove the charge from each and every electrified particle. Instead it is postulated that the lightning stroke accomplishes only a rough neutralization by introducing a tree-like pattern of electric charge into the cloud having an opposite sign to the charge originally within the cloud. This figure shows how the lightning stroke that had jumped into a negative region of the cloud might appear within the cloud.

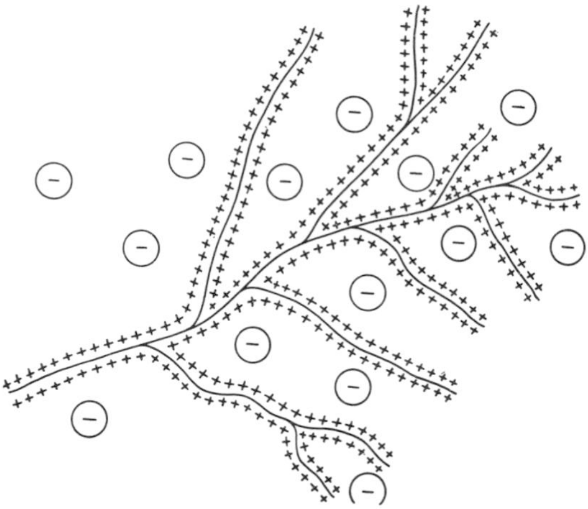

It is suggested that the fast ions released into the cloud by the lightning stroke (of positive polarity in this case) move out into the cloud under the influence of the electrical forces acting on them.

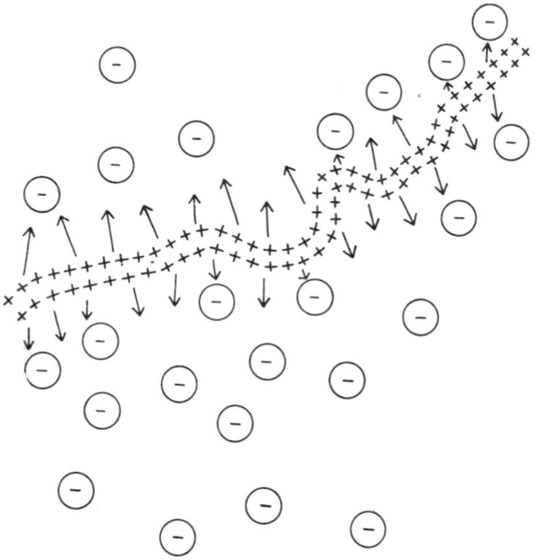

After moving only a short distance the fast ions encounter the oppositely charged cloud particles to which they became attached, thus first neutralizing their original charge (negative in this case) and then giving them a far stronger opposite charge (positive in this case).

Fig. 5(a). Sketches illustrating the mechanism for the charging of cloud droplets near a lightning channel.

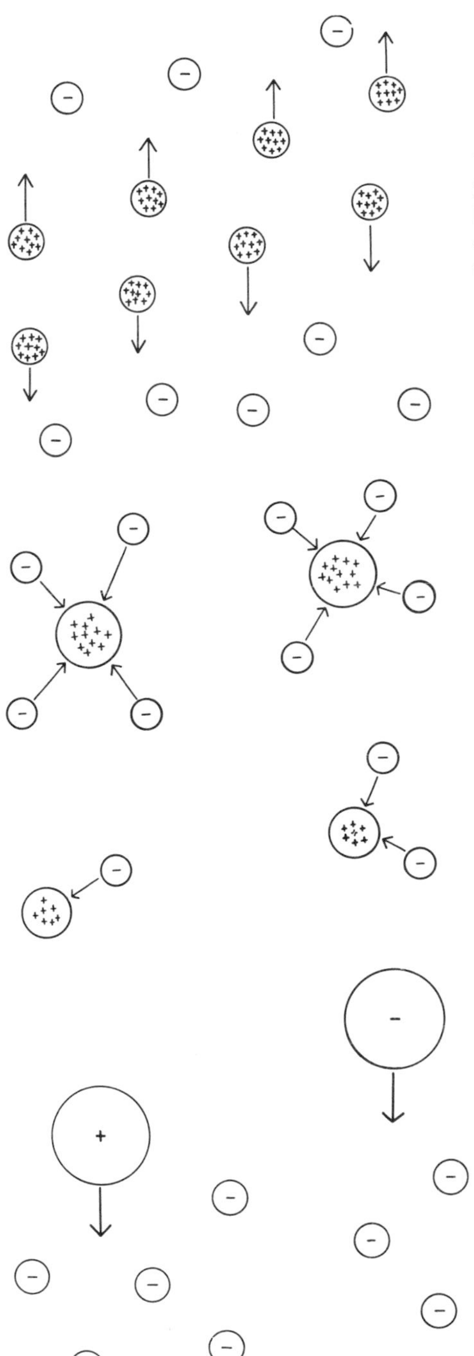

Under the influence of electrical forces the cloud particles that have become highly charged as the result of the lightning then start to move through the cloud.

As the result of their movement in the electric field and their strong electrical attraction for the oppositely charged cloud particles, the highly electrified cloud particles produced by the lightning undergo many collisions with cloud particles thereby losing charge and increasing in mass.

Finally, when the electrified cloud particles resulting from the lightning have undergone many collisions they will have lost their charge and become big enough that they now fall at appreciable velocities under the influence of gravity. They continue to increase in size as they fall by further collisions with cloud droplets. These big falling drops make up the gushes of rain following the lightning.

Fig. 5(b). Sketches illustrating the motion of highly charged droplets in a cloud of oppositely charged ones and the subsequent collisions and coalescences to form the rain gush after lightning.

them that prevents their direct contact during the brief time of the collision. Because sufficiently intense fields to prevent rebounds were assumed to be absent early in the development of the cloud, many investigators minimized the importance of coalescence and have considered alternative processes by which precipitation may be formed. When Wegener (1911), Bergeron (1935) and Findeisen (1938) suggested that much of the precipitation falling on the earth starts out as snow, pursuit of the coalescence process seemed less important. Since 1946, much effort has gone into seeding clouds with ice-forming nuclei in efforts to initiate the formation of snow within them (Section 6). Efforts of this nature aimed at augmentation of rainfall by cloud seeding have had limited (and often questioned) success due, probably in part, to the actual importance of the coalescence mechanism in natural precipitation processes.

Langmuir (1948a) treated the physics of collisions and coalescence for unelectrified droplets quantitatively by assuming that every direct contact resulted in coalescence and by ignoring the droplet deformation and rebound effects. He proposed a *chain reaction mechanism* for the formation of larger cloud droplets that overtake and collide with more slowly falling ones according to an aerodynamically derived inertial "collection-efficiency". The times that are required for some larger cloud droplets to grow into raindrops by this process were calculated to be of the order of two hours. Laboratory simulations of Langmuir's process by a number of investigators confirmed his predictions approximately and supported his assumptions although the electrical conditions within natural cumuli were not necessarily reproduced.

We have found that electrical energies of the order of 10^{-14} J may be necessary to "nucleate" the onset of coalescence between colliding neutral droplets in the air by a local disruption of the air film between the droplets. Because of contact differences of potential, it is difficult to avoid supplying such energies to simulated rain droplets in laboratory experiments. On the other hand, in the natural atmosphere the electrical energies available to condensed cloud droplets as a result of ionization, conduction or from dipoles induced in the fine-weather field are often less than the critical coalescence energy. This may account for the observed stability of many liquid-water clouds that do not readily form precipitation particles. Clouds can persist for hours and produce no radar echo at a Z level of detection of less than 10^{-2} mm^6 m^{-3}. This "stability" immediately disappears when a cloud has become electrified; soon afterwards the entire cloud has reflectivities appreciably above the radar threshold.

The electric forces exerted between charged and adjacent uncharged droplets are often sufficient to allow coalescence on their collision (Plumlee,

1964; Colgate, 1967; Jennings, 1976). Further, the migration of charged droplets in the presence of external electric fields can increase significantly the frequencies of their collisions with other particles; these two effects may explain some of the observed, initially rapid growths of cloud droplets.

In addition to rain, thundercloud precipitation includes hail, graupel (i.e. snow pellets) and ice particles in various forms. Hail in many New Mexican storms occurs in slender shafts that can be recognized with a vertically scanning radar before it reaches the earth. This hail almost invariably follows nearby lightning and often arrives in a "burst" covering the ground over an area only a few hundred metres in diameter. Hail particles falling from New Mexican clouds have a characteristic pyramidal shape with a curved lower surface beneath a pyramid of bubbly ice similar to that reported by Bentley (1932). The maximum diameter for this hail is usually about 5 or 6 mm while the vertical dimension is often 9 or 10 mm. Much larger hail has been observed elsewhere; hail stones with masses in excess of 3·4 kg have been reported (Borovikov and Khrgian, 1963).

Many investigators have inferred that large hail has been recycled several times as it is carried up on different updraughts in severe thunderstorms. Our radar evidence indicates that in New Mexican storms the small hail falls directly out of the cloud and acquires its growth on a single pass through the region having a high liquid-water concentration. When captive balloons are brought down after being flown near the 0 °C isotherm in these clouds, graupel are often found to have collected on the horizontal fins. These particles, when falling freely, usually melt and turn into raindrops before they arrive at the earth.

As a thundercloud begins to dissipate, the character of the echo displayed on the indicator of a vertically scanning radar changes. The echoes higher than the level of the 0 °C isotherm become less intense, as rain falls out and the remaining cloud is converted into snowflakes. Just below the 0 °C level a radar "bright band" frequently appears. This is produced by falling snow which, when it begins to melt, provides large, reflective targets for the radar. This signature is usually absent during the developing phases of a thundercloud when the cloud is composed mostly of liquid-water droplets but toward the end of the storm it appears, showing the presence of snow aloft. The precipitation intensity at the surface of the earth decreases to 1 to 3 mm h^{-1}, which persists for 15 or 20 minutes as the cloud overhead subsides. After the larger snow particles have fallen out and melted, the rain usually ceases. A veil of more slowly falling ice crystals is often left aloft in a cirrostratus overcast.

Newton (1966, 1968) has estimated budgets for water in a number of thunderclouds and reports that for midwestern storms about 40 to 45% of the water vapour entering the cloud falls to the earth as precipitation, while an equivalent fraction is evaporated into the downdraught created by the falling precipitation. The remainder is converted into a cirrus cloud of ice crystals aloft or leaves as uncondensed vapour.

Over the western United States marginal thunderclouds often produce little or no precipitation at the earth's surface (Fuquay, 1962). These clouds, common in early summer, often turn into ice crystals early in their development. Many of them produce a few lightning discharges and a trace of rain before dissipating. Evaporation of precipitation in the subcloud air is common over New Mexico with losses amounting to as much as 1 cm of precipitation.

3.3 Electrification

The earth carries a net negative charge of the order of 5×10^5 C which produces a downwardly directed atmospheric electric field with an intensity of about 0·13 kV m^{-1} at its surface. An equivalent amount of positive electricity resides as a distributed space charge, mostly in the lower atmosphere, so that the strength of the fine-weather atmospheric electric field decreases with altitude. As a result of the vertical electric field, the upper atmosphere is at a mean potential of about +300 kV relative to the earth.

Convective clouds develop in the atmospheric fine-weather electric field and often produce electrical disturbances of their own. Most, but not all, electrified clouds appear to accumulate a net positive charge in their tops and a larger negative charge in their lower regions. The fields from these charges become sufficiently intense to reverse the fine-weather field over and beneath the cloud and to cause currents that maintain the negative charge on the earth against atmospheric ion conduction currents.

The first electrification associated with a developing cloud can often be detected from an aircraft flying over the vigorous turrets rising from a cumulus congestus 3 or 4 km deep. At this time the cloud droplets are usually still liquid (although the temperature of the air near the cloud top may be as low as −15 °C). When the cloud top slumps downward, the early electrical excursion disappears and then recurs as a new turret develops. Similar excursions have been detected over warm clouds (Moore and Vonnegut, 1973).

A radar observer beneath a developing cloud can almost invariably detect a precipitation echo before a significant electrical perturbation is shown by a surface electric field meter. The initial electrification of a cloud, as measured from the earth beneath, is usually a reversal of the fine-weather field polarity with an exponentially increasing perturbation: the time required for the cloud's external electric field to increase by a factor of Naperian e is often about 2 minutes.

A typical electric field recording from beneath a thundercloud is shown in Fig. 4. The electric field is shown here as a potential gradient in the common convention of atmospheric electricity: dominant positive charge overhead produces a positive potential gradient.

When the surface field strength exceeds 1·5 to 2 kV m^{-1} well-exposed objects with small radii of curvature begin to release point-discharge ions with positive polarity into the air under the influence of the field (Chapter 4). Light rain arriving at the earth in this period frequently carries a positive charge as though it had captured some of the positive point-discharge ions moving upward from the surface and returned them to earth. This effect is widely observed: strip chart records of the surface electric field strength and the electric current carried down by precipitation commonly show a "mirror image" relation whenever point-discharge ions are emitted (Simpson, 1949; Chalmers, 1967).

After a period of about 8 minutes, the surface field strength caused by the cloud overhead increases to values of the order of 3 kV m^{-1} and dielectric breakdown and lightning flashes occur within the cloud. It should be noted, however, that many clouds never reach the lightning stage, even though they develop precipitation and strong electric fields. Coincident with the discharge, the surface electric field recording makes a large and abrupt discontinuous excursion back to the fine-weather polarity, as though negative charge had abruptly been removed from overhead. The field record then returns more slowly to its original "foul-weather" trend at an exponentially decaying rate as the cloud's electrification continues.

Recent measurements of the electric field above the earth's surface with balloon-borne meters during this period show a different behaviour (see Fig. 6). The field strengths just before a discharge and the magnitudes of the field changes caused by lightning are much larger aloft than at the surface. The direction of the electric field often does not change significantly, i.e. there is no polarity reversal, and the recovery after a discharge shows a linear increase until the next discharge occurs (Winn and Byerley, 1975, 1976).

After lightning occurs, the intensity of the cloud's radar-echo often increases greatly and, as already discussed, a gush of rain or hail falls from the cloud base a few minutes later.

As the precipitation shaft nears the earth, the surface electric field directly beneath often makes a smooth change back to fine-weather polarity to intensities of 10 kV m^{-1} or so. The charge carried on the precipitation arriving at the earth also reverses polarity, changing from positive to negative during the arrival of the rain gush. The negative charge on the precipitation cannot easily be used to explain the change in electric field as the polarities are wrong for the observed time sequence. The field change towards fine weather is almost completed before the gush arrives. Therefore, if negative charges on the approaching precipitation made significant contributions to the surface field, they would be expected to increase the field excursion in the foul-weather sense instead of reversing it. The observed field excursions suggest that a locally dominant positive charge is appearing over the observatory; this effect is often used as indicating a lower positive charge in the thundercloud. In our observatory, the sequence is known, in shorthand, as the GEAWP or the FEAWP, the gradient (or the field) excursion associated with precipitation.

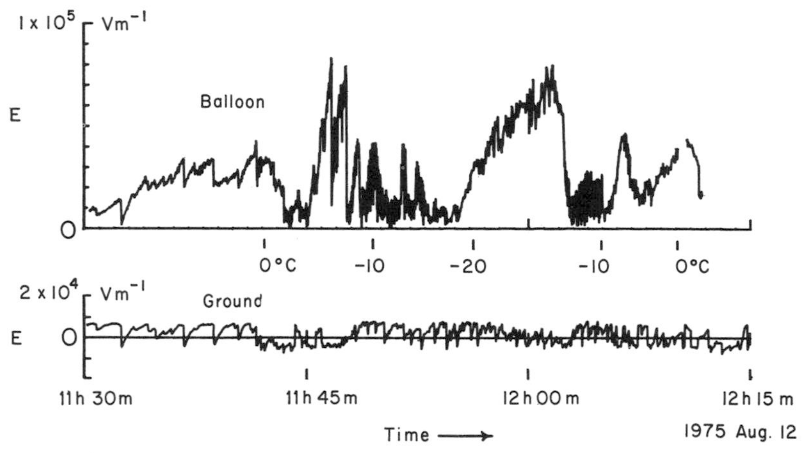

Fig. 6. Simultaneous electric field records from a mountain top and from an instrumented free balloon rising into a thundercloud over the mountain top. Note that the electric field is much stronger aloft than it is at the mountain top. This is due to the balloon proximity to the cloud charges and because it is above the point-discharge ions released at the earth's surface that limit the electric field there. The recovery curves after lightning are also of interest: aloft they are linear whereas at the earth their regeneration is limited by the emission of point-discharge ions when the fields are strong. (From Winn and Byerley, 1976.)

To continue the description of the events beneath a thundercloud: as the gush decays away, the surface field returns to the foul-weather polarity (with values of up to 12 or 15 kV m^{-1}) again indicating the presence of negative charge in the cloud. As the storm grows, lightning flashes frequently and more torrential rain falls from the cloud but any cause-and-effect relations become obscured. Abrupt changes in the electric field record accompany each lightning flash while smooth reversals to fair-weather field polarity frequently precede the arrival of a burst of new precipitation.

Finally, the repeated lightning ceases and the rain dwindles away. One last lightning discharge may occur and often it is anomalous: instead of bringing negative charge to earth it brings positive charge. In this period the vertically scanning radar usually displays the characteristic return from snow throughout the upper parts of the cloud with a horizontal "bright-band" just below the level of the 0 °C isotherm. The electric field records from beneath the centre of the storm now show a typical "end of storm oscillation" (see Fig. 6). As the oscillation starts the field makes another smooth reversal back to strong, fine-weather values while the rain ceases. A continuing and prevalent downdraught can be observed as the cloud subsides. After 5 or 10 minutes the electric field reverses again to foul-weather polarity but now with diminished intensity. The field returns after a few minutes to fine-weather values as the cloud remnants evaporate and the storm ends.

The time sequence reported above is often repeated but we do not understand all of the processes that must be operating in and around the cloud to produce it.

4. Electrical Conditions around Thunderclouds

4.1 Conduction

The lower atmosphere conducts electricity, although poorly in normal situations. This conduction rises as the result of the motion of charged particles under the influence of electrical forces. Under fine-weather conditions, the charge carriers of primary importance are the singly charged clusters of molecules of either positive or negative polarity called small ions. These are produced primarily by ionization from cosmic rays and radioactivity. Small ions are lost to the atmosphere by collisions between ions of opposite polarity which neutralize each other, by attachment to atmospheric aerosols and by contact with the earth. In the free atmosphere, $\mathrm{d}n/\mathrm{d}t$, the rate of change for an ionic concentration with a given polarity is given by relations of the form

$$\mathrm{d}n/\mathrm{d}t = \gamma - \alpha np - \beta N_0 n - \ldots,$$

where

γ is the volume rate of ion production,

α is a coefficient describing volume rate of the recombination of ions of opposite polarity,

p is the concentration of ions with the polarity opposite to n,

β is a coefficient describing the volume rate of attachment of ions with neutral aerosol particles with concentration N_0.

The equilibrium ionic concentrations are calculated from differential equations of this form.

Ions migrate under the influence of the Coulomb forces in a direction such as to weaken the electric field. When the density of the resulting electric current J is linearly proportional to the strength E of the atmosphere's electric field, the conduction is termed *ohmic* with $J = \lambda E$ where λ, the electric conductivity of the atmosphere, is defined by $\lambda = \sum_i^\infty n_i q_i K_i$ with n_i the concentration of the ith class of charge carrier, q_i the charge on the carrier and K_i the ion carrier mobility, i.e. its velocity per unit electric strength.

The conductivity of the air near the earth's surface depends on several ionizing processes: cosmic rays, radioactivity in the earth's crust and radioactive aerosols and gases in the atmosphere. This property, therefore, varies as a function of height above the earth's surface, the stability of the atmosphere, and with local pollution and radioactivity. Over the oceans, which cover most of the planet, terrestrial radioactivity is small and the conductivity of air near but above the surface is found to be of about $1{\cdot}5 \times 10^{-14}\,\Omega^{-1}\,\mathrm{m}^{-1}$ for each polarity. In clean air the conductivity increases with altitude as a result of increasing cosmic-ray-produced ionization. Kraakevik (1957) found in measurements over Greenland that the total conductivity for altitudes between 0·2 and 6 km could be represented by $\lambda(z) \approx 30 \times 10^{-14} \exp[(z-6{,}000)/3{,}100]$ per Ω m.

The flow of current in the atmosphere as a result of the electric field usually acts to dissipate the field with a time constant $\tau_r = \varepsilon/\lambda$, where ε is the permittivity of the medium, usually $8{\cdot}85 \times 10^{-12}\,\mathrm{F\,m^{-1}}$. Accordingly the charge on the earth should be neutralized by conduction from the atmosphere with a time constant of about 600 s. The observations showing that the earth's charge is not being neutralized indicate that, on the average, charge is being supplied to the earth at the same rate at which it is lost by conduction. Evidence suggests that electrified clouds are serving as the primary electrical generators and drive charging currents against the existing electric fields to maintain the earth in its electrified state.

Cloud and aerosol affect electrical properties of the atmosphere in several ways. One of the more important is to reduce the mean free distance that an ion can migrate in the local electric field before it becomes attached to the particulate matter thus shortening its life on the average. The length $\bar{l}_0$ of the mean path of an ion in a cloud of polarized but uncharged, spherical droplets of radius a is given by $\bar{l}_0 = (3\pi a^2 N_c)^{-1}$, where N_c is the number concentration of droplets.

The ion balance relation applicable to weakly electrified clouds is often given in the form

$$\frac{\mathrm{d}n}{\mathrm{d}t} = \gamma - \alpha np - 4\pi a N_c\, Dn - \frac{EK}{\bar{l}_0} n,$$

where D is the ionic diffusion coefficient. At equilibrium $\mathrm{d}n/\mathrm{d}t = 0$ and often $n \approx p$ so that

$$n = \begin{cases} \sqrt{\dfrac{\gamma}{\alpha}} \quad \text{for } \alpha\gamma \gg \left(\dfrac{EK}{2\bar{l}_0} + 2\pi a N_c\, D\right)^2, \\ \gamma\left(\dfrac{EK}{\bar{l}_0} + 4\pi a N_c\, D\right)^{-1} \quad \text{for } \alpha\gamma \ll \left(\dfrac{EK}{2\bar{l}_0} + 2\pi a N_c\, D\right)^2. \end{cases}$$

For the latter case, when $EK/\bar{l}_0 \gg 4\pi a N_c\, D$, $n \rightarrow \gamma \bar{l}_0/EK$. In this limit with strong fields, $J \rightarrow 2q\gamma\bar{l}_0$, which is independent of the electric field strength and therefore non-ohmic (Krehbiel, 1967, 1969; Phillips, 1967; Klett, 1971).

The normal attachment of ions to cloud particles, as a result of their diffusion and field-induced migration, causes clouds to become even poorer conductors of electricity than the clear air surrounding them. Measurements of conduction within the bases of active thunderclouds give values for the electrical relaxation in excess of 4,000 s (Rust and Moore, 1974). Because clouds are poor conductors of electricity, large amounts of free charge can accumulate there and culminate in lightning.

When the field strength is very great, point discharge may take place at the surfaces of cloud particles, greatly increasing the rate of ion production and the flow of leakage charges within the cloud. Kasemir *et al.* (1976) have used this phenomenon in efforts to suppress lightning by large scale releases of conducting fibres into thunderclouds.

Freier (1962) and Evans (1969) have inferred conductivities in natural thunderclouds at least 20-fold greater than those in the adjacent clear air but these values seem unlikely in view of (a) the resulting great leakage that would be imposed on the electrical generator in the cloud, that already is difficult to explain, and (b) all other measurements made inside clouds, to date, indicate that they are normally poor conductors of electricity and therefore good repositories of free charge.

The length of the ionic mean free path changes when ions move within a cloud of charged droplets. Then

$$\bar{l} = \bar{l}_0\left(1 \pm \left|\frac{\bar{Q}}{\bar{Q}_m}\right|\right)^{-2}, \tag{1}$$

where $\bar{Q}$ is the average charge on each cloud particle and $\bar{Q}_m = 12\pi a^2\, \varepsilon_0\, E_0$.

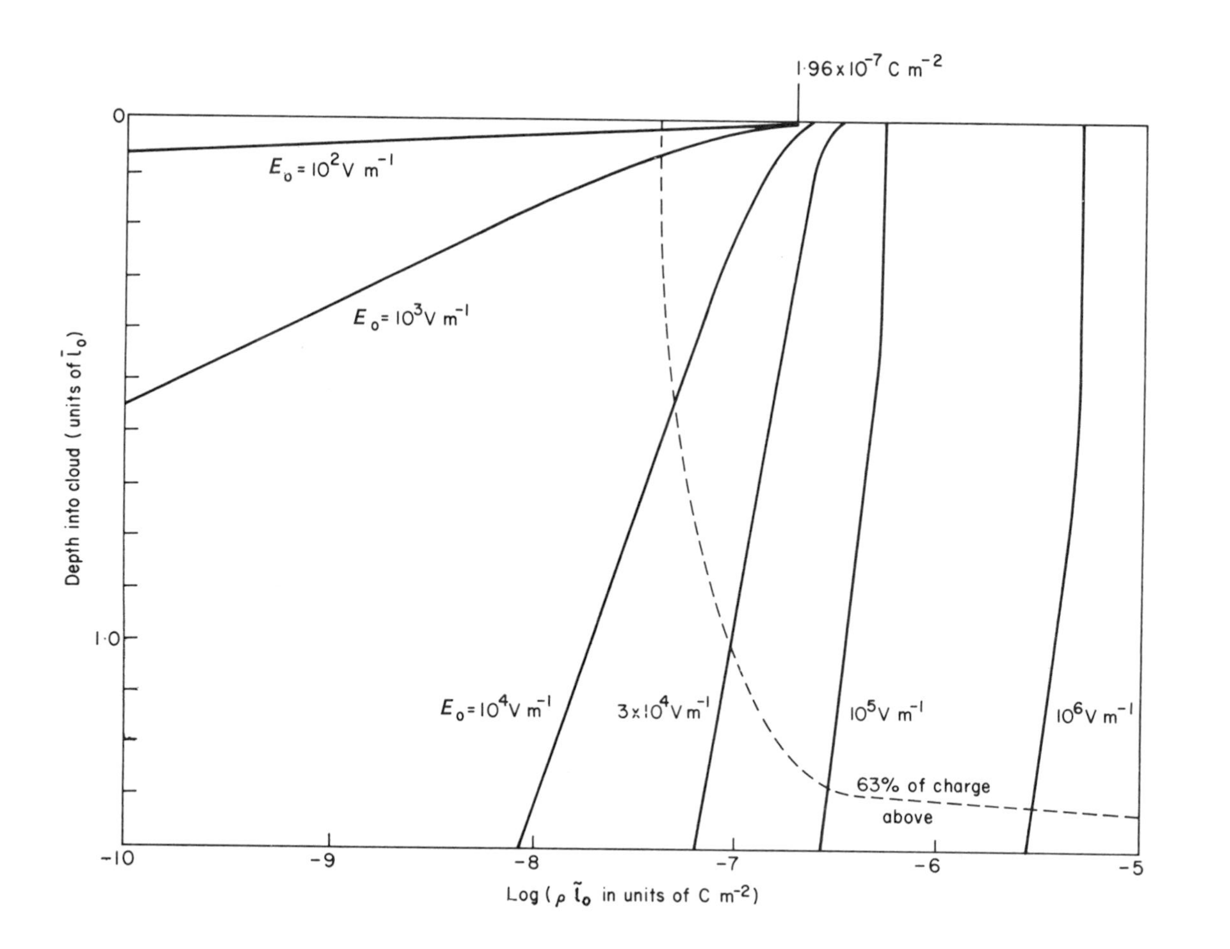

1·96 x 10^-7 C m^-2
0
$E_o = 10^2$ V m^-1
$E_o = 10^3$ V m^-1
$E_o = 10^4$ V m^-1
3 x 10^4 V m^-1
10^5 V m^-1
10^6 V m^-1
63% of charge above
1·0
Depth into cloud (units of $\bar{l}_o$)
-10
-9
-8
-7
-6
-5
Log ($\rho\,\tilde{l}_o$ in units of C m^-2)

If $\bar{Q}$ has the same polarity as the migrating ion, the negative sign is chosen in Equation (1).

Under these conditions the cloud particles have high collision cross-sections for oppositely charged ions and small cross-sections for similarly charged ions. Until breakdown occurs, however, the principal source of high mobility ions is cosmic radiation which is insufficient to make natural clouds highly conductive.

Another significant effect probably takes place near the boundaries of an electrified cloud immersed in a more conducting atmosphere. Ions having the same polarity as the locally effective charge in the cloud are repelled while those having the opposite polarity are attracted into the cloud. After entering the cloud these ions become attached quickly to cloud particles and thus effectively are immobilized as charge carriers. In a motionless electrified cloud the process continues at an exponentially decreasing rate, with a time constant τ_r, until sufficient charge has been accumulated to neutralize externally the cloud's electric field. The region in the outer portion of a cloud containing the neutralizing charge is sometimes called the "screening layer" and has been discussed by Grenet (1947), Lecolazet (1948), Vonnegut (1953) and Brown *et al.* (1971). We calculate that, in strong electric fields, a cloud with an ionic mean free path of $\bar{l}_0$ accumulates its maximum charge concentration in the outer one fourth of an ionic mean free path. At equilibrium this maximum charge concentration, ρ_m, is given by

$$\rho_m = \frac{4\varepsilon_0 E_0}{e^2 \bar{l}_0}$$

where $e = 2{\cdot}71828 \ldots$. At equilibrium the charge in the first $1{\cdot}4\bar{l}_0$ neutralizes all but $1/e$ of the cloud's electric field. In weaker fields, diffusion becomes dominant in attaching field-attracted ions to droplets near the cloud surface so the screening layer becomes thinner but the concentrations of deposited charge remain high (of the order of $2 \times 10^{-7}\, \bar{l}_0$ C m^{-2}). A plot of the computed distribution of screening-layer space charge under the two regimes is shown in Fig. 7. The time constant for space charge accumulation in the screening layer is calculated to be one half the field relaxation time constant given by the conductivity of the surrounding clear air.

Fig. 7. Calculated distributions of space charge concentration in the screening layer formed around a motionless, electrified cloud with an ionic mean free path length of l_0. In electric fields stronger than 3×10^4 V m^{-1}, peak charge concentrations of $0{\cdot}5413\varepsilon_0\, E_0/l_0$ are attained as a result of geometric capture of ions by the polarized droplets. In weaker fields, diffusion processes cause attachment of ions to droplets near the cloud boundary thus producing a thin screening layer of concentrated charge. These distributions were calculated for 6μm diameter cloud droplets.

Indications of the presence of screening layers around actual clouds are provided by Gunn (1955) and by Vonnegut *et al.* (1962). Laboratory studies of the effect have also been made by Eden and Vonnegut (1965) and by Colgate (1967).

The formation of a screening layer is a charge-concentrating process and it produces the highest concentrations and gradients of charge that have yet been identified in thunderclouds. Several speculations (Grenet, 1947; Vonnegut, 1953; Phillips, 1967) have been advanced on the role of this concentrated charge in thundercloud electrification but insufficient measurements exist to permit any quantitative evaluations.

4.2 Point-discharge Currents

The electric fields around exposed and elevated conductors often become so intense that the surrounding air becomes ionized and an electric current flows to weaken the field. These flows, called *point-discharge currents* (Chapter 4), develop beneath thunderclouds from trees, bushes and structures when the field strengths exceed 2 kV m^{-1} or so depending on the object's exposure. In thunderstorm electric fields of 10^4 V m^{-1}, currents of the order of a few microamperes flow from each isolated exposed point. Under these conditions, we have observed current densities of up to 10 nA m^{-2} to flow into the air from a small forest of potted and isolated trees 2 m high and with about 2 m between each.

Wilson (1925) has shown that these point-discharge currents act to limit the strength of the electric field near the earth and that in the presence of these currents the strength of the field beneath a widespread storm should increase with altitude approximately as

$$E(z) = \sqrt{\left(\frac{2Jz}{\varepsilon_0 K} + E_0^2\right)}.$$

Kasemir (private communication) has reported electric field strengths in excess of 100 kV m^{-1} at cloud base about 1000 m above Langmuir Laboratory where the surface field strength was no greater than 10 kV m^{-1}. If all of this field change were due to the Wilson effect, the average point-discharge current densities required would not have exceeded 10 nA m^{-2}.

Positive-ion space charge concentrations of the order of 2×10^{-9} C m^{-3} have been measured 15 m above the surface at Langmuir Laboratory by use of sensitive electric field meters within a screen wire Faraday cage. As has been suggested earlier (Smith, 1951), this space charge probably is responsible for the temporary reversals of the electric field that are usually observed immediately after a nearby lightning flash neutralizes negative charge aloft.

Surface densities of positive-ion charge amounting to more than 10^{-7} C m^{-2} in the subcloud layer have been inferred from the field changes caused by lightning. The dominant transfers of charge beneath thunderclouds over land are those released by point-discharge (Chalmers, 1967) rather than by precipitation or lightning.

The very strong electric fields (10^5 V m^{-1} or more) that have been found beneath thunderclouds over water surfaces (Toland and Vonnegut, 1976) must be due in part to the lesser amount of point-discharge emission from water surfaces. As a result the cloud's electric field is screened less over water than over land.

4.3 Charge Transported by Precipitation

The collection of precipitation in a shielded Faraday funnel and the measurement of the charge that it carries is a useful technique for the determination of precipitation current density (Simpson, 1949; Chalmers, 1967). Measurements at the earth's surface usually indicate that the charges on the precipitation carry the same polarity as that of the space charge in the subcloud layer through which the particles have fallen: under fine-weather fields, with positive charge in the lower air, the polarity of the precipitation arriving at the earth is positive. As the cloud becomes weakly electrified and the surface electric field changes direction, an electrode layer of negative space charge forms beneath the cloud and falling precipitation soon begins to arrive with negative charge. As the cloud overhead increases its electrification, its electric field becomes more intense and an upward emission of positive point-discharge ions commences from exposed conductors on the earth. The precipitation polarity becomes positive within a minute or two and the well-known mirror-image relation appears between the electric field and the rain-current density (Simpson, 1949).

Rust and Moore's (1974) simultaneous measurements at the surface and near cloud-base level show that precipitation often reverses its polarity on falling through the subcloud air; they conclude that Wilson ion-capture by the precipitation beneath the cloud probably accounts for this effect. The precipitation charge that they collected at cloud base with captive balloons almost invariably had the polarity of the charge that created the local field aloft.

Subsequent measurements with Faraday funnels on top of free balloons rising through active thunderclouds show much larger current densities in midcloud than Rust found at cloud base (60 nA m^{-2} versus Rust's maxima of about 9 nA m^{-2}). The maximum precipitation space-charge concentrations aloft were of the order of 8×10^{-9} C m^{-3}. Results from a recent flight are

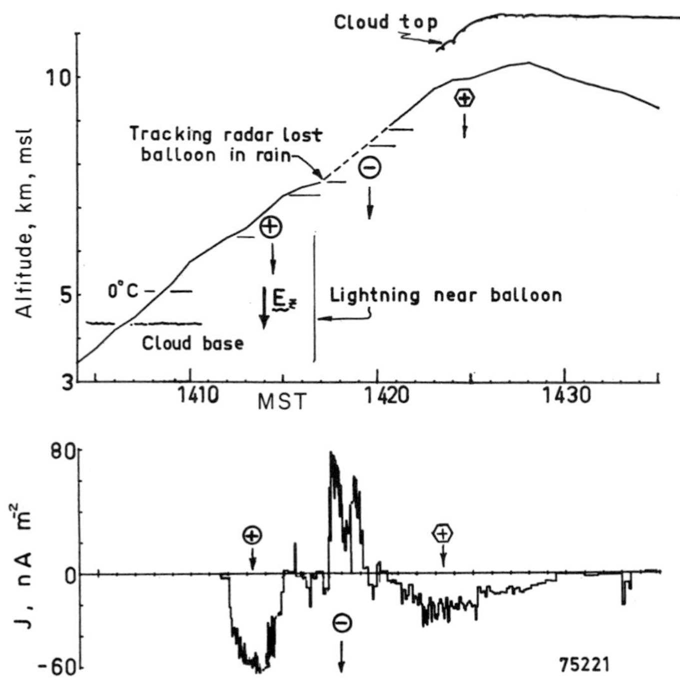

Fig. 8. The density of the electric current carried downward by charges on falling precipitation in a thundercloud. A Faraday cup mounted on top of a free balloon was used to catch precipitation for measurements of its current density. The balloon was launched into an isolated thundercloud that developed over our laboratory on 9 Aug. 1975 and rose almost vertically to an altitude in excess of 9 km MSL. This storm was of the usual polarity and produced nine lightning flashes between 14.10 and 14.25 MST. The radar record indicates that after 14.20 MST, the precipitation in the upper regions of the cloud was in the form of snow.

Our polarity convention is that positive charge carried downward (i.e. in the $-\hat{z}$ direction) constitutes a *negative* current. The time integral of this current density record gives a net transport downward of 12 μC m^{-2} of positive charge on the precipitation.

shown in Fig. 8. In these flights more positive charge than negative was lowered by the precipitation; the net downward flux of positive charge in one flight was 12×10^{-6} C m^{-2} over a 20-minute period with the maximum rate occurring when the balloon was in heavy rain at an altitude of 5,700 m.

4.4 Charge Distribution in Thunderclouds

Because most thunderclouds appear to have positive charge in their upper regions and a somewhat larger quantity of negative charge lower in the cloud, they have often been represented as a dipole with positive charge uppermost superimposed on a negative monopole in their lower regions. Their actual charge distributions are much more complicated than is shown in this simple model (see Chapter 6). Field meters on aeroplanes usually detect a field reversal when flown over the top of developing thunderclouds indicating the effect of positive charge below the aircraft although Fitzgerald (1965) has reported regions with negative charge around the positively charged, upwelling tops of some severe storms. Beneath thunderclouds the electric field is usually directed upward as though there were an appreciable negative charge within the lower regions of the cloud. However, when lightning or downdraughts occur the surface field reverses indicating temporarily the presence of positive charge overhead. Winn and Byerley's (1975, 1976) measurements of the electric field in clouds do not show this reversal; often the direction of the electric field is changed little by lightning. These observations suggest that much of the positive charge detected beneath thunderclouds may arise as point-discharge currents flowing from the earth under the influence of negative charge in the cloud base.

Although it is premature to present a sketch representing the actual distribution of charges around a thundercloud in view of the lack of detailed information, our best guess for a mature thundercloud is shown in Fig. 9. We hold the opinion that much of the net charge in thunderclouds probably resides on cloud particles: strong electric fields are often observed at great distances from precipitation in clouds; lightning and thunder are often produced in regions of thunderclouds free from intense precipitation. It may be reasonable to expect that much of the cloud's charge resides on cloud particles since the cloud-particle surface-area per unit volume of air often exceeds the concentration of precipitation-particle surface-area by a hundredfold or more. Further, far more of the mass of the cloud's condensed water is initially in the form of small droplets and ice crystals than in precipitation. The sources of the principal charges within the cloud, however, have not been determined.

It has been suggested that the upper negative charges reported by Fitzgerald (1965) may have come from screening layer charges transported by the cloud motions, but this has not been established.

The damped oscillation in the electric field commonly observed beneath a dissipating thundercloud during its final subsidence supports the charge distribution shown in Fig. 9. At this time, our measurements (Moore *et al.*, 1959) indicate that downdraughts from the centre of the cloud displace the lower negative charges outwards and expose regions of positive charge above to field meters beneath the cloud. In our observatory this phenomenon is known as the EOSO: the "end of storm oscillation". It seems to have been

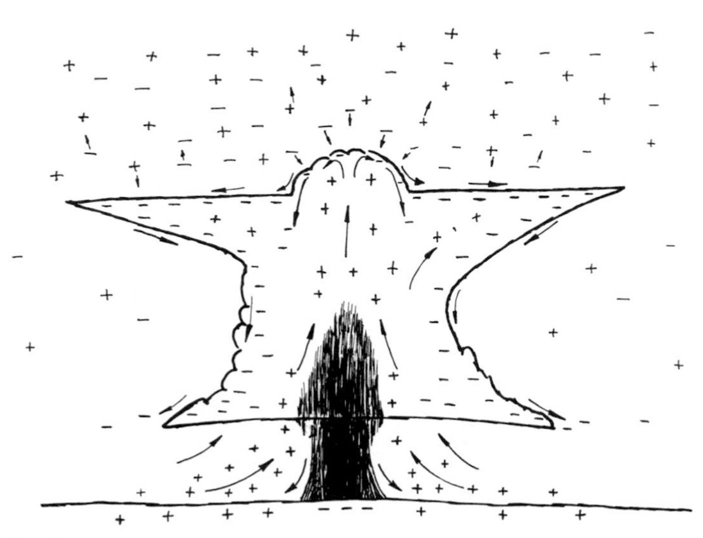

Fig. 9. A schematic view of the probable distribution of charges in and around a thundercloud. This is a speculative view of the charge distribution; adequate information on the distributions is not yet available. We do know that point-discharge ions are dominant features of the subcloud air. Similarly the existence of strong electric fields in the appreciably conductive air at the cloud top supports Wilson's view of a current flowing there. Our guesses as to the interior charges are based on electric field measurements outside the cloud and on a few soundings.

reported first by Franklin in 1753 (Schonland, 1952). A similar field excursion associated with precipitation-driven downdraughts (the FEAWP) has long suggested that a lower positive centre of charge exists in the precipitating portions of the cloud. In our view (Moore *et al.*, 1959) and as Malan (1963) has suggested, the lower centre of positive charge may arise because of positive point-discharge ions carried into the cloud on updraughts. These will produce a noticeable effect at the earth's surface whenever some of the surrounding negative charge is displaced outwards by downdraughts. A sketch of this behaviour is shown in Fig. 10. The small final excursion in the electric field overhead during the EOSO may be due to the remaining negative charge trapped in the screening layers of the subsiding cloud.

One of the remarkable features noted in Winn's rocket probes of thunderclouds is the scarcity of the regions with intense electric fields. In 90 rocket penetrations of thunderclouds, fields stronger than 100 kV m^{-1} were observed on seven occasions and those stronger than 400 kV m^{-1} were observed only twice (see also Chapter 4.6). An equally interesting feature of his measurements is that very strong electric fields were confined to rather small volumes. Winn *et al.* (1974) inferred, for one of the field maxima, a localized charge region with a diameter of the order of 700 m that contained a net charge of about 5 C. The calculated mean concentration of space charge in this volume was of the order of 30 nC m^{-3}. It was at an altitude of about 5,800 m above sea level in a cloud whose base was at about 4,000 m and top about 9,000 m.

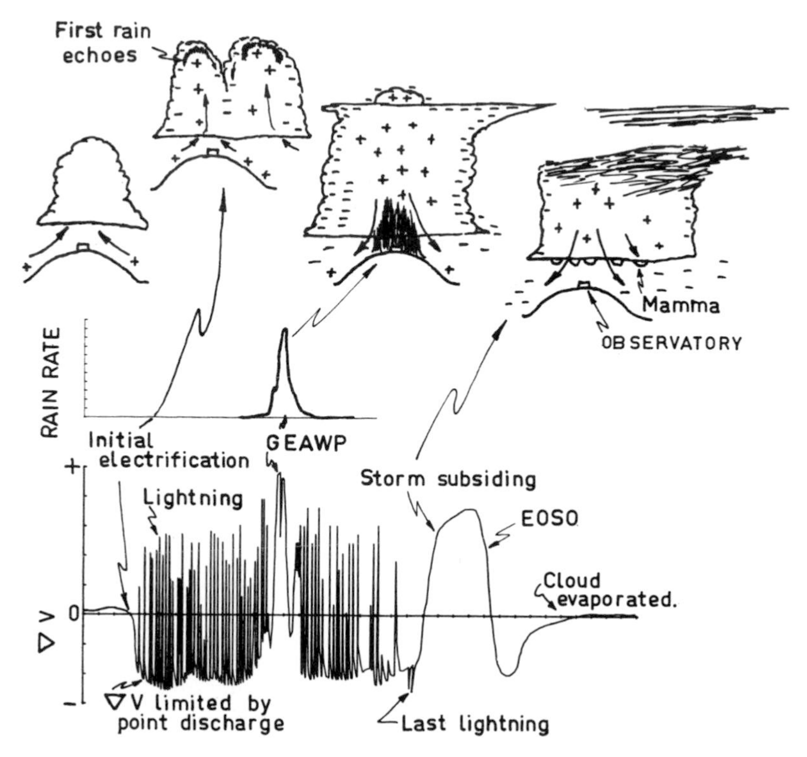

Fig. 10. Sketches illustrating our hypotheses to explain the potential gradient changes beneath an active thundercloud.

We observed that as a cloud became electrified, negative charge was accumulated in the vicinity of the lower part of the cloud (both within and without) and positive charge was accumulated aloft. When a strong downdraught occurred, the lower, negative charge was displaced to the sides. Thus the effect of the upper positive charge could be observed from beneath the cloud, for a partial "window" was created into the cloud by downdraughts as the superposition of the effects of various charges is decreased by the downdraught displacement.

In the same manner, on flying above cumuli our pilot could detect positive charge beneath him whenever a cloud turret grew. We have suggested that an electrical screening layer on the cloud surface, neutralizing any inner charge, becomes displaced to the sides when the cloud top grows. The radial motion of the cloud top here may create another "window" into the cloud.

These displacements suggest that moving air transports large quantities of charge that must be quite significant in the electrical budget of a thunderstorm.

The direction and polarity of the charge could not be determined with the rocket-borne instrumentation. In other measurements Winn and Byerley (1975) have estimated the total amount of charge in a New Mexican thundercloud to be of the order of 140 C.

4.5 The Electric Field Intensities Necessary to Produce Lightning

The maximum strengths reported for electric fields in thunderclouds are slightly in excess of 4×10^5 V m^{-1}. These measurements, corrected for the distortion and the charge of the instrument, were obtained by Winn and his associates (1974) with a spinning rocket that used the displacement currents between electrodes behind dielectric windows to sense the exterior field. In one of the studies the rocket was launched into the region of a thundercloud (that had been producing lightning at 1-minute intervals) about 1 minute after the discharge. The field strengths at the rocket's extremities undoubtedly were much greater than those sensed in the cylindrical regions of the instrument but, despite the local field enhancements, no indications of a resulting discharge appear on the telemetry records.

Electric field meters recessed in a spherical housing were carried by a captive balloon into thunderclouds by Clark (1971) who used a dielectric tether for the balloon. He repeatedly found $1{\cdot}3 \times 10^6$ V m^{-1} field strengths on the surface of his sphere at altitudes of 4 km; these values appeared to approach a limit imposed by local breakdown at his instrument possibly due to disruption of water collected by the instrument.

Dawson and Warrender (1973) have supported simulated raindrops with diameters as large as 3·8 mm in a vertical wind tunnel and applied electric fields of the order of 10^6 V m^{-1} without causing sparks or local breakdown. The resulting field strengths at the electrical poles of the drops probably approached 3×10^6 V m^{-1}.

Dawson (1969) found that the minimum field strength E_{pc} necessary to cause positive corona from a single raindrop in an electric field at pressures less than 650 mb is

$$E_{pc} \approx 703P(\sigma/r_0)^{\frac{1}{2}}/T,$$

where

P is the atmospheric pressure,

σ is the water surface tension,

r_0 is the equivalent spherical radius of the drop,

T is the temperature and all units are S.I.

His results indicate that, in regions of clouds where all the particles are liquid, field strengths in excess of 900 kV m^{-1} may be needed to initiate electric discharges. Griffiths and Latham (1974) after a study of the onset of corona from ice crystals concluded that field strengths of the order of 400 to 500 kV m^{-1} were sufficient to start discharges from ice in the central regions of thunderclouds, as further discussed in Chapter 4.6.

The field strengths that are necessary to propagate a streamer appear to be less than those necessary to initiate the discharge. Phelps (1971, 1974) has triggered positive streamers, that propagated as laboratory lightning-like discharges, by allowing 3 mm diameter drops to fall through a uniform field of 4×10^5 V m^{-1} into a region where the field was concentrated about three-fold by the presence of a hemispherical conductor. In weaker fields the streamers triggered in this manner propagated for short distances and died away exponentially as they left their source. These measurements were all made at atmospheric pressure near sea level; in more recent work (Griffiths and Phelps, 1976) the critical field strengths were found to vary approximately with the atmospheric pressure raised to the 3/2 power.

5. Cloud Electrification Mechanisms

5.1 Precipitation-powered Processes

At the present time there is no consensus on the mechanisms by which thunderclouds become electrified. An appreciation for the current state of cloud electrification explanations can be obtained by reading the expositions and exchanges published by Vonnegut (1963, 1965), Mason (1971, 1972, 1976) and Moore (1974, 1976). There are at least two different categories of explanations: one uses the falling precipitation particles to separate charges within clouds and the other depends on the convective motions of cloudy air to transport externally derived charges in a somewhat organized fashion.

It has long been widely assumed, without evidence, that negative charge is selectively separated and transported downwards in thunderclouds by falling precipitation particles. Modern proposals to explain how this may be accomplished began with Elster and Geitel's (1885, 1913) induction charge transfer process shown in Fig. 11 in which charge is postulated as being separated in elastic collisions between cloud particles and falling precipitation polarized by a vertical electric field. Although many other interactions involving precipitation have been suggested in the past 90 years to explain thundercloud electrification, most of them have been shelved as unworkable for various reasons. One of them, the selective capture of ions by raindrops polarized in an electric field (Wilson, 1929) appears, however, to provide an adequate explanation for the charging of precipitation beneath thunderclouds. On the other hand, however, this mechanism has been found inadequate for the production of intense fields in electrified clouds for two reasons: there are insufficient ions in cloudy air and this ion-capture mechanism depends on the precipitation fall velocity being greater than the ion velocity in the local electric field; in fields stronger than 10^5 V m^{-1}, many ions move faster than the precipitation.

In recent years, Sartor (1967), Mason (1972), Levin and Scott (1975) and others have again advanced the Elster–Geitel mechanism with the transfer of polarization charges to rebounding cloud droplets as the principal explanation of charge separation in clouds. They assume that the electric field is always directed vertically and ignore the frequently observed, horizontal components of the field (Winn *et al.*, 1971, 1974, 1975) which will oppose the postulated charge separation.

The Elster–Geitel process has been studied in a wind tunnel by Aufdermaur and Johnson (1972) who instrumented a simulated hail pellet and caused supercooled cloud droplets to collide with it along an externally applied electric field. They found that, while most of the collisions resulted in accretion of the water droplets by the hail pellet causing its growth, a fraction (from 0·1 to 1%) of the colliding droplets escaped and carried away some of the lower polarization charge as postulated by Elster and Geitel. The hail pellet was maintained at ground potential through the measuring circuit so that no determination was possible of the equilibrium charge that might be acquired by a pellet in this manner. The droplets that rebounded and escaped were found to be somewhat smaller than they were originally; this suggested to Aufdermaur and Johnson that the escaping droplets were ones that had struck the hail pellet near its edges in a grazing collision. If this is generally the condition for droplet escape, it places another severe limit on the net amounts of charge transferred in the process since the polarization charge varies with the cosine of the polar angle and approaches a minimum at the drop edge, i.e. its equator, where most of the rebounds occur. The applicability of the Elster–Geitel mechanism to thunderstorm charge separation is currently being debated in scientific journals.

The lack of supporting experimental data is a major difficulty with all precipitation-powered mechanisms (excepting Wilson's beneath the thundercloud). There are no good observations showing that falling precipitation is carrying sufficient charge in opposition to the local electric fields to account for the accumulation of electrical energy in the cloud in the observed time sequence. Thus far the observations indicate that the fall of charged precipitation is dissipating rather than increasing the electrical energy of the cloud.

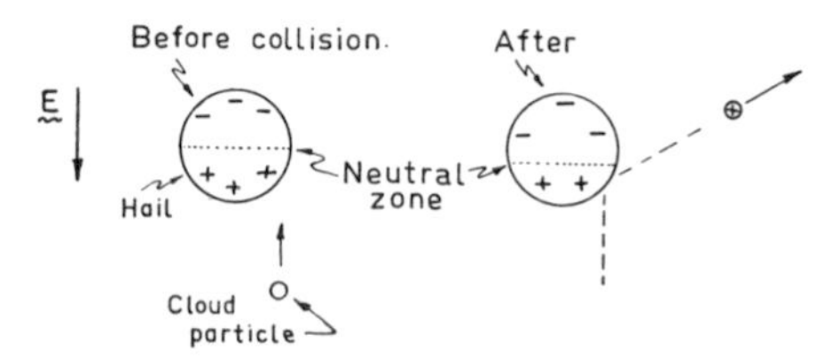

Fig. 11. Elster–Geitel model for the charging of a neutral precipitation particle, polarized in a vertically directed electric field, by elastic collision with a neutral cloud droplet that carries away some of the lower polarization charge. This sketch is drawn in the reference frame of the falling precipitation particle. The polarization of the cloud droplet and its deflection around the precipitation particle by the flow streamlines are not shown in this figure for brevity.

5.2 Convective Electrification Mechanism

In the mechanism proposed by Grenet (1947) and later, independently, by Vonnegut (1953), the convective motions of the cloud transport and accumulate electrified cloud particles in opposition to forces imposed by the electric field. The resulting accumulations of positive and negative charge increase the intensity and extent of existing electric fields and thereby their energy. Both the positive and negative electric charges required for such a process are derived from outside of the cloud. The conduction current, of the order of an ampere, known to flow from the clear upper atmosphere above the cloud, brings small negative ions to the cloud's upper surface where they rapidly become immobilized by attachment to cloud particles. The resultant shallow layer of dense negative space charge formed on the cloud surface is then caused by the convective circulation to accumulate as pockets of charge within the cloud. On the surface of the earth beneath the cloud, positive ions released at the rate of about an ampere by point-discharge are carried by updraughts into the cloud, where they become attached to cloud particles and accumulate as charged regions in the upper part of the cloud.

This process, having positive feedback, is self-intensifying. It therefore requires the existence of electrification in the form of an electric field or space charge to get it started during the initial phase of the cloud's development. This priming electrification is available from several possible sources, such as fair-weather electricity, electrified ocean spray, blowing dust, charged precipitation, or a nearby cloud that has already become electrified. The process is shown schematically in Fig. 12.

The convective electrification process has been tested experimentally by the artificial releases of space charge of anomalous polarity into subcloud air resulting in the anomalous electrification of the cloud (Vonnegut *et al.*, 1962;

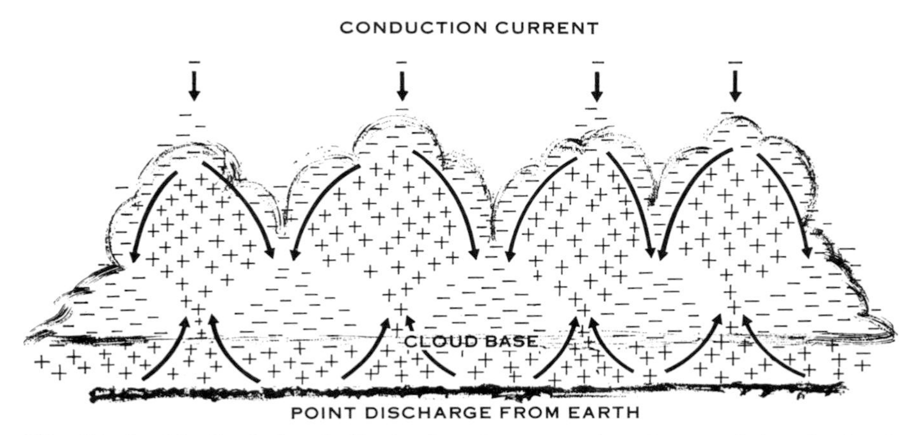

Fig. 12. An idealized sketch illustrating the convection theory of thundercloud electrification (after Vonnegut, 1963). In this mechanism, external charges drawn to the cloud are caught in the convective overturn and carried by the motions of the air against the local electric fields to increase the cloud's electrical potential energy. The ions carrying the Wilson current at the cloud top become attached to cloud particles and are carried downward by the compensating downdraughts. Point-discharge ions beneath the cloud are carried toward the cloud top by updraughts, and attract more negative ions to the cloud top from the conductive clear air around thus causing positive feedback in the cloud electrification.

Moore, 1974). The principal impediments to the evaluation of the convection explanation are the difficulties of demonstrating the positive feedback and our lack of knowledge of the motions of air in and around thunderclouds. Efforts to test these ideas are continuing with several approaches: numerical modelling, artificial releases of space charge and observations of natural thunderclouds. Criticisms and defenses of this explanation for thundercloud electrification may be found in Coroniti (1965) and Dawson (1974).

5.3 Explanations Based on the Contact Differences of Potential

Recently proposed mechanisms of cloud electrification by Wåhlin (1973) and Takahashi (1973) deserve a brief mention. These mechanisms are based on laboratory experiments demonstrating that test samples of various materials or drops suspended in a Faraday cage and connected to the cage through an electrometer reach some equilibrium potential when they are ventilated with ionized air. By extrapolation from these experiments the authors argue that particles in a cloud under similar conditions will acquire charges leading to electrification. Griffiths in an unpublished comment on Wåhlin's paper and Griffiths and Vonnegut (1975) commenting on Takahashi have raised the question as to whether the results of these experiments can properly be applied to the behaviour of particles in clouds.

5.4 Requirements for a Satisfactory Explanation of Cloud Electrification

A single mechanism, perhaps one of those described above, may be the primary agency responsible for the electrification of all thunderstorms. However, many scientists active in the field now incline to the view expressed by Schonland (1953), Chalmers (1954) and others that several processes are capable of giving rise to lightning. If we recognize that no single explanation will probably be able to satisfy them all, we can draw up a set of requirements for a satisfactory understanding of thunderstorms as others have done in the past (Mason, 1953; Schonland, 1953; Wormell, 1953; Chalmers, 1954). The mechanism or mechanisms must offer qualitative and quantitative explanations for the following facts.

(a) The initial electrification of a developing cloud often increases exponentially, e-folding every two minutes or so (Reynolds and Brook, 1956).

(b) The production of lightning in an average thunderstorm occurs at the rate of several flashes per minute (Chapter 14.2.3.2), requiring an average charging current of the order of an ampere.

(c) The lightning flash produces a dipole moment change of the order of 100 C km with a charge transfer of the order of tens of coulombs (Wormell, 1953; Chapters 6.3.4 and 9.5). The electric dipole of the thunderstorm usually has positive charge above and negative charge below, but in some clouds the polarity is reversed (Hughes, 1965). The dipole destroyed by intra-cloud lightning discharges usually departs from the vertical—in some cases by as much as 90° (Reynolds *et al.*, 1957; see also Chapter 6).

(d) The electrification process is capable of generating electric fields within the cloud stronger than 400 kV m^{-1} (Winn *et al.*, 1974).

(e) The electrification process is capable of producing space charges in excess of 20 nC m^{-3} (Vonnegut, 1963; Winn *et al.*, 1974).

(f) In order to give strong electrification and lightning, the cloud must have a depth of at least 3 or 4 km (Moore *et al.*, 1960).

(g) Both strong convective activity and the presence of falling precipitation appear to be necessary but not sufficient conditions for the production of lightning. There is, however, an observed close association between the development of the electric field and the rapid vertical development of the cloud (Reynolds *et al.*, 1957).

(*h*) A cloud giving a rainfall of as little as 3 mm h^{-1} can produce lightning (Moore, 1965, *et seq.*).
(i) The intensity of electrification as indicated by the frequency of lightning discharges bears little relation to the intensity of the present precipitation or the amounts or intensities of the previous precipitation (Watt, 1931; Moore *et al.*, 1964).
(j) Strong electrification can occur in the absence of ice within the cloud (Marchand, 1903; Foster, 1950; Moore *et al.*, 1960; Pietrowski, 1960).
(k) Strong electrification and lightning can occur in regions of the cloud where the temperature is so low that only ice can be present (Fuquay, 1962).
(l) Very tall thunderstorms produce far more frequent lightning than storms of ordinary height (Vonnegut, 1963).
(m) Above the cloud strong electric fields are observed primarily over penetrative convective cells (Vonnegut *et al.*, 1966).
(n) The electric fields are often much higher within the cloud than in the surrounding clear air, the field intensity increasing at the cloud boundary (Gunn, 1955).
(o) With the exception of triggered discharges (see Chapters 5.3.2 and 15.5.5), lightning always originates within the cloud.
(p) The ratio between intra-cloud lightning events and cloud-to-ground lightning events is extremely variable (Chapter 14.4). Some clouds produce only intra-cloud lightning while others produce only cloud-to-ground lightning.
(q) During the dying stage of the storm when both convection and precipitation have virtually ceased, there is often a strong negative electric field on the ground under the cloud that can persist for 10 minutes or more. Sometimes this may be accompanied by an abnormally energetic lightning discharge lowering positive instead of the usual negative charge (Moore *et al.*, 1959, Chapter 5.5.9).

6. Modification of Thunderclouds

The size and energies of thunderclouds make our efforts to control them seem feeble. We are also hampered by a lack of knowledge of their many processes, so that at present we often do not know whether the application of the methods available to us will be beneficial or harmful. Several techniques have been developed that can affect thunderstorm behaviour markedly. Among these is the release of ice-nucleating agents into clouds to hasten the onset of ice-crystal formation. This technique known as "cloud seeding" often uses solid carbon dioxide or small, dispersed silver iodide crystals to initiate the formation of snow crystals. Cloud seeding has been employed to release the heat of fusion of the cloud water thereby increasing its convective energies (Langmuir, 1948b; Simpson, 1967). Intensive cloud seeding has been used by the U.S. Forest Service in an effort to convert much of the cloud water into ice crystals with the aim of reducing lightning discharges to earth and thereby decreasing the incidence of forest fires (Fuquay, 1974).

Hail-suppression attempts use a similar approach: silver iodide nuclei are released with concentrations in excess of 100 active particles per litre of air beneath developing clouds. The usual aim is to glaciate the cloud so that little supercooled water is available for capture by falling hail embryos. Over New Mexico the early conversion of thunderclouds into ice-crystal clouds appears to be a feasible means of decreasing the total amount of precipitation that falls to the earth and therefore many residents oppose the use of cloud seeding in this region.

Many attempts have been made to suppress lightning and some approaches appear to be worthy of further study. The technique that presently offers the greatest promise is one suggested by Kasemir and Weickmann (1965): the release of large quantities of thin conducting fibres, about 10 cm in length, into the updraughts beneath growing thunderclouds. These fibres are carried by the updraughts into the cloud; in the presence of intense electric fields they ionize the air around their tips and allow the air to conduct an electric current that moves so as to decrease the local electric field. Releases of 2 kg of fibres have reportedly reduced the rate at which lightning flashes in the treated clouds to about one quarter of that in control clouds. The lightning that did occur probably originated in regions of the clouds that were not affected by the chaff. The effects of large scale chaff releases on the precipitation formation and on severe electrical storms have not yet been evaluated.

Convective clouds may be primed with release of artificially produced point-discharge ions into the subcloud air from which they grow (Vonnegut *et al.*, 1962). Space charge released at the surface has been detected in updraughts at cloud level 2 km above the surface within 500 s or less after release. The introduction of negative space charge into the subcloud air has caused some convective clouds to become weakly electrified with a dipole polarity opposite to the usual sense; appreciable quantities of positive charge were observed in the base of one of the clouds. Several thunderclouds that developed over our net during the emission of negative charge appear to have maintained their initially inverted polarity as they become fully electrified: negative point discharges in excess of the artificial supply were drawn from the net by the storms as they grew. No data on the polarity and nature of the associated lightning were obtained but these experiments show that large scale charge releases are feasible and that they have some effect on cloud electrification.

Experiments to increase convection by blackening the surface of the earth have been repeatedly suggested over the past decade (Black, 1963). Similarly, there are a number of indications that urban complexes cause a local "heat island" and they modify cloud behaviour as do islands surrounded by a warm ocean (Changnon, 1969).

7. Summary

Despite the advances of our knowledge in recent decades, made possible by the modern tools of aircraft, radar and electronics, the thunderstorm remains a major meteorological mystery. As yet, we have no satisfactory understanding of the mechanisms by which these violent convective overturnings of the atmosphere produce rain, hail, electrical effects and tornadoes. The lack of good information on the convective motions within, about and beneath the storm that is necessary to understand the growth of cloud and precipitation particles and the movement of the charged particles responsible for electrification continues to be a major barrier. Equally important missing data, that are indispensable to an understanding of cloud electrification, are the population of cloud and precipitation particles in the various regions of the cloud and the sign and magnitude of the electric charges that they carry.

Obtaining the required information will be a difficult task that will require the concerted efforts of many scientists. Until this knowledge becomes available, the origin and role of thunderstorm electricity is certain to remain controversial, and efforts to predict or control these storms will be problematic.

Acknowledgements

Thanks to our many collaborators and associates for the help and suggestions that they contributed to the work reported here. Among the many are: J. Hughes, M. Brook, P. Wyckoff, D. McCaig, J. Cook, P. R. Leavitt, J. Machado, E. Vrablik, D. Ching, A. Botka, A. G. Emslie, C. R. Holmes, W. P. Winn, J. Longmire, J. Cobb, F. Johnson, J. W. Bullock, J. Shortess, H. O. McMahon, B. Stein, H. Survilas, H. F. Eden, F. Mallahan, Sally Marsh, P. Sanders, D. Clark, W. D. Rust, E. F. Corwin and our wives.

These studies were made possible by the support of the U.S. National Science Foundation under its Meteorology and its RANN programs and by the Atmospheric Sciences Program of the U.S. Office of Naval Research.

References

Aufdermaur, A. N. and Johnson, D. A. (1972). Charge separation due to riming in an electric field. *Q. Jl R. met. Soc.* **98**, 369–382.

Battan, L. J. and Theiss, J. B. (1970). Measurements of vertical velocities in convective clouds by means of pulsed-Doppler radar. *J. atmos. Sci.* **27**, 293–298.

Bénard, H. (1901). Les tourbillons cellulaires dans une nappe liquide transportant de la chaleur par convection en régime permanent. *Annls Chim. Phys.* **23**, 62.

Bentley, W. A. (1932). Conical snow. *Science, N.Y.* **75**, 383.

Bergeron, T. (1935). On the physics of clouds and precipitation. *Proc. 5th Assembly U.G.G.I., Lisbon* **2**, 156.

Black, J. F. (1963). Weather control: use of asphalt coatings to tap solar energy. *Science, N.Y.* **139**, 226–227.

Borovikov, A. M. and Khrgian, A. Kh. (1963). "Cloud Physics", p. 172. Translated by Israel Program for Scientific Translations, Jerusalem.

Brown, K. A., Krehbiel, P. R., Moore, C. B. and Sargent, G. N. (1971). Electrical screening layers around charged clouds. *J. geophys. Res.* **76**, 2825–2835.

Browning, K. A. (1965). Some inferences about the updraft within a severe local storm. *J. atmos. Sci.* **22**, 669–677.

Browning, K. A. and Donaldson, R. J. (1963). Airflow and structure of a tornadic storm. *J. atmos. Sci.* **20**, 533–545.

Browning, K. A. and Ludlam, F. H. (1962). Airflow in convective storms. *Q. Jl R. met. Soc.* **88**, 117–135.

Byers, H. R. and Braham, R. C., Jr (1949). "The Thunderstorm", p. 53. Washington, D.C., U.S. Government Printing Office.

Chalmers, J. A. (1954). Atmospheric electricity. *Rep. Prog. Phys.* **17**, 125.

Chalmers, J. A. (1967). "Atmospheric Electricity", 2nd ed. p. 515. Pergamon Press, Oxford.

Changnon, S. A., Jr (1969). Recent studies of urban effects on precipitation in the U.S. *Bull. Am. met. Soc.* **50**, 411–421.

Clark, D. (1971). Balloon-borne Electric Field Mills for Use in Thunderclouds. M.S. Thesis, New Mexico Institute of Mining and Technology, Socorro, New Mexico.

Colgate, S. A. (1967). Enhanced drop coalescence by electric fields in equilibrium with turbulence. *J. geophys. Res.* **72**, 479–487.

Colson, D. V. (1960). High level thunderstorms of July 31, August 1, 1959. *Mon. Weath. Rev.* **88**, 279, 285.

Coroniti, S. C. (1965). "Problems of Atmospheric and Space Electricity", p. 616. Elsevier, Amsterdam.

Davies-Jones, R. P. and Kessler, E. (1974). Tornadoes, *in* "Weather and Climate Modification" (W. N. Hess, Ed.), Ch. 16. John Wiley and Sons, New York.

Dawson, G. A. (1969). Pressure dependence of water drop corona onset and its atmospheric importance. *J. geophys. Res.* **74**, 6859–6868.

Dawson, G. A. and Warrender, R. A. (1973). The terminal velocity of raindrops under vertical electric stress. *J. geophys. Res.* **78**, 3619–3620.

Dawson, G. A. (1974). Lightning modification, *in* "Weather and Climate Modification" (W. N. Hess, Ed.), pp. 596–604. John Wiley and Sons, New York.

Ebert, C. H. V. (1963). The meteorological factor in the Hamburg fire storm. *Weatherwise* **16**, 70–75.

Eden, H. F. and Vonnegut, B. (1965). Laboratory modelling of cumulus behavior in a gaseous medium. *J. appl. Met.* **4**, 745–747.
Elster, J. and Geitel, H. (1885). Über die Elektrizitätsentwicklung bei der Regenbildung. *Annls Phys. Chemy.* **25**, 121–131.
Elster, J. and Geitel, H. (1913). Zur Influenztheorie der Niederschlagselektrizität. *Phys. Z.* **14**, 1287–1292.
Emmons, G., Haurwitz, B. and Spilhaus, A. F. (1950). Oscillation in the stratosphere and high troposphere. *Bull. Am. met. Soc.* **31**, 135–138.
Evans, W. H. (1969). Electric fields and conductivity in thunderclouds. *J. geophys. Res.* **74**, 939–948.
Few, A. A. (1974). Lightning sources in severe thunderstorms, pp. 387–390. Preprints of the American Meteorological Society Conference on Cloud Physics (Tucson), Oct. 21, 1974.
Findeisen, W. (1938). Die kolloidmeteorologischen Vorgänge bei der Niederschlagsbildung. *Met. Z.* **55**, 121.
Fitzgerald, D. R. (1965). Measurement techniques in clouds *in* "Problems of Atmospheric and Space Electricity" (S. C. Coroniti, Ed.), pp. 199–211. Elsevier, Amsterdam.
Flora, S. D. (1953). "Tornadoes of the United States", p. 194. University of Oklahoma Press, Norman.
Foster, H. (1950). An unusual observation of lightning. *Bull. Am. met. Soc.* **31**, 140–141.
Freier, G. (1962). Conductivity of the air in thunderstorms. *J. geophys. Res.* **67**, 4683–4691.
Fujita, T. T. (1975). New evidence from April 3–4, 1974 tornadoes. Reprinted from preprint volume, Ninth Conference on severe local storms, Oct. 21–23, 1975, Norman, Okla. Published by American Meteorological Society, Boston, Mass.
Fuquay, D. M. (1962). Mountain thunderstorms and forest fires. *Weatherwise* **15**, 149–152.
Fuquay, D. M. (1974). Lightning modification, *in* "Weather and Climate Modification" (W. N. Hess, Ed.), pp. 604–612. John Wiley and Sons, New York.
Graham, H. E. (1952). A fire-whirlwind of tornadic violence. *Weatherwise* **5**, 59, 62.
Grenet, G. (1947). Essai d'explication de la charge électrique des nuages d'orages. *Extrait annls Géophys.* **3**, 306–307.
Griffiths, R. F. and Latham, J. (1974). Electrical corona from ice hydrometeors. *Q. Jl R. met. Soc.* **100**, 163–180.
Griffiths, R. F. and Phelps, C. T. (1976). The effects of air pressure and water vapour content on the propagation of positive corona streamers and their implications to lightning initiation. *Q. Jl R. met. Soc.* **102**, 419–426.
Griffiths, R. F. and Vonnegut, B. (1975). Comments on "Electrification of condensing and evaporating liquid drops". *J. atmos. Sci.* **32**, 226–227.
Gunn, R. (1955). Comment *in* Proceedings on the conference on atmospheric electricity. *Geophys. Res. Pap.* **42**, 242.
Harland, W. B. and Hacker, J. L. F. (1966). "Fossil" lightning strikes 250 million years ago. *Adv. Sci. Lond.* **22**, 663–671.
Hiser, H. W. (1973). Sferics and radar studies of south Florida thunderstorms. *J. appl. Met.* **12**, 479–483.
Hissong, I. N. (1926). Whirlwinds at oil tank fire, San Luis Obispo, California. *Mon. Weath. Rev.* **54**, 161–163.

Hooke, R. (1664). Royal Society Manuscript 29, reprinted in "Early Science in Oxford" by R. T. Gunther, Vol. VI, p. 180. Oxford University Press.
Hughes, J. (1965). Discussion *in* "Problems of Atmospheric and Space Electricity" (S. C. Coroniti, Ed.), pp. 51, 153, 154. Elsevier, Amsterdam.
Jennings, S. G. (1976). Electrical charging of water drops in polarizing electric fields. *J. Electrostat.* **1**, 15–25.
Jones, R. F. (1952). Turbulence in relation to radar echoes from cumulonimbus clouds. Proc. Third Radar Weather Conference, p. A37. McGill University.
Kasemir, H. W. and Weickmann, H. K. (1965). Modification of the electric field of thunderstorms. Proc. International Conf. on Cloud Physics, Japan, pp. 519–523.
Kasemir, H. W., Holitza, F. J., Cobb, W. E. and Rust, W. D. (1976). Lightning suppression by chaff seeding at the base of thunderstorms. *J. geophys. Res.* **81**, 1965–1970.
Klett, J. D. (1971). Ion transport to cloud droplets by diffusion and conduction, and the resulting droplet charge distribution. *J. atmos. Sci.* **28**, 78–85.
Kraakevik, J. H. (1957). The Electrical Conductivity and Current Density in the Troposphere, p. 126. Ph.D. Thesis, University of Maryland.
Krehbiel, P. R. (1967). Electrical conduction in clouds. Study Paper for Physics 531, New Mexico Institute of Mining and Technology, Socorro, New Mexico.
Krehbiel, P. R. (1969). The ionic conduction current in thunderstorms. (Abstract.) *EOS, Trans. Am. geophys. Un.* **50**, 618.
Langmuir, I. (1948a). The production of rain by a chain reaction in cumulus clouds at temperatures above freezing. *J. Met.* **5**, 175–192.
Langmuir, I. (1948b). Studies of the effects produced by dry ice seeding of stratus clouds. General Electric Res. Lab. Rpt, RL–140, p. 26, *in* "The Collected Works of Irving Langmuir" (C. G. Suits, Ed.), Vol. 11, pp. 74–100. Pergamon Press, New York (1962).
Lecolazet, R. (1948). Les phénomènes électriques à la frontière de deux atmosphères différentes. *Annls Geophys.* **4**, 181–192.
Le Grand, A. (1680). Historia naturae, *in* "Variis Experimentis et Ratiociniis Elucidata", 2nd ed., p. 273. London.
Levin, Z. and Scott, W. D. (1975). Polarization charging may produce a large electric field before the radar echo maximum. *J. geophys. Res.* **80**, 3918–3923.
Levine, J. (1959). Spherical vortex theory of bubble-like motions in cumulus clouds. *J. Met.* **16**, 653–662.
Lewis, W. and Perkins, P. J. (1953). Recorded pressure distribution in the outer portion of a tornado vortex. *Mon. Weath. Rev.* **81**, 379–385.
Lucretius, T. (58 B.C.). Great meteorological phenomena, *in* "De Rerum Natura", Book VI, pp. 255–261. Translated by W. E. Leonard, 1957. E. P. Dutton and Co., New York.
Malan, D. J. (1963). "Physics of Lightning", p. 87. The English Universities Press Ltd., London.
Malkus, J. S. (1954). Some results of a trade cumulus cloud investigation. *J. Met.* **11**, 220–237.
Malkus, J. S. (1960). Recent developments in studies of penetrative convection, *in* "Cumulus Dynamics" (C. E. Anderson, Ed.), pp. 65–84. Pergamon Press, New York.
Marchand, E. (1903). Études sur les nuages. *Bull. Soc. Ramond, Toulouse* (*Explors pyren.*), **8**, 12.

Mason, B. J. (1953). On the generation of charge associated with graupel formation in thunderclouds. *Q. Jl R. met. Soc.* **79**, 501–509.
Mason, B. J. (1971). "The Physics of Clouds." p. 671, 2nd ed. Clarendon Press, Oxford.
Mason, B. J. (1972). The Bakerian Lecture, 1971, The physics of the thunderstorm. *Proc. R. Soc.* A **327**, 433–466.
Mason, B. J. (1976). In reply to a critique of precipitation theories of thunderstorm electrification by C. B. Moore. *Q. Jl R. met. Soc.* **102**, 219–240.
Montgomery, F. C. (1956). Some observations on the tornado at Blackwell, Oklahoma, 23 May 1955. *Weatherwise* **19**, 97, 101.
Moore, C. B. (1965). Charge generation in thunderstorms, *in* "Problems of Atmospheric and Space Electricity" (S. C. Coroniti, Ed.), pp. 257, 259. Elsevier, Amsterdam.
Moore, C. B. (1974). "An assessment of thundercloud electrification mechanisms". Proc. of the International Conference on Atmospheric Electricity, Garmisch-Partenkirchen (in the press).
Moore, C. B. (1976). Reply to B. J. Mason "In reply to a critique of precipitation theories of thunderstorm electrification by C. B. Moore". *Q. Jl R. met. Soc.* **102**, 219–240.
Moore, C. B. and Vonnegut, B. (1973). Reply to P. R. Brazier-Smith, S. G. Jennings and J. Latham, "Increased rates of rainfall production in electrified clouds". *Q. Jl R. met. Soc.* **99**, 779–786.
Moore, C. B., Vonnegut, B. and Emslie, A. G. (1959). Observations of thunderstorms in New Mexico. Report on Contract NONR–1684 (Arthur D. Little, Inc.).
Moore, C. B., Vonnegut, B., Stein, B. A. and Survilas, H. J. (1960). Observations of electrification and lightning in warm clouds. *J. geophys. Res.* **65**, 1907–1910.
Moore, C. B., Vonnegut, B., Machado, J. A. and Survilas, H. J. (1962). Radar observations of rain gushes following overhead lightning strokes. *J. geophys. Res.* **67**, 207–220.
Moore, C. B., Vonnegut, B., Vrablik, E. A. and McCaig, D. A. (1964). Gushes of rain and hail after lightning. *J. atmos. Sci.* **21**, 646–665.
Newton, C. W. (1966). Circulations in large, sheared cumulonimbus. *Tellus* **18**, 699–713.
Newton, C. W. (1968). Convective cloud dynamics—a synopsis. Proc. of the International Conference on Cloud Physics, Toronto, pp. 487–498.
Newton, C. W. and Newton, H. R. (1959). Dynamical interactions between large convective clouds and environment with vertical shear. *J. Met.* **16**, 483–496.
Phelps, C. T. (1971). Field-enhanced propagation of corona streamers. *J. geophys. Res.* **76**, 5799–5806.
Phelps, C. T. (1974). Positive streamer system intensification and its possible role in lightning initiation. *J. atmos. terr. Phys.* **36**, 103–111.
Phillips, B. B. (1967). Ionic equilibrium and the electrical conductivity in thunderclouds. *Mon. Weath. Rev.* **95**, 854–862.
Pietrowski, E. L. (1960). An observation of lightning in warm clouds. *J. Met.* **17**, 562–563.
Plumlee, H. R. (1964). Effects of electrostatic forces on drop collision and coalescence in air. Charged Particle Res. Lab. Report No. CPRL–8–64, Dept. of Electrical Engr, Engr Experiment Station, University of Illinois and Illinois State Water Survey, Urbana, Ill.

Rayleigh, Lord (1879). The influence of electricity on colliding water drops. *Proc. R. Soc.* **28**, 406–409.
Raymond, D. (1975). A model for predicting the movement of continuously propagating convective storms. *J. atmos. Sci.* **32,** 1308–1317.
Reynolds, S. E. and Brook, M. (1956). Correlation of the initial electric field and the radar echo in thunderstorms. *J. Met.* **13**, 376–380.
Reynolds, S. E., Brook, M. and Gourley, M. F. (1957). Thunderstorm charge separation. *J. Met.* **14**, 426–436.
Roach, W. T. (1967). On the nature of the summit areas of severe storms in Oklahoma. *Q. Jl R. met. Soc.* **93**, 318–336.
Rust, W. D. and Moore, C. B. (1974). Electrical conditions near the bases of thunderclouds over New Mexico. *Q. Jl R. Met. Soc.* **100**, 450–468.
Sartor, J. D. (1967). The role of particle interactions in the distributions of electricity in thunderstorms. *J. atmos. Sci.* **24**, 601–615.
Saunders, P. M. (1962). Penetrative convection in stably stratified fluids. *Tellus* **14,** 177–194.
Schonland, B. F. J. (1950). "The Flight of Thunderbolts", p. 127. Clarendon Press, Oxford.
Schonland, B. F. J. (1952). The work of Benjamin Franklin on thunderstorms. *J. Franklin Inst.* **253**, 375–392.
Schonland, B. F. J. (1953). "Atmospheric Electricity", 2nd ed., pp. 80–82. Methuen and Co., London, John Wiley and Sons, New York.
Scorer, R. S. (1958). "Natural Aerodynamics", p. 270, Pergamon Press, New York.
Simpson, G. C. (1949). Atmospheric electricity during disturbed weather. *Geophys. Mem. Lond.* **84**, 1–51.
Simpson, J. (1967). An experimental approach to cumulus clouds and hurricanes. *Weather* **22**, 95–114.
Smith, L. G. (1951). In discussion on cloud physics. *Q. Jl R. met. Soc.* **77**, 683.
Steiner, R. and Rhyne, R. H. (1962). Some measured characteristics of severe storm turbulence. 17 pp. National Severe Storms Project, Report No. 10, U.S. Weather Bureau, Washington, D.C.
Takahashi, T. (1973). Electrification of condensing and evaporating liquid drops. *J. atmos. Sci.* **30**, 249–255.
Telford, J. W. and Warner, J. (1962). On the measurement from an aircraft of buoyancy and vertical air velocity in cloud. *J. atmos. Sci.* **19**, 415–423.
Thomson, J. (1881). On a changing tesselated structure in certain liquids. *Proc. Glasg. phil. Soc.*
Toland, R. B. and Vonnegut, B. (1977). Measurement of maximum electric field intensities over water during thunderstorms. *J. Geophys. Res.* **82**, 438–440.
Vaughan, O. H., Jr and Vonnegut, B. (1976). Luminous electrical phenomena in Huntsville, Alabama tornadoes on April 3, 1974. *Bull. Am. met. Soc.* **57**, 1220–1224.
Vonnegut, B. (1953). Possible mechanism for the formation of thunderstorm electricity. *Bull. Am. met. Soc.* **34**, 378.
Vonnegut, B. (1960). Electrical theory of tornadoes. *J. geophys. Res.* **65**, 203–212.
Vonnegut, B. (1963). Some facts and speculations concerning the origin and role of thunderstorm electricity. *Met. Monogr.* **5**, 224–241.
Vonnegut, B. (1965). "Thundercloud electricity". *Discovery* **26**, 12–17.

Vonnegut, B. and Moore, C. B. (1958). Giant electrical storms, *in* "Recent Advances in Atmospheric Electricity", pp. 339–411. Pergamon Press, London.

Vonnegut, B. and Moore, C. B. (1960). A possible effect of lightning discharge on precipitation formation process, *in* "Physics of Precipitation", pp. 287–290. Monogr. No. 5, Am. Geophys. Un.

Vonnegut, B. and Weyer, J. R. (1966). Luminous phenomena in nocturnal tornadoes. *Science* **153**, 1213–1220.

Vonnegut, B., Moore, C. B., Semonin, R. G., Bullock, J. W., Staggs, D. W. and Bradley, W. E. (1962). Effect of atmospheric space charge on initial electrification of cumulus clouds. *J. geophys. Res.* **67**, 3909–3922.

Vonnegut, B., Moore, C. B., Espinola, R. P. and Blau, H. H., Jr (1966). Electric potential gradients above thunderstorms. *J. atmos. Sci.* **23**, 764–770.

Vul'fson, N. I. (1957). Compensation downdraughts caused by developing cumulus clouds. *Izv. AN SSSR, seriya geofiz.* 1, **94**, 103.

Wåhlin, L. (1973). A possible origin of atmospheric electricity. *Found. Phys.* **3**, 459–472.

Watt, R. A. W. (1931). Some problems of modern meteorology, 3. The present position of theories of the electricity of thunderstorms. *Q. Jl R. met. Soc.* **57**, 17–26.

Wegener, A. (1911). Kerne der Kristallbildung, *in* "Thermodynamik der Atmosphäre", pp. 94–98. J. A. Barth, Leipzig.

Wilk, K. E. (1963). National Severe Storms Project, aircraft and radar observations. Severe Local Storms Conference, Urbana, American Meteorological Society pp. 1–6 (and private communication available from the senior author).

Wilson, C. T. R. (1925). The electric field of a thundercloud and some of its effects. *Proc. phys. Soc. Lond.* **37**, 32D–37D.

Wilson, C. T. R. (1929). Some thundercloud problems. *J. Franklin Inst.* **208**, 1–12.

Winn, W. P. and Byerley, L. G. (1975). Electric field growth in thunderclouds. *Q. Jl R. met. Soc.* **101**, 979–994.

Winn, W. P. and Byerley, L. G. (1976). "Vertical and horizontal components of electric fields in thunderclouds—1975 measurements" (in preparation).

Winn, W. P. and Moore, C. B. (1971). Electric field measurements in thunderclouds using instrumented rockets. *J. geophys. Res.* **76**, 5003–5017.

Winn, W. P., Schwede, G. W. and Moore, C. B. (1974). Measurements of electric fields in thunderclouds. *J. geophys. Res.* **79**, 1761–1767.

Wormell, T. W. (1953). Atmospheric electricity; some recent trends and problems. *Q. Jl R. met. Soc.* **79**, 3–38.

Ziv, A. and Levin, Z. (1974). Thunderstorm electrification: cloud growth and electrical development. *J. atmos. Sci.* **31**, 1652–1661.

4. Point-discharge

J. LATHAM and I. M. STROMBERG

University of Manchester, England

1. Introduction

This chapter is concerned with electric discharges occurring in the atmosphere which are not so energetic and violent as lightning, although they are intimately linked to the latter phenomenon. The topic divides naturally into two parts, point-discharge at the ground and corona discharges from hydrometeors within clouds.

Point-discharge from protuberances at the earth's surface has been a subject of study since the time of Benjamin Franklin. It is of importance from a number of viewpoints. First, it may contribute significantly to the overall electrical budget of the atmosphere (Wormell, 1930). Indeed some workers have estimated that point-discharge is of primary importance in this respect. However, these estimates have been based upon extrapolations of isolated localized measurements onto the global scale, and therefore cannot be regarded as truly representative. For this reason a discussion of the global electrical budget will not be given. A second possible role for point-discharge is in influencing the growth of the electric field within thunderclouds. Proponents of precipitation-based theories of thunderstorm electrification have tended to regard the introduction of point-discharge currents into the bases of thunderclouds as a dissipative process acting to inhibit the growth of the electric field. Some workers have ascribed great importance to this effect. However, their estimated current flow was based on no direct evidence whatsoever and in view of the facts that the current depends on the field *at the ground* and that it might take at least 10 minutes for the ions to travel from the earth's surface to the cloud base it is difficult to imagine that field growth would be seriously inhibited by point-discharge; one such aspect is discussed in Chapter 17.4.3. The interesting possibility that air motions may distribute point-discharge ions in such a way as to cause the electric field to grow within a thundercloud is discussed in Chapter 3.4.2 and thus need not be treated here. A third possibly important role for point-discharge is that its occurrence at the ocean surface may have been responsible for prebiotic synthesis of organic compounds in the primitive atmosphere of the earth. Experiments have shown that weak electrical discharges are capable of synthesizing amino acids and other products in a simulated primitive atmosphere, and recent studies have shown that two mechanisms exist by which corona discharges can occur at the ocean surface. This location is important because the necessary transportation and storage of the organic products beneath the ocean surface would be greatly facilitated. A further role of point-discharge at the earth's surface is to modify the charge distribution on precipitation as it falls from cloud to ground. Although the reality of this phenomenon has been established for a long time no sufficiently comprehensive studies have been made, and it will not be further discussed.

Corona discharges from hydrometeors in the highly electrified interiors of thunderclouds is a fairly recent topic of intensive study. Its importance resides mainly in the fact that a lightning stroke is probably triggered by corona from protuberances on hydrometeors. Recent work has established the conditions under which corona can be produced from individual raindrops, pairs of colliding raindrops, ice hydrometeors and melting hailstones. A second possibly significant role of corona discharges within thunderclouds is a corollary of the first. Under conditions when these discharges do not result in lightning the electrical conductivity of the air will increase locally thus possibly inhibiting further field growth in these regions. A discussion of recent work on this topic is presented in Section 6.

2. The Mechanism of Point-discharge

Most of the existing information on the properties of corona discharge has been obtained by studying discharges on metal points but it has been shown that the nature of these discharges is the same whether they occur on metal points, water drops or on trees. A brief description is now given of the salient properties of corona discharge. A more detailed treatment of this topic is given in Loeb (1965).

For a metal point in air the discharge process is initiated in a small volume of air close to the tip, where the electric field is sufficiently enhanced to permit ionization of the gas molecules by collision with electrons which have accelerated in the electric field, during their mean free path, to a point where their kinetic energy exceeds the ionization potential of the gas. This action releases further electrons which can act in the same way. Thus a process of cumulative ionization occurs which is known as an electron avalanche and is of fundamental importance in all forms of corona discharge. It will be seen that an initial electron is required for an avalanche to start. This may be provided by cosmic-ray activity or as a result of radioactive decay.

In the case of a positive point in air the discharge starts with short pulses known as onset streamers. The streamers advance away from the point by virtue of their own space charge field which enhances the applied field in one direction. Photoionization of the O_2 molecules ahead of the tip initiates further avalanches. Provided that the electron avalanches reach a critical size the streamer will continue to propagate out into the low field region, where the ambient value is too low to initiate avalanches on its own. The pulsed nature of this discharge is caused by the accumulated space charge left behind the propagating tip of the streamer. This inhibits further streamers until it is cleared by the action of the electric field. It should be noted that the electron avalanches propagate in the direction of increasing field strength. At fields above the onset value the space charge distribution will ultimately reach a state of dynamic equilibrium with the field and the discharge will be in the form of a steady glow with no pulses. With further increases of the applied field the space charge is cleared more rapidly and pulsed breakdown streamers occur which lead to spark breakdown if they cross the entire point-plane gap (see Chapter 7.4).

For a negative point, electron avalanches propagate away from the tip in the direction of decreasing field strength. In air, electron attachment can occur on oxygen molecules and negative ions are formed in the lower field region. This constitutes space charge which acts to inhibit further avalanches until it is cleared by the field and so negative discharges are also in the form of pulses, known as Trichel pulses. The frequencies of these pulses range from 1 kHz at onset fields up to about 100 kHz at higher fields. In appearance the discharge is similar to the positive glow discharge, producing a luminous region confined closely to the tip of the electrode.

3. Point-discharge from Artificial Points

Early studies of point-discharge from metal points showed that appreciable currents occur, even from low points, in the strong potential gradients of storms and showers, but that a considerable height is needed in fine weather. Wilson (1920) suggested that point-discharge might be an important factor in the transfer of charge between clouds and the earth. This suggestion led directly to the work of Wormell who used elevated metal points and Schonland who used an electrically insulated living tree. Schonland (1928) using a small tree, supported on insulators and connected to ground through a galvanometer, measured point-discharge currents during thunderstorms in South Africa. The net charge transferred to ground was estimated by integration of the current records and found to be nearly always negative and, on average, 20 times greater than the charge brought down by lightning flashes. He found that the charge brought to ground by precipitation is negligible and obtained a figure of 2.1 A for the average total current below a thunderstorm. It is generally thought that Schonland's figure is an over-estimate because the tree used in his experiment was in a more exposed position than surrounding trees in the area, but it seems probable that point-discharge plays a major role in the transfer of charge between thunder-storms and the earth.

Table I

Ratios of negative to positive point-discharge currents measured at the ground by various investigators

Observer	Place	Ratio of −ve to +ve	Duration
Wormell (1927, 1930)	Cambridge (England)	2·0	3 years
Whipple and Scrase (1936)	Kew (England)	1·7	2 years
Immelmann (1938)	Pretoria (S. Africa)	2·8	2 years
Yokouti (1939)	Japan	2·1	1 year
Gerashimova (1939)	Elbrus (Caucasus)	1·6	14 years
Chiplonkar (1940)	Colaba (India)	2·9	1 year
Perry *et al.* (1942)	Nigeria	2·86	9 months
Lutz (1944)	Munich (Germany)	2·0	3 years
Chalmers and Little (1947)	Durham (England)	1·36	8 months
Allsopp (1952)	S. Africa	2·2	—
Michnowski (1957)	Poland	1·5	6 storms
Sivaramakrishnan (1957)	Poona (India)	2·0	1 year

Wormell (1927, 1930) measured the ratio of the amounts of negative to positive charge brought to ground by point-discharge using an elevated metal point and found it to be 2 : 1. A more accurate value of 1·7 : 1 was obtained by Whipple and Scrase (1936) over a measurement period of two years at Kew in England. Measurements of the point-discharge ratio have been made in various parts of the world and the ratio of negative to positive charge transferred has been shown to vary from about 1·5 to 2·9. It can be seen from Table I (Chalmers, 1967) that the higher values seem to occur in the tropical regions.

Point-discharge has also been investigated using tethered balloons and kites. Davis and Standring (1947) used a tethered balloon at various heights up to 2,400 m and found an increase in point-discharge current with height, potential gradient at the ground and windspeed. The effect of windspeed on a discharging point is now well established. The point-discharge current will increase with windspeed by an amount given by the equation of Large and Pierce (1957)

$$I = a(V - V_0)(W^2 + c^2 V^2)^{\frac{1}{2}},$$

where a, c and V_0 are constants, I is the point-discharge current, V is the potential difference between the point and its surroundings and W is the windspeed. When W is zero the equation reduces to

$$I = A(V - V_0)\, V,$$

where A is a constant, which is effectively the formula of Whipple and Scrase (1936).

Point-discharge frequently occurs on aircraft wherever the potential difference between the aircraft and its surroundings exceeds the onset potential (see Chapter 21.5.5). This can typically occur when an aircraft flies through or near storm clouds or through heavy precipitation below cloud base. A study of the various ways in which aircraft can become charged, leading in extreme cases to the initiation of corona, has been made by Coroniti *et al.* (1952).

Rust and Moore (1974), using rockets capable of measuring the electric field vector and the charge on hydrometeors, have found that the sign of the charge on precipitation leaving a thunderstorm cloud can be reversed by the time the precipitation reaches the ground (see Chapter 3.4.3). This effect is explained by assuming that the falling precipitation intercepts ions produced by point-discharge, rising towards the cloud base.

4. Point-discharge at the Ground

The most commonly occurring natural objects from which point-discharge currents occur in relatively weak electric fields are trees. However, in higher fields point-discharge occurs from less protuberant objects such as grassland.

Attempts to simulate point-discharge from trees by performing laboratory experiments on single metal points, or arrays of points, have not been very informative since the threshold fields and the current-voltage characteristics have not been accurately reproduced.

However, some useful experiments have been performed on growing trees. Milner and Chalmers (1961) used by-passing electrodes in the trunk of a tree to measure currents directly with a galvanometer. Chalmers (1962), using a similar system, reported that during a thunderstorm a tree equipped with by-passing electrodes showed anomalous electrical behaviour. When the potential gradient dropped below the onset value, the current in the tree did not fall immediately to zero but decayed with a time constant of approximately 90 s. This phenomenon has been examined theoretically and experimentally by Ette (1966a) who suggests that the conductivity of a living tree arises chiefly from the sap, which can be regarded as a heterogeneous electrolyte. When metal electrodes are inserted into the tree, electric double layers form around them and, depending on the electrode area in contact, double-layer capacitances of several microfarads may exist. A change in the potential difference between the electrodes disturbs the chemical equilibrium around the electrodes, and the system reaches a new equilibrium state at a rate governed by ionic diffusion at the electrodes. The fact that tree sap contains ions of various sizes and mobilities leads to the existence of a number of different electrical relaxation times for a given tree-electrode system. Ette (1968) has devised an experimental arrangement which largely overcomes the above-mentioned measurement problems by using two mercury electrodes installed 2 m apart on each tree trunk. A third electrode sited close to the crown of each tree enables a discharge current from a laboratory source to be injected into the tree for calibration purposes. Prior to calibration the optimum resistive load across the by-passing electrodes is determined to reduce residual currents below a limiting value and suppress relaxation effects for upwards of 60 minutes. By injecting a known current into the tree, the by-passing efficiency of the electrode system is determined and this is typically of the order of 0·05%. Another problem associated with the measurement of point-discharge currents is that displacement currents caused by rapid potential gradient changes occurring during a thunderstorm can be around several micro-amperes and will be indistinguishable from the point-discharge current. Lines of force, which would have ended on an area of the earth's surface, end instead on the tree. This area, S_e, is the effective area of the tree. For a potential gradient changing at the rate dF/dt at a given time, the magnitude of the displacement current in the tree is given by

$$i_D = \varepsilon_0 \frac{dF}{dt} S_e,$$

where ε_0 is the permittivity of free space. Ette (1966b) calculated the effective area by assuming that a tree could be treated as a vertical half-prolate spheroid. For a tree of height 18 m and a maximum horizontal spread of 9 m, in a potential gradient changing at a rate of 500 V m^{-1} s^{-1}, the displacement current is 5·1 μA. Thus during periods of violent lightning activity with the associated rapid field changes, quite erroneous point-discharge data may be obtained.

It has been found by Stromberg (1971a) that both positive and negative point-discharge currents occurring in tree branches are of a pulsed nature whereas it is known that for metal discharge points negative currents are often not pulsed. He devised a measuring technique which overcomes the problems of displacement currents and spurious electrode currents mentioned previously. Capacitative electrodes wrapped around the top branches of a tree enable the point-discharge pulses to be detected. The system responds only to pulses with a fast rise-time and therefore is not affected by displacement currents. As the electrical connection to the tree is purely capacitative, there are no spurious electrode currents. The repetition frequency and amplitude of the pulses give the sign and magnitude of the current flowing down the tree. Due to the relatively high impedance to ground, a tree can act as a radio antenna and Stromberg experienced some difficulty in eliminating radio signals. This was achieved ultimately by using a pulse detector which responds only to asymmetric or pulsed wave-forms and rejects, by integrating to zero, any sinusoidal signals.

Maund and Chalmers (1960) suggested a further method by which the difficult problem of measuring point-discharge currents from a living tree might be resolved. It involves the measurement of the potential gradient, upwind and downwind of a tree, during a period of point-discharge. The space charge so produced should cause a reduction in the potential gradient at the ground downwind of the tree. Ogden (1968) showed that the space charge produced by a point-discharge current I would give a potential gradient reduction downwind of

$$\Delta E \approx \frac{I}{2\pi\varepsilon_0 Wh}\{1 + d/(d^2+h^2)^{\frac{1}{2}}\},$$

where W is the wind speed, h the height of the tree, d the distance downwind and ε_0 the permittivity of free space. He also obtained a limited amount of experimental data which showed statistically significant reductions in potential gradient downwind of a clump of trees if the upwind values were greater than about 1,000 V m^{-1}. The largest reduction observed was 300 V m^{-1} which corresponds to a point-discharge current of 0·42 μA. This technique, although attractive in principle, could not be used for the determination of point-discharge currents over large areas because the upwind and downwind surface potential gradients would not generally be identical in the absence of point-discharge.

The most extensive and probably the most representative point-discharge current measurements beneath thunderstorms that have been made to date

Table II
Discharge data for palm trees, grass blades and metal points (Ette and Utah, 1973)

Period	No. of discharge events	−ve Charge Q_- (mC)	+ve Charge Q_+ (mC)	Net charge $Q_- - Q_+$ (mC)	−ve Discharge duration T_- (min)	+ve Discharge duration T_+ (min)	Total discharge duration $T_+ + T_-$ (min)	Q_-/Q_+	T_-/T_+	$I_- = \frac{Q_-}{T_-}$ (μA)	$I_+ = \frac{Q_+}{T_+}$ (μA)	Net mean current (μA)
18-m palm, 1970												
April–May	13	96·15	20·97	75·18	1250	410	1660	4·6	3·1	1·28	0·85	0·76
June–July	15	58·18	23·40	34·78	808	481	1289	2·5	1·7	1·20	0·81	0·45
October–November	10	56·80	9·16	47·64	683	244	927	6·2	2·8	1·40	0·63	0·86
Overall	38	211·13	53·53	157·50	2741	1135	3876	3·9	2·4	1·28	0·79	0·68
18-m palm, 1971												
March–April	13	46·80	20·20	26·60	705	451	1156	2·3	1·6	1·11	0·75	0·39
May–June	20	57·93	31·76	26·17	752	523	1275	1·8	1·4	1·26	1·01	0·34
July–October	17	78·65	21·13	57·52	770	417	1187	3·7	1·8	1·70	0·85	0·81
Overall	50	183·38	73·09	110·29	2227	1391	3618	2·5	1·6	1·37	0·88	0·51
13-m palm, 1971												
February–March	12	30·40	14·06	16·34	484	242	726	2·2	2·0	1·05	0·97	0·38
April–May	15	35·13	10·68	24·45	500	191	691	3·3	2·6	1·17	0·93	0·59
June–September	14	53·11	11·57	41·54	599	230	829	4·6	2·6	1·48	0·84	0·84
Overall	41	118·64	36·31	82·33	1583	663	2246	3·3	2·4	1·25	0·91	0·61
10-m point, 1971												
March–May	18	25·03	15·14	9·89	1091	617	1708	1·7	1·8	0·38	0·41	0·10
June–July	15	17·67	9·53	8·14	604	387	991	1·9	1·6	0·49	0·41	0·14
Overall	33	42·70	24·67	18·03	1695	1004	2699	1·7	1·7	0·42	0·41	0·12
7-m point, 1971												
February–March	11	3·99	1·23	2·76	555	194	749	3·2	2·9	0·12	0·11	0·061
April–May	11	2·97	1·10	1·87	249	152	401	2·7	1·6	0·20	0·12	0·078
June–September	9	7·96	1·74	6·22	445	137	582	4·6	3·2	0·37	0·21	0·18
Overall	31	14·92	4·07	10·85	1249	483	1732	3·7	2·6	0·20	0·14	0·10
5-m point, *1971*												
May–July	12	3·41	1·75	1·66	487	265	752	2·0	1·8	0·12	0·11	0·037
Grass, 1971												
March–June	12	0·155	0·039	0·116	226	107	333	4·0	2·1	0·011	0·006	0·0058

appear to be those of Ette and Utah (1973). Their measurements, summarized in Table II, were made at Ibadan in Nigeria over two storm seasons and were based on data from 88 thunderstorms. Currents were measured, using the electrode by-passing arrangement described previously, from two isolated palm trees, heights 13 and 18 m, three metal points of heights 5, 7 and 10 m and also an area of grass transplanted to cover a large rectangular plate of area 4·2 m^2 mounted flush with the ground on four all-weather insulators. By comparing the discharge data from the trees and metal points with those from the grass they were able to calculate the corresponding "effective areas" of the trees and points. Using these values together with the net mean currents, a mean annual discharge-current density from thunderstorms of $(1 \cdot 2 \pm 0 \cdot 2) \times 10^{-9}$ A m^{-2} was obtained for the two-year period covered by the measurements. It may be inferred from this figure that a typical storm cloud of area 50 km^2 maintains a mean current of about 0·6 A below it—in general agreement with the measurements of Gish and Wait (1950) for storms in the U.S.A. From the discharge data obtained by Stromberg (1971b), Table III, the mean discharge current for a tree inside a spruce plantation at Lanehead, England, was $-0 \cdot 084$ μA. This yields, with an average spacing of 6 m between the trees, a mean discharge-current density of $(2 \cdot 3 \pm 0 \cdot 6) \times 10^{-9}$ A m^{-2}.

Table III

Comparison between point discharge in two spruce trees in a plantation for July and August 1968 (Stromberg, 1976)

	Tree inside		Tree at edge	
	Positive	Negative	Positive	Negative
Mean current (μA)	0·15	−0·18	0·15	−0·14
Total duration (min)	9	22	15	40
Total charge transferred (μC)	84	−240	132	−332
Net charge transferred (μC)	−156		−200	
Ratio of negative to positive charge transferred	2·9		2 5	

5. Corona at the Ocean Surface

Whilst considerable data exist on the electrical activity over land surfaces, conditions that occur beneath thunderclouds over the oceans are not known. It is to be expected, however, that the electric fields may attain higher values than over land due to the relative difficulty of obtaining corona from water. There appear to be three processes by which corona discharges may occur in the vicinity of the ocean surface—drop splashing, bubble bursting and raindrop collisions. All these mechanisms involve the production of points at a liquid surface and are not significantly affected by the purity of the water.

When a drop falls into water and splashes, a jet of liquid is ejected vertically upwards. It has a diameter close to that of the original drop and may rise several centimetres above the water surface. If such splashing events occur in the presence of an electric field, E, the jet is pulled out further by the field. It has been shown by Phelps *et al.* (1973) that when E achieves a critical value, E_c, the tip of the jet becomes pointed and a corona occurs (Figs 1 and 2). For drops in the millimetre size range, of radius r (mm), the critical field ($kV\ cm^{-1}$) is given by the empirical relationship

$$E_c(r) = (1{\cdot}9/r^{\frac{7}{2}}) + (1{\cdot}78 \pm 0{\cdot}04).$$

For drops of $r = 3$ mm, about the largest size existing in natural rainfall, $E_c \approx 180\ kV\ m^{-1}$. The production of corona by colliding raindrops, which may occur in uniform fields as low as $250\ kV\ m^{-1}$, is discussed in detail in Section 6.

Air forced into the oceans by wave motion or precipitation rises to the surface in the form of bubbles. When a bubble bursts at the surface, a jet of liquid rises at high velocity from the bottom of the bubble crater and achieves a height of several millimetres before ejecting a number of drops. The tip of the jet formed when a bubble bursts is likely to be a particularly effective site for corona emission. Consequently, Latham (1975) performed experiments in which bubbles produced under water from a fine capillary tube rose to the water surface and burst in the presence of a vertical electric field. Corona were initiated in fields exceeding about $260\ kV\ m^{-1}$. A more detailed investigation of the variation of E_c with bubble size and the charges transferred has yet to be made.

Electric discharges are very efficient in synthesizing amino acids and other organic compounds in conditions which simulate the atmosphere of the primitive earth. Consequently, they may have been of considerable importance in the origin of life. Corona emission from pointed objects is probably the most effective form of electric discharge involved in prebiotic synthesis (Miller and Orgel, 1974). Corona occurring in the vicinity of the oceans would be particularly important since accumulation within the ocean of organic products produced by the discharges would be facilitated. Thus the three processes by which corona discharges may occur at or near the ocean surface—drop splashing, bubble bursting and raindrop collisions—may well have been important in biogenesis, as pointed out by Latham (1975).

Fig. 1. Liquid jet with a conical tip in negative corona produced when a water drop of radius 2·2 mm falls 2·5 m into deep water in a vertical electric field of 270 kV m^{-1}. The maximum width of the jet is about 5 mm. (From Phelps *et al.*, 1973.)

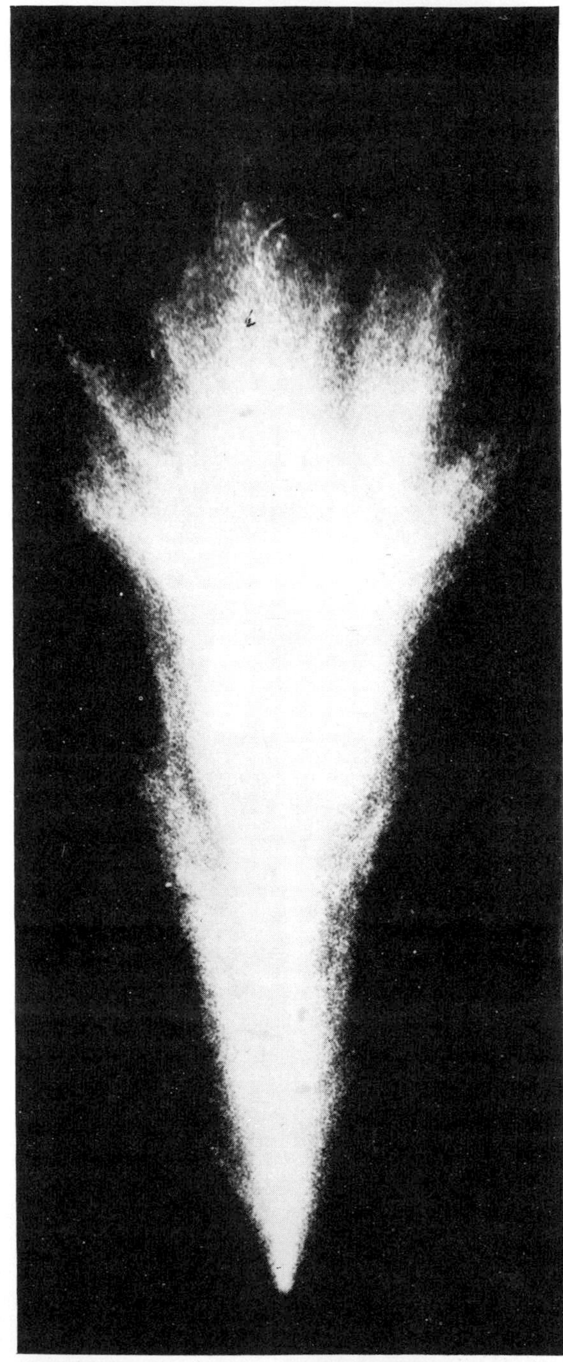

Fig. 2. Positive corona discharge from the tip of the water jet for the conditions specified in Fig. 1. (From Phelps *et al.*, 1973.)

Insufficient evidence exists at present to permit an assessment of the relative importance of these three processes.

Phelps *et al.* (1973) measured the amount of charge transferred per splashing event, and found this to be of the order of 0·1 μC. The current pulses as displayed on an oscilloscope were typically 3 to 20 ms in duration. For drops at terminal velocity in the radius range 1 to 3 mm, the critical field $E_c(r)$ was found to be inversely proportional to the momentum of the drop. The charge transferred was proportional to the excess field $[E-E_c(r)]$.

When the surface field beneath oceanic thunderclouds rises significantly above the threshold value, corona will constitute the most powerful mechanism of charge production. Corona from splashing or bubble bursting will cause a general reduction in the surface field until it is reduced to the threshold value.

6. Corona Discharges within Thunderclouds

The problem of lightning initiation within thunderclouds has provoked considerable discussion in recent years. The majority of investigators have favoured the suggestion that lightning is triggered by the emission of positive corona from the surface of a raindrop, highly deformed by strong electric fields. In this situation the surface field at the regions of high curvature will be appreciably magnified compared with the ambient value. This possibility has been studied in considerable detail in a series of exhaustive and elegant experiments by Dawson (1969) and Richards and Dawson (1971).

Dawson (1969) measured the corona-discharge onset-potential of water surfaces of radius from 0·22 to 1·46 mm over a range of pressures corresponding to altitudes in the standard atmosphere of 1 to 13 km (see also Chapter 3.4.5). He found that at low pressures the electrical stress is relieved by means of pure corona from an undisrupted water surface. However, at high pressures, the corona is triggered by the production of a liquid point at the water surface, which occurs as a consequence of electrohydrodynamical disruption. In the low pressure régime the onset field varies inversely with pressure in the classical manner. At the high pressures the corona onset field is determined by the condition for surface disruption E_d, which in its simplest form depends only on the surface tension σ and radius r of the drop, as demonstrated by Taylor (1964), and is thus independent of pressure. Dawson observed the transition from the surface disruption mode of discharge to pure corona for both positive and negative surfaces. The positive transition occurs at a pressure around 470 mb (4·5 km) and the negative transition around 340 mb (7·0 km). Between these two points the products of discharges are mostly positive ions and negative droplets. Furthermore, the different onset potentials for the two polarities cause any drop exposed to a high enough field, regardless of direction and pressure, to be left with a large negative charge. Dawson points out that these results should be true generally, regardless of the source of the surface electric field.

In an effort to simulate as precisely as possible conditions which might exist in a thundercloud, Richards and Dawson (1971) conducted an experimental study of the onset of instability and corona in water drops of raindrop size falling at their terminal velocity in the presence of vertical electric fields. The instability proceeds by the formation of a point at the upper surface of the drop from which corona is emitted, modifying the net charge on the drop. For uncharged drops of radius in excess of 2 mm the field required to produce instability exceeded that predicted by the Taylor criterion ($E_d(r/\sigma)^{\frac{1}{2}} = 1{\cdot}6$) and was essentially independent of radius at about 950 kV m^{-1}. If the drops carried very high charges of the appropriate polarity the instability (and therefore corona) field was reduced to about 550 kV m^{-1}, and if the field was inclined at about 45° to the vertical—thereby providing a particularly favourable configuration for the occurrence of instability—it was even lower.

However, as pointed out by Richards and Dawson, even under the most advantageous conditions the corona-onset fields measured in these experiments were much higher than the maximum fields that have been recorded in thunderstorms. Extensive measurements, conducted by Gunn (1948), Chapman (1958), Fitzgerald (1969) and particularly by Winn and Moore (1971), have failed to reveal fields in excess of about 400 kV m^{-1} (see also Chapter 3.4.4), while the large scale fields in active thunderclouds are generally around several tens of kilovolts per metre. Thus it appears probable that corona accompanying the disruption of individual raindrops under the influence of strong electric forces cannot be responsible for lightning initiation.

Richards and Dawson suggested that the collision of a pair of raindrops within a thundercloud may produce, momentarily, a grossly deformed object whose shape is particularly conducive to corona onset in relatively weak electric fields. This possibility has been investigated experimentally by Crabb and Latham (1974). Using a specially designed wind tunnel, measurements were made of the electric field values required to produce corona from a pair of water drops, of radii 2·7 and 0·65 mm, colliding with a relative velocity of 5·8 m s^{-1}. For central collisions the larger drop went into the "bag-mode" while for more glancing collisions a liquid filament, typically several times the larger drop radius in length, was drawn out between the separating drops, and eventually collapsed. The values of the corona onset field decreased with increasing values of L—the length of the combined drop-pair in the direction of the field at the moment of separation—which itself increased as the collisions became less central. High-speed photographs used to determine values of L also revealed that the tip of the filament was sharply pointed at the time that corona occurred. The onset field ranged from about 500 kV m^{-1} for head-on collisions to about 250 kV m^{-1} for glancing collisions producing the longest filaments. Both positive and negative corona pulses were detected each time a collision resulted in a discharge but these were few in number because of the transience of the exaggerated liquid shape responsible for the occurrence of corona. Each pulse carried a charge of about 10^{-10} C. Subsidiary experiments with model drops reinforced the conclusion that the corona resulted initially from a point on the drop surface which had been disrupted by mechanical or electrical forces. Calculations by Crabb and Latham indicated that the emission of positive corona from colliding raindrops can readily occur in thunderclouds, and thereby possibly trigger lightning, in fields as low as 350 kV m^{-1}. This value is considerably lower than that associated with other natural triggering processes so far studied.

An alternative source of corona in relatively weak electric fields is the surface of ice hydrometeors. This possibility was completely discounted in the literature until recently, mainly because of the studies of Bandel (1951) who failed to obtain corona discharges from ice particles. However, it was re-examined by Griffiths and Latham (1974a) in a series of experiments in which artificial hailstones or ice crystals grown from the vapour were mounted on fine insulating supports in an electric field. The dimensions of these particles were typically several millimetres. They found that at temperatures above −18 °C—for pure ice—vigorous corona occurred when the field was raised to a critical onset value which was well defined for a particular ice specimen. Values of the onset field were determined as a function of the size, shape, purity, orientation and the surface features of the particles, their initial charge, pressure and temperature. All of these parameters were found to exercise a significant effect on the onset field, E_c. E_c was found to be inversely proportional to pressure, in accordance with classical theory. E_c decreased with increasing dimension of the particle in the direction of the field vector, and surface features on a scale of tens of micrometres were found to have a more profound effect on the onset field than the overall shape of the specimen. A series of experiments, pursued in greater detail by Griffiths and Latham (1974b), showed that the failure of ice to sustain a corona discharge at temperatures below a well defined critical value was a consequence of the fact that the surface electrical conductivity, which falls with decreasing temperature, was too low. The application of an initial charge to the specimen, of a magnitude expected to exist on ice hydrometeors within thunderclouds, produced a redution in E_c ranging from about 10 to 20%. A detailed examination of the influence of the applied charge on the form of the corona and on the onset field has been conducted by Griffiths (1975). It was concluded from these various studies that the onset fields for corona emission from hailstones or snow crystals in the central regions of thunderclouds are probably in the range 400 to 500 kV m^{-1}. A similar range for E_c was found by Reed (private communication) in experiments on the production of corona from melting hailstones.

Although the lowest values of E_c so far measured in conditions representative of those occurring in thunderclouds are associated with the collision of raindrops—and it seems probable that this process is responsible for triggering the lightning discharge in many situations—the values of E_c for ice hydrometeors are sufficiently low that corona from ice may be the triggering mechanism in other circumstances. Much more information on the temperature location of the initiatory discharge in thunderstorms is required before this question can be definitely resolved.

The forms of the corona discharges from ice were studied in some detail by Griffiths and Latham (1974a). They were found to be identical with those accompanying the application of a potential to a metal point. For negative surfaces the types of discharge occurring as the potential was raised were: Trichel pulses, negative glow and spark. For positive surfaces the corresponding types were: onset streamers with burst pulses, positive glow, breakdown streamers and spark. The positive and negative corona discharges occurred simultaneously at opposite ends of the freely suspended ice particles. The continuous current measured when the corona onset field was just achieved was typically about 0·1 μA. The current was found to be independent of the pressure and the shape, size and orientation of the ice specimen. It increased with increasing field strength above E_c until, at the streamer-to-glow transition, there was a sudden decrease. As the field was increased further the current again rose until breakdown streamers and the subsequent spark appeared. Griffiths and Latham presented an interpretation of the characteristics of the streamers in terms of the model of propagation propounded by Dawson and Winn (1965) and extended by Phelps (1971).

The experimental measurements of corona-onset fields and currents from ice hydrometeors made by Griffiths and Latham were used by Griffiths *et al.* (1974) to calculate the normalized electrical conductivity—the ratio of the in-cloud value to the clear-air value at the same altitude—within highly electrified clouds. These calculations were based on the ionic balance equations derived by Phillips (1967), which take account of the immobilization of ions on the cloud particles as a consequence of diffusion and electrostatic capture. Since temperatures warmer than −18 °C are required in order to produce significant corona from ice this process will occur in thunderclouds in regions where high concentrations of supercooled water droplets exist. Thus the conductivity will be very sensitive to both the liquid-water content and the concentration of corona sites. The calculations suggest that corona from ice provides a copious supply of positive and negative ions which causes the normalized conductivity to increase from between 10^{-3} and 10^{-4} for fields just below E_c to greater than 1 when the onset field is achieved. In this situation the altitude z also plays an important role because E_c decreases quite rapidly with decreasing pressure. For example, with a typical distribution of temperature and pressure within a thundercloud, corona will be initiated from hailstones at $E_c = 720$ kV m^{-1} if $z = 2{\cdot}5$ km (0 °C) and 540 kV m^{-1} at $z = 5{\cdot}5$ km (-18 °C): the corresponding values of E_c for snowflakes are 540 and 390 kV m^{-1} respectively. At $z = 2{\cdot}5$ km the normalized conductivities are slightly higher than and lower than 10 respectively, as L is increased from 1 to 5 g m^{-3} with an assumed concentration of corona sites of 1 m^{-3}.

Corona discharges occurring at the surfaces of precipitation particles within highly electrified clouds can act either to increase the conductivity and leakage currents, thereby opposing further field growth, or to initiate a lightning stroke. The latter process will occur if the high field region surrounding a particle in corona extends over a sufficient volume for the positive streamers to drain a critical amount of charge, the magnitude of which is still under debate. Phelps (1971) has shown that the field strength required to maintain positive streamer propagation is somewhat less than the values of E_c determined for ice particles, although it exceeds the lowest values of E_c reported for colliding drops. Thus the distribution of field becomes a crucial factor in determining whether a lightning stroke will be initiated. These arguments suggest that if the region of very intense field is highly localized, corona currents can greatly increase the conductivity in that region, and possibly inhibit the onset of lightning. If, however, the intense field exists over a large volume of the cloud a corona discharge will tend to trigger a lightning stroke.

References

Allsop, H. L. (1952). An investigation of point-discharge currents. *S. Afr. J. Sci.* **48**, 244–246.

Bandel, H. W. (1951). Corona from ice points. *J. appl. Phys.* **44**, 984–985.

Chalmers, J. A. (1962). Point-discharge currents through a living tree during a thunderstorm. *J. atmos. terr. Phys.* **24**, 1059–1063.

Chalmers, J. A. (1967). "Atmospheric Electricity", 2nd ed. Pergamon Press, Oxford.

Chalmers, J. A. and Little, E. W. R. (1947). Currents of atmospheric electricity. *Terr. Magn. atmos. Elect.* **52**, 239–260.
Chapman, S. (1958). "Recent Advances in Atmospheric Electricity". pp. 277–286, Pergamon Press, New York.
Chiplonkar, M. W. (1940). "Measurement of point discharge current during disturbed weather at Colaba. *Proc. Indian Acad. Sci.* A **12**, 50–56.
Coroniti, S. C., Parziale, A. J., Callahan, R. C. and Patten, R. (1952). Effect of aircraft charge on airborne conductivity measurements. *J. geophys. Res.* **57**, 197–205.
Crabb, J. A. and Latham, J. (1974). Corona from colliding drops as a possible mechanism for the triggering of lightning. *Q. Jl R. met. Soc.* **100**, 191–202.
Davis, R. and Standring, W. G. (1947). Discharge currents associated with kite balloons. *Proc. R. Soc.* A **191**, 304–322.
Dawson, G. A. (1969). Pressure dependence of water-drop corona onset and its atmospheric importance. *J. geophys. Res.* **74**, 6859–6868.
Dawson, G. A. and Winn, W. P. (1965). A model for streamer propagation. *Z. Phys.* **183**, 159–171.
Ette, A. I. I. (1966a). Anomolous electrical behaviour of a tree "looking in" between electrodes. *J. atmos. terr. Phys.* **28**, 285–294.
Ette, A. I. I. (1966b). Measurement of electrode by-passing efficiency in living trees. *J. atmos. terr. Phys.* **28**, 295–302.
Ette, A. I. I. (1968). Point discharge current distribution below storm clouds. *J. Geomagn. Geodet.* **20**, 423.
Ette, A. I. I. and Utah, E. U. (1973). Studies of point-charge characteristics in the atmosphere. *J. atmos. terr. Phys.* **35**, 1799–1809.
Fitzgerald, D. R. (1969). Electric field and precipitation structure near convective cloudtops. *Trans. Am. geophys. Un.* **50**, 169–180.
Gerashimova, T. W. (1939). Atmospheric-electric measurements on Elbrus. *Isv. Akad. Nauk SSSR.* Ser. *Geogr. Geofiz.* **13**, 4–5.
Gish, O. H. and Wait, G. R. (1950). Thunderstorms and the earth's general electrification. *J. geophys. Res.* **55**, 473–484.
Griffiths, R. F. (1975). The initiation of corona discharges from charged ice particles in an electric field. *J. Electrostatics* **1**, 3–13.
Griffiths, R. F. and Latham, J. (1974a). Electrical corona from ice hydrometeors. *Q. Jl R. met. Soc.* **100**, 163–180.
Griffiths, R. F. and Latham, J. (1974b). A new method for the measurement of the surface electrical conductivity of ice. *J. met. Soc. Japan* **52**, 238–242.
Griffiths, R. F., Latham, J. and Myers, V. (1974). The ionic conductivity of electrified clouds. *Q. Jl R. met. Soc.* **100**, 181–190.
Gunn, R. (1948). Electric field intensity inside of natural cloud. *J. appl. Phys.* **19**, 481–484.
Immelmann, M. N. S. (1938). Point discharge currents during thunderstorms. *Phil. Mag.* **25**, 159–163.
Large, M. I. and Pierce, E. T. (1957). The dependence of point-discharge currents on wind, as examined by a new experimental approach. *J. atmos. terr. Phys.* **10**, 251–257.
Latham, J. (1975). Possible mechanisms of electrical discharge involved in biogenesis. *Nature* **256**, 34.
Loeb, L. B. (1965). "Electrical Coronas." University of California Press, Berkeley.

Lutz, C. W. (1944). Über den Beitrag der Spitzenentladung zur Aufrechterhaltung der negativen Erdladung. *Beitr. Geophys.* **60**, 9–16.
Maund, J. E. and Chalmers, J. A. (1960). Point discharge from natural and artificial points. *Q. Jl R. met. Soc.* **86**, 85–90.
Michnowski, S. (1957). Point discharge in the interchange of electric charge between earth and the atmosphere. *Acta Geophys. Pol.* **5**, 123–124.
Miller, S. L. and Orgel, L. E. (1974). "Origins of Life on Earth." Prentice-Hall, New Jersey.
Milner, J. W. and Chalmers, J. A. (1961). Point discharge from natural and artificial points (II). *Q. Jl R. met. Soc.* **87**, 592–596.
Ogden, T. L. (1968). Apparent point discharge at a clump of trees. *Q. Jl R. met. Soc.* **94**, 598–600.
Perry, F. R., Webster, G. H. and Baguley, P. W. (1942). The measurements of lightning voltages and currents in Nigeria. *J. Instn elect. Engrs* **89**, 185–209.
Phelps, C. T. (1971). Field enhanced propagation of corona streamers. *J. geophys. Res.* **76**, 5799–5806.
Phelps, C. T., Griffiths, R. F. and Vonnegut, B. (1973). Corona produced by splashing of water drops on a water surface in a strong field. *J. appl. Phys.* **44**, 3082–3086.
Phillips, B. B. (1967). Ionic equilibrium and the electrical conductivity in thunderclouds. *Mon. Weath. Rev.* **95**, 854–862.
Richards, C. N. and Dawson, G. A. (1971). The hydrodynamic instability of water drops falling at terminal velocity in vertical fields. *J. geophys. Res.* **76**, 3445–3455.
Rust, W. D. and Moore, C. B. (1974). Electrical conditions near the bases of thunderclouds over New Mexico. *Q. Jl R. met. Soc.* **100**, 450–469.
Schonland, B. F. J. (1928). The interchange of electricity between thunderclouds and the earth *Proc. R. Soc.* A **118**, 252–262.
Sivaramakrishnan, M. V. (1957). Point discharge current, the earth's electric field and rain charges during disturbed weather at Poona. *Indian J. Met. Geophys.* **8**, 379–390.
Stromberg, I. M. (1971a). Some characteristics of point discharge pulses from metal points and tree branches. *J. atmos. terr. Phys.* **33**, 473–484.
Stromberg, I. M. (1971b). Point discharge current measurements in a plantation of spruce trees using a new pulse technique. *J. atmos. terr. Phys.* **33**, 485–495.
Taylor, G. I. (1964). Disintegration of water drops in an electric field. *Proc. R. Soc.* A **280**, 383–397.
Whipple, F. J. W. and Scrase, F. J. (1936). "Point discharge in the electric field of the earth", *Geophys. Mem., London.* **68**, 1–20.
Wilson, C. T. R. (1920). Investigations on lightning discharges and on the electric field of thunderstorms. *Phil. Trans. R. Soc.* A **221**, 73–115.
Winn, W. P. and Moore, C. B. (1971). Electric field measurements in thunderclouds using instrumented rockets. *J. geophys. Res.* **76**, 5003–5017.
Wormell, T. W. (1927). Currents carried by point discharge beneath thunderclouds and showers. *Proc. R. Soc.* A **115**, 443–55.
Wormell, T. W. (1930). Vertical electric currents below thunderstorms and showers. *Proc. R. Soc.* A **127**, 567–90.
Yokouti, K. (1939). Point discharge in the atmospheric electricity. *J. met. Soc. Japan* **17**, 73–75.

5. The Earth Flash

K. BERGER

*Zollikon, Switzerland**

1. Early Investigations

During the 150 years after Franklin's introduction of the lightning conductor (see Chapter 2.6), lightning protection was largely guided by the more or less fantastic ideas of individual workers. It was round about 1900 that real scientific progress was initiated by the use of lightning photography on a moving film or plate by Hoffert (1889) in England and Walter (1902) in Germany. About the same time Pockels (1897) began to measure the amplitudes of lightning currents and this led to the development of the magnetic link. A third branch of research was concerned with the measurement of electric fields, the first results being reported by Wilson (1916).

A strong impetus was given to lightning research in the second decade of this century by the needs of the electricity supply industry to ensure continuity of supply during thunderstorm periods. Technical necessity thus led to scientific research, the principal methods of which may be briefly outlined.

1.1 Magnetic-link Measurements

As shown in Chapter 22.4 the risk of back flashover on a transmission tower is a function of the severity of the lightning current. Magnetic links were therefore installed in large numbers on the towers and earth wires of high-voltage transmission lines and, later, on tall chimneys and towers. The limitations of magnetic links and the important results obtained from their analysis are reviewed in Chapter 9.3.2.1.

* Formerly Swiss Technical University, Zurich, Switzerland.

1.2 The Fulchronograph

Wagner and McCann (1940) utilized the magnetic link in an attempt to determine the time variation of the lightning current. On the circumference of a rotating disk they fitted a large number of small steel laminations which were subjected to the magnetic field of the lightning current to be recorded. At best, a resolution of about 30 μs could be achieved.

1.3 The Klydonograph

The klydonograph was developed by Peters (1924) by using the phenomenon of Lichtenberg figures (Lichtenberg, 1778). It consists in essence of a point/plate arrangement with intervening photographic plate. The physical properties of Lichtenberg figures were elucidated by Toepler (1917). Their diameter indicates the amplitude of the applied voltage, their pattern determines the polarity and one of their main advantages is that voltage impulses lasting no more than a small fraction of a microsecond produce clearly defined figures. However, measuring errors may exceed ±30% (Toepler, 1920).

Installations of klydonographs on American and European transmission systems required potential dividers to reduce overvoltages of the order of 1 MV to between 4 and 20 kV. Inadequacy of the dividers then available and the great sensitivity of the klydonograph to short impulsive spikes produced erroneous results. The klydonograph is today only of historical interest but it was further developed in the form of the cine-klydonograph and the similar teinograph (Griscom, 1960; Griscom *et al.*, 1965). These instruments are also subject to serious errors.

1.4 The Cathode-ray Oscillograph

The cathode-ray oscillograph was first utilized in 1924 to record atmospheric overvoltages on transmission lines (Norinder, 1925; Berger, 1928, 1930). Observation of earth flashes adjacent to high-voltage lines which caused no trip-out led to the conclusion that induced surges on such lines were harmless (Fortescue, 1930, and Chapter 23). Three-phase oscilllograms of lightning surges on overhead lines soon established that, on lines of 60 kV and above, flashovers were caused exclusively by direct strikes (Berger, 1936). Following this important result, the interest switched to the recording of lightning currents, particularly on the Empire State Building in New York and Mount San Salvatore in Switzerland.

2. Lightning Recording at a Distance

2.1 General

Recording at a distance is mainly concerned with field measurements and lightning photography. The literature on these aspects is so extensive that reference may merely be made to the comprehensive treatment by Uman (1969).

2.1.1 Field Measurements

The electric field-changes shortly before and during the development of a lightning discharge are recorded by antennae and by field mills. The frequency response of both devices is restricted, at its lower limit, by the leakage resistance of their insulation. At its upper range, the frequency response is theoretically unlimited for the antenna. In the field mill it is restricted by the number of segments and the speed of rotation (see also Chapter 13.2.2). The field mills used by Schonland and Malan (Malan, 1963) have an inherent frequency of 1 kHz and thus a time resolution of 1 ms. The corresponding limits of the field mills on Mount San Salvatore are 1,150 Hz and 0·85 ms. The upper limit can be readily increased by adding together three measurements which are displaced in time and which make it possible to eliminate the shift of the zero line (Berger, 1972).

The variation with time of the photographic image of an earth flash together with the resulting electric field and field-change records is shown schematically in Fig. 1. Two different types of earth flash are shown, Kitagawa *et al.* (1962) named them respectively, discrete and hybrid flashes. A discrete flash consists of a sequence of separate discharges which produce sudden light and current impulses. In contrast, a hybrid flash consists of a multiple discharge, at least one component of which exhibits prolonged light emission. Each new illumination of the lightning channel causes the electric field to increase suddenly while, between these sudden changes, the field remains either constant or only increases very slowly. In a hybrid flash the field increases strongly throughout the duration of the light emission.

In accordance with generally accepted nomenclature (see also Chapter 6.5) the following terminology is used for the different components of field-change illustrated in Fig. 1:

L-change: due to leader,
R-change: due to return stroke,
C-change: due to continuing discharge,
K-change: short, sharp pulse during interval between component strokes,

J-change: slow field-change during interval between component strokes,
M-change: short, sharp pulse during continuing *C*-change.

Following Schonland's terminology the entire lightning discharge is called a *flash* and the individual components *strokes*. A multiple lightning flash thus consists of several strokes. This requires a time resolution of several milliseconds. A faster resolution of several microseconds enables each stroke to be resolved into its leader and return strokes.

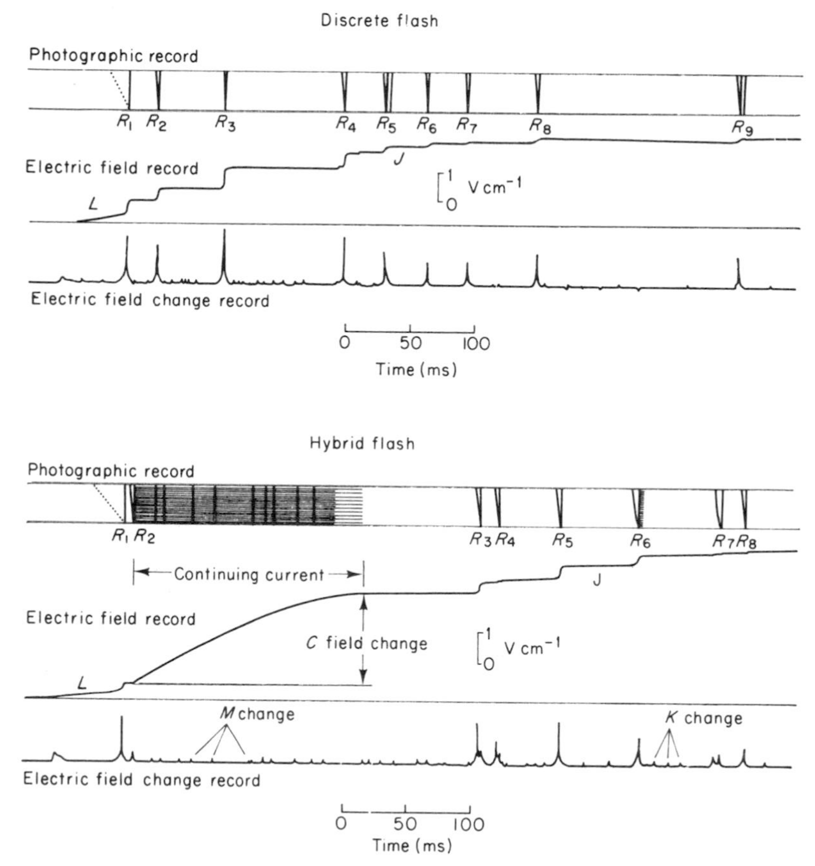

Fig. 1. Schematic representation of simultaneous electric field and field-change records, positive deflections upwards (after Kitagawa *et al.*, 1962).

Schonland has already noted that the fast field-changes due to individual strokes invariably have the same direction and correspond to the discharge of a negative cloud to earth. In contrast to the field-changes caused by intra-cloud discharges (see Chapter 6.3), earth flashes produce no polarity reversal at greater distances. Figures 2(a) to (c) show the time variation of the electric field at three distances, namely at 15 km and 3 km, as typical for the results obtained by Schonland (1956) and at the point of strike as recorded by Berger (1973) on Mount San Salvatore. At 15 km distance the movement of negative charge results in a reduction of the negative field. At a distance of 3 km, and much more so at the point of strike, the downward moving leader causes the negative field to increase up to the commencement of the return stroke.

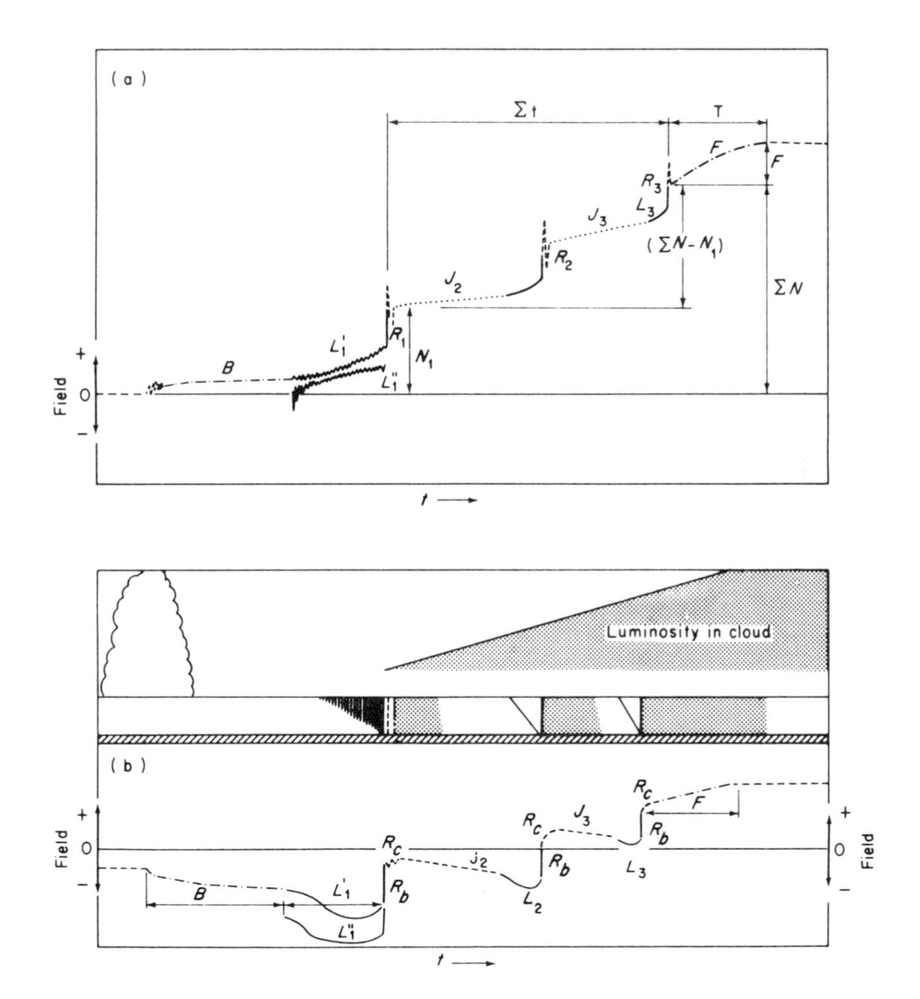

Fig. 2. Electric field variation due to an earth flash. (a) At 15 km distance. (b) At 3 km distance. (*Continued on pp.* 124 *and* 125.)

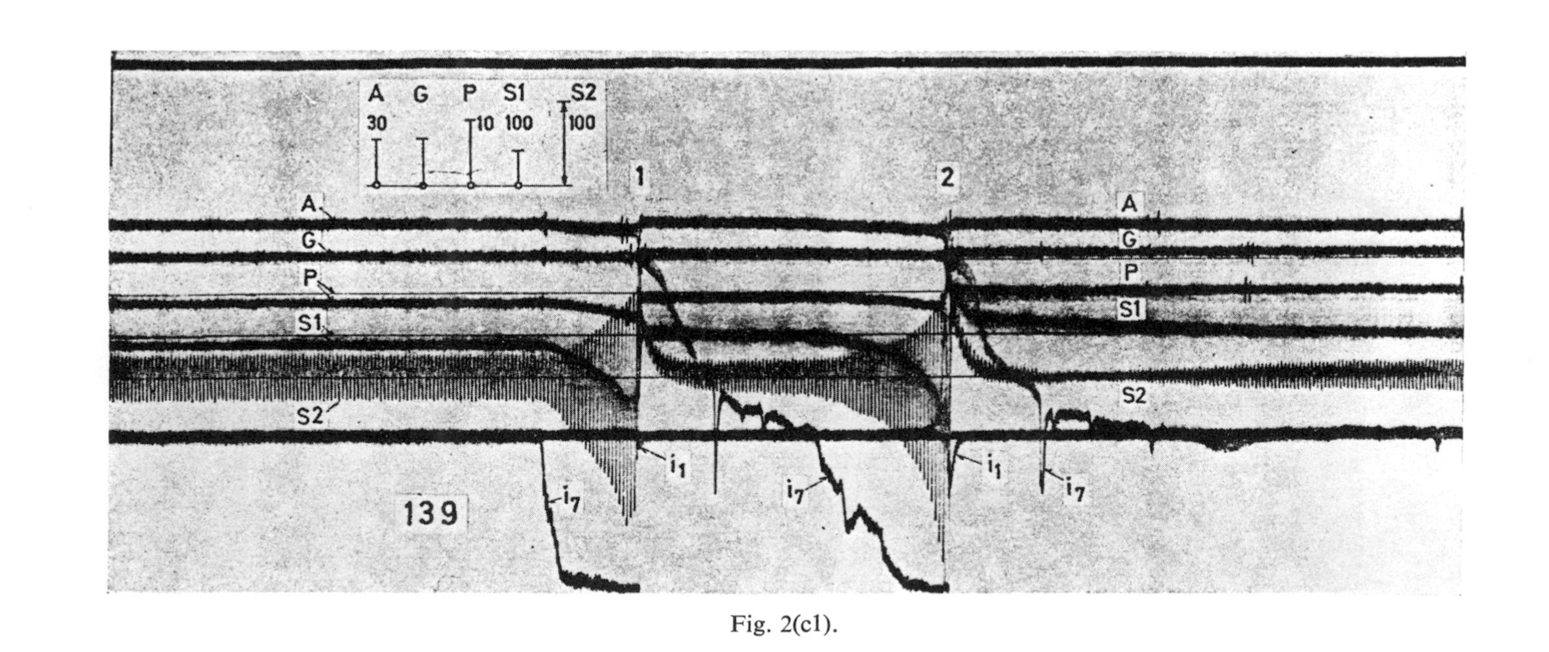

Fig. 2(c1).

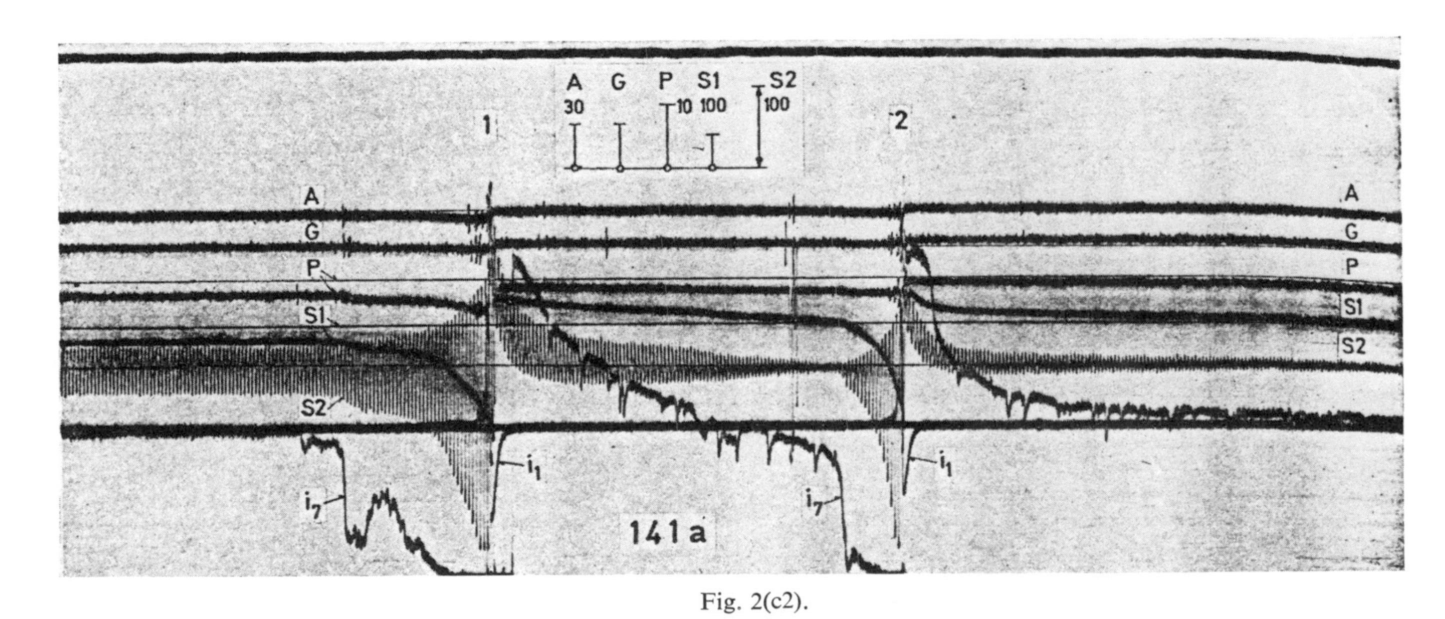

Fig. 2(c2).

Fig. 2 (*cont.*). (c) Near a two-stroke flash (curves $S1$ and $S2$—the scales refer to 100 kV m^{-1}). A: electric field at Agra; G: electric field at Gemmo; P: electric field at Pugerna; $S1$: electric field at Salvatore Church; $S2$: electric field at side of tower 1; i_1: antenna current, range 1 A; i_7: antenna current range 7 mA; 1: first stroke; 2: second stroke; frequency $S2$ —1,150 Hz.

Measurements at San Salvatore, Fig. 2(c), show, as a rule, a change of polarity due to the return stroke and this disappears within several tens of milliseconds.

As mentioned earlier, the position of the zero line in records of the field by means of antennae or simple field mills is, as a rule, unknown. In Fig. 2(b) at 3 km distance a negative initial field is assumed to correspond with the field due to a close discharge from a negative cloud. On the other hand, it is assumed in Fig. 2(a), at 15 km distance, that the fields due to the lower negative and the higher positive cloud discharges just cancel; the field is therefore assumed to start at zero. In Fig. 2(c) the position of the zero line S1 is correct (Berger, 1973). An important result of all three records is the observation that the development of the leader is preceded by a slow field-change (indicated in Fig. 2(a) by *B*). This proves conclusively that the leader develops in the cloud charge as the result of discharges which prepare its onset. The field-change designated by *F* which follows the last component stroke may be assumed to be due to a subsequent slow neutralization of charges within the cloud. Figure 2(c) also shows the displacement current from the top of the tower (lightning-current antenna); it is recorded with a high sensitivity of 7 mA for full deflection and is designated as i_7. In this way fast field-changes can be recorded with great accuracy, particularly the onset of discharges within the cloud, the start of the leader and the progression of the leader towards earth.

2.1.2 Lightning Photography

Initially, the original Boys' camera (Boys, 1926) was employed for photographic studies of the lightning discharge. Later modified cameras with rotating film were used (Schonland, 1956; Malan, 1950, 1963; Berger and Vogelsanger, 1966). All Schonland's photos were obtained near Johannesburg, i.e. in a sub-tropical area and they invariably show the discharge of negative clouds. The leader of the first stroke is stepped while subsequent strokes are preceded by continuously developing dart leaders. Only when the time interval between two subsequent strokes is exceptionally long are fine steps observed in subsequent leaders. A leader originating in a negative cloud carries negative charge to earth and is termed a negative stroke and the current it produces is defined as a negative current (see Section 3.2).

Figure 3 shows a lightning strike to tower No. 2 on San Salvatore (see Section 4.3) which has three component strokes. The stepped leader of the first stroke strikes the tower below its tip (Berger, 1955), the two subsequent strokes have dart leaders and the upward streamer from the tip of the tower indicates an intense local field gradient.

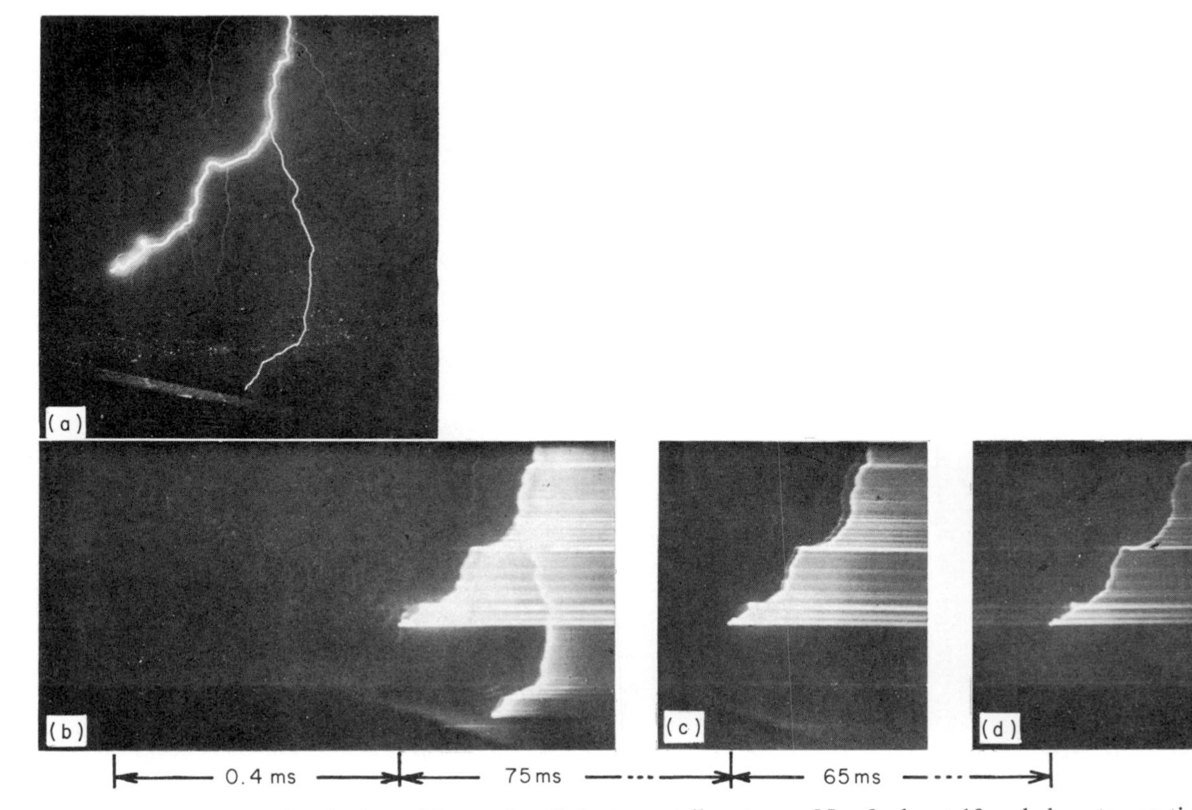

Fig. 3. Downward negative flash to Mount San Salvatore strikes tower No. 2 about 10 m below tower tip. (a) Still photo. (b) First stroke with stepped leader. (c) and (d) Subsequent strokes.

While lightning flashes photographed in South Africa invariably showed discharges from negative clouds with downward leaders as illustrated in Fig. 3, upward leaders were first recorded from the Empire State Building. The interpretation of lightning photographs and the associated currents is presented in Section 4.3.

2.2 Temporal Characteristics of Lightning Flashes

2.2.1 Duration of Lightning Flashes

Table I lists the durations of lightning flashes in different parts of the world. The mean duration is notably shorter on San Salvatore than in South Africa or New Mexico. The reason for this and other differences may be geographical or meteorological although the dispersion of the frequency and other

Table I

Duration of lightning flashes, recorded in South Africa (Malan, 1956), New Mexico, U.S.A. (Kitagawa *et al.*, 1962) and Switzerland (Berger, 1970)

		Malan	Kitagawa *et al.*			Berger		
No.	Duration of flashes (ms)	530 Flashes (%)	36 Discrete flashes (%)	36 Hybrid flashes (%)	72 Flashes (mean value) (%)	58 Downward flashes (%)	245 Upward flashes (%)	303 Flashes (mean value) (%)
0	0	0	0	0	0	0	0	0
1	0–50	11·5	0	0	0	26	5	9
2	0–100	23·5	6	0	3	43	26	29
3	0–150	36	16·5	0	8	53	43	45
4	0–200	50	22	0	11	63	60	61
5	0–250	62	28	0	14			
6	0–300	74	37	6	21	80	78	78
7	0–350	81·5	47·5	11	29			
8	0–400	86	53	16·5	35	86	91	90
9	0–450	89	55	28·5	41·5			
10	0–500	92	67	42	54·5	93	95	95
11	0–550	93·5	72	50	61			
12	0–600	95·5	81	53	67	96	97	97
13	0–800	98	92	78	85	98	99	99
14	0–1,000	98	95	80	92	100	100	100
15	0–1,200	98	97·5	95	97·5			
16	0–1,400	98	98	97·5	99			
17	0–1,600	98	99	97·5	99			
18	0–1,800	100	100	100	100			

parameters of the lightning discharge, measured over many years, is rather large even in the same locality. Malan (1956) determined the variation of the flash duration as a function of the number of component strokes from 530 flashes. The result, illustrated in Fig. 4(a), shows that, e.g., the median

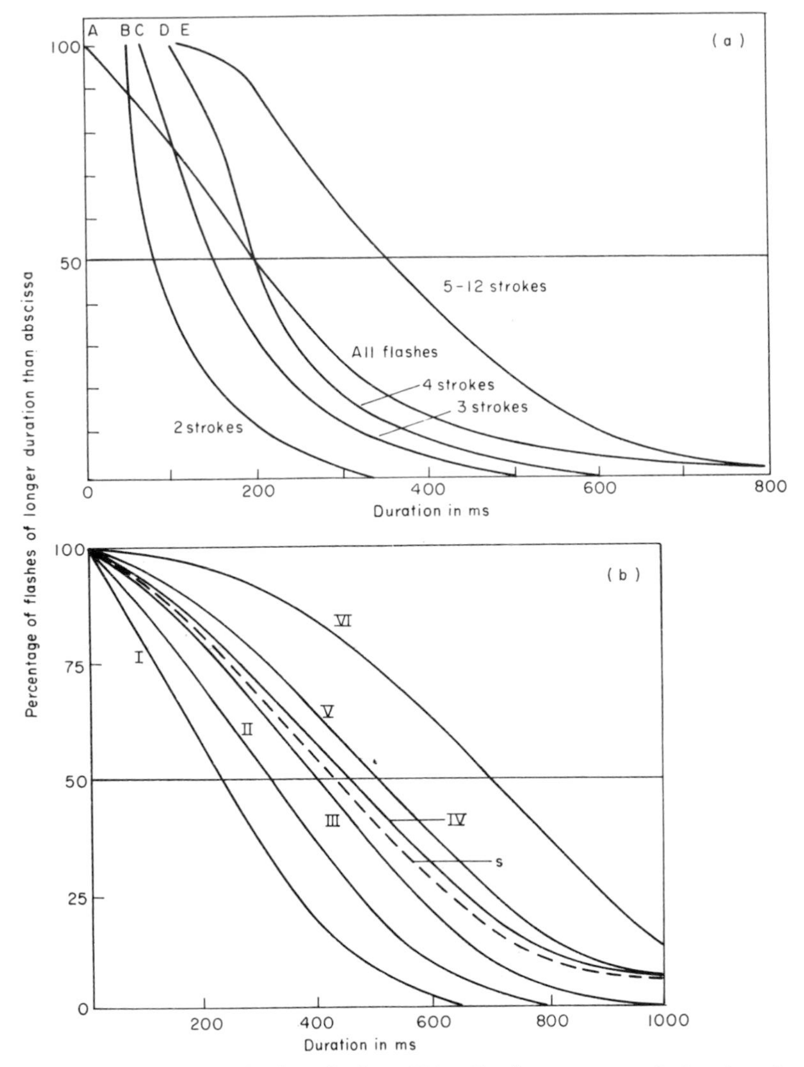

Fig. 4. Duration of lightning flashes. Distribution curves of the duration of lightning discharges with different numbers of component strokes. (a) Curves for different numbers of strokes per flash, in South Africa (Malan, 1956). (b) Curves measured in Asherbaidshan, U.S.S.R. (Kulijew, 1976). I to V: flashes with 1 to 5 strokes; VI: flashes with 6 to 23 strokes; s: all flashes.

duration of 80 ms for 2 strokes increases to 350 ms for a flash with 5 to 12 component strokes. For earlier results see Bruce and Golde (1941) where a survey of the state of knowledge prior to 1941 is given; later results are given by Kulijew (1976) for the Southern Caucasus, see Fig. 4(b).

2.2.2 Number of Component Strokes

Information on the number of component strokes in multiple flashes is given in Tables IIa, IIb, IIc. Malan (1956, Table IIa) drew attention to the uncertainty in deriving the number of strokes from lightning photographs. Depending on the blackening of the photographic film by continuing strokes, the superimposed impulses may be counted as *M*-changes (pulses within continuing discharges) or as component strokes. Thus a slowly rotating film showed 48 traces of light which were counted as component strokes while the same lightning discharge, photographed on a faster moving film, indicated only 16 traces which could be assessed as strokes. The remaining 32 consisted of short light pulses within continuing strokes and must therefore be interpreted as *M*-changes. From 530 flashes Malan derived 14 component strokes as the maximum in a particular flash.

From a photographic analysis of 36 discrete flashes Kitagawa *et al.* (1962), Table IIb, obtained a maximum of 22 strokes and, from 36 hybrid flashes, 26 strokes lasting 1·93 s. In one case Workman *et al.* (1960), also operating in New Mexico, photographed an exceptional lightning flash, lasting 2 s with 54 component strokes, 26 of which showed leader and return strokes. On San Salvatore (Berger, 1955), direct current measurements gave a maximum of 17 component strokes lasting 1·8 s. Values obtained from direct lightning

Table IIa

Number of component strokes within 530 flashes (Malan, 1956)

Number of strokes per flash	Percentage within 460 multiple flashes	Percentage within all 530 flashes
1	—	13
2	22	19
3	20·5	18
4	23	20
5	14	12
6–14	20·5	18
	100	100

Table IIb

Number of flashes with a given number of strokes per flash (Kitagawa *et al.*, 1962)

Number of strokes per flash	Number of discrete flashes	Percentage of 36 discrete multiple flashes	Percentage of 36 discrete +5 single flashes	Number of hybrid flashes	Percentage of 36 hybrid multiple flashes	Percentage of 36 hybrid +5 single flashes
1	5	—	12	5	—	12
2	11	31	26·5	3	8·5	7·3
3	3	8·5	7·3	5	14	12
4	2	5·5	4·9	5	14	12
5	—	—	—	1	2·8	2·5
6	6	16·5	14	5	14	12
7	1	3	2·5	5	14	12
8	2	5·5	4·9	3	8·5	7·3
9	—	—	—	1	2·8	2·5
10	—	—	—	3	8·5	7·3
11	2	5·5	4·9	—	—	—
12	2	5·5	4·9	—	—	—
13	2	5·5	4·9	1	2·8	2·5
14	1	2·8	2·5	1	2·8	2·5
15	1	2·8	2·5	2	5·5	4·9
16	—	—	—	—	—	—
17	1	2·8	2·5	—	—	—
18	—	—	—	—	—	—
19	—	—	—	—	—	—
20	—	—	—	—	—	—
>20	2	5·5	4·9	1	2·8	2·5
	36/41	100	100	36/41	100	100
Mean value: strokes/flash		260/36 = 7·2	265/41 = 6·5		253/36 = 7·0	258/41 = 6·3

Table IIc

Number of flashes with a given number of strokes per flash (Berger, 1972)

Number of strokes per flash	Number of flashes	Percentage within 242 multiple flashes	Percentage within all 242+784 flashes
1	784	—	76·5
2	79	33	7·7
3	49	20	4·8
4	20	8	1·9
5	24	10	2·3
6	14	6	1·4
7	10	4	1
8	7	3	0·7
9	5	2	0·5
10	6	3	0·6
>10	28	11	2·7
	784+242 = 1,026	100	100
Mean value: strokes/flash		4·8	1·9

current measurements on San Salvatore (Berger, 1970, Table IIc) include both downward and upward flashes.

As indicated by Tables IIa to IIc, multiple flashes are particularly prevalent in New Mexico while single-stroke flashes predominate in Switzerland. This difference is clearly due to the large number of upward discharges on San Salvatore which are very frequently single strokes.

2.2.3 Time Intervals between Strokes

In the South African investigations (Malan, 1956) the most frequent or modal time interval between individual strokes varied between 30 and 50 ms, with a minimum value of 15 ms and a maximum of 700 ms. Kitagawa and Kobayashi (1958, Table III) determined time intervals between strokes from field registrations. During four thunderstorm periods, 7 multiple flashes were recorded with 10 or more component strokes and particularly long time intervals. This led to the conclusion that the time interval increases with increasing number of component strokes.

Pierce (1955) recorded the field-changes produced by earth flashes up to 200 km away. He found that, up to about 50 km, slow electrostatic field-changes occurred in about 25% of all multiple flashes, while in the remaining

Table III

Time intervals between strokes (Kitagawa and Kobayashi, 1958)

Flash No.	Flash duration (ms)		Time intervals between strokes (ms)										
			Stroke order										
			1	2	3	4	5	6	7	9	9	10	Mean
1	1088		48	66	50	66	56	46	175	165	416	—	121
2	862		45	19	45	38	36	243	45	49	200	144	86·4
3	1256		31	16	41	67	48	107	41	188	228	439	120·6
4	871		20	72	108	82	105	46	193	67	118	60	87·1
5	605		18	25	37	31	37	41	72	81	77	186	60·5
6	749		28	29	34	35	55	73	84	38	73	300	74·9
7	847		38	47	36	46	116	54	62	139	114	195	84·7
1/7	899	Mean	32·6	39·1	50·1	52·1	64·7	87·1	96·0	103·8	175·0	220·7	90·3

75% of cases the field remained constant between component strokes. This subdivision roughly agrees with the fraction of multiple flashes containing continuing currents.

3. Types of Lightning Discharge

3.1 Definitions

As shown in Chapters 3.4.4 and 6.3.2 the lower parts of a thundercloud are usually negatively charged while the higher region carries positive charges. The polarity of the prevailing cloud charge defines the polarity of the lightning current. Thus, the discharge of a positive cloud charge to earth is called a *positive flash* while the discharge of a negative cloud is termed a *negative flash*. In a positive flash a positive current flows from cloud to earth, in a negative flash the current is described as negative.

A *downward leader* progresses from the cloud to earth. In an *upward leader* the direction is from earth to the cloud. An *upward connecting leader* is a discharge from earth (or an earthed object) which meets a downward leader and discharges it to earth.

No such agreement in terminology exists for the definition of polarity of the electric field below a charged cloud. In the context of this book the electric field below a positive cloud is defined as positive. A *positive field* is thus directed from the cloud (as the origin of the field) to earth. The discharge of a positive cloud then reduces the field and causes a *negative field-change*; the discharge of a negative cloud causes a *positive field-change*.

The polarity of a leader may be defined either by the polarity of the electric charge involved or by that of the resulting current. Either a *downward leader* lowers positive charge from the cloud by means of a positive current or it lowers a negative charge by means of a negative current. Such a leader therefore has always the same polarity of charge and current.

An *upward leader*, on the other hand, has opposite polarities of charge and current. This fact is of importance for the discharge of a positive cloud as discussed in Section 3.2. In this chapter the *polarity of a leader* is defined by the polarity of its charge and not by that of its current. The same rule applies to the polarity of an upward *streamer*, whether this is followed by a lightning stroke or not.

3.2 Description of Lightning Types

The leader stroke is invariably initiated at a point of high electric field strength. This can occur either between positive and negative charge centres in a thundercloud or between a negative (or positive) charge centre in the cloud and its induced countercharge in the ground.

The first case starts as an intra-cloud flash between the charges of opposite polarity. Depending on the height of the cloud above ground and the transient field change between cloud and earth, such a discharge can be confined to a pure intra-cloud flash or it can proceed towards ground, thus producing an earth flash.

The second case leads to a leader stroke from the cloud charge to its induced charge in the ground. The highest field strength can then arise either at the lower boundary of the cloud or on a very tall earthed object. This, in turn, leads either to a downward flash from cloud to earth or to an upward leader which develops from the earthed object towards the cloud.

Four different types of lightning flash have therefore to be considered and these are sketched in the upper half of Fig. 5. This schematic illustration merely shows the polarity of the cloud charge which directly participates in the flash, without reference to the distribution of other charges in the cloud—this aspect is discussed in Chapters 3.4.4 and 6.3.2. A leader stroke can be followed by a return stroke or not; this statement may seem self-evident but it is important. The predominant case of the downward leader from a negative cloud being followed by a return stroke (as described by Schonland) is widely accepted as *the* picture of the earth flash. Thus the picture of a lightning flash without a return stroke came only to the fore as the result of observations on such tall structures as the Empire State Building and San Salvatore. The measurements on San Salvatore then led to an extension of the picture of the lightning discharge in the sense that an upward leader can progress so far into a cloud as to initiate a return stroke downwards.

A comprehensive catalogue of the different forms of a lightning discharge therefore requires a further subdivision of the four leader types illustrated in Fig. 5. into flashes with and without return strokes. Following the method first introduced by Berger (1975), a flash without a return stroke is designated by the letter *a* and a flash having a return stroke by *b*. The eight possible types of lightning flash are therefore defined as follows:

Type 1a (Fig. 5). The discharge begins with a downward leader from a negative cloud. This is predominantly the case over open country without very tall objects. The leader is negatively charged and its current is negative. If it does not reach earth (air discharge), no return stroke ensues; it constitutes a cloud discharge, as discussed in Chapter 6.

Type 1b (Fig. 5). When the negative downward leader reaches the earth, the very fast upward moving return stroke develops and this discharges the leader and part of the cloud charge to earth. This sequence can occur once,

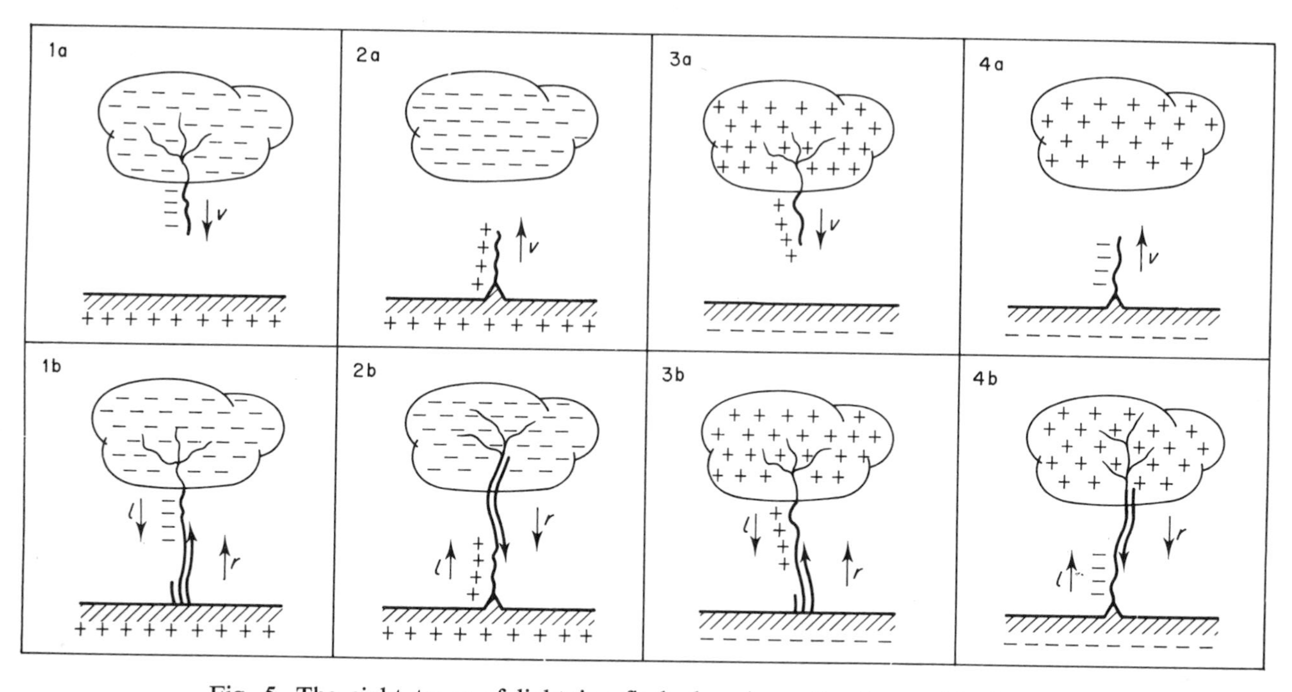

Fig. 5. The eight types of lightning flash, based on directions of leader and return strokes. l: leader; r: return stroke; v: direction of propagation.

producing a single-stroke flash, or it can be repeated, producing a multi-stroke flash.

Designation: negative downward flash (exclusively examined by Schonland).

Type 2a (Fig. 5). The discharge is initiated by an upward leader from a tall earthed object (tower or mountain top), progressing towards a negative cloud. The leader is charged positively (the top of the tower constitutes the anode); it is a positive leader. The charge flowing to earth, and thus the currents in the tower are negative.

Designation: positive upward leader/negative continuous flash.

Type 2b (Fig. 5). Initial stage as for type 2*a*, followed by subsequent strokes, each having a downward leader and upward return stroke as type 1*b*.

Designation: positive upward leader/negative multiple flash (investigated on Empire State Building and San Salvatore).

Type 3a (Fig. 5). This corresponds to type 1*a* but under a positive cloud. Both the charge on, and the current in the leader are positive. Since the leader does not reach earth, a displacement current flows in the ground. This is an intra-cloud flash as discussed in Chapter 6.

Type 3b (Fig. 5). When the positive downward leader reaches the earth, it gives rise to a positive upward return stroke by which the leader and part of the cloud are discharged. This type is very rare in mountainous regions. Only one photograph has been secured of a leader to the shore of Lake Lugano and no single case has been recorded in 15 years on San Salvatore; see also type 4*b*.

Designation: positive downward flash.

Type 4a (Fig. 5). From a tall earthed object an upward leader is initiated under a positive cloud. The leader is negatively charged (the tip of the structure constitutes the cathode). The charge flowing to earth is positive and so is the current which, being of long duration, is described as a *continuous* current. This type of flash was first observed on San Salvatore (Berger, 1966).

Designation: negative upward leader/positive continuous current flash, or, in short, positive continuous-current flash (mountain type).

Type 4b (Fig. 5). This type is initiated as type 4*a* but the upward leader is followed, after 4 to 25 ms, by an exceptionally severe positive downward discharge which must be regarded as a return stroke. This type was also first observed and described on San Salvatore (Berger and Vogelsangar, 1966) and was discussed in greater detail by Berger (1975). The negative upward leader progresses in the form of a very long "connecting leader" into an existing intra-cloud flash which, first, caused the transient field by which the upward leader was initiated and, later, is discharged to earth through this "connecting leader".

Designation: negative upward leader/positive impulse-current flash, or in short, positive impulse current flash (mountain type).

Lightning type $4b$ seems to exclude flashes of type $3b$ in mountainous areas or near tall structures because of high field concentration on the ground by which a long upward connecting leader is initiated before a downward leader approaches.

Figure 5 is oversimplified. In reality, when a downward leader approaches the earth, an upward streamer is often initiated from the ground or from a projection and this progresses towards the tip of the downward leader in the form of an upward leader as described in greater detail in Chapter 17. Simultaneously several additional streamers may sometimes develop from earth, extending for several meters into the air as illustrated in Fig. 3, Chapter 17.

4. Results of Lightning Research and Recording Stations

4.1 Instruments for Recording Lightning Currents

Three Instruments have been employed to record lightning currents.

4.1.1 Measuring Shunt

The first method to record the wave shape of the lightning current utilizes a precision shunt of resistance R to measure the potential drop which a current i produces across its terminals. The shunt consists of a thinly walled tube made of manganin, constantan or a similar material or of several concentric tubes, electrically connected in series, through which the lightning current is discharged. In the centre line of this arrangement, a measuring cable is positioned to record the potential drop (Berger, 1965, 1967).

This method requires concentration of the full lightning current in the shunt. This can be achieved by installing the shunt directly, and without any bypass, below a lightning rod, a ring or an antenna on top of a high structure or in a down conductor, provided this carries the full lightning current.

4.1.2 Induction Coil

An inductive loop or coil can be used to measure the induced voltage $u = M\,\mathrm{d}i/\mathrm{d}t$ or the integral of u with respect to time and thus the lightning current i. This presupposes that the mutual inductance M between the discharge path of the current i and the coil is accurately known. A toroidal coil or pulse transformer is particularly suitable for this purpose (Hentschke and Illgen, 1975, and Chapter 13.4).

A non-ferrous, long toroidal coil with many turns, arranged on an insulating core, is installed around the current-carrying object, e.g. a television tower. Such a coil is known in German literature as a "Rogowski-coil". It has the remarkable property that its induced voltage is proportional to the variation with time of the current which is discharged through it, independently of where the partial currents penetrate the coil. The pulse transformer differs from the toroidal coil by having a magnetic core for high frequencies; in power engineering it corresponds to the current or voltage transformer. Unfortunately, the frequency range of all these devices is limited in both directions. They are suitable for the measurement of impulse currents but not of continuing and continuous currents.

4.1.3 The Cathode-ray Oscillograph

The cathode-ray oscillograph is the only device at present capable of determining the wave shape of the lightning current. However, attempts have recently been made to record the current wave shape in digital form and to retrace it from the stored information. Since several points have to be measured within a front duration of 1 μs, a high recording rate must be achieved. Registration of the wave shape on magnetic tape also requires a high frequency response since it can only be accomplished by frequency modulation. Furthermore, the tape must have a sufficiently long life. Up to now the problem has not been solved although it would constitute a simple and cheap method of registration.

4.2 The Empire State Building

The first investigations on the 380 m high Empire State Building (McEachron, 1939) employed the crater-lamp oscillograph, magnetic links and, at about 780 m distance, rotating cameras.

Photographs taken with the Boys' camera showed for the first time upward leaders emanating from the top of the building as defined in Section 3.2 as type 2 and as illustrated schematically in Fig. 5. Branches from the main lightning channel point upwards for the first upward strokes but downwards for downward strokes. The following values were found for the rate of propagation of the component discharges:

stepped upward leader	0·25 m μs^{-1}
downward leader in subsequent stroke following upward leader	12·1 m μs^{-1}
return stroke following upward leader	61·7 m μs^{-1}.

Out of 53 photographs, 36 showed upward stepped leaders, 2 produced downward leaders, while 2 others had upward leaders of the dart type, namely without steps. The remainder were not clear. All subsequent strokes started with dart leaders. Mean time intervals between the steps of upward leaders were between 20 and 30 μs, with extremes of 20 μs and 100 μs respectively.

The second important result of these investigations is that all oscillograms indicated negative currents. The upward leader was thus invariably positive towards a negative cloud. Later investigations on San Salvatore showed that positive leaders produce much less light than negative leaders. This explains why McEachron faced much greater difficulties in securing clear photographs of leaders than Schonland who recorded exclusively negative leaders. The only photograph reproduced by McEachron (1939) shows a few fine points at the leader tip and this confirms a positive leader according to experience on San Salvatore.

In contrast to these latter recordings, the Empire State Building experienced no positive lightning currents of types 3 and 4 in Fig. 5. This is likely to be simply due to the small number of 27 component strokes recorded with the crater lamp oscillograph and 11 magnetic link records.

4.3 Mount San Salvatore

4.3.1 Recording Equipment

The station on Mount San Salvatore is on Lake Lugano, latitude 45°59′ north, longitude 8°57′ east. Its peak is 914 m above sea level and 640 m above the lake. Its position is marked by a cross in Fig. 6(a), in which the few points where upward flashes were photographically recorded are also indicated. Figure 6(b) indicates the points where downward flashes were recorded by night photography. Where the exact points of strike were not visible the numbers of strikes are entered. Figure 6(b) clearly shows that downward flashes may occur anywhere.

The first tower, erected in 1943, was made of wood with four insulated stays. Its height was 60 m with an additional steel conductor of 10 m length. The current shunt was installed near a recording room at the foot of the tower. In 1958 it was replaced by a 60 m transmitting tower of the Swiss Post Office to which was added a platform for the shunt and a lightning rod extending 70 m above ground. In this way the second shunt had a notably higher frequency limit than the first, with a response time of 16 ns. The construction of this shunt is described by Berger (1965, 1967). A second tower was erected in 1950 to obtain photographs from a distance of 365 m and to increase the number of records. Its shunt is of the same construction as that on the other tower.

The wave-shapes of lightning currents were determined by recording oscillographically the potential drop across the shunt via a special cable leading into the recording room. Following the early use of a Dudell oscillograph with eight galvanometers, a special cathode-ray oscillograph was installed in 1958. It has four beams for the recording of currents in both towers and two time deflections with a resolution of 0·5 μs. Tripping is accomplished (Berger, 1972) down to about 20 A.

Since 1950 eight Leica cameras have been used to cover the entire panorama for night photography. Speeds of film rotation, initially 50 ms^{-1}, varied between 1 and 27 ms^{-1}. For photographs of the panorama the cameras with stationary films can be actuated by remote control either singly or together. The rotating cameras are duplicated to avoid losses during the changing of films. An additional photographic station at a distance of 3·2 km records direct strikes to both towers. Spectroscopic photography is described in Chapter 8.

Since 1967 the electric field has been registered by two types of field mill, installed respectively at the tower and at distances of several kilometres. The principle of the design of a field mill is discussed in Chapter 13.2.2.1 and the novel construction used on San Salvatore is described by Berger (1972). By periodically discharging the rotating vanes, its lower frequency response could be practically reduced to zero, unless it was subjected to excessively strong and gusty wind. A resulting measuring error can be detected in the oscillogram.

The first simple field mill was installed about 18 m above ground on one leg of tower No. 1. The four improved field mills were installed:

(i) above the viewing platform of a church on the mountain top,
(ii) at about 2·5 km to the east (Pugerna, P),
(iii) at about 3·45 km to the north (Gemmo, G) and
(iv) at about 3·3 km to the west (Agra, A).

The values measured by field mills (i) to (iv) were transmitted to the recording station on Mount San Salvatore by direct telephone cable at a frequency of 13·5 kHz. All field mills were operated from the recording station by remote control. The five results were recorded permanently on moving magnetic tapes and were erased after 5 s. When a direct stroke occurred to San Salvatore, the stored record was transmitted to, and recorded by, a Dudell oscillograph with an upper frequency limit of 4 kHz. This method has the advantage that all five field curves could be continuously observed on the oscillographic screen. Direct strikes which did not hit either tower could still be recorded in this manner, if the field curves showed interesting features.

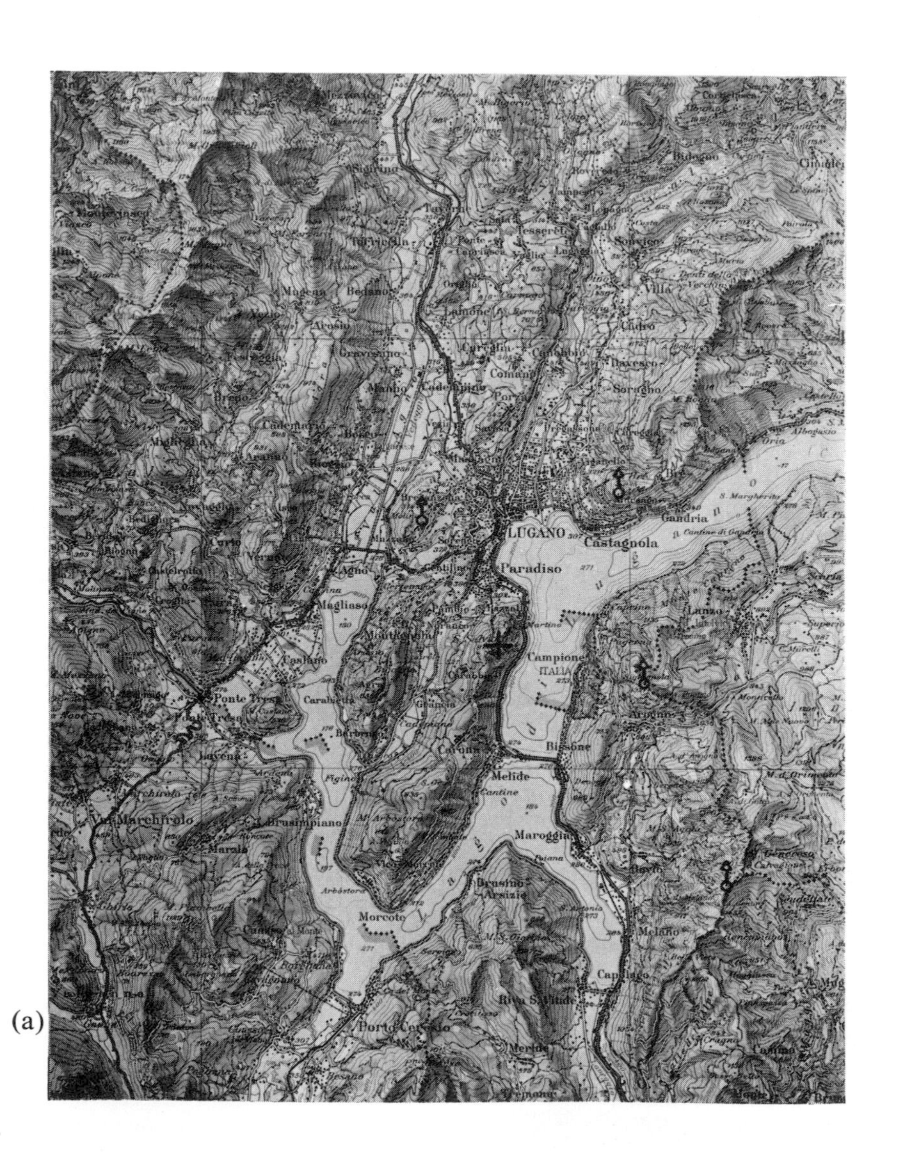

LUGANO
Castagnola
Paradiso
Gandria
Campione
ITALIA
Bissone
Melide
Maroggia
Morcote
Ponte Tresa
Agno
Magliaso
Caslano
Lavena
Brusimpiano
Arogno
Lanzo
Melano
Capolago
Riva S. Vitale
Tesserete
Sonvico
Villa
Cadro
Davesco
Soragno
Bidogno
Bedano
Gravesano
Manno
Cademario
Carona
Figino
Arbostora
Cantine
S. Margherita

(a)

Fig. 6. Lightning flashes to the surroundings of Mount San Salvatore in the period 1955 to 1965. (a) Map of Lugano with San Salvatore (cross). Dots indicate origins of upward flashes. (b) Same map with indication of downward flashes. Numbers indicate flashes to points where ground is not visible from Mount San Salvatore. Different symbols indicate different years within the observation period 1955 to 1965. (Reproduced with the permission of the Eidgenössische Landestopographie.)

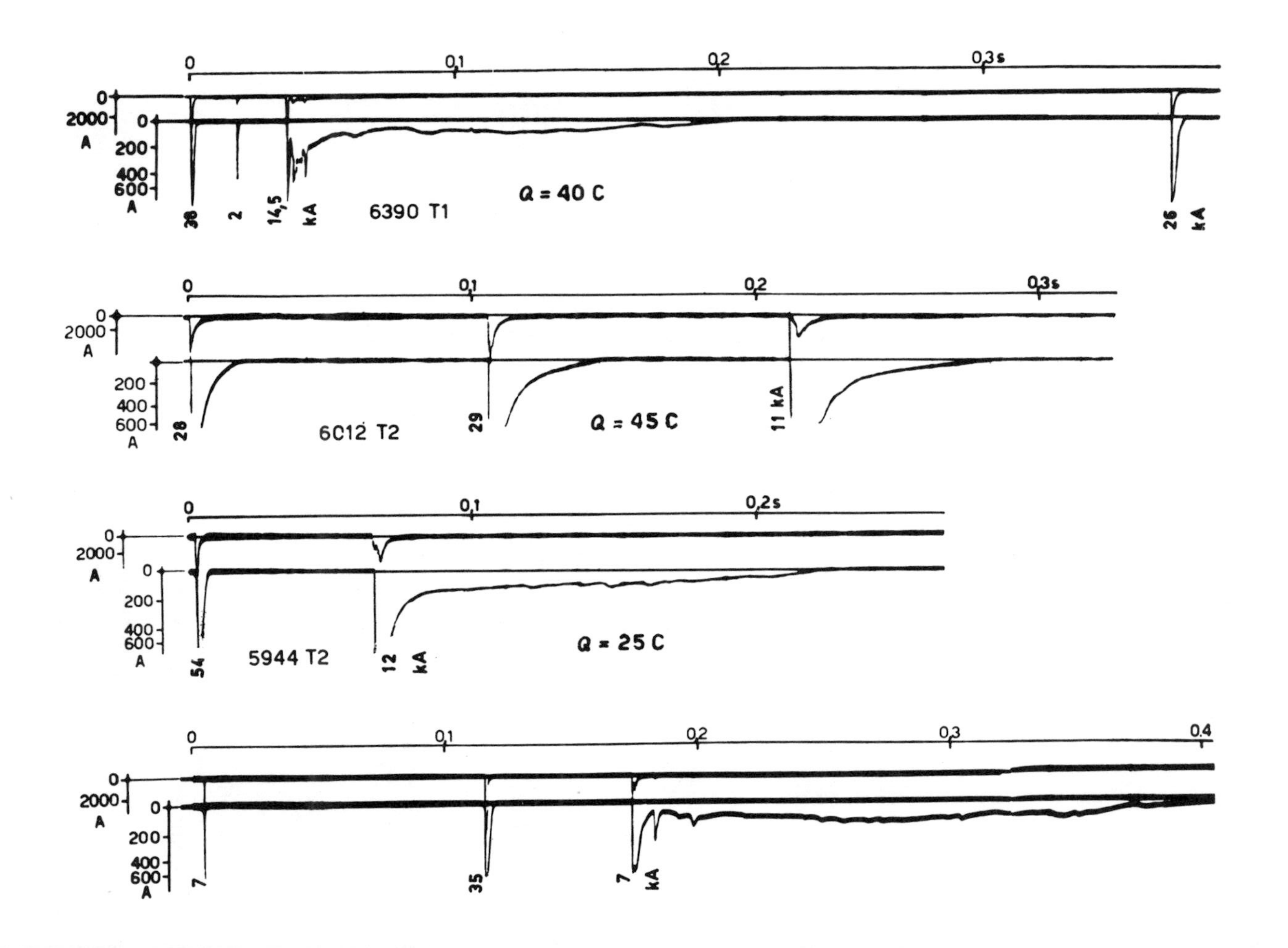
0
0,1
0,2
0,3s
0
2000
A
0
200
400
600
A
38
2
14,5
kA
6390 T1
Q = 40 C
26
kA
0
0,1
0,2
0,3s
0
2000
A
200
400
600
A
28
6012 T2
29
Q = 45 C
11 kA
0
0,1
0,2s
0
2000
A
0
200
400
600
A
54
5944 T2
12
kA
Q = 25 C
0
0,1
0,2
0,3
0,4
0
2000
A
0
200
400
600
A
7
35
7
kA

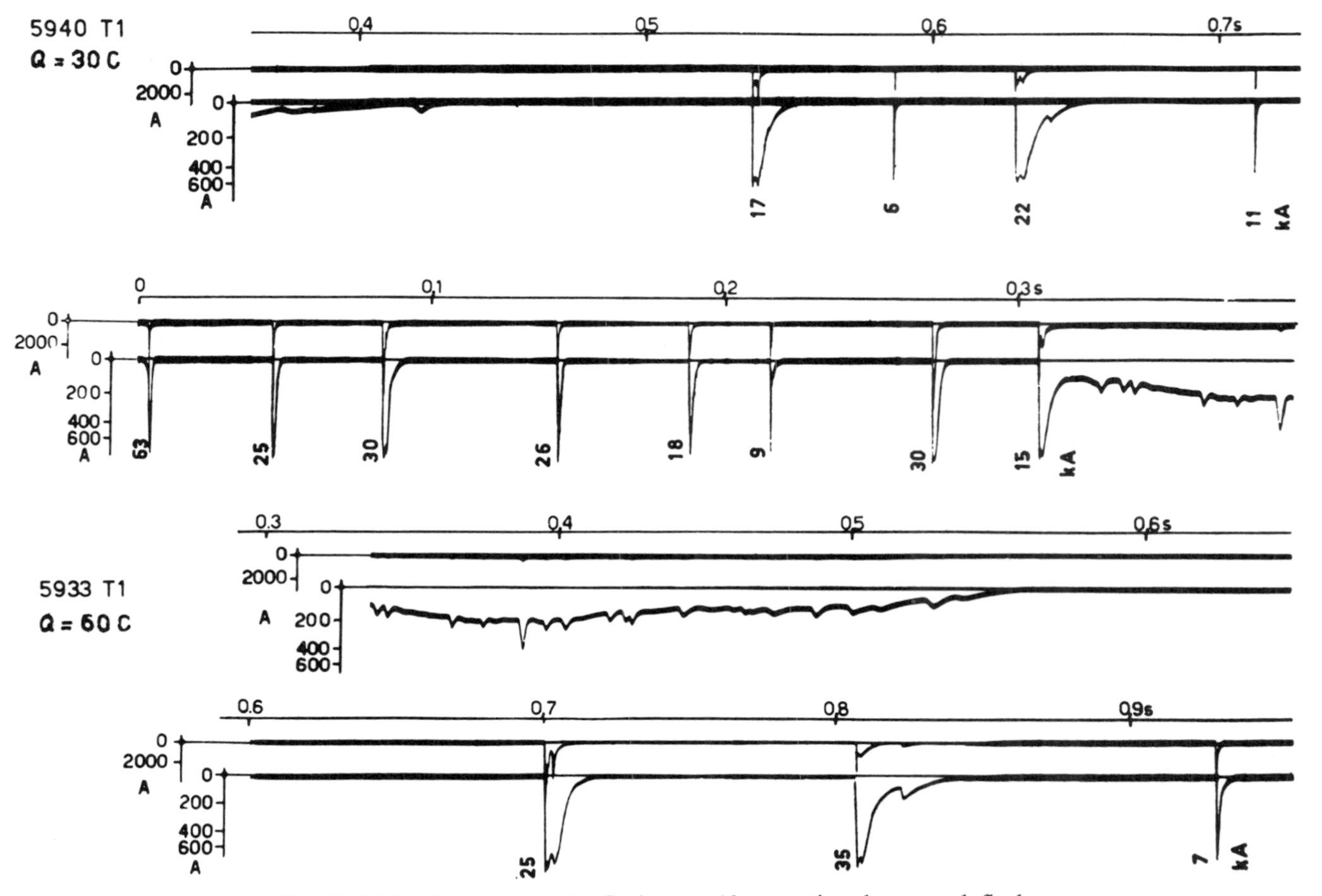

Fig. 7. Lightning currents to flash type 1*b*; negative downward flash.

4.3.2 Lightning Current Characteristics

During the recording period 1946 to 1954, 204 negative flashes and 57 positive flashes with 390 negative strokes and 59 positive strokes were recorded oscillographically (Berger, 1955). Between 1955 and 1963, 324 negative flashes and 46 positive flashes were recorded (Berger, 1965) and between 1963 and 1974, there were 732+106 negative and 102+27 positive flashes (Berger, 1972).

In what follows the typical characteristics of the four types of lightning discharge as defined in Section 3.2 are described.

Type 1: negative downward leader—negative flash.

The current in Fig. 7 begins with an impulse amounting to several tens of kiloamperes. This is frequently followed by successive strokes, some of which have continuing currents of the order of 100 to 200 A lasting hundredths or tenths of a second. The initial current impulse is due to the fact that the current is not noticeable before the downward leader contacts the ground. Typical wave shapes of first currents are reproduced in Fig. 8 (see also Chapter 9.4.1.1).

Type 2: positive upward leader—negative flash.

The example in Fig. 9 shows a continuous current of several hundred amperes lasting several tenths of a second; the shape is irregular with small superimposed impulses. In other cases, these impulses may reach 10 kA (Fig. 10). The negative continuing current flows through the upward leader to earth so that it is positively charged and assumes the character of a streamer of great length and long duration.

Figure 11 illustrates the equally frequent occurrence of a multiple flash in which the continuous current is followed, after a no-current time interval, by a negative impulse current. This can be repeated several times (subsequent strokes). The amplitudes of these subsequent currents are mostly between 10 and 30 kA and, usually, lower than those in the first stroke.

Types 3 and 4: positive cloud—positive flashes; general comments.

The subdivision of positive flashes in downward and upward discharges encounters difficulties. In the first place, only one fast rotating-camera photograph was secured of a positive downward stroke, Fig. 12 (see Section 4.3.3). Secondly, as explained in Section 3.2, the frequent type 4 must be described as a negative upward leader and positive downward stroke.

Out of 129 positive flashes between 1963 and 1971, a total of 102 comprised positive upward discharges, namely upward leaders without subsequent discharges of the cloud. These are to be defined as *Type 4a*. The remaining 27 positive upward leaders were followed by intense discharges from the cloud. In Table III of Berger and Vogelsanger (1966) they are listed as positive

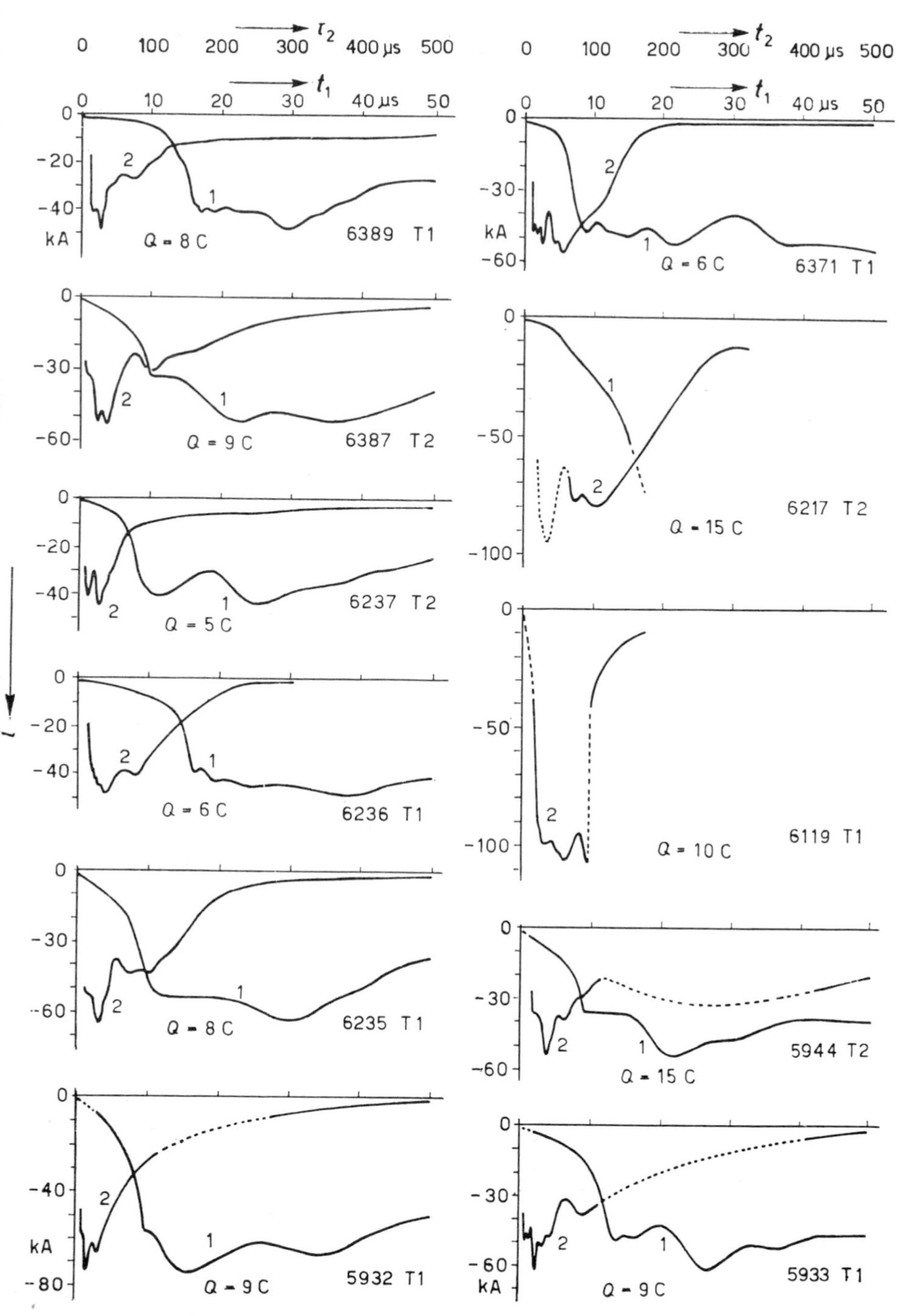

Fig. 8. Negative impulse currents in flashes type 1*b*.

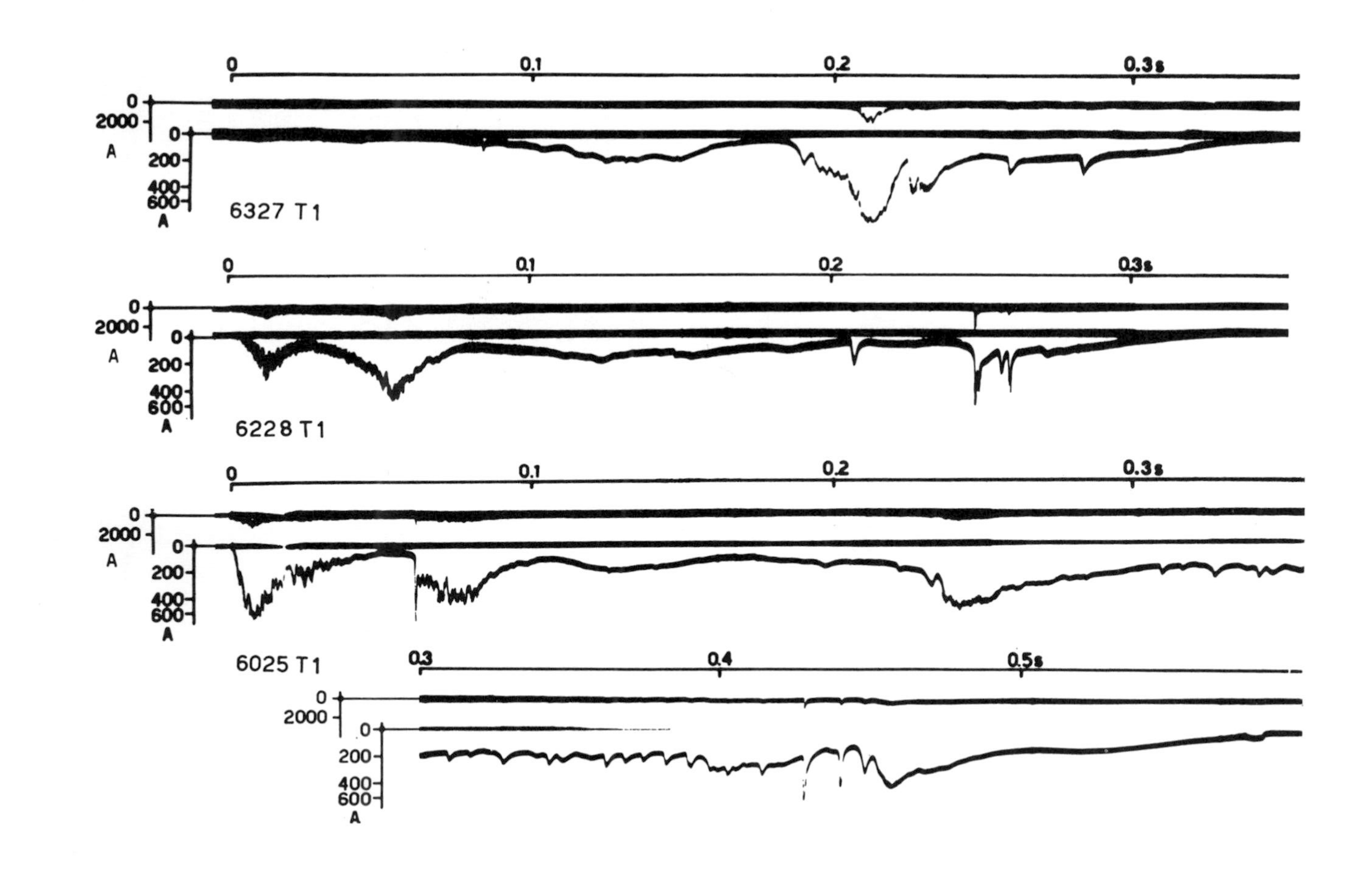

6327 T1
6228 T1
6025 T1
0
0.1
0.2
0.3s
0.3
0.4
0.5s
0
2000
A
200
400
600

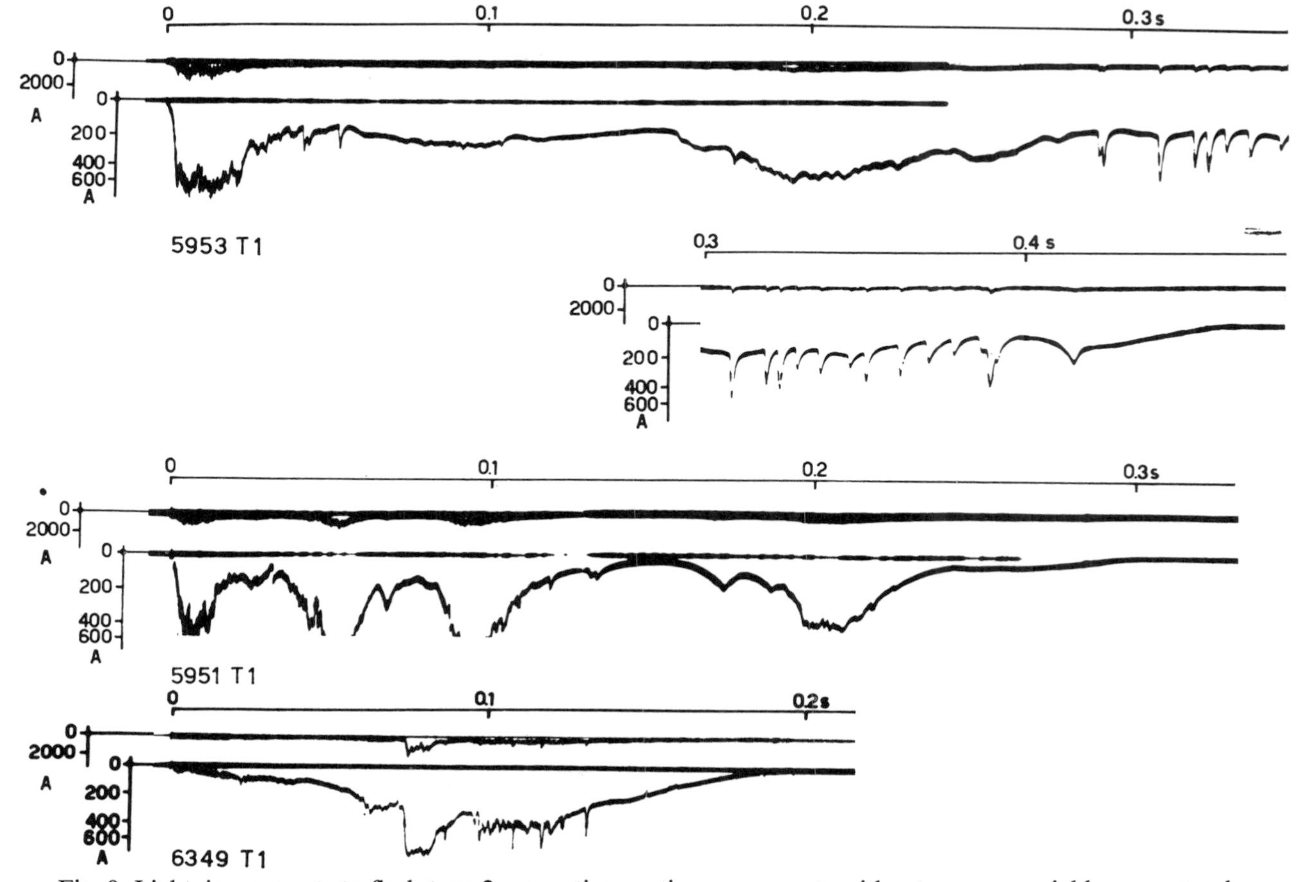

Fig. 9. Lightning currents to flash type $2a$; negative continuous currents without any appreciable current pulses.

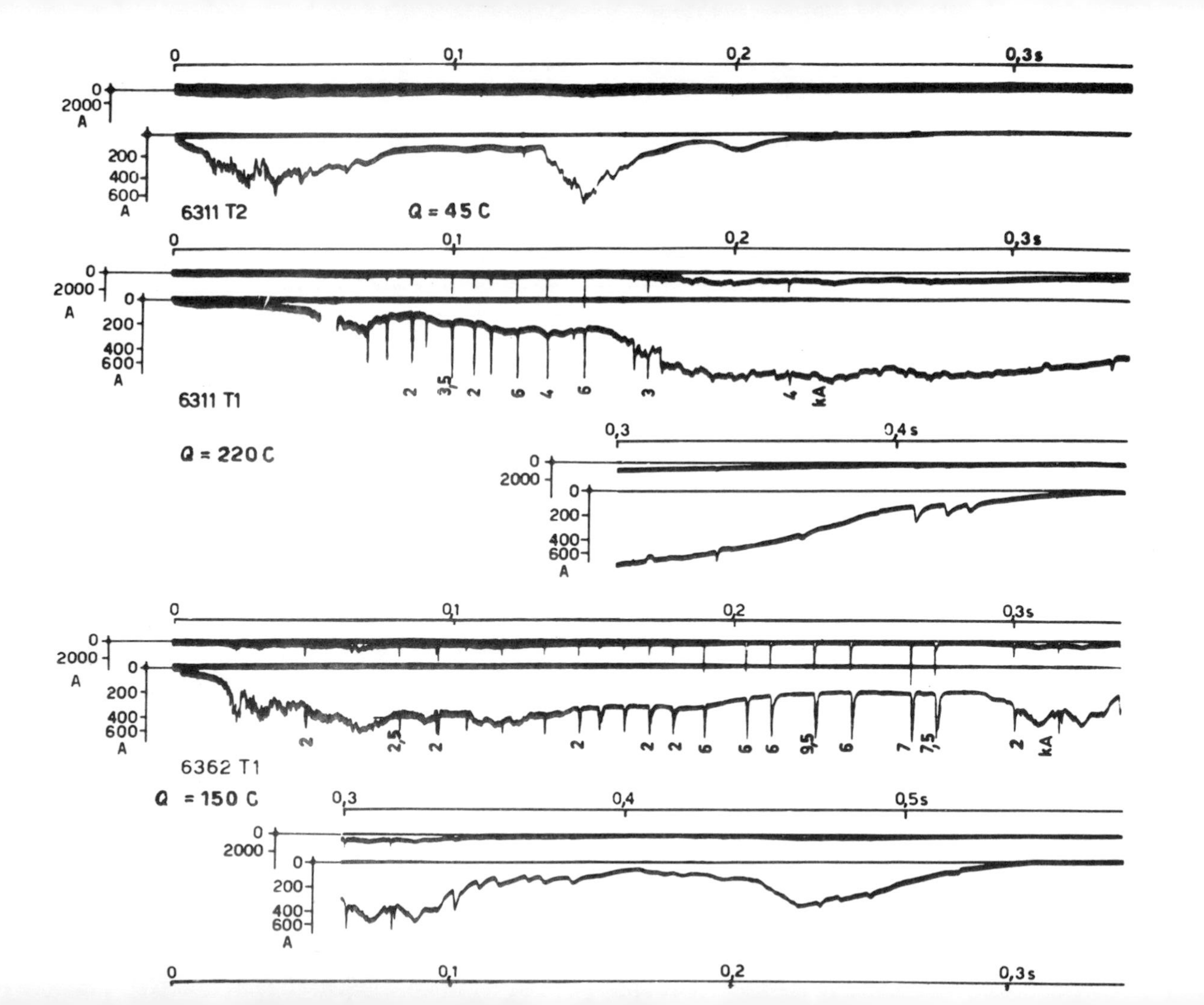
6311 T2
Q = 45 C
6311 T1
Q = 220 C
6362 T1
Q = 150 C
kA

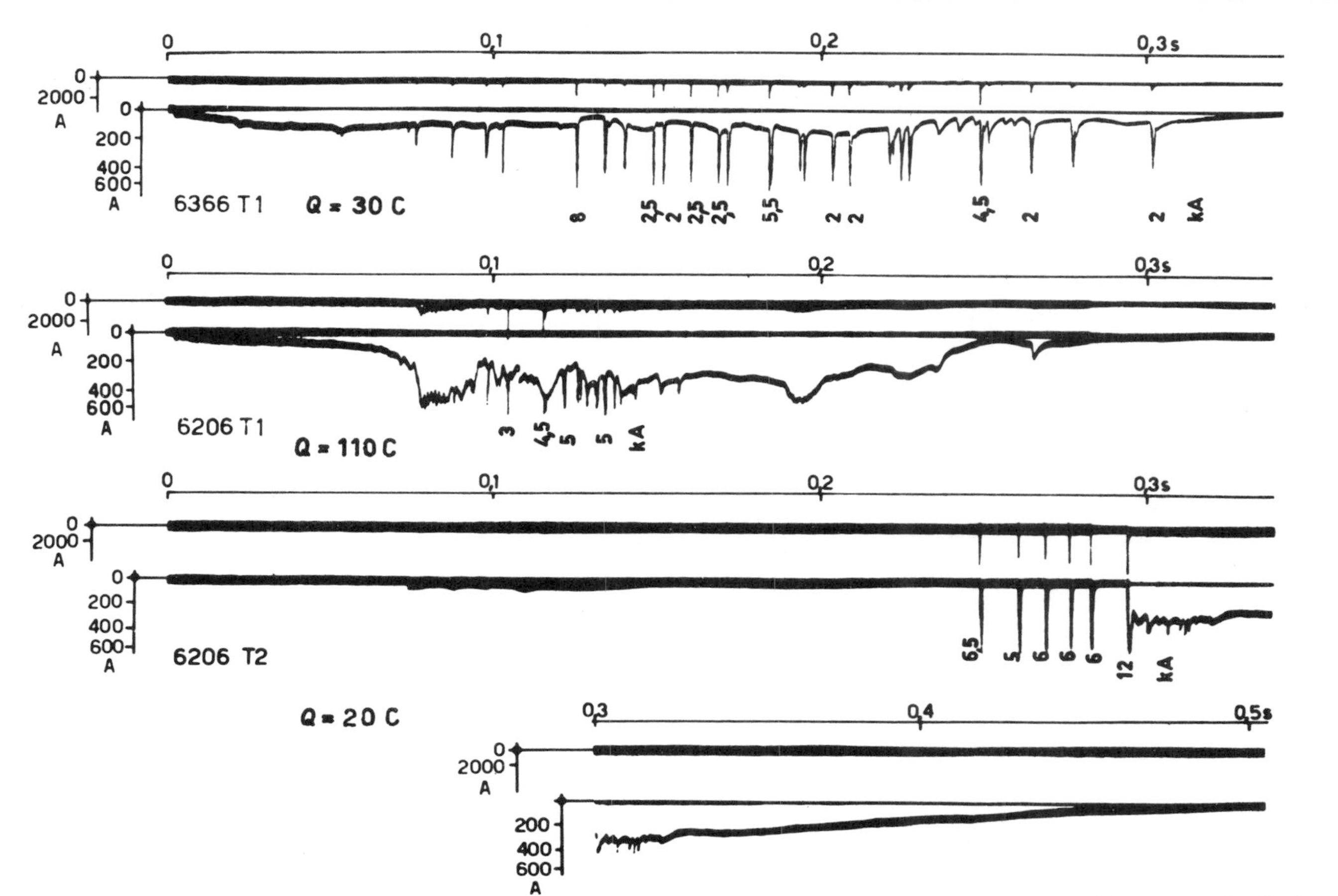

Fig. 10. Lightning currents to flash type $2a$, with superimposed current pulses up to 10 kA, upward leader/single negative flash.

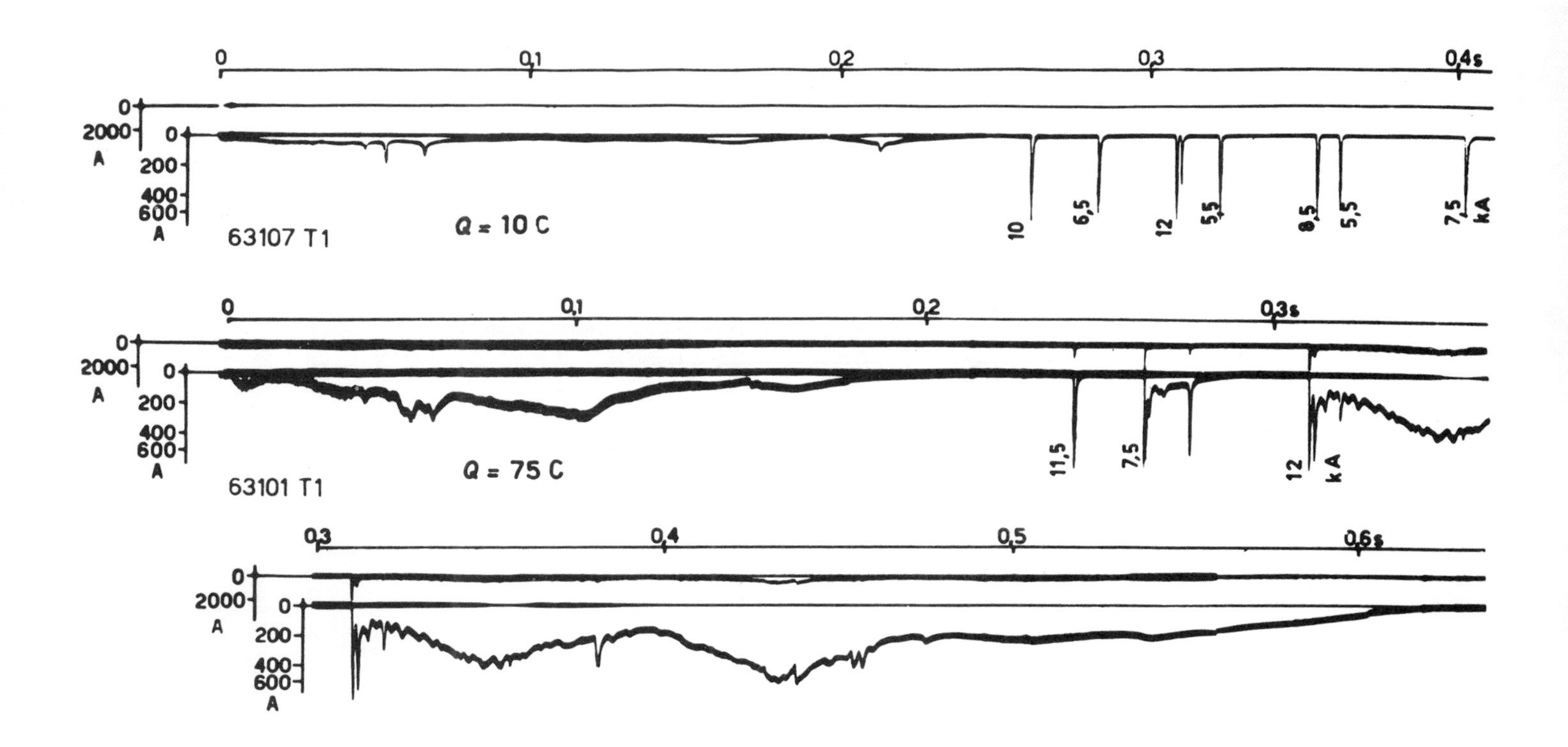
63107 T1
Q = 10 C
0
0,1
0,2
0,3
0,4s
2000
A
0
200
400
600
A
10
6,5
12
5,5
8,5
5,5
7,5
kA
63101 T1
Q = 75 C
0,3s
11,5
7,5
12
kA
0,4
0,5
0,6s

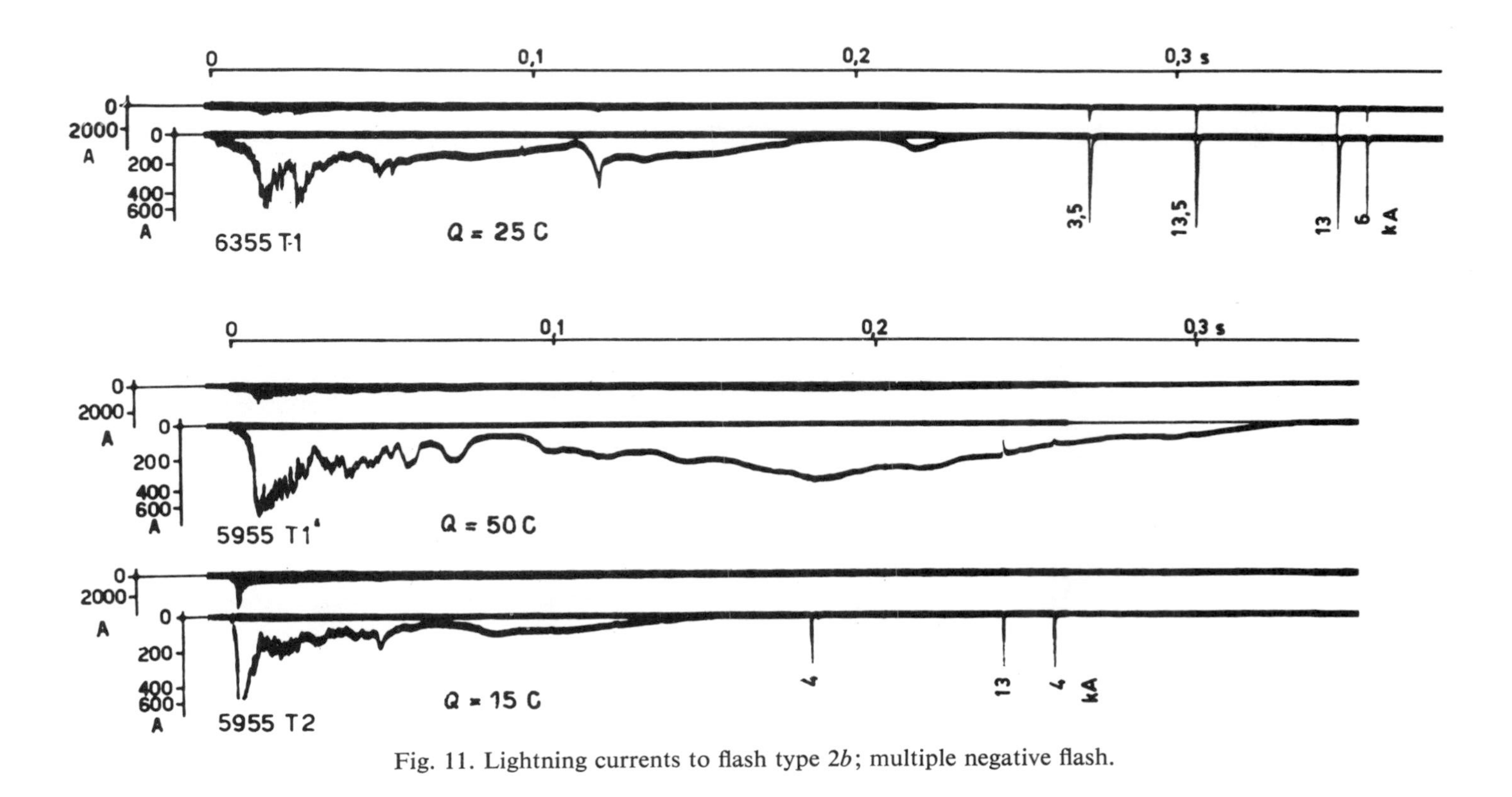

Fig. 11. Lightning currents to flash type 2*b*; multiple negative flash.

1900 m
1 ms
(a)

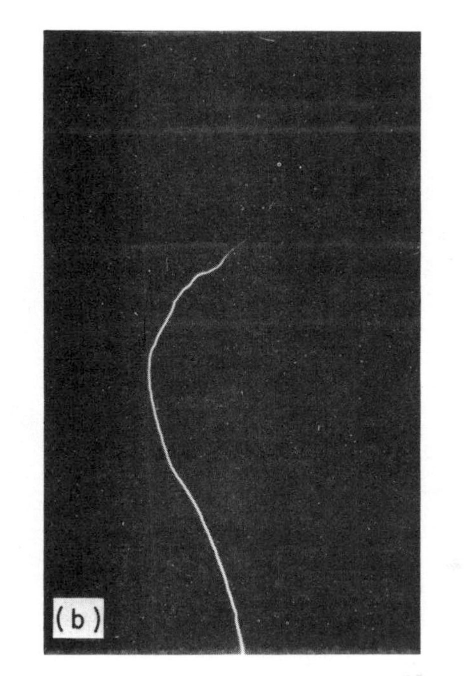

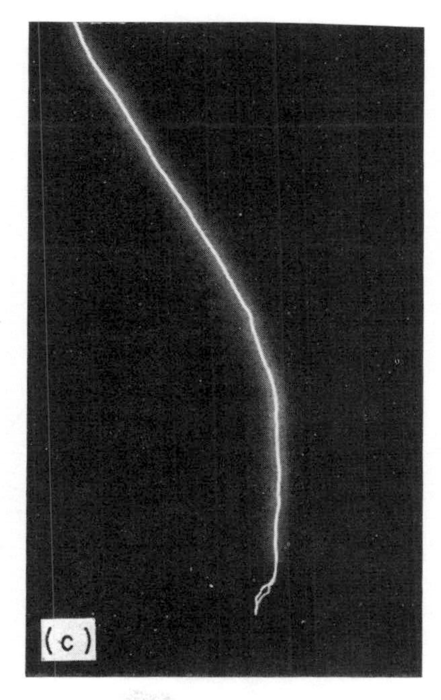

Fig. 12. Single example of a positive flash to Campione near Mount San Salvatore, flash type 3*b*. (a) Leader photograph from 3·3 km distance. (b) Same flash on stationary film, from 3·3 km. (c) Same flash on stationary film, from tower No. 1. (Note the "loop" between downward and upward connecting leaders.)

downward flashes, instead of the more exact description of upward leader—downward flash. They are defined as *Type 4b*.

Type 4b was first observed on Mount San Salvatore. As compared with negative flashes this type represents, in thermal and mechanical stresses, by far the most severe lightning discharges.

Type 3: positive downward leader—positive flash.

During the period 1955 to 1965, when current measurements as well as leader photography were carried out, only a single such leader was recorded (Fig. 12). This involved a strike to the shore of Lake Lugano at Campione. No current oscillogram is therefore available.

Type 4: The subdivision into types 4*a* and 4*b* is explained above. The important type, 4*b*, is represented by Fig. 13. As indicated by the field record, the leader begins to develop as a result of intense field-changes which are due

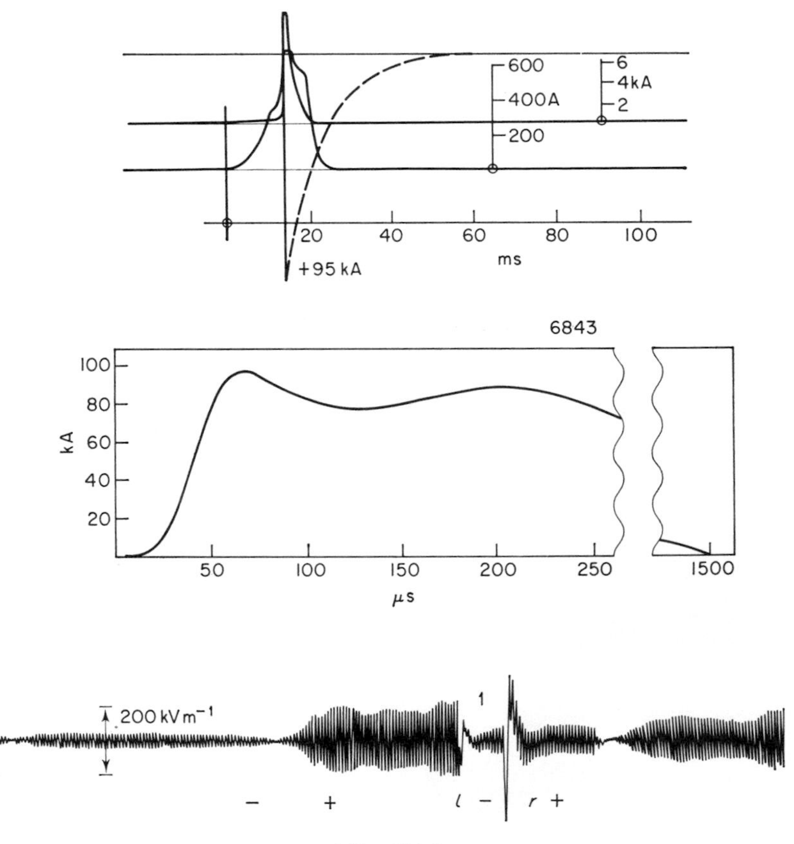

Fig. 13(a).

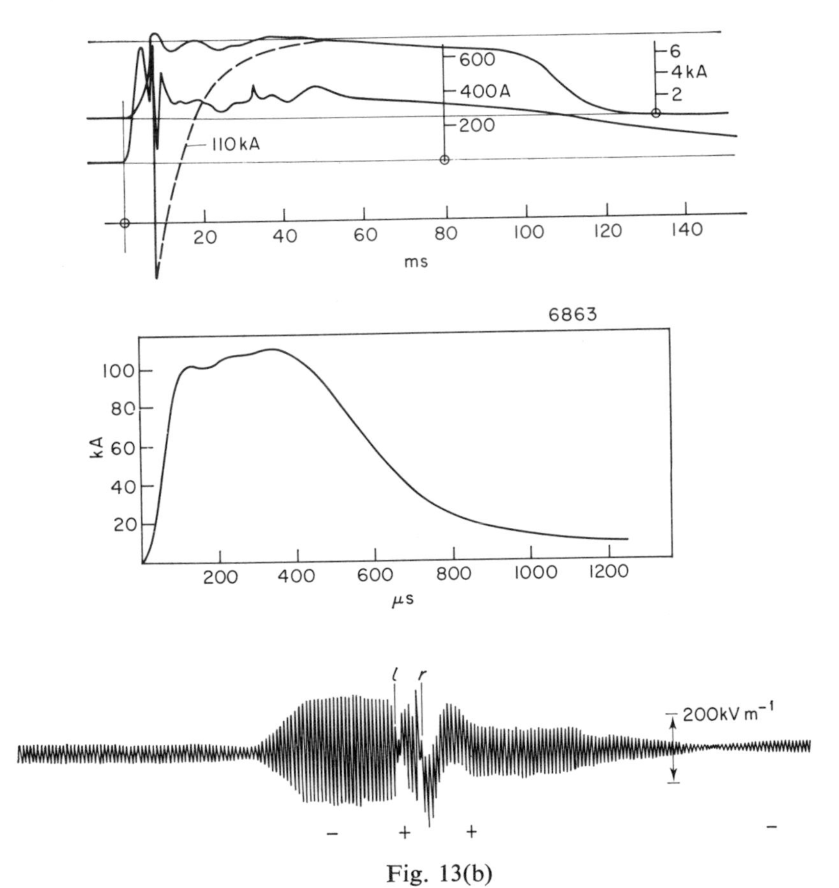

Fig. 13(b)

Fig. 13. Positive impulse currents in flash type 4*b*. (a) Without continuing current. (b) With long continuing current.

to a preceding intra-cloud discharge. The leader is designated by *l*, the intense current impulse, which constitutes a return stroke, by *r* (Berger, 1975). Between 1959 and 1973, a total of 48 flashes of type 4*b*, i.e. 3·2 p.a., were recorded oscillographically, their mean amplitude being 55 kA. The duration of the leader is plotted as a function of the subsequent impulse-current amplitude in Fig. 14.

4.3.3 Photography

Each type of flash is illustrated by only one photograph; for other examples see Berger and Vogelsanger (1966) and Berger (1975).

Type 1 is illustrated in Fig. 15. The negative downward leader reaches an upward streamer at point A. This is confirmed by the branch point B since

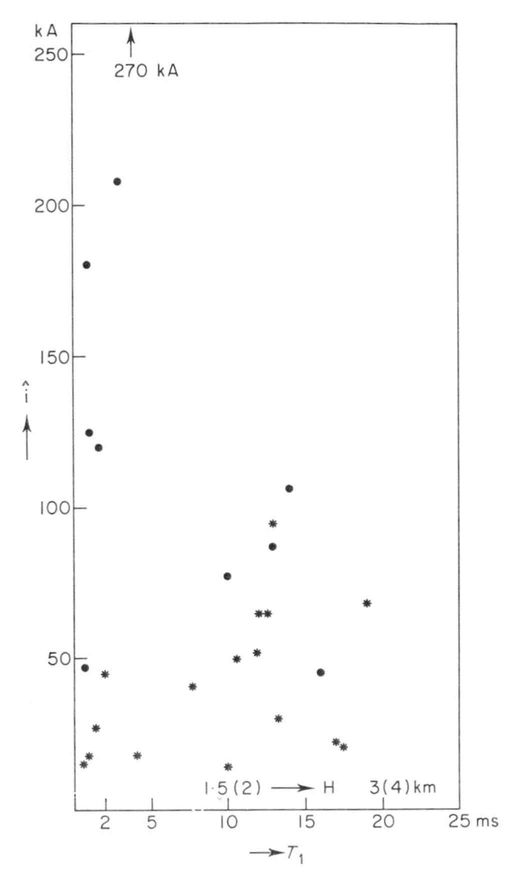

Fig. 14. Duration of upward negative leaders and subsequent positive impulse currents of type *4b*. T_1: duration of leader; H: approximate height of cloud discharge; i: return-stroke current.

it is well established that all branches of a leader assume a strong light intensity when the return stroke develops.

Type 2 is shown in Fig. 16. Within 1 ms two positive upward leaders developed from the two towers, first from the top of tower No. 1 and then from that of tower No. 2. The tips of the faint positive leaders are characterized by bright spots. These are never clearly defined but are always diffuse. Despite the frequency of occurrence of these discharges the low light intensity made it impossible to secure many photographs.

Type 3 is represented by a single photograph, Fig. 12. As already mentioned in Section 4.3.2, this strike occurred to the shore of Lake Lugano at a distance of about 2 km from the mountain top. The positive leader shows no clearly

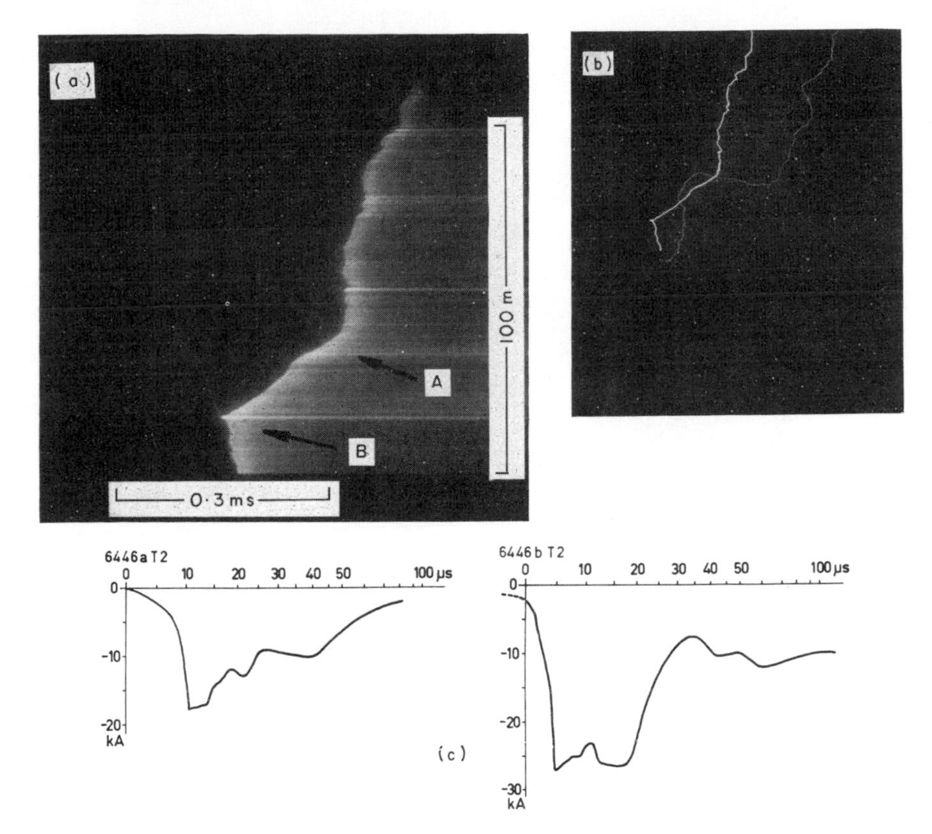

Fig. 15. Photograph of flash type 1*b* to tower No. 2. (a) Downward leader progression to junction point A with invisible upward connecting leader. B: visible branch of the upward connecting leader. (b) Photo on stationary film showing another flash which ends below the tower tip. (c) Lightning currents in both flashes.

defined trace but a variable, though continuous, luminosity over the entire space. The bright leader tips suggest a continuous progression of the leader, similar to type 2. In Fig. 12(b) the point of strike is hidden but this is clearly visible in Fig. 12(c). It is interesting that this shows the loop formation by which the junction with an upward connecting leader is typified (see Chapter 17.2.1).

For type 4 discharges a large number of good photographs are available showing the progression of a negative upward leader, there are many more than for the more frequent type 2. In these photographs the step formation of the negative leader from the tower top can be studied in great detail; the steps are similar to those described by Schonland for negative downward leaders. Figure 17 shows an example for type 4*a* without a return stroke

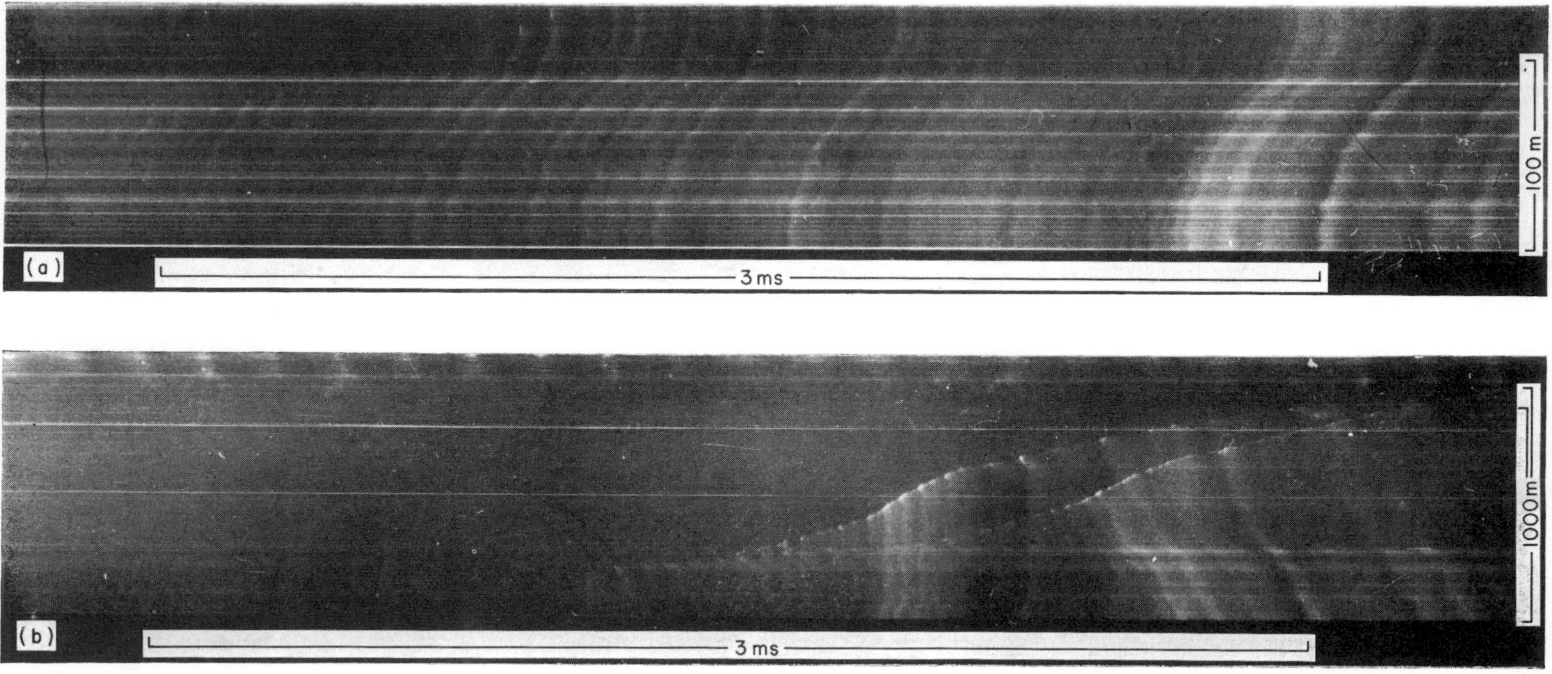

Fig. 16(a, b)

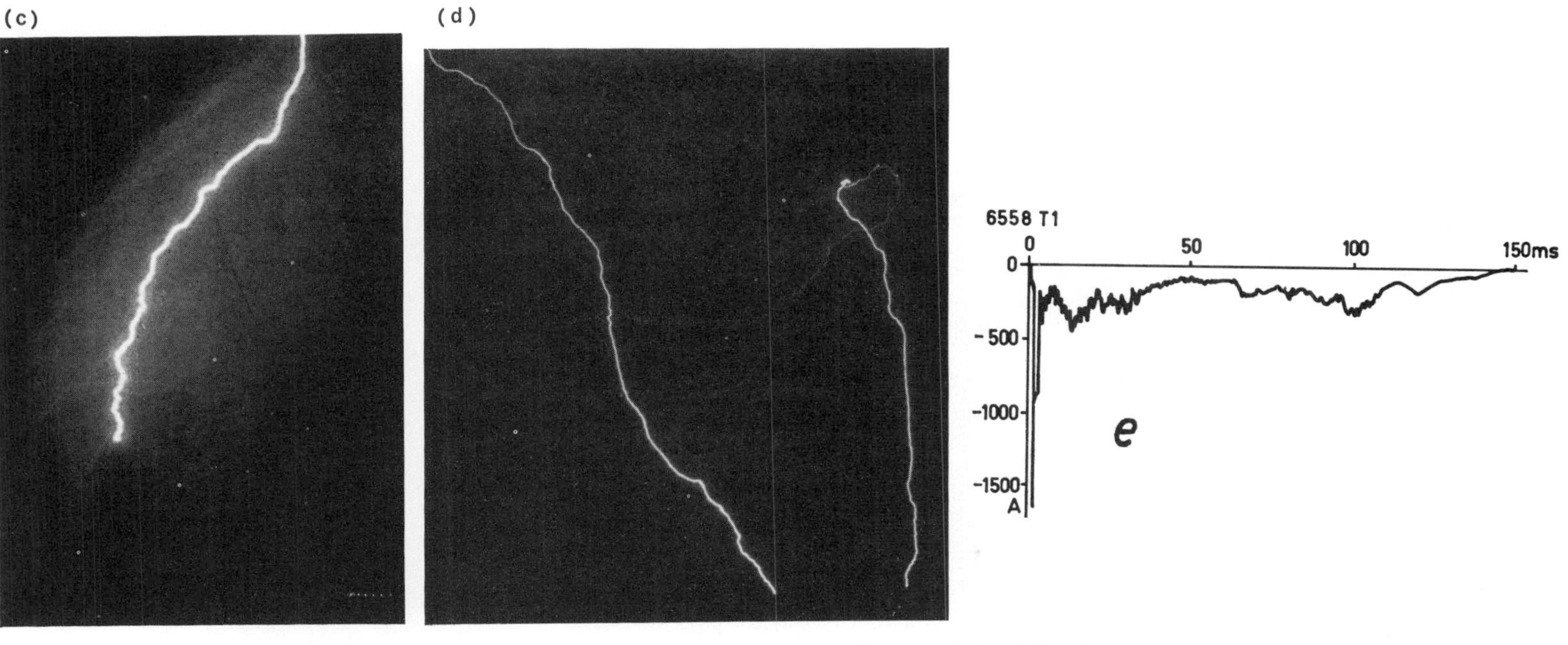

Fig. 16. Photograph of flash type 2*a* to towers Nos. 1 and 2. (a) Leader on tower No. 2, photographed from tower No. 1. (b) Leaders on towers Nos. 1 and 2, photographed from 3·3 km distance. (c) As (a) but on stationary film. (d) As (b) but on stationary film. (e) Lightning current in tower No. 1.

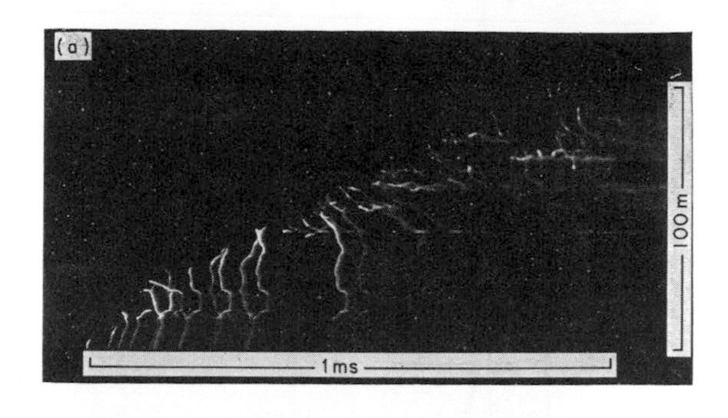

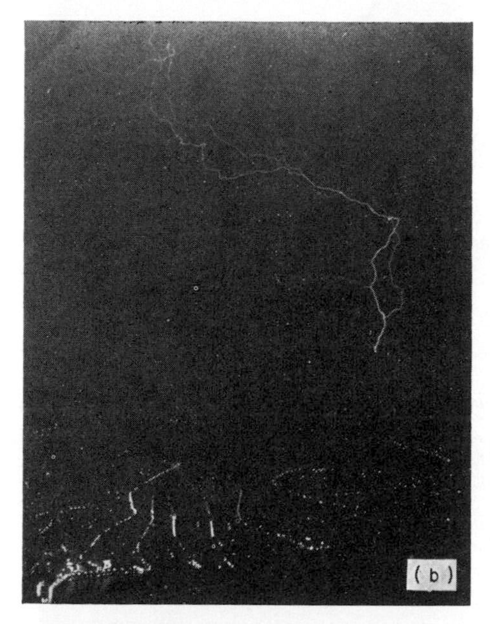

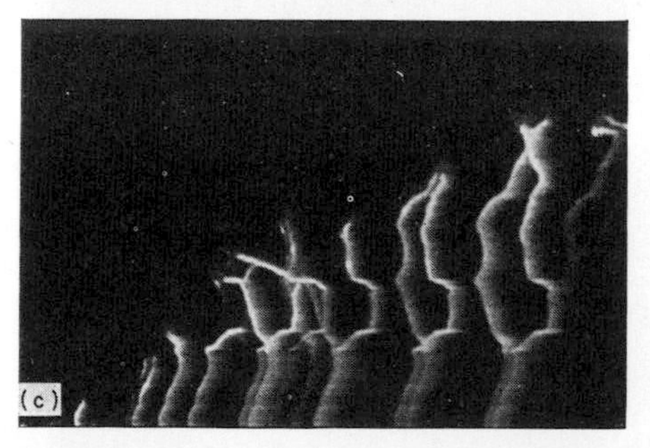

Fig. 17. Photograph of flash type 4*a*. (a) and (c) Leader from tip of tower No. 2. (b) Photograph on stationary film from tower No. 1.

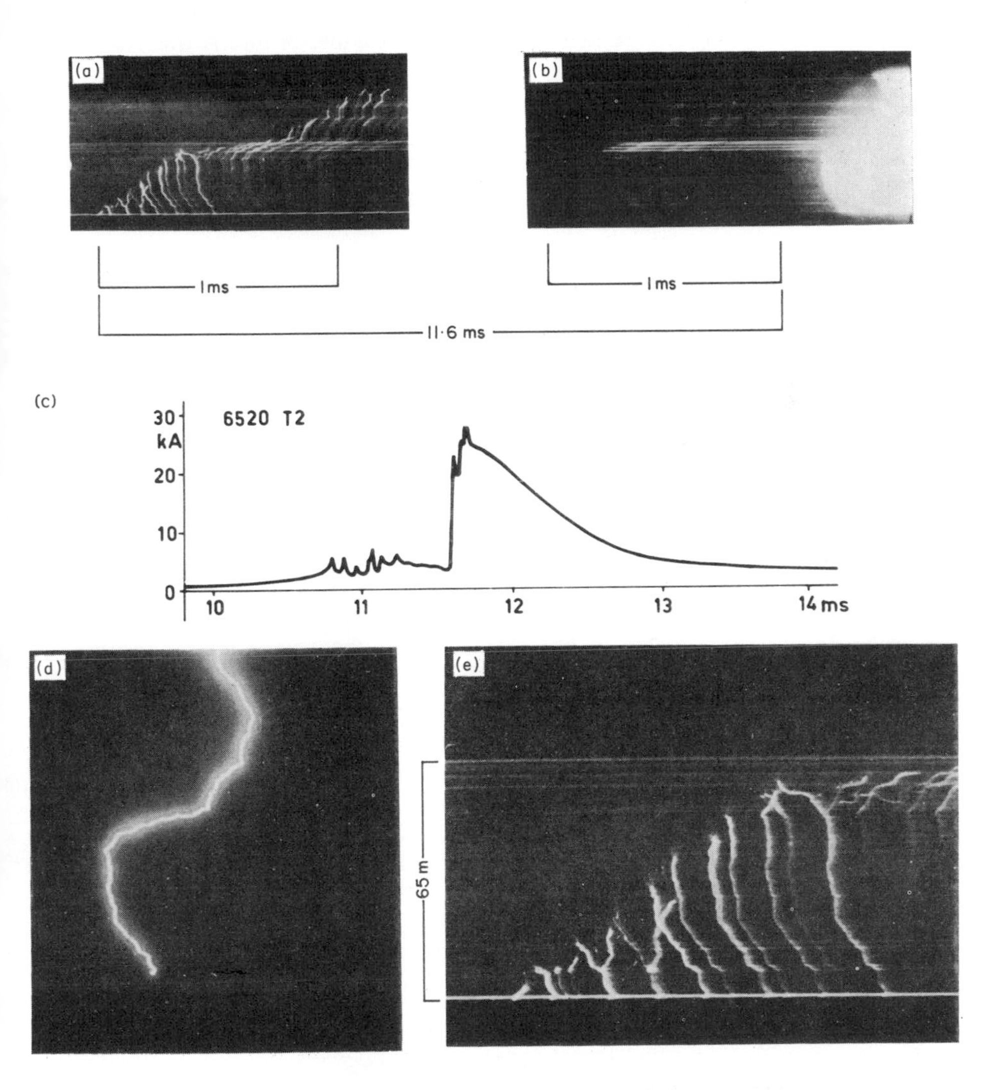

Fig. 18. Photograph of flash type 4*b*. (a) First and (b) last millisecond of leader formation on tower No. 2 before initiation of downward return stroke. (c) Current of leader and return stroke. (d) Same flash on stationary film, photographed from tower No. 1. (e) Detail from (a), corona in front of leader tip.

while Fig. 18 is typical for type 4*a* without a return stroke while Fig. 18 is typical for type 4*b* with a return stroke, after 11·6 ms, of 28 kA and 700 μs time to half-value.

The characteristics of the different types of leader are listed in Table IV.

Table IV

Propagation characteristics of positive and negative leaders on Mount San Salvatore (Berger and Vogelsanger, 1966)

P_q	P_i	Type of flash	n	H (m)	v mms^{-1}	T_{St} (μs)	H_{St} (m)
−	−	1*b*	4	0–100	185–220	40–52	8–10
−	−	1*b*	14	0–1,300	85–440	29–47	3–17
				0–1,750	65		
			1	1,750–2,000	700	41	29
				2,000–2,350	1,060	47	50
			8	0–110	120–190	33–50	4·5–8
−	+	4	3	250–1,200	110–450	40–47	5–18
			1	20–110	870–1,150	4–6·5	3·5–7·5
−	+	4	6	0–55	85–140	34–47	3–6
−	+	4	1	540–900	2,200	55	120
			4	40–110	40–75	65–110	4–8
+	−	2	7	110–≈500	130–490	45–115	8–27
			7	~500–1,150	105–970	40–115	12–40
				320–920	2,400	–	–
+	+	3*b*	1	920–1,660	1,700	–	–
				1,660–1,870	360	–	–

P_q Polarity of leader charge q
P_i Polarity of leader current i and cloud charge
n Number of evaluated photographs
H Height above tower tip or earth
v vertical velocity of leader propagation
T_{St} time interval between steps
H_{St} vertical step length
} mean of 5–30 steps

4.3.4 Field Registrations

In Fig. 2(c), deflections 1 and 2 are due to two partial strokes to tower No. 1; *S*2 is recorded by the simple field mill and *S*1, *A*, *G*, *P* show the new field records obtained at the four recording stations listed in Section 4.3.1. The scales indicated give the field strengths directly at the field mills in kV m^{-1}. Curves i_1 and i_7 show the currents discharged into the top of the tower, with sensitivities of 1 A and 7 mA respectively for full deflection to the edge of the film. Differences in curves *S*1 and *S*2 are due to the installations on tower

No. 1 and on the church near that tower. The highly sensitive current record i_7 shows the displacement current between the cloud and the antenna on tower No. 1. This record gives a clear indication of the initiation of a discharge inside the cloud which precedes the onset of the visible leader (see Chapter 6.4). It also shows that, between two component strokes, the current to the antenna exceeds about 10 mA only during the approach of a downward leader. The upward discharges are preceded by irregular field-changes which, as shown in Fig. 2(c), repeatedly change polarity. The field-change begins near Gemmo, position *G*. Both upward discharges then increase the field about Agra, position *A*. The registration of the field at the point of strike indicates beyond doubt that upward discharges invariably develop in consequence of cloud discharges which produce large transient field concentrations on towers erected on mountains. This is the reason why type 4 may be termed the "mountain-type" of discharge.

4.4 Other Recording Stations

4.4.1 U.S.A. Stations

McCann (1944) reported the results from 25 stations during the years 1940 to 1944. Thirteen of these flashes, of which three showed upward leaders, were recorded by fulchronograph on the Cathedral of Learning in Pittsburgh. Two independent current ranges of 50 to 1,000 A and 1 to 30 kA were used. Negative currents up to 30 kA were recorded. At the lower end, continuing currents exceeding 50 A for periods of 0·1 s were observed.

Reference to the cine-klydonograph has been made in Section 1.3. The results obtained with this device have been justifiably rejected by Wagner (1967) and are therefore not discussed further.

4.4.2 Italian Research Stations

The first station was built in 1969 at Foligno, 42°15′ north, 12°11′ east, and this was followed by a further similar station at the foot of the telecommunication tower on Monte Orsa, 45°55′ north, 8°52′ east (Garbagnati and Lo Piparo, 1970a, 1973). Lightning strikes to the towers were photographed and currents were measured through shunts between a special antenna and the top of the mast (Garbagnati and Lo Piparo, 1970b). The oscillograph is tripped at 3 kA but upward discharges are not invariably recorded. Photographic records indicate that most discharges start with upward leaders from the 40 m high towers (Garbagnati *et al.*, 1975).

Out of 27 discharges which could be analysed, 17 gave negative currents of type 1 with a maximum of 67 kA and 25 μs duration. A positive current record of 170 kA with more than 500 μs time to half-value must be classified as type 4. Further statistical results are given in Chapter 9.

4.4.3 Rocket Stations

The suggestion to initiate a lightning discharge artificially by firing a rocket carrying a fine steel wire towards a thundercloud is due to Newman (1963). This method is described in Chapter 21.3.

Another station employing rockets was opened in 1973 at Saint Privat sur l'Allier in central France (Gary and Fieux, 1974). The rockets are fired either from the ground or from a research mast about 1100 m above sea level, situated 45°0′ north, 3°30′ east. The firing circuit is actuated when the field strength at ground level reaches 13 to 20 kV m^{-1}. Results obtained are included in Chapter 9. The highest current amplitudes are obtained when the rocket initiates a lightning strike at heights between 70 and 220 m.

The heights to which rockets have to be fired to cause a strike are, on some occasions, surprisingly low, namely a few hundred metres. It seems doubtful whether the parameters of such artificially induced flashes are the same as those of natural discharges; this applies at least to the first component stroke since this corresponds to an upward stroke of type 2. On the other hand, the conditions governing subsequent strokes should be the same for artificial and natural lightning discharges.

4.4.4 U.S.S.R. Station

The 537 m high television tower in Moscow is situated at 55°45′ north, 2°40′ east. In 4·5 thunderstorm seasons it was struck 143 times (Gorin *et al.*, 1972), i.e. 32 times p.a., with a maximum of 12 strikes in a single thunderstorm. Up to the time of reporting 83 strikes have been photographed and 41 component strokes have been recorded oscillographically by means of a toroidal coil surrounding the uppermost part of the tower. The highest current measured was 46 kA, front durations varied between 1 and 10μs and times to half value between 20 and 70μs. Roughly one-half of all strikes were negative. Some currents were derived from magnetic field records made 400 m distant from the tower. However, no details have been published on the frequency responses of these measuring devices. Most strikes (49) were upward flashes and some of these started 12 to 36 m below the tower top. Two downward flashes also struck the tower 200 and 300 m below the top. The authors report that the tower, in addition to attracting strikes to itself, appears to be responsible for more earth flashes within 1 km than at distances between 1 and 3 km. This would indicate a negative "protective effect" of the tower.

4.4.5 South African Station

In 1973 a 60 m high insulated mast was erected on a low hill near Pretoria (Eriksson, 1974), 25°50′ south, 28°0′ east (see Chapter 13.4). Current measurement is effected by means of a wide band width gapped-core pulse transformer. The first 6 strikes produced negative currents between 11 and 64 kA of which two were multiple flashes; there were 3 downward flashes, 1 was upward and 2 uncertain. The wave shapes of three component strokes exhibit front lengths exceeding 20 μs for second strokes and a total duration of more than 200 μs for a first component stroke. These measurements are being continued with particular respect to providing information on the striking distance, see Chapter 17.3.

5. Physical Picture of the Earth Flash

5.1 Charge Distribution in a Thundercloud

The first results concerning the charge distribution in a thundercloud were obtained by Simpson and Scrase (1937). They found that the higher regions of the cloud carry positive charges while the lower regions are negatively charged with a concentrated small positive charge near the lower frontal region of the cloud. The intensity and importance of this low positive charge are still open to doubt since it seems more likely to constitute a uniform layer of low charge density below the negative cloud. Present thoughts on the charge distribution are discussed in Chapters 3.4.4 and 6.3.2.

Charge measurements on San Salvatore show irregular changes in polarity and charge density (Berger, 1955), indicating both negative and positive charge pockets when an ion counter traverses fog cells during a thunderstorm. Figure 19 shows a typical record of the number of charge carriers as measured by an ion counter after Israel (1929) together with the point-discharge currents recorded during the same period in the two towers on the mountain. The record suggests a rather irregular charge distribution in the lower regions of the cloud. The highest ion concentration is observed when the station on San Salvatore dips into cloud fog. It must be concluded that electric charges reside on fog particles.

The vertical charge distribution determined by Schonland (1956) and Malan (1963) in South Africa seems to be characteristic of genuine heat thunderstorms. Cold-front and warm-front storms seem to be characterized by many vertical cells, rapidly rotating winds and charge pockets. Under these conditions subsequent strokes have pronounced horizontal components (see also Chapter 6) as frequently observed, particularly for upward flashes.

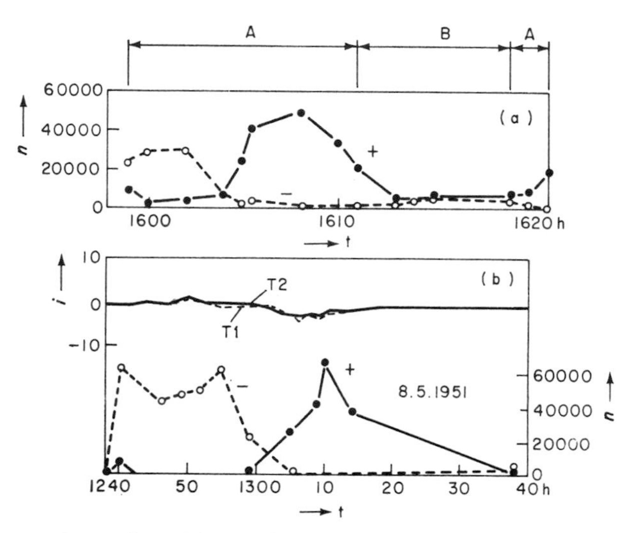

Fig. 19. Ion content of positive and negative polarity during thunderstorm on Mount San Salvatore. (a) A: Laboratory is dipping in fog. B: Laboratory is out of fog, in dry north-wind. n: number of heavy ions per cm^3; t: time. (b) $T1$ and $T2$ point-discharge currents i (in mA) on towers Nos. 1 and 2.

A highly promising multistation method to determine the location of the charge centres involved in multiple earth flashes has been described by Krehbiel *et al.* (1974). This method is discussed in Chapter 6.4.3. In principle, multistation recording of field changes requires flat surrounding country. Furthermore, five or more recording points produce an excess amount of information and this results in a certain scatter of the derived values. Two earth flashes recorded in New Mexico were analysed by Krehbiel and his colleagues. The charges involved were found to be located at above 5 km above the terrain and successive strokes had substantial horizontal components.

5.2 Point-discharge

The mechanism of point-discharge is described in Chapter 4.2. On a high tower, such as on Mount San Salvatore, point-discharge may last from a few minutes to several hours. The discharge is so weak that it can only be photographed by an exposure of several minutes and in complete darkness. At a short distance, it can be detected by a slight hissing noise. When lightning strikes, this noise suddenly ceases and gradually reappears within several seconds.

On San Salvatore, the point-discharge currents from the two towers were continuously recorded during thunderstorm periods. They reached 3 mA. During the summer months of 1971, currents exceeding 0·3 mA were registered for 1,203 min or about 20 h. As shown in Table V point-discharge is more

Table V

Duration of point-discharge currents exceeding 0·3 mA

Month	May	June	July	Aug.	Sept.	Oct.	Nov.	May/Nov.
Duration (min)	663	103	12	180	—	—	245	1,203

pronounced in spring than in late summer. This may be due to the lower height of thunderclouds in spring and late autumn or to annual variation. In 80 to 90% of all cases, point-discharge is negative; it occurs under a negative cloud and conveys negative charge to earth. However, in the course of several minutes, the current reverses polarity repeatedly (Berger, 1955). This even applies in the absence of lightning flashes as seen from Fig. 20. The total charge transported through the towers during the 20 h in 1971 amounted to 72 C which constitutes a small fraction of the charges involved in the lightning strikes to these towers.

A less well-known form of point-discharge occurs at the tower tip at the instant of a lightning flash nearby or in the cloud. Under otherwise silent conditions, a whip-like noise emanates from the tip of the tower. It is caused by so-called impulse corona which is too weak to be visible. Acoustically it is very pronounced and is indicative of the intense transient field-changes on a tall object due to lightning flashes within a radius of several kilometres. In the extreme, they initiate upward leaders.

As shown by Fig. 20 the occurrence of intense point-discharge does not justify the conclusion that it will be followed by a lightning discharge within a few minutes. This can follow but it is by no means certain. This fact needs to be considered in the design of lightning-warning devices (see Chapter 15.5).

5.3 Transition from Corona to Leader Arc

In the transition from an electron avalanche to the arc regime in a leader the following physical processes are involved:

(a) ionization at the leader tip, i.e. creation of new electrons and positive ions;

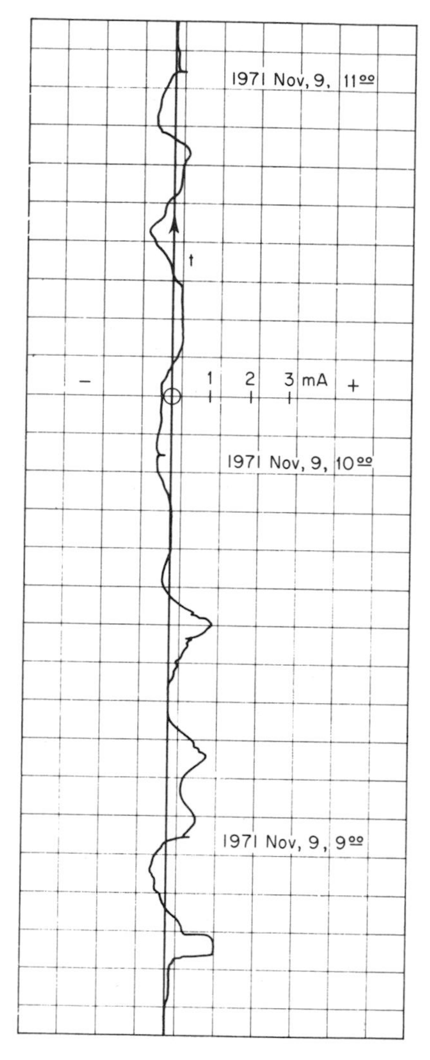

Fig. 20. Point-discharge current on tower No. 1 without formation of flashes.

(b) retraction of electrons backwards by the positive channel charge, just behind a negative front;
(c) attachment of newly created electrons to gas molecules resulting in the formation of negative ions;
(d) recombination of electrons and positive ions in the retraction space behind the front;

(e) diffusion of electrons and ions radially from the ionizing channel;
(f) energy transmission from electrons and ions to neutral gas molecules (gas heating);
(g) transition from electron conductivity to arc conductivity (thermal ionization).

In contrast to items (a) to (d), items (e) and (f) have received no great attention in the literature.

Concerning item (e) the following observations may be made:

The diffusion as a function of time is given by

$$\rho^2 = 4Dt,$$

where (Knoll *et al.*, 1935)

ρ = radius of expanded channel,
D = diffusion constant = $\frac{1}{3}\lambda v$,
λ = mean free path of ions or electrons,
v = mean velocity of ions or electrons,
t = time in seconds since beginning of the arc.

(i) For N_2-ions in normal gas density, the following values apply:
$\lambda_i = 8 \times 10^{-5}$ cm; $v_i = 500$ m s^{-1}; therefore $D = 1{\cdot}33$ cm^2 s^{-1} and $\rho_i^2 = 4 \times 1{\cdot}33t$.
First example: $t = 50$ μs (step time) $- 2\rho_i = 0{\cdot}32$ mm.
Second example: $t = 0{\cdot}5$ s (flash duration) $- 2\rho_i = 3{\cdot}2$ cm.

(ii) For free electrons in normal gas density:
$\lambda_e = 4 \times \sqrt{(2)}\,\lambda_i$; $v_e = (\lambda_e/\lambda_i)^{\frac{1}{2}} \times (m_i/m_e)^{\frac{1}{2}}$; $v_i = 380 v_i$; therefore $D_e = 716\lambda_i v_i = 2{,}866$ cm^2 s^{-1} and $\rho_e^2 = 4 \times 2866t$.
First example: $t = 50$ μs (step time) $- 2\rho_e = 1{\cdot}5$ cm.
Second example: $t = 5$ μs (current front time) $- 2\rho_e = 4{\cdot}8$ mm.

The foregoing calculation presupposes constant air density around the electron avalanche. This constitutes a close approximation for the case of glow-to-arc transition but not for the transition from a leader to the arc regime. This latter case is examined in Section 5.6.

Concerning the above item (f), Aleksandrov (1969) calculated the heating of the gas in a spark channel as a function of the electron collisions with gas molecules and the resulting formation of a *stem* in a brush discharge. Now it can be shown that the transmission of kinetic energy from particles produced by ionization (electrons and ions) to neutral gas molecules is due more to ion collision and less to collision with electrons. This is true because of Newton's law, despite the fact that electrons have a mean free path and energy which is $4\sqrt{2}$ times higher. Attachment of electrons to gas molecules leads to the replacement of highly mobile charge carriers by heavy negative ions which heat the gas much more effectively than electrons. The energy released during the process of attachment appears essentially in the form of photons by whose short-wave radiation the surroundings are excited or ionized. In this way further electron avalanches are produced in the heated gas. This results in longer free paths for electrons and therefore higher ionization and further heating by ions, leading ultimately to an instability and the formation of an arc channel, the conductivity of which is based on thermal dissociation and ionization.

The transition from corona discharge to the lightning leader involves the interaction of a series of individual effects. It is too complex to be open to simple calculation. The optical spectrum is discussed in Chapter 8.3.3.2.

Toepler (1917, 1925) was the first to investigate glow-to-arc transition. His first law describes the transient resistance of a corona filament (streamer) which he assumes to be inversely proportional to its impulse charge. Thus

$$R_w = kl/Q,$$

where

R_w = resistance of corona filament in Ω,

k = Toepler constant = 0·3 to $0{\cdot}8 \times 10^{-3}$, best value 5×10^{-4},

l = length of corona filament in cm,

Q = electric charge during spark formation in C.

This law enables the transient resistance of the discharge channel or the discharge voltage to be calculated as a function of time from infinity to zero.

Toepler's second law defines the transition between positive glow and negative arc characteristics. When a critical impulse charge has been reached in a corona filament, its voltage drop begins to decrease with increasing charge. This critical charge is about 1 e.s. unit, namely $\frac{1}{3} \times 10^{-9}$ C. The corresponding resistance is about 1·45 MΩ cm^{-1} and the current about 20 mA, given by 30 kV cm^{-1} divided by 1·45 MΩ cm^{-1}.

In a uniform field the duration of transition or breakdown time T_F of a gap of a cm and a breakdown voltage $U_F = E_F \times a$ is given by Toepler as

$$T_F = 2\pi ka/U_F = 2\pi k/E_F = 2\pi/\alpha.$$

For example, for $a = 1$ cm, $E_F = 30$ kV cm^{-1}, $T_F \approx 1$ μs. The k values depend to a certain extent on the discharge circuit: $T_F = (4 \text{ to } 8)(1/\alpha) = \frac{2}{3}$ to $\frac{4}{3}$ μs, with a mean value of 1 μs.

In the non-uniform field in front of the leader tip and around the leader channel the filament current must flow through its capacitance. The significance of Toepler's second law lies in the fast reduction of voltage along the filament when a multiple of the critical charge of 1 e.s. unit passes through it. The critical time t_c is approximately:

$$1 \text{ e.s. unit} = \tfrac{1}{3} \times 10^{-9}\,\text{C} = 20\,\text{mA} \times t_c,$$

thus

$$t_c = \tfrac{1}{60}\,\mu\text{s}.$$

For a critical mean current of 20 mA, the critical charge is reached after 17 ns. For a higher current, transition occurs more quickly; the corresponding frequencies range from 50 MHz upwards. Toepler's laws thus also confirm radio noise at very high frequencies.

5.4 The Stepped Leader

Stepped leaders appear as bright lines at the tips of the negative leaders of first strokes as shown in Figs 3(b), 15 and 17. In contrast, positive leaders exhibit no bright steps but show bright spots at the tips of a faint channel of variable luminosity, see Figs 12(a) and 16(b).

Toepler (1920) compared these leader steps with the steps in his "gliding discharges" in Lichtenberg figures. When such a filament has extended over several centimetres in length, it is suddenly transformed into a good conductor, a bright streamer. The length of the filament before its transition depends on its capacitance. On the insulating surface on which a gliding discharge is formed, this capacitance is a multiple of that of a filamentary discharge of a lightning leader in space. However, Toepler's law cannot be extrapolated to explain the length of the lightning leader step of some 5 to 50 m.

Loeb and Meek (1940) and Raether (1964) have independently advanced a valuable theory of the breakdown process of spark discharges, based on the effect exerted by positive charge in an ionized channel. At the tip of an avalanche, the electrons are subjected to two fields: the static field between the electrodes and the field of the positive-ion charge behind the head of the avalanche. If the latter field exceeds the former, the resultant field either vanishes or changes direction. Electrons then fall back into the positive channel charge, the local field collapses and strong recombination takes place. This entails intense photo ionization in all directions and emission of visible light. When the progress of the head of the avalanche has been arrested, the highly concentrated charges progress back along the ionized channel and, as a result of further ionization, breakdown occurs between the electrodes.

The foregoing theory can be extended to non-uniform fields, provided that the field strength throughout the region is known. For very long sparks, like lightning, Loeb and Meek (1940) proposed a modified theory. The voltage drop along the leader increases by the attachment of electrons to neutral molecules, by recombination of positive and negative ions and by diffusion of ions. If the voltage drop is high enough, a "potential wave" progresses from the cloud to the tip of the leader where it produces renewed ionization and thus a new step.

Artificial arcs of about 100 A do not show a quasi-periodical change in resistivity such as to produce "potential waves". Such potential waves therefore need an independent pulse signal, which is probably not situated along the arc channel but at its tip where corona is transformed, over a certain length, into an arc. The basic problem thus lies in this transition process which has not yet been quantitatively resolved.

The experimental work of the Les Renardières Group (1974) on *positive* rod/plane gaps has failed to show a prevailing process of the stepped leader. In this respect there is no contradiction to the lightning discharge since positive leaders likewise show no steps. In contrast, as described in Chapter 7.4.3, the most recent work on *negative* rod/plane gaps of 4 to 7 m has confirmed the occurrence of "mid-gap streamers" with reillumination back towards the negative rod electrode (Gruber *et al.*, 1975). If it can be shown that such mid-gap streamers also develop at greater distances from the negative electrode, the resulting picture would approach that of the lightning discharge.

Observations by Stekolnikov and Shkilev (1963), as well as those by the Les Renardières Group, have confirmed that step formation disappears in the long spark if the rate of increase of the applied voltage is high. This corresponds to the velocity of propagation of subsequent strokes which is higher than that of first strokes.

Future theories to explain the step formation of leaders should take into account that well-developed steps occur in the first stroke. Subsequent strokes with time intervals of less than about 100 ms are invariably preceded by dart leaders which progress much faster than first leaders. When the time interval between successive strokes exceeds some 100 ms, subsequent leaders exhibit faint steps.

Future progress in this field should be based on the following numerical values (Schonland, 1956; Kitagawa and Kobayashi, 1958):

velocity of electrons at the leader tip	$1{\cdot}4$ to 18×10^6 cm s^{-1}
velocity of propagation of first leader	1 to 8×10^7 cm s^{-1}
velocity of propagation of subsequent leaders	1 to 20×10^8 cm s^{-1}
step length of first leader	10 to 200 m, mean 25 m
time interval between steps of first leader	10 to 100 μs, mean 50 μs
velocity of propagation of return stroke	2 to 11×10^9 cm s^{-1}.

Note that, on San Salvatore, the step length of first leaders varied between 3 and 50 m. Positive leaders have velocities up to 24×10^7 cm s^{-1}.

5.5 Transition from Leader to Return Stroke

As a positive or negative leader approaches the ground, the electric field increases until breakdown takes place. According to Wagner (1967), a point of discrimination and a corresponding "striking distance" (Golde, 1947) can be determined by extrapolation of laboratory measurements on rod/rod or rod/plane gaps. Such test results suggest an average critical field strength of about 500 kV m^{-1}. If the potential V of the tip of the leader is known, the striking distance S can be calculated from $S = V/E$. A leader potential of 50 MV thus suggests a striking distance of 100 m (see Chapter 17.3).

In comparison with the laboratory spark, the striking distance of a lightning discharge is notably longer. In consequence, the counter-discharges from earth do not only take the form of corona discharges but high-intensity streamers. One of these can progress to establish contact with the downward leader. The point of contact may be termed the junction point. If the upward streamer, sometimes called a "connecting leader", starts at the same instant that the downward leader reaches the striking distance, the junction point should be at about half that height. Connecting leaders can be recognized in lightning photographs by the direction of branching. Calculation of striking distances, assuming a mean breakdown field of 500 kV m^{-1}, does not take into account the observation that, for high switching surges of negative polarity, the variation of voltage with gap distance becomes non-linear when the gap length exceeds 5 to 10 m. This effect is well known in high-voltage testing with voltages above about 5 MV. The mean breakdown field therefore decreases to below 500 kV m^{-1} and striking distances may be considerably longer than indicated by linear calculation.

The return stroke begins at the junction point. It is characterized by a rapid increase of the current in the downward leader and in the upward connecting leader. As a first approximation, let the downward leader be replaced by a resistanceless conductor of surge impedance Z and the connecting leader by a good conductor (breakdown spark) of inductance L. From a knowledge of the leader potential V, the ultimate current i and the initial rate of rise of current di/dt can then be calculated. Thus

$$i = V/Z \quad \text{and} \quad di/dt = V/L.$$

Taking $V = 50$ MV, $Z = 500\ \Omega$, $L = 100\ \mu H$ (for a height of 50 m of the junction point), the

$$i = 10^5 \text{ A} \quad \text{and} \quad di/dt = 500 \text{ kA } \mu s^{-1}.$$

Even in subsequent strokes the values for di/dt rarely exceed 120 kA μs^{-1} and in first strokes 32 kA μs^{-1} (see Table I, Chapter 9). The foregoing assumptions must therefore be regarded as rough approximations and the resulting numerical values as applying to the special assumptions only.

The return stroke starts from the junction point in the form of a wave travelling upward along the leader which has been charged before to a high potential. It can be visualized as a wave discharging the negatively charged leader or, alternatively, as an upward moving positive wave by which the negative leader charge is neutralized. Connection to earth is effected through the upward connecting leader. The comparatively small leader current thus increases in a very short time from an order of magnitude of 100 A to an arc current of many kiloamperes. An arc consists of a mixture of positive and negative ions and electrons which are the result of thermal ionization due to a gas temperature of several 1,000 K (Knoll *et al.*, 1935). Due to their small mass, the current is conveyed practically exclusively by electrons. A stationary arc has a relatively low gradient of 20 to 60 V cm^{-1}. It is therefore electrically neutral. There is a strong tendency for electrons to combine with atoms or molecules and thus to form ions. The number of free electrons is determined by current and electron mobility. A sudden increase of the current in an arc requires an increased field strength or more free electrons.

An increase in the field strength has two effects, firstly because of increased velocity of the electrons and, secondly, because of an increased number of electrons, due either to the electron emission from ions or to Townsend ionization, ultimately up to impulse breakdown.

The mechanism of arc formation for normal-frequency currents is well established. For currents of several tens of amperes, peaks occur in each half-cycle of the arc voltage. They produce new electrons and reheat the arc (Berger and Pichard, 1956).

These investigations show that the free electrons disappear in the arc within 1 μs if this is shorted within a fraction of that time. An immediate breakdown now requires approximately the same breakdown voltage as an unionized gas channel of the same temperature.

The laws of Toepler were applied by Lundholm (1957) to calculate the steepness and velocity of propagation of lightning surges. Independently, Wagner (1963) applied Maxwell's field equations to calculate the velocity of the return stroke from the measured impulse energy and the impulse voltage required to establish a high-current arc.

In place of Toepler's spark constants used by Lundholm, Wagner used the energy in W s cm^{-1} as obtained oscillographically from investigations of the impulse spark or arc (Wagner *et al.*, 1958). In contrast to a wave propagated with the velocity of light in a loss-free space, a retarded wave releases energy. Wagner equates this with the energy required to heat the arc in front of the wave charging the return stroke (see also Chapter 17.3.2). This leads to a relation between the current amplitude and the velocity of the return stroke.

The results of Lundholm and Wagner are reproduced in Fig. 21. In this calculation the return-stroke channel was simulated by a cylinder of 3 cm radius, with a corona shell of 9 m radius for a lightning current of 100 kA.

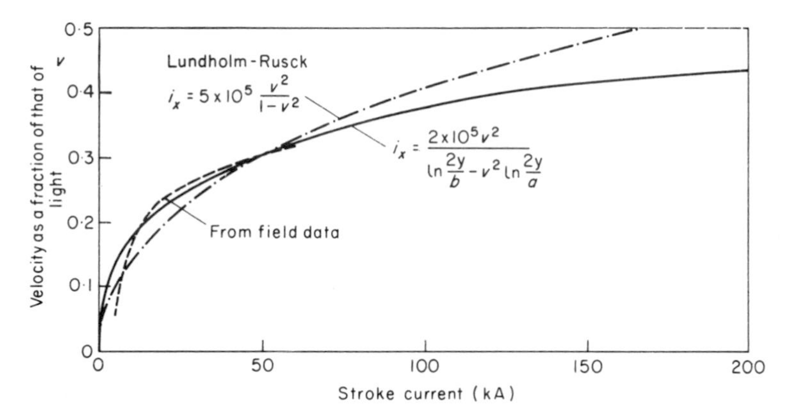

Fig. 21. Variation of return-stroke velocity with stroke current. v: velocity as a fraction of that of light; i: stroke current in kA.

The reduced velocity of propagation of the return stroke, as compared with the velocity of light, leads Wagner to derive a value of 3,000 Ω for the transient surge impedance in the front of the channel of the return stroke against the value of 300 Ω which represents the surge impedance of the return-stroke wave behind its front.

In principle, the transient phenomena in the heating of the connecting leader to an arc of many kiloamperes produce a delay and reduction in current steepness. This explains why the measured values of the rate of current steepness in the return stroke are lower by a factor of about 20 (for first currents) and 5 (for subsequent currents) as compared with the approximate estimates for a perfect no-loss lightning channel.

No measurements are at present available concerning the relation between current amplitude and velocity of the return stroke. The "field data" used by Wagner were calculated by Rusck (1958) by comparing the frequency distribution curves of lightning-current amplitude and return-stroke velocity. If the resulting relationship were to be confirmed by direct measurements, an agreement with Wagner's (1963) theory would indicate that the energy impulse of 2 W s cm^{-1} kA^{-1} does not depend on the initial temperature of the arc channel. This impulse was determined for room temperature at the beginning, whereas it is the leader temperature in the transition from the leader to arc. This would suggest that most of the impulse energy is used for ionization rather than for gas heating.

Information is available on the duration of the "final jump" (see Chapter 7.4.2.2), namely the time which elapses between the instant when the first corona filaments reach the counter-electrode and the instant of breakdown or arc formation.

The Les Renardières Group (1974) has shown that the duration of the final jump for positive rod/plane gaps is remarkably constant. For spacings from 2 to 15 m it is about 20 μs. The breakdown voltage U_B of the air gap between the leader tip and the plane at the start of the final jump is found to be the same as though the leader was a metal conductor, namely,

$$U_B = 0{\cdot}44S + 0{\cdot}15 \quad \text{MV}$$

when S is measured in metres. For negative rod/plane gaps of 1 to 4 m spacing, the duration of the final jump is 3 to 5 μs. These values agree approximately with the front durations of positive and negative lightning strokes; see Table I in Chapter 9. This suggests that, for both polarities, the front duration and steepness are functions of the final jump and that there is no basic difference between the long rod/plane gap and the natural lightning stroke.

5.6 The First Return Stroke

Return strokes below negative clouds always proceed from earth to the cloud. A particular return stroke below a positive cloud is discussed in Section 5.9.

The velocity of first return strokes lies between 2 and 11×10^9 cm s^{-1} (Schonland, 1956). Statistical information on front duration, front steepness, electric charge and $\int i^2 \mathrm{d}t$ is given in Chapter 9 in which the shape of the current is also discussed. The voltage drop along the return channel is about 60 V cm^{-1} or very similar to that of the 100 A arc of the leader, as mentioned in Section 5.5. The uniform voltage drop along the return stroke or the leader does not imply a uniform charge distribution. This is governed by the electromagnetic field in the air which, in turn, is determined by the stroke current and its variation with time.

Travelling-wave theory enables the stroke current i to be calculated from the leader potential V if the lightning channel is replaced by a metal conductor of surge impedance Z. Thus:

$$i = V/Z, \quad Z = (L/C)^{\frac{1}{2}}$$

and the velocity

$$v = 1/(LC)^{\frac{1}{2}}$$

where L and C are respectively the inductance and capacitance per m of a long line. Therefore

$$L = 2\times 10^{-7}\ln(D/r)\ \mathrm{H\ m^{-1}},$$

$$C = 2\pi\varepsilon/\ln(D/r_q)\ \mathrm{F\ m^{-1}} \quad \text{and}$$

$$\varepsilon = 8{\cdot}84\times 10^{-12},$$

$$D = \text{length of conductor},$$

$$r_i = \text{radius of conductor},$$

$$r_q = \text{radius of corona shell} = r_i \text{ for a metal conductor.}$$

As a first approximation, we take:

$$L = 2\ \mu\mathrm{H\ m^{-1}},\quad C = 10\ \mathrm{pF\ m^{-1}},\quad V = 50\ \mathrm{MV},$$

$$Z = 450\ \Omega,\quad i = 110\ \mathrm{kA},\quad v = 3\times 10^{8}\ \mathrm{m\ s^{-1}}.$$

Following Wagner's (1963 and 1967) suggestion, we accept:

$$r_i = 3\ \mathrm{cm} \quad \text{and} \quad D = 300\ \mathrm{m}.$$

This gives a second approximation

$$L = 18{\cdot}4\times 10^{-7}\ \mathrm{H\ m^{-1}},\quad C = 12{\cdot}1\times 10^{-12}\ \mathrm{F\ m^{-1}},\quad Z = 390\ \Omega,$$

$$v = 2{\cdot}12\times 10^{8}\ \mathrm{m\ s^{-1}} \quad \text{or} \quad v/c = 0{\cdot}707,$$

if c is the velocity of light.

A third approximation, due to Wagner (1963 and 1967), is based on the thermal energy or retarded voltages of the wave front. This picture is the only one which is physically acceptable and alone explains the low value of the velocity of the return stroke. Wagner's assumption of a value of 3,000 Ω for the surge impedance of the front of the return stroke is difficult to understand but his values for L and C are the same as quoted above. Once the arc regime is established the surge impedance is determined by the interchange of magnetic energy $i^2L/2$ and $u^2C/2$. It is only during the current front that a supplementary voltage impulse must be postulated to heat the ionized channel during its transition from the leader current to the return-stroke current. This transient voltage impulse acts like a transient resistance, inserted for the duration of the current front. It is shorted when the current has reached its final value when the voltage drop in the arc reverts to its stationary value of about 60 V cm^{-1}. In my view, it is therefore physically not sound to speak of a "surge impedance" of 3,000 Ω of the return stroke since this value is transient and effective only for the duration of the current front.

After the full current has been attained on earth, it travels upwards at 10 to 30% of the velocity of light. When it has reached the upper end of the leader, it is reflected in a similar manner as a surge reaching the open end of a transmission line. The current is therefore reflected, doubling progressively the positive charge on the channel. When the reflected current wave reaches the earth, the entire channel is essentially without current but positively charged. At this instant, the current may be extinguished at the earth. However, having regard to damping in the cloud and along the channel, the reflected current amplitude will be lower than that of the initial return current. Furthermore, the corona shell of the leader channel will conduct part of its charge back into the channel and to earth. As a result, the current at the earth will not fall to zero after twice its travelling time. If the velocity of propagation of the return stroke is taken as 10 to 30% of that of light, twice travelling time becomes 100 to 330 μs for a channel length of 5 km, or 60 to 200 μs for 3 km. The median impulse duration of a first current is 75 μs and its total duration varies between about 200 and 500 μs. This is more than the above travelling time, as would be expected, considering damping by corona and the consequent lengthening of the travelling wave. The picture of a highly damped travelling wave to represent a return stroke thus seems justified in principle.

A problem which is not only of scientific but also of general interest is the diameter of a lightning stroke. To avoid misunderstanding, a definition of diameter is first required.

Three different diameters must be distinguished:

(a) the diameter of the corona shell, based on the ionization sheath;
(b) the diameter of the visual stroke, based on optics;
(c) the diameter of the arc channel, based on conductivity and current density.

As to (a), the diameter of the corona shell has never been measured. Its calculation is based on the stroke potential or the charge density in the leader, on the air density and on the mean field strength in the ionized corona shell. Approximate values may vary from a few metres to 20 or 40 m (see Section 5.10).

Concerning (b), the visual diameter is of interest if the whole spectrum is known. Orville succeeded in photographing not only the spectrum of strokes but also that of leaders with a time resolution down to 2 to 5 μs (Chapter 8.3.3.2). The prevailing spectral wave-lengths are the same in both cases and this is a further proof for their common arc character.

An extensive optical analysis of a stroke by Orville *et al.* (1974) reveals a fairly constant optical diameter of 6·5 cm during the first 1·6 ms of the flash.

Earlier results are summarized in Table VI. Measurements of time-resolved spectra constitute the only means for calculating the temperature and ion density in the stroke.

Table VI

Diameter of the lightning return stroke

	Experimental (cm)	Method
Schonland (1937)	15–23	Photographic[a]
Schonland (1950)	<5	Fulgurites in sand
Evans and Walker (1963)	3–12	Photographic[b]
Hill (1963)	0·03–0·52	Fulgamites
Uman (1964)	0·2–0·5, 2–3·5	Holes in fibreglass bonnets
Taylor (1965)	0·05–0·3, 1–8	Tree trunk damage
Jones (1968)	0·1–0·3	Discharge craters in aluminium
Orville *et al.* (1974)	6–7	Photographic[c]
	Theoretical (cm)	
Braginskii (1958)	0·3–2	Spark discharge model
Oetzel (1968)	0·1–8	Electrical circuit model
Plooster (1971)	0·33–1·76	Spark discharge model

[a] Images uncorrected for overexposure.
[b] Time-resolved images corrected for overexposure.
[c] Time-resolved images of one stroke corrected for overexposure.

For (c), the diameter of the electric arc includes all longitudinal current filaments within the optical diameter. If the bell-shaped distribution curves of temperature and current density are taken into account, the effective arc diameter is found to be much smaller than the optical diameter. Its extent depends essentially on the duration of the impulse current.

5.7 The Multiple Flash

The picture developed in the preceding section indicates that after twice the travelling time of the current in the return stroke, the lightning channel must be positively charged. At the instant when the current first reaches the upper end of the channel, the local field at this point changes from negative to positive polarity with respect to the surrounding space charges. This increased field causes a "junction streamer" (see Chapter 6.5.1) which progresses towards an untapped charge centre (Bruce and Golde, 1941).

Two alternative developments may now follow. If the decaying current maintains its connection with earth, the junction streamer feeds the conducting channel and a continuing current will flow to earth. Small current pips may occur as a consequence of the junction streamer tapping small charge centres, see Figs 7 and 9. If, on the other hand, the current has been extinguished at ground level, a new leader may develop from a higher charge centre downward in the same manner as for the first leader. This second leader follows the still hot and ionized channel within the persisting corona shell. Its velocity of propagation is one order of magnitude higher than that of a first leader; it is a dart leader. Contact with earth is established as in the first leader but there is no connecting streamer.

Figure 2(c) indicates in curves *S*1 and *S*2 opposite (positive) polarity of the charge on the channel immediately following the first return stroke. This transient field disappears within 10 ms and a negative field is slowly re-established by the development of the junction streamer until the downward moving dart leader increases the field into the second return stroke.

5.8 Subsequent Strokes

Subsequent strokes have front times of 0·5 to 1 μs and their decay is regular and more or less exponential. Why do they differ from first strokes? In the pause following a first stroke the channel temperature decreases. Following calculations by Uman and Voshall (1968), the central temperature of a 2 cm diameter channel falls from 8,000 K to 2,000 K, or from 8,000 K to 3,300 K for a 4 cm channel, during the interval between successive strokes of 40 ms. The air density may therefore be one-tenth that in a first stroke, the mean free path and the extension of electron avalanches would be ten times longer and the velocity of propagation of the dart leader would be about three times greater. This temperature effect is supported by an effect exerted by the high radial temperature gradient. This latter implies a high radial gradient of the mean free path of electrons which acts like a reflecting wall and which exerts an axial guiding effect. A third acceleration effect is caused by a residual ion charge within the corona shell produced by the first stroke. This shell was created by the first leader to equalize the radial field and to neutralize free space charges by branches. The dart leader has therefore not to feed another corona shell or any branches. Branches were discharged by the first stroke. In consequence, a subsequent stroke may be treated as a linear conductor with better approximation than a first stroke.

Persisting ionization from the first return stroke is sometimes assumed to be responsible for the high velocity of dart leaders. Such an effect is not confirmed by experiments made by the Swiss High Voltage Research

Committee (Berger and Pichard, 1956). A steep impulse voltage of 150 kV was superimposed on a 50 Hz arc of about 300 cm length and carrying 10 A, at the instant when the sinusoidal current had decreased to +6·7 A before passing through zero. It was presumed that the superimposed impulse current would not materially affect the arc voltage which was 37 kV. Contrary to expectation the voltage rose to 150 kV and then collapsed after 40 μs. The measured breakdown voltage and its time delay were found to agree with those of a non-ionized air gap at an arc temperature of, say, 3,000 K. This result suggests that a "dying" arc contains an insufficient amount of electrons but a large number of ions. These are responsible for the stationary arc current but are incapable of ionization. If this result is applicable to the lightning discharge, persistent ionization of the channel cannot affect the development of subsequent strokes.

The wave-shape of subsequent strokes is regular and more or less exponential (see Fig. 10, Chapter 9). The absence of branches makes it more convenient to treat subsequent strokes as travelling waves on the lightning channel. The short front time of 0·5 to 1 μs is of special interest. A main explanation may be that the dart leader finds a hot channel reaching all the way to earth. Branching has never been observed.

5.9 Discharge of Positive Clouds

The rare occurrence of positive earth flashes is due to the great height of positive charge centres in the thundercloud. There is a further reason for the small number of type 3 flashes to tall structures or mountains. When an intra-cloud flash lowers positive charge, the electric field at a tall structure may initiate an upward leader from its tip, thus causing a type 4 flash. Above the earth, only an upward leader will be visible and this often produces an upward streamer (or connecting leader) towards a cloud discharge. This is the reason why the only type 3 flash recorded at Mount San Salvatore struck the lake shore as shown in Fig. 12.

In contrast, type 4, and particularly type 4*b*, flashes constitute by far the most severe and interesting type of lightning flash. One is tempted to call these "giant flashes" because of their exceptionally large $\int i^2 dt$ values of up to $1{\cdot}5 \times 10^7$ A^2 s as compared with 5×10^5 A^2 s for first negative strokes (see also Chapter 9.5.3). An extreme value of $2{\cdot}2 \times 10^7$ A^2 s was experimentally confirmed in Switzerland by Meister (1973).

The integral $\int i^2 dt$ is responsible for the thermal and mechanical effects of lightning. This has been confirmed on San Salvatore by the fact that the measuring shunt was destroyed by type 4*b* flashes. It appears that these giant flashes may have to be taken into account for the protection of tall telecommunication towers, rockets and other structures on mountain tops.

A specialized problem concerning flashes of types 1*b* and 3*b* is associated with the meaning of the term "striking distance" as defined in Chapter 17.3. Let it be accepted that the striking distance is equivalent to the final jump in the long spark discharge as suggested in Section 5.5. If extrapolation of the equation $S = V/E$ in Section 5.5 is permissible, a leader potential of 20 MV would cause a breakdown at a height of 46 m above the earth. Experience in very-high-voltage laboratories places such an extrapolation very much in doubt.

As is shown in Chapter 7.4.2.2, the actual length of a positive spark leader is considerably longer than the shortest sparkover distance. Any mathematical basis for calculating the striking distance for flashes of types 1*b* and 3*b* is therefore questionable. Some realistic information may, however, be derived from the loops in lightning photographs mentioned in Section 4.3.4 and from sharp kinks in the lightning channel above tall structures, similar to those mentioned in Chapter 17.2.1. No such problem arises in connection with types 2 and 4.

5.10 Stroke Potential and Energy in Lightning Flash

The potential with respect to earth of a thundercloud cannot be determined, since knowledge of the distribution of positive and negative charges in the cloud is inadequate (see Chapters 3.4.4 and 6). However, the voltage U_L between the upper end of a leader and earth shortly before it reaches earth can be estimated from known data. It is the sum of the voltage drop U_l along the leader and the voltage U_t of the tip of the corona shell.

The gradient along the leader can be taken with reasonable certainty as 60 V cm^{-1} (see Section 5.5). The voltage U_l thus becomes:

$$U_l = 30 \text{ MV for a 5 km lightning channel} \quad \text{and}$$

$$U_l = 18 \text{ MV for a 3 km lightning channel.}$$

The voltage U_t of the corona tip of the leader can be derived approximately from a knowledge of the corona shell. At the outer surface of this shell the field must be $E = 30$ kV cm^{-1}. The charge density Q in the leader is

$$Q = 2r\pi E \text{ C m}^{-1}.$$

This charge density Q on the leader can be calculated from the minimal leader current $i = 100$ A and the minimal leader velocity $v = 10^5$ m s^{-1}. Thus

$$Q = i/v = 10^{-3} \text{ C m}^{-1} \quad \text{or} \quad 1 \text{ C km}^{-1}.$$

On the other hand, the recorded median value of the leader charge is 5 C, a value which corresponds with the figure of 1 C km^{-1}.

The radius of the corona shell at ground level is therefore

$$R = Q/2\pi\varepsilon E = 6 \text{ m}.$$

In the corona shell the field is limited by the breakdown value of 30 kV m^{-1}. The voltage U_t will therefore not exceed $U_t = 18$ MV. So for an average charge of 5 C on the leader, the total voltage becomes

$$U_L = 30+18 = 48 \text{ MV for a 5 km lightning channel} \quad \text{and}$$

$$U_L = 18+18 = 36 \text{ MV for a 3 km lightning channel.}$$

The radial field strength on the corona shell is presumably somewhat lower than 30 kV cm^{-1}. If it is taken as 20 kV cm^{-1}, the radius of the corona shell would be between 6 and 9 m. On the other hand, the voltage of the corona shell would not be materially changed.

If the calculation was to be based on the statistical 5% *value* of the leader charge, namely 20 C or 400 A for the leader current with $v = 10^7$ cm s^{-1}, the radius of the corona shell would be $R = 24$ m, $U_K = 24 \times 3 = 72$ MV and

$$U_L = 30+72 = 102 \text{ MV for a 5 km channel} \quad \text{and}$$

$$U_L = 18+72 = 90 \text{ MV for a 3 km channel.}$$

For comparison it may be mentioned that Moore (1974) mentions values of between 10 and 20 m for the radius of the corona shell.

These approximate calculations should not be accepted as accurate. However, they indicate that the lightning potential lies in the order of magnitude from 20 to 100 MV.

The *energy* of an earth flash is stored in the electrostatic field between the cloud charge and earth and is given by $\frac{1}{2}QU$. Taking a mean value for negative clouds of $U = 50$ MV and a mean charge of $Q = 8$ C, the energy in an average flash becomes 2×10^8 Ws or 55 kWh; this is equivalent to 48 kcal, or about 4 kg of oil, per flash (see also Chapter 9.5.3). Values for positive flashes may be ten times higher.

The frequency of lightning discharges is examined in Chapter 14. If it is assumed that there are 100 negative flashes per second over the entire globe (Brooks, 1925) and that one-half of these go to earth, the mean *power* of lightning flashes over the globe becomes:

$$P = 50 \times 2 \times 10^8 = 10^{10} \text{ Ws s}^{-1} \quad \text{or} \quad 10{,}000 \text{ MW}.$$

If 5% of positive flashes were taken into account, the foregoing value would have to be increased by 50%. However, this will be discounted. The power transmitted to 1 km^2 of the earth's surface is on average:

$$P = 10^{10}/5 \times 10^8 = 20 \text{ W km}^{-2}.$$

This is more than one million times smaller than the solar radiation reaching the earth.

On the basis of 3 earth flashes per km^2 per year, a mean value for Switzerland, the following values result:

$$\text{energy—}2 \times 10^8 \times 3 = 6 \times 10^8 \text{ W s km}^{-2} \text{ yr}^{-1} = 167 \text{ kW h km}^{-2} \text{ yr}^{-1}$$

$$\text{power—}6 \times 10^8/(8{,}760 \times 3{,}600) = 19 \text{ W km}^{-2}$$

This is the answer to those who wish to utilize lightning for an energy supply. The mean power per km would be insufficient to maintain an electric light.

In contrast, a lightning discharge of 30 kA and a voltage drop of 60 V cm^{-1} exerts a power per km of channel of

$$P = 3 \times 10^4 \times 6 \times 10^6 = 180{,}000 \text{ MW km}^{-1}$$

By comparing this figure with the output of an atomic power plant, the enormous *power* of a lightning discharge becomes evident. On the other hand, the modest *energy* of an average lightning flash is merely equivalent to that involved in the explosion of 4 kg of oil.

6. Outlook

In the foregoing presentation of present knowledge of the earth flash several aspects have been mentioned on which further information is required.

These may be summarized as follows.

Simultaneous time-resolved photo-multiplier pictures of downward and upward connecting leaders ("last step").

Physical clarification of the corona–arc transition within the striking distance ("final jump").

Physical and mathematical elucidation of the formation of steps in first and subsequent strokes, taking into account the different air temperatures in the channel.

Significance and value of the "surge impedance" of the return stroke.

Localization and frequency of flashes to the surroundings, within several kilometres of high TV towers.

Verification of the "negative protective value" of lightning discharges triggered by high towers and rockets.

Direct measurements of the correlation between stroke current and velocity of return strokes.

Comparison of direct measurement of stroke current in towers with electric and magnetic field measurements at distances over open country.
Experimental research on the electric breakdown strength of decaying arcs, using very steep impulse voltages as a probe.

References

Aleksandrov, G. N. (1969). Besonderheiten der Entwicklung der Funkenentladung in langen Luftfunkenstrecken. *Z. tech. Phys.* **39** (4), 744–747.

Berger, K. (1928). Weiterentwicklung des KO von Dufour. *Bull. schweiz. elektrotech. Ver.* **19**, 292–301.

Berger, K. (1930). Ueberspannungen in elektr. Anlagen. *Bull. schweiz. elektrotech. Ver.* **21**, 77–108.

Berger, K. (1936). Resultate der Gewittermessungen in den Jahren 1934/35. *Bull. schweiz. elektrotech. Ver.* **27**, 145–163.

Berger, K. (1955). Messungen und Resultate der Blitzforschung der Jahre 1947–1954 auf dem Monte San Salvatore. *Bull. schweiz. elektrotech. Ver.* **46**, 291–301, 405–424.

Berger, K. (1967). Novel observations on lightning discharges: results of research on Mount San Salvatore. *J. Franklin Inst.* **283**, 478–525.

Berger, K. (1970). Summary about duration of flashes and stroke intervals. Conf. int. grands Res. Elect. W.G. 33–01.

Berger, K. (1972). Methoden und Resultate der Blitzforschung auf dem Monte San Salvatore 1963–71. *Bull. schweiz. elektrotech. Ver.* **63**, 1403–1422.

Berger, K. (1973). Oszillographische Messungen des Feldverlaufs in der Nähe des Blitzeinschlages auf dem Monte San Salvatore. *Bull. schweiz. elektrotech. Ver.* **64**, 120–136.

Berger, K. (1975). Development and properties of positive lightning flashes at Mount San Salvatore. Culham Conf. Section I2.

Berger, K. and Pichard, R. (1956). Experimentelle und theoretische Untersuchung der Erdschluss-Ueberspannungen und -Lichtbögen. *Bull. schweiz. elektrotech. Ver.* **47**, 485–517.

Berger, K. and Vogelsanger, E. (1965). Messunger und Resultate der Blitzforschung der Jahre 1955–1963 auf dem Monte San Salvatore. *Bull. schweiz. elektrotech. Ver.* **56**, 2–22.

Berger, K. and Vogelsanger, E. (1966). Photographische Blitzuntersuchungen der Jahre 1955–1963 auf dem Monte San Salvatore. *Bull. schweiz. elektrotech. Ver.* **57**, 591–620.

Boys, C. V. (1926). Progressive lightning. *Nature.* **118**, 749–750.

Braginskii, S. I. (1958). Theory of the development of a spark channel. *Sov. Phys.* (English translation) **34**, 1068–1074.

Brooks, C. E. P. (1925). The distribution of thunderstorms over the globe. *Met. Office geophys. Memoirs* **3** (24), 147–164.

Bruce, C. E. R. and Golde, R. H. (1941). The lightning discharge. *J. Inst. elect. Engrs.* **88**, Part II, 487–505.

Eriksson, A. J. (1974). The measurement of lightning and thunderstorm parameters. C.S.I.R.—Special Report, Elek. 51, Pretoria.

Evans, W. H. and Walker, R. L. (1963). High-speed photographs of lightning at close range. *J. geophys. Res.* **68**, 4455–4461.

Fortescue, C. L. (1930). Direct strokes—not induced surges—chief causes of high-voltage line flashovers. *Electl. J.* **27**, 459–462.

Garbagnati, E. and Lo Piparo, G. B. (1970a). Shunts per la misura delle correnti di fulmine. (Shunts for lightning current measurement.) *Energia elett.* **47**, 254–259.

Garbagnati, E. and Lo Piparo, G. B. (1970b). Stazione sperimentale per il rilievo delle caratteristiche dei fulmini (Research station for determining lightning characteristics.) *Elettrotecnica* **74**, 388–397.

Garbagnati, E. and Lo Piparo, G. B. (1973). Nuova stazione automatica per il rilievo delle caratteristiche del fulmine. *Energia elet.* **50**, 375–383.

Garbagnati, E., Giudice, E., Lo Piparo, G. B. and Magagnoli, U. (1975). Rilievi delle caratteristiche dei fulmini in Italia. Risultati ottenuti negli anni 1970–1973. (Survey of the characteristics of lightning stroke currents in Italy, results obtained in the years from 1970 to 1973.) Report E.N.E.L. R5/63–72. *Elettrotecnica* **57/62**, 237–249.

Gary, C. and Fieux, R. (1974). The Experimental Station of St Privat d'Allier. Conf. int. grands Res. Elect. W.G. 33–01.

Golde, R. H. (1947). Occurrence of upward streamers in lightning discharges. *Nature.* **160**, 395.

Gorin, B. N., Lewitow, V. I. and Skiljew, A. K. (1972). Blitzmessungen am Ostankino-Fernsehturm in Moskau. *Élektrichestvo.* **2**, 24–29.

Griscomb, S. B. (1960). The cine-klydonograph—transient waveform recorder. *Trans. Am. Inst. elect. Engrs, III. Power App. Syst.* **79**, 603–610.

Griscomb, S. B., Caswell, R. W., Graham, R. E., McNutt, H. R., Schlomann, R. H. and Thornton, J. K. (1965). Five-year field investigation of lightning effects on transmission lines. *Trans. Inst. elect. electron Engrs. Power App. Syst.* **84**, 257–280.

Gruber, G., Hutzler, B., Jouaire, J. and Riu, J. P. (1975). Long sparks in negative polarity (in the press). Report to Electricité de France.

Hentschke, S. and Illgen, M. (1975). Messung extremer Blitzströme an Fernmeldetürmen. *Elektrotech. Z. Ausg. A.* **96**, 352–355.

Hill, R. D. (1963). Determination of charges conducted in lightning strokes. *J. geophys. Res.* **68**, 1365–1375.

Hoffert, H. H. (1889). Intermittent lightning flashes. *Phil. Mag.* **28**, 106–109; also in *Proc. phys. Soc.* **10**, 176–180.

Israel, H. (1929). Ein transportables Mesgerät für schwene Ionen. *Z. Geophys.* **5**, 342–350.

Jones, R. G. (1968). Return stroke core diameter. *J. geophys. Res.* **73**, 809–814.

Kitagawa, N. and Kobayashi, M. (1958). Distribution of negative charge in the cloud taking part in a flash to ground. *Pap. Met. Geophys., Tokyo* **9**, 99–105.

Kitagawa, N., Brook, M. and Workman, E. J. (1962). Continuing currents in cloud-to-ground lightning discharges. *J. geophys. Res.* **67**, 637–647.

Knoll, M., Ollendorff, F. and Rompe, R. (1935). "Gasentladungstabellen." Springer, Berlin.

Krehbiel, P., McCrory, R. A. and Brook, M. (1974). The determination of lightning charge location from multistation electrostatic field change measurements. Proc. Conf. on Cloud Phys., Tucson, Arizona.

Kulijew, D. A. (1976). Oszillographische Blitzuntersuchungen in der Gebirgsgegend Aserbaidshans. 13th Internat. Blitzschutzkonferenz, Venezia.

Les Renardières Group (1972). Research on long air gap discharges at Les Renardières. *Electra* **23**, 53–157.
Les Renardières Group (1974). Research on long air gap discharges at Les Renardières. *Electra* **35**, 49–156.
Lichtenberg, G. C. (1778). Novo methodo naturam ac motum fluidi electrici investigandi. *Societatis Regiae Scientiarum Gottingensis.* T **8**, 168–180.
Loeb, L. B. and Meek, J. M. (1940). The mechanism of spark discharge in air at atmospheric pressure. I and II. *J. appl. Phys.* **11**, 438–447 and 459–474.
Lundholm, R. (1957). Induced overvoltage surges on transmission lines. *Chalmers tek. Högsk. Handl.* **188.**
Malan, D. J. (1950). Appareil de grand rendement pour la chronophotographie des éclairs. *Rev. Opt.* **29**, 513–523.
Malan, D. J. (1956). The relation between the number of strokes, stroke intervals, and the total durations of lightning discharges. *Geofis. pura appl.* **34**, 224–230.
Malan, D. J. (1957). The theory of lightning photography and a camera of new design. *Geofis. pura appl.* **38**, 250–260.
Malan, D. J. (1963). "Physics of Lightning." The English Universities Press Ltd, London.
McCann, G. D. (1944). The measurement of lightning currents in direct strokes. *Trans. Am. Inst. elect. Engrs* **63**, 1157–1164.
McEachron, K. B. (1939). Lightning to the Empire State Building. *J. Franklin Inst.* **227**, 149–217.
Meister, H. (1973). Dynamische Zerstörung einer Erdleitung durch einen Blitz. *Bull. schweiz. elektrotech. Ver.* **64**, 1631–1635.
Moore, C. B. (1974). An assessment of thundercloud electrification. Conf. on Atm. Elec. Garmisch-Partenkirchen, *in* "Electrical Processes in the Atmosphere" (H. Dolezalek and R. Reiter, Eds). Steinkopff, Darmstadt.
Newman, M. M. (1963). Use of triggered lightning to study the discharge process, *in* "Problems of Atmospheric and Space Electricity" (S. C. Coronity, Ed.), pp. 482–490. Elsevier, Amsterdam.
Norinder, H. (1925). Recherches sur la nature des décharges électriques des orages. Conf. int. grands Res. Elect. report No. 82.
Oetzel, G. N. (1968). Computation of the diameter of a lightning return stroke. *J. geophys. Res.* **73**, 1889–1896.
Orville, R. E., Helsdon, J. H. and Evans, W. H. (1974). Quantitative analysis of a lightning return stroke for diameter and luminosity changes as a function of space and time. *J. geophys. Res.* **79**, 4059–4067.
Peters, J. F. (1924). The klydonograph. *Electl Wld* **83**, 769–773.
Pierce, E. T. (1955). Electrostatic field-changes due to lightning discharges. *Q. Jl R. met. Soc.* **81**, 211–228.
Plooster, M. N. (1971). Numerical model of the return stroke of the lightning discharge. *Physics Fluids* **14**, 2124–2133.
Pockels, F. (1897). Über das magnetische Verhalten einiger basaltischer Gesteine. *Ann. Physik Chem.* **63**, 195–201.
Raether, H. (1964). "Electron Avalanches and Breakdown in Gases." Butterworth, Washington, D.C.
Rusck, S. (1958). Induced lightning overvoltages on power transmission lines with special reference to the overvoltage protection of low-voltage networks. *Trans. R. Inst. Technol.* **120.**

Schonland, B. F. J. (1937). The diameter of the lightning channel. *Phil. Mag.* **37**, 503–508.

Schonland, B. F. J. (1950). "The Flight of Thunderbolts," p. 63. Oxford University Press, New York.

Schonland, B. F. J. (1956). The lightning discharge, *in* "Encyclopaedia of Physics," Vol. 22, pp. 576–628. Springer Verlag, Berlin.

Simpson, G. C. and Scrase, F. J. (1937). The distribution of electricity in thunderclouds. *Proc. R. Soc.* A **161**, 309–353.

Stekolnikov, I. S. and Shkilev, A. V. (1963). Long positive spark development on a ramp voltage wave. *Dokl. Akad. Nauk SSSR, Tekn. Fiz.* **151**, 837–840, 1085–1088.

Taylor, A. R. (1965). Diameter of lightning as indicated by the tree scars. *J. geophys. Res.* **70**, 5693–5695.

Toepler, M. (1917). Über den inneren Aufbau von Gleitbüscheln und Gesetze ihrer Leuchtfäden. *Beibl. Annln Phys.* **53**, 217–234.

Toepler, M. (1920). Messungen und Beobachtungen an Polbüscheln gleitender Entladung. *Phys. Z.* **21**, 706–711.

Toepler, M. (1925). Funkenkonstante, Zündfunken und Wanderwelle. *Arch. Elektrotech.* **14**, 305–318.

Uman, M. A. (1964). The diameter of lightning. *J. geophys. Res.* **69**, 583–585.

Uman, M. A. (1969). "Lightning." McGraw-Hill Book Co., New York.

Uman, M. A. and Voshall, R. E. (1968). Time interval between lightning strokes and the initiation of dart leaders. *J. geophys. Res.* **73**, 497–506.

Wagner, C. F. (1963). Relation between stroke current and velocity of the return stroke. *Trans. Inst. elect. electron. Engrs Power App. Syst.* 609–617.

Wagner, C. F. (1967). Lightning and transmission lines. *J. Franklin Inst.* **283**, 558–594.

Wagner, C. F. and McCann, G. D. (1940). New instruments for recording lightning currents. *Trans. Am. Inst. elect. Engrs* **59**, 1061–1068.

Wagner, C. F., Lane, C. M. and Lear, C. M. (1958). Arc drop during transition from spark discharge to arc. *Trans. Am. Inst. elect. Engrs, III Power, App. Syst.* 242–247.

Walter, B. (1902). Ein photographischer Apparat zur genaueren Analyse des Blitzes. *Physik. Z.* **3**, 168–172.

Wilson, C. T. R. (1916). On some determinations of the sign and magnitude of electric discharges in lightning flashes. *Proc. R. Soc.* A **92**, 555–574.

Workman, E. J., Brook, M. and Kitagawa, N. (1960). Lightning and charge storage. *J. geophys. Res.* **65**, 1513–1517.

6. The Cloud Discharge

M. BROOK and T. OGAWA

New Mexico Institute of Mining and Technology, Socorro, New Mexico, U.S.A., and Kyoto University, Kyoto, Japan

1. Introduction

1.1 General Remarks

In spite of the fact that the great preponderance of lightning discharges is confined to the thundercloud itself, research into the mechanism of the intra-cloud lightning discharge has lagged behind the more spectacular flashes to ground. In his review article "The lightning discharge" Schonland (1956) devoted a total of 18 lines to the cloud discharge, with perhaps a similar number of lines to air discharges. Such biased attention to ground flashes persists to this day. This is undoubtedly due to the fact that lightning damage to power lines, tall buildings and old churches, the loss of valuable timber through lightning-initiated forest fires, and the destruction of human and animal life each provide strong economic motivations for the study of lightning strokes to ground. In addition, where photographs of lightning channels can be used to supplement electrical measurements of ground strokes, no technique for obtaining in-cloud photographs of lightning is presently available.

For the atmospheric physicist and meteorologist, studies of intra-cloud and cloud-to-ground lightning are equally important in that they are both necessary in improving our crude models of thunderstorm charge distributions, and, hopefully, for obtaining an improved understanding of how charge is separated in clouds.

In-cloud discharges may involve both positive and negative distributions of considerable extent. It is a curious fact, therefore, that both Malan and Schonland, whose brilliant researches into progressive lightning uncovered the fascinating sequence of events associated with the ground flash as we know it today, chose to speak almost exclusively of negative charge in the cloud, the so-called "negative column", without much concern for the disposition of an equal amount of positive charge which must also have been separated at the same time (see, e.g., Malan and Schonland, 1953).

A knowledge of the distribution of electric charge in a thundercloud is of singular importance in estimating the role of electric forces in influencing cloud element growth. Although electric forces may well rival hydrodynamic forces under appropriate conditions (their existence has been inferred from *in situ* measurements), a coherent, even rudimentary picture of the important interactions has not yet emerged (see Chapter 3). Only recently have *in situ* electric field measurements in cloud become credible, but the thunderstorm environment is a most hostile one, and results are slow in coming. Thus, electric field-changes measured on the ground are still our main source of information about the distribution of electric charge in clouds.

In this chapter we confine our attention to cloud discharges, i.e. those categories of lightning usually referred to as intra-cloud, intercloud and air discharges. We shall make no distinction between the three categories.

In the mathematical expressions which appear in the following sections we use the symbols given in Table I.

Table I
Symbols

	Electric field (E) and field-change (ΔE)	Charges	Heights	Distance	Electric moment (M) and moment change (ΔM)
Cloud	E	Q_1, Q_2	H, h	D_1, D_2	M
Cloud discharges	ΔE_1	q	H_1, h_1	D_{11}, D_{12}	ΔM_1
K changes	ΔE_2	q_2	H_2, h_2		ΔM_2

1.2 Methods of Investigation

The investigation of cloud-discharges relies primarily upon surface electric field and field-change measurements. The measurements are useful in constructing models (not necessarily unique) of charge distributions, of charge transport (as in streamer propagation), and in estimating charge and current magnitudes with the help of additional data. Auxiliary data of considerable value are obtained from still-camera and moving-film photographs of streamers when they emerge from the clouds; from luminosity measurements with light sensors (Malan, 1955; Kitagawa and Kobayashi, 1958); from thunder measurements (Holmes *et al.*, 1971; Nakano, 1973; Teer and Few, 1974, and Chapter 11) which yield estimates of discharge energy and in-cloud channel lengths and orientations; from radar studies of the appearance and decay of ionized channels (Hewitt, 1957); and from high-resolution direction finding at VHF (≈ 250 MHz) to pinpoint sources of electromagnetic emission (Proctor, 1974b, and Chapter 10.3.3).

Most of the electrostatic measurements to be discussed were made with field mills and with flat-plate antennae ("slow" and "fast" antennae) (see Chapter 13). Although field mills are useful for continuous recording of the total electric field, their slow time response is unsuited to the study of the finer structures. The slow antenna and the fast antenna have been most useful in this respect, especially in providing the microscopic view of events necessary to distinguish between in-cloud flashes and flashes to ground (Kitagawa and Brook, 1960).

The flat-plate antenna and associated measuring circuit normally consists of a disk-shaped conductor of effective area A set flush with the earth and connected to a charge amplifier. A change, ΔE, in the electric field terminating on the plate is accompanied by a change in the induced charge $\Delta Q = \varepsilon_0 A \, \Delta E$, where ε_0 is the permittivity of free space. An operational amplifier maintains the plate at ground potential by charging the feedback capacitor, C, such that a potential difference ΔV appears across it and hence at the output. Thus $\Delta V = \Delta Q/C = \varepsilon_0 A \, \Delta E/C$. Sensitivity is adjusted by changing the value of C, i.e. one tries to keep the ratio $\Delta E/C$ as large as is consistent with the dynamic range of the amplifier.

In practice, a resistor is placed across the capacitor to provide a time constant long enough to reproduce accurately the field-changes under study, but short enough to bring the output voltage back to a value near zero between the events of interest. In this way, maximum dynamic range is utilized for the field changes. In the case of the slow antenna, time constants from 4 to 12 s have been used for studying the totality of events in a flash. For the study of individual pulses, a fast antenna with time constant between 70 and 300 μs has proved useful.

2. The Electric Field and Charge Distribution

2.1 Observed Fields

Charges neutralized in a cloud flash need to be considered in the context of the total charge distribution. To this end, we reproduce in Fig. 1 a classification of 99 thunderstorm electric field patterns observed during the period

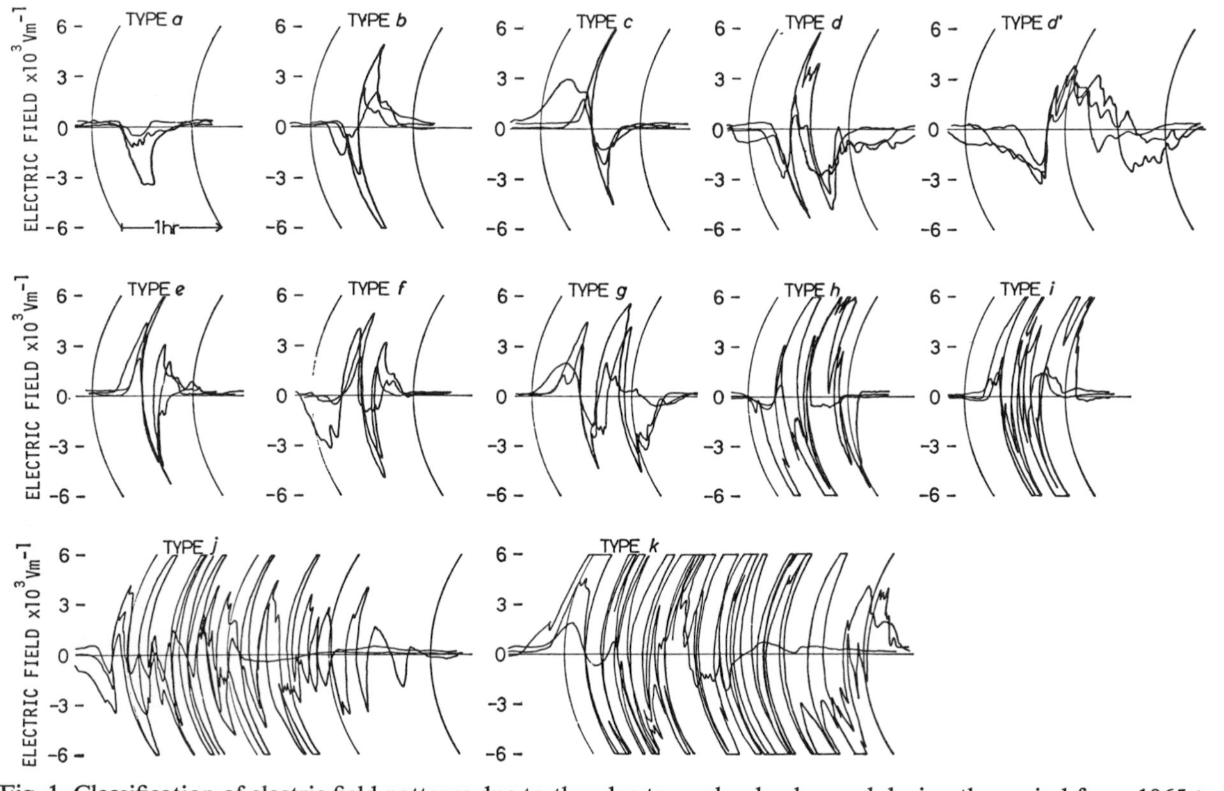

Fig. 1. Classification of electric field patterns due to thunderstorm clouds observed during the period from 1965 to 1967 in Kyoto. Type d' differs from type d only in time scale (from Huzita and Ogawa, 1976a).

1965 to 1967 in Kyoto (Huzita and Ogawa, 1976a). Twelve patterns from type *a* to *k*, including type *d'*, are shown in the figure (types *d* and *d'* differ only in their time scale). Field variations which appear frequently are: negative first, then positive, then negative and finally positive (type *f*: 13%). Type *j* is the variation which appears most often (16%), but this type appears to be due to the successive appearance of the simpler types, *b*, *d* and *d'* and *f*. The seven types, *a*, *b*, *d* and *d'*, *f*, *h* and *j*, constitute 70% of all the field variations measured during this period.

Figure 1 is typical of electric field variations with time observed by many workers. The records do not explicitly separate the variation in electric field with time from the variation with distance from cloud to observer. In fact, Fig. 1 implies a space stationary behaviour which is difficult to justify against the known fact that the thunderstorm cell duration is typically from 25 to 45 min, and that storms move with a velocity which may range from 2 to 20 m s^{-1}. On the basis of many observations, including aircraft measurements over thunderstorms (Gish and Wait, 1950), the observed electric field of a thunderstorm in first approximation appears to be consistent with a bipolar charge distribution in which the positive charge is uppermost. The variations shown in Fig. 1 are generally derivable from such a distribution but with some important qualifications:

(a) the charges are not usually equal;
(b) the charges are not generally vertically distributed;
(c) other charges of lesser magnitude, such as a low-lying, smaller, positive charge below the negative one are sometimes present (Simpson and Robinson, 1941).

2.2 The Electric Field Produced by Simple Charge Distributions

A point or spherically symmetric charge Q at height H above a flat, conducting earth will produce a vertical electric field at the surface given by

$$E = \frac{1}{4\pi\varepsilon_0}\left(\frac{2QH}{(D^2+H^2)^{3/2}}\right) \text{ V m}^{-1}, \tag{1}$$

where $D = (x^2+y^2)^{1/2}$ is the horizontal distance from the charge to the observer. All quantities are expressed in SI units. The sign convention adopted here (as throughout the book) is that in which a positive charge aloft causes a positive field on the ground. The quantity $M = 2QH$ is the dipole moment of the charge at height H and its image charge in the earth.

It is obvious that the point charge or monopole produces a field E which is inconsistent with the fields shown in Fig. 1. Equation (1) indicates that E is either only positive or only negative depending upon the sign of Q. No dependence of polarity on distance is predicted.

We next consider the bipolar distribution. Let the upper positive charge be Q_1, and the lower negative charge be $-Q_2$. The vertical electric field at the surface is then

$$E = \frac{1}{4\pi\varepsilon_0}\left(\frac{2Q_1 H}{(D_1{}^2+H^2)^{\frac{3}{2}}} - \frac{2Q_2 h}{(D_2{}^2+h^2)^{\frac{3}{2}}}\right), \tag{2}$$

where $H > h$. Since the assumed point-charges in this model are not necessarily one above the other, D_1 and D_2 are distinct.

Equation (2), in most cases, provides a qualitative fit to the observed data. In a Cartesian coordinate system, each charge magnitude Q with its coordinates x, y, z constitute four unknowns; thus eight independent, simultaneous field measurements, E_i, would suffice to solve a system of simultaneous equations

$$E_i = \frac{1}{4\pi\varepsilon_0}\left(\frac{2Q_1 z_1}{[(x_1-x_i)^2+(y_1-y_i)^2+z_1{}^2]^{\frac{3}{2}}} - \frac{2Q_2 z_2}{[(x_2-x_i)^2+(y_2-y_i)^2+z_2{}^2]^{\frac{3}{2}}}\right)$$

$$i = 1 \text{ to } 8, \tag{3}$$

where x_i, y_i are the coordinates of the measuring stations.

In practice, attempts to calculate the Q's and their locations in cloud from measured E_i values have generally proved disappointing. The reason, of course, is that simultaneous measurements of E at widely separated stations will usually include contributions from local space-charge distributions as well as from charges embedded in the cloud.

It is not difficult to write down equations for more complex models such as linear, disk-shaped, or cylindrical charge distributions. We shall not deal with such models because existing uncertainties in measurements have not justified their use. Distributions involving point charges in numbers greater than two are easily constructed using Equation (1) and the superimposition principle.

3. Electric Field-changes

3.1 Observed Data

Ogawa and Brook (1964) presented a good example of electric field and field-change data for a well-defined isolated thunderstorm which passed within 4 km of the observing station. This storm moved in an approximately straight line as a single unit, and was active for a period of more than 2½ hours. Excellent visibility made it possible to identify clearly the type of discharge (cloud-to-ground, or intra-cloud). Thunder could be heard for almost 1 hour, and the absence of other storms made it possible to obtain a good field mill record of the total thunderstorm electric field. The slow antenna yielded a continuous record of field-changes from both ground and cloud discharges throughout the storm passage. The net field-changes, ΔE_1, accompanying the visually observed cloud flashes are given in Fig. 2. Also shown is the electric field, E, observed with a field mill, after all disturbances due to the individual discharges had been smoothed. To supplement the time variations, the approximate distance to the storm estimated from time-to-thunder measurements is also shown. The distances do not necessarily correspond to the individual cloud discharges which occurred in the same time period. About 600 field changes due to cloud discharges were recorded from 19.17 to 20.27 MST.

Figure 2 is a good illustration, for cloud flashes, of the dependence of the sign of the field-change upon distance. ΔE_1 tends to be positive for near flashes and negative for distant flashes. Evident in Fig. 2 is the reciprocal behaviour between E and ΔE_1, i.e. when the field E is negative, the field-change ΔE_1 is positive and vice versa. Also to be noted is that, in spite of the large scatter, ΔE_1 changes sign from positive to negative somewhat earlier than E changes from negative to positive. Thus a smoothed curve of the average ΔE_1 values would cross the time axis at about 19.36, whereas E changes polarity at 19.40.

3.2 The Cloud Flash Model

The bipolar model represented by Equation (2) provides good agreement with the general features of the observed E field variation in Fig. 2. It is reasonable to proceed one step further and to assume that portions of the main bipolar charges will be involved in the cloud discharge. For example, one possible model consists of equal positive and negative charges of magnitude q (less than either of the charges Q_1 or Q_2 of Equation 2) which are neutralized by the discharge. The field-change would then be predicted by

$$\Delta E_1 = -\frac{1}{4\pi\varepsilon_0}\left(\frac{2qH_1}{(D_{11}{}^2+H_1{}^2)^{\frac{3}{2}}}-\frac{2qh_1}{(D_{12}{}^2+h_1{}^2)^{\frac{3}{2}}}\right), \tag{4}$$

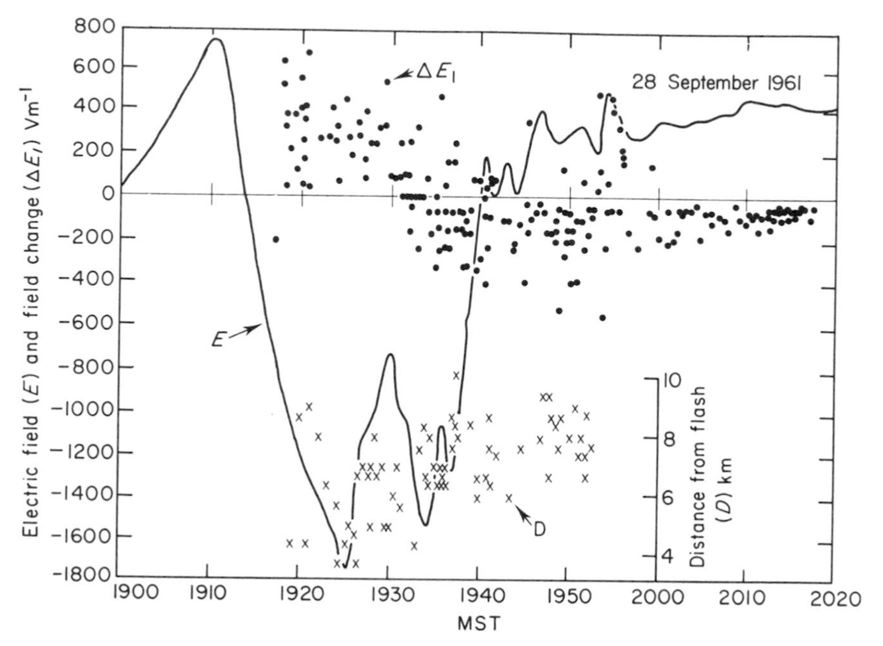

Fig. 2. Electric field E (solid curve) and electric field-change ΔE_1 (dots) plotted as a function of time for the isolated thunderstorm of 28 Sept. 1961, observed in Socorro, New Mexico. The approximate distance to the storm at any given time is shown by the crosses (from Ogawa and Brook, 1964).

where H_1 and h_1 are the heights of the positive and negative charges, $\pm q$, respectively.

One must be careful not to take these models too literally. A measured field-change, ΔE_1, reflects a change in the electric moment M of the cloud, for which the uniqueness of a chosen model cannot be established. A change in electric moment can be produced, for example, by neutralizing two equal and opposite charges, or, alternatively, by distorting an initially spherically symmetric charge distribution into an elongated one such as a line or a cylinder. The latter will not involve two charges of opposite sign. Yet the change in electric moment can be represented by models which *effectively* either remove charges or neutralize charges or move charges up or down or horizontally by the proper amount. In this sense, the representation of a field-change by the simple model involving the disappearance of two charges, for example, is justified if consistent with observations. Equation (4) will therefore be used, where applicable, to model total field-changes due to cloud discharges. Obviously, multistation measurements of field-changes can lead to improved models by demanding consistency with a greater number of simultaneous observations.

3.3 Heights of the Charges

We first consider the bipolar charges of Equation (2) to constitute a vertical dipole, i.e. $Q_1 = Q_2 \equiv Q$ and $D_1 = D_2 \equiv D$. The horizontal distance from a point on the surface directly beneath the dipole to a point on the surface where the electric field is zero, the so-called reversal distance is then given by

$$D_0^2 = (H \cdot h)^{\frac{2}{3}} (H^{\frac{2}{3}} + h^{\frac{2}{3}}). \qquad (5)$$

Note that D_0 is independent of the magnitude of the charge Q; it depends only upon the location of the charges.

A similar analysis is possible for the electric field-change produced by the discharge of a vertical dipole with equal charges q^+ and q^- located at heights H_1 and h_1 respectively. We then find that the measured ΔE_1 of Equation (4) will be zero at the horizontal distance D_{01} given by

$$D_{01}^2 = (H_1 h_1)^{\frac{2}{3}} (H_1^{\frac{2}{3}} + h_1^{\frac{2}{3}}). \qquad (6)$$

Mean values of D_0 and D_{01} derived from Fig. 2 are 7·5 and 7·0 km, respectively; i.e. $D_0 > D_{01}$. Thus, the ratio $D_0/D_{01} > 1$ is a constraint which limits the values of H_1 and h_1 for a given pair of values H and h. The condition $D_0/D_{01} > 1$ may be written

$$\{(Hh/H_1 h_1)^{\frac{1}{3}} [(H^{\frac{2}{3}} + h^{\frac{2}{3}})/(H_1^{\frac{2}{3}} + h_1^{\frac{2}{3}})]^{\frac{1}{2}}\} > 1. \qquad (7)$$

One consistent set of values limits the discharge to the region between the main dipole; i.e. $H_1 < H$ and $h_1 \geqslant h$.

Malan (1956) studied photographs of discharges in transparent thunderclouds and concluded

(a) the negatively charged region of a thundercloud reaches an altitude of at least 8 km above ground (9·8 km MSL);
(b) the negatively charged region has a horizontal extent of about 8 km;
(c) the positively charged region is situated above 8 km altitude.

Estimates of the height of the flashing region of cloud have been made by Mackerras (1968) using photographs and time-to-thunder measurements in

Brisbane. For 24 thunderstorms observed between 1964 and 1967 he finds two height ranges where flashes occur; high intra-cloud flashes typically range from about 4 to 12 km while the more usual cloud flashes range from 1 to 8 km.

Takagi (1961), using a field mill and an antenna system of 1 ms time constant, reported that the positive and negative charges associated with a cloud discharge are located at altitudes of 7 to 11 km and 3 to 6 km respectively, and the mean inclination of the axis connecting the charges is generally less than 30°.

Workman and Holzer (1942), Workman *et al.* (1942), Reynolds and Neill (1955) and Tamura *et al.* (1958) calculated cloud-charge heights from multi-station field and field-change measurements. These results will be discussed in the next section.

The most recent results relating to cloud-charge heights have been made with arrays of microphones to locate thunder sources for the reconstruction of lightning channels in cloud. Nakano (1973), from a small sample of discharges, finds the lightning channel in cloud to be approximately horizontal and located at a height of about 8 km. Teer and Few (1974) reconstructed 20 intra-cloud flashes and found the channels to have considerable horizontal extent, average $11{\cdot}0 \pm 4{\cdot}7$ km, with a ratio of horizontal to vertical extent of about 3 to 1. The mean height of the reconstructed channels was 5 km, ranging from a low point of 4 km to a maximum height of 6 km.

The large spread in calculated and measured heights of cloud flashes by various investigators does not, in our opinion, indicate discrepancies which need to be resolved; rather, they indicate that cloud flashes occur over a large altitude range with a similarly large range in horizontal extent. Teer and Few (1974) have shown how the channel-length distribution varies at four locations from a maximum in the distribution occurring at 5 km in length at Socorro, New Mexico, to a maximum at 14 km in length at Houston, Texas.

Although Smith (1957), Takeuti (1965) and others have found reasonable agreement with the vertical dipole model (Smith found that 80% of his data fit a vertical dipole), the work of Ogawa and Brook (1969), Nakano (1973) and Teer and Few (1974) clearly indicates that intra-cloud flash channels, and hence inferred charge distributions, must quite often have large horizontal components. These results are not generally inconsistent with the model represented in Equation (4), which allows for large horizontal or vertical separation of the equivalent bipolar charges. Obviously, single-station field-change measurements are not very useful in making reliable inferences about the horizontal and vertical extent of the equivalent charge distribution. Yet most of our estimates come primarily by way of such inferences.

3.4 Values of Charge and Moment Change

Figure 2 is again useful for illustrating the relationship between the magnitude of the field-change, ΔE_1, and the distance, D from observer to storm. The closest distance of approach was about 3·5 km. As the storm receded, the field-changes reached a maximum value of about 600 V m^{-1} and reversed sign reaching a value of about -200 V m^{-1} at a distance of about 8 km. The vertical extent of the channel in Socorro storms was found by Teer and Few (1974) to be approximately 2 km. Using this value for $H_1 - h_1$ in Equation (6) with $D_{01} \approx 7$ km we find the charge heights $H_1 \approx 6$ km and $h_1 \approx 4$ km. We can then calculate the charge q in Equation (4) with $\Delta E_1 = -200$ V m^{-1} at $D_{11} = D_{12} \approx 8$ km. The calculation gives the average charge $q \approx 28$ C. The equivalent moment change $\Delta M_1 = 2q(H_1 - h_1)$ is calculated to be about 110 C km.

Equation (3), which was developed on the bipolar charge model for the measurement of electric fields at i stations, can be extended to apply to field-change measurements by writing ΔE_i for E_i, and by setting $Q_1 = Q_2 \equiv q$. The latter condition implies the discharge of an equivalent pair of charges of equal magnitude and opposite sign, but not necessarily one above the other. We then have

$$\Delta E_i = -\frac{1}{4\pi\varepsilon_0}\left(\frac{2qz_1}{[(x_1-x_i)^2+(y_1-y_i)^2+z_1^2]^{\frac{3}{2}}} - \frac{2qz_2}{[(x_2-x_i)^2+(y_2-y_i)^2+z_2^2]^{\frac{3}{2}}}\right), \quad i = 1 \text{ to } 7. \tag{8}$$

In this case, seven simultaneous measurements of ΔE_i will suffice to determine the charges and their coordinates.

Workman and Holzer (1942) made a set of measurements with eight electric field-change stations in Albuquerque, New Mexico, during the summer of 1939. They analysed 32 field-changes of which 16 were attributed to cloud flashes. Their results gave an average height of $5{\cdot}8 \pm 0{\cdot}43$ km for the upper positive charge and a height of $4{\cdot}7 \pm 0{\cdot}32$ km for the lower negative. The mean vertical distance between the two charges was thus 1·1 km, while q

was found to be 32 ± 18 C on the average. The average electric moment change was 70 C km. Workman *et al.* (1942) constructed an improved system for electric field-change measurement during the summer of 1940 which had a time constant of 1 s and a resolution of 10 ms. They used a 16 mm film run at a speed of 0·32 cm s^{-1} to record the field-changes. Of 312 strokes, 187 were analysed using a mechanical analogue system, and solutions were obtained from 60% of the observed data. The results showed equivalent charges of from 0·3 to 100 C neutralized in cloud discharges, and the average vertical distance between the charges was 0·6 km, the upper positive being at a height of 5·8 km, and the lower negative at 5·2 km on the average. The horizontal separation between the charges varied between 1 and 10 km, the average separation being 3 km.

Reynolds and Neill (1955) also made simultaneous field-change observations at 12 stations near Socorro, New Mexico, using a technique similar to the one developed by Workman *et al.* (1942). In addition, a 3 cm radar provided precipitation echo data and two time-lapse cameras were operated to photograph the visible cloud. Reynolds and Neill reported data from 35 cloud discharges, in which, for 28 cases, the positive charge was located above the negative charges. The average positive charge centre was at $5{\cdot}5 \pm 1{\cdot}4$ km and the average negative charge centre at $5{\cdot}1 \pm 1{\cdot}0$ km above terrain which was 2·3 km above MSL. The average vertical separation of the charges was 0·47 km, while the horizontal separation was essentially negligible at $0{\cdot}19 \pm 0{\cdot}13$ km. The charges neutralized ranged from 1·3 to 63 C with a mean value of 21 ± 14 C. Thus the average electric moment change for a cloud flash was 21 C km.

For observations taken far from the storm, the heights H_1 and h_1 in Equation (4) can be neglected compared to the distance $D_{11} \approx D_{12} = D$. Then the field-change can be written

$$\Delta E_1 = -\frac{1}{4\pi\varepsilon_0}\frac{2q(H_1 - h_1)}{D^3} \tag{9}$$

and the change in electric moment can be estimated as

$$\Delta M_1 = 2q(H_1 - h_1) = -4\pi\varepsilon_0 \Delta E_1 D^3. \tag{10}$$

As seen in Fig. 2, the electric field-changes are negative at great distances, so that ΔM_1 in Equation (10) must be positive.

Pierce (1955a) made extensive observations in Cambridge, England, of the electric field-changes produced by lightning discharges at distances within which the electrostatic field was still dominant. A Wilson sphere (see, e.g., Schonland, 1953, or Malan, 1963) was used as the sensing antenna. Simultaneous observations were made with two instruments, one with the capillary electrometer and the other with an oscilloscope in order to relate his measurements to the earlier measurements of Wilson (1920) and Wormell (1939). Pierce (1955a) was thus able to compare electric moment change observations taken in 1939, 1946 and 1947. The magnitude of field-changes, for distances of 25 to 250 km from the discharge, obeyed the expected inverse cube relations (Equation 9), in which the median magnitude corresponded to a change in electric moment of 110 C km of either sign.

Later evidence based on the cathode-ray oscilloscope measurements suggested that the 110 C km figure may have been an overestimate for both positive and negative changes. In 1947 and 1949, oscillograph records of over 1,500 electrostatic field-changes were obtained. Of these, 293 field-changes observed for two storms at distances of 50 to 90 km from Cambridge were used to estimate the average total change in moment. For slow positive field-changes which were generally supposed to involve the downward movement of negative charge in flashes which did not reach the earth, the average moment change was 45 C km. For slow negative field-changes due to air or intra-cloud discharges, which equivalently either lower positive charge or raise negative charge, the average moment change was 81 C km. Since the cloud-discharge field-change is negative for distant storms, the slow negative field-changes may be appropriate for the estimation of cloud-discharge moment change.

There is always some question regarding the degree of confidence which can be assigned to the interpretation, without ambiguity, of field-changes due to cloud-to-ground flashes in the absence of direct visual confirmation. In this regard, Pierce was careful to categorize his field-changes according to polarity and he did not try to distinguish between intra-cloud and cloud-to-ground flashes.

Mackerras (1968) obtained 100 C km for the median value of electric moment change from electric field observations made in Brisbane, Australia.

In Japan, a summer thunderstorm research project was executed near Maebashi in the Kanto Area in the period from 1940 to 1944. Based on a five-station measurement of sudden changes of antenna–earth current due to lightning discharges, in which galvanometers were used in the ballistic mode, Hatakeyama (1949) found from the 1940 and 1941 data that almost all the cloud discharges were vertical. Most of the intra-cloud discharges over a circular cylinder 2 km in radius extending from 4 km to 8 km above ground and an upper positive charge distributed over the cylinder extending from 8 km to 12 km. From the analyses of data obtained in 1942 to 1944 it was found that the neutralized electric charge in the intra-cloud discharge depends on the stage of storm development, i.e. the charge was generally more than 100 C and as high as 300 to 400 C during the developing stage of the storm, and was less than 40 to 60 C and usually less than 10 C during the decaying stage. Hatakeyama (1949) also mentioned that the distance between the neutralized charges was larger in the developing stage, e.g. 2 to 3 km for 100 C of the neutralized charge, and smaller in the decaying stage, e.g. about 1 km for 10 to 20 C of neutralized charge. He also mentioned that horizontal discharges most often occur in the decaying stage of the thunderstorm. These results were also reported by Hatakeyama (1958) in a revised version, in which other data were also included. The average neutralized charge was found to be 90 C for the 17 discharges analysed and the charge ranged from 50 to 152 C with 55% of the charges lying between 70 and 100 C. For a critical comment on the foregoing values see Chapter 9.5.1.1.

Tamura (1949) studied electric field-changes using rotating field mills in Kyoto, and Beppu in Kyushu, and found that the average electric moment change for intra-cloud discharges was 60 C km for positive polarity, and 100 C km for negative polarity field-changes. Also, from simultaneous observations at three stations and from statistical analyses of the data obtained at one station, Tamura (1949) analysed the transition of the electric field sudden change and the succeeding recovery curve with distance and deduced probable modes of discharges. These were that the discharge occurred either in an upper dipole (upper positive and lower negative charge), in a lower dipole (lower negative and lower positive charge), or in both upper and lower dipoles simultaneously: the corresponding moment changes might be 40, 40 and 80 C km respectively.

Tamura *et al.* (1958) performed an eight-station field-mill measurement during the summer of 1957 in Kyoto, similar to the earlier studies of Workman and Holzer (1942). The results showed that intra-cloud discharge moment changes were between 7 and 72 C km with the most frequent value being 20 C km. The height at which the intra-cloud discharges occurred ranged from 4 km to 7 km and the most frequent height was 6 km. Tamura *et al.* (1958) showed that cloud discharges may occur in regions of the thundercloud different from where ground discharges occur, the charges involved in cloud flashes being located at lower levels.

Takeuti (1965) reported from three-station measurements that the moment change in the cloud discharge often exceeded 100 C km and that the discharge was normally several kilometres in length.

In Singapore, Wang (1963a) studied net electrostatic field-changes produced by tropical lightning during the normal winter rainy season in 1950 to 1951. He used a Wilson sphere connected to a capillary electrometer, and found that tropical thunderclouds are of positive polarity and the mean of 39 field-change reversal distances observed for cloud discharges was 11·2 km with a median of 11·5 km. Later (in 1952 to 1953), the mean value, using an oscillographic technique, was found to be about 10 km (Wang, 1963b). Wang obtained a mean electric moment change for cloud discharges of about 200 C km by extrapolating the uncorrected moments of negative field-changes to great distances. He found the charge neutralized to be about 15 C, the probable height of the negative charge was about 4·8 km, ranging between 2 and 8·5 km, and the height of the positive charge at about 8·2 km, ranging between 6 and 11 km.

In situ measurements of electric fields and field-changes inside clouds have been made by Winn and Moore (1971) and Winn and Byerly (1975) with small rockets and balloons. Supporting surface measurements underneath the clouds included field mills and radar. Combined in-cloud and surface measurements were used (Winn and Byerley, 1975) to obtain an approximate value for the total quantity of charge in a region of cloud just before it was partially discharged by lightning. In two cases, charges were estimated to be −120 and −160 C, at a height of between 5·2 to 7·0 km MSL. The amounts of charge neutralized by the flashes were between −37 and −50 C for the first case and between −47 and −66 C for the second. These measurements constitute a new method with much promise, despite some weakening assumptions.

As must be evident from the above descriptions of investigations of different workers during different periods and in different geographical locations, the reported average electric charge and moment change in the cloud discharge vary from 10 to 100 C, and from 20 to 400 C km, respectively. The results obtained by Hatakeyama (1949, 1958) using the ballistic galvanometer, which is an integrating instrument, yielded the largest value of charges neutralized. These results are reminiscent of the magnetographic measurements made by Hatakeyama (1936), Meese and Evans (1962) and Nelson (1968) on the charge transfer in cloud-to-ground lightning flashes, also using integrating ballistic-type instruments (see Chapter 9.5.1.1). These latter measurements again gave much larger charge values than obtained by other workers using other methods.

As a final comment on the large spread of values obtained by many workers studying intra-cloud discharges, we must remark that most authors provide the reader with little insight into the accuracy of their measurements, and do not indicate whether or not the results can be or have been used to test the validity of the assumed model, be it a monopole, dipole, bipole or more complex distribution of charge. Obviously, *all* reasonable charge distributions will look like monopoles or dipoles if the observer is sufficiently far away from the storm. A plot of the data of field-change against distance will always approach the inverse cube dependence as expressed in Equation (9). The inverse cube fit to the data is therefore *not* a sensitive test for distances much beyond the reversal distance, a test which has been used by many authors to justify the use of the dipole model. Two remarks are appropriate:

(a) single-station measurements over many storms at near and far distances do not appear to be adequate for testing the validity of models;

(b) multiple, simultaneous measurements may provide an appropriate set of constraints within which the adequacy of models can be judged.

Naturally, the more constraints, the higher will be the probability that a fitted model is the appropriate one. One such example is the analysis given by Jacobson and Krider (1976) for multiple-station measurements of ground-stroke field-changes from a network of 25 field mills. It is hoped that future studies will be designed to provide a much needed assessment of the consistency of measurements with the validity of assumed models.

4. Initial Cloud Discharge Streamer

4.1 Classification of Electric Field-change Patterns

Electric field-changes due to all types of lightning discharge (cloud and ground) are often classified in earlier work as simple or complex changes, slow or rapid changes, and either positive going or negative going. These classifications are confusing and, useful as they may have been, should be abandoned in favour of a grouping based separately upon the known characteristics of ground discharges and cloud discharges. Indeed, much of the need for the earlier classifications was overcome through the use of high time resolution recording of electric field-changes as measured, for example, with a slow or fast antenna, and correlated with direct visual or photographic observations (see Chapter 5.4.3).

A discussion of the fine structure of field-changes has been given by Kitagawa and Brook (1960). Electric field-changes from approximately 5,000 lightning discharges of all types were observed with high time resolution on both electric field and electric field-change meters, and correlated with simultaneous visual observation of stroke type and photoelectric recordings of luminosity fluctuations. On the basis of these correlations criteria were developed which allow for positive identification of cloud and ground discharges.

The cloud discharge field-change may be divided into three parts, an initial part, a very active part, and a final or *J*-type part. An examination of the final or *J*-type part discloses a relatively quiescent period, throughout which short duration step changes occur approximately 10 ms apart. These step changes (*K* changes) constitute the major contribution to the positive going or negative going final part of the cloud discharge.

Ogawa and Brook (1964) gave data on field-change patterns of cloud discharges observed throughout the storm of Fig. 2. Figure 3 illustrates the four basic types of intra-cloud field-change variations. For clarity, only a representative fraction of the discharges are plotted on the right, and their occurrences in numbers and in time are plotted on the left. In Fig. 3, the general trend in the slope and value of each field-change curve labelled type I is positive; in type II the slope is first positive and then reverses sign to negative, while the value of the field remains positive. The curves labelled type III differ from type II only in that the final value of the field becomes negative. In the type IV curves, both the slope and the value of the field-change are negative. A distinction between adjacent types on the basis of the slow field-change record is sometimes difficult to make. A better method for distinguishing between types of field-change, especially between types I and II, requires a more detailed examination of high-speed oscillographic records (fast antenna). If the polarity of the *K* changes is used as a basis for classification, it is relatively easy to distinguish between types I and II field-changes.

It is not simple to correlate these four types with specific distances, since there is a considerable overlap. Nevertheless, a type I field-change is usually observed at distances within 6 km of the storm. Types II and III field-changes may be observed over the range of 4 to 10 km. Type IV field-changes are, in general, produced by cloud discharges occurring at distances greater than 8 km.

The type I field-changes were observed only during a short period in the early stages of the storm. The total duration of this type of discharge was surprisingly short, often as short as 100 ms and seldom exceeding 300 ms. The average duration of the other three types was about 500 ms.

The curves of cloud flash field-changes will be divided into two parts, the initial streamer process and the later *K* change process. They will be discussed separately.

4.2 The Initial Streamer

It was considered by Pierce (1955a, b) that the shape of the field-change produced by an intra-cloud discharge can be interpreted in terms of vertical

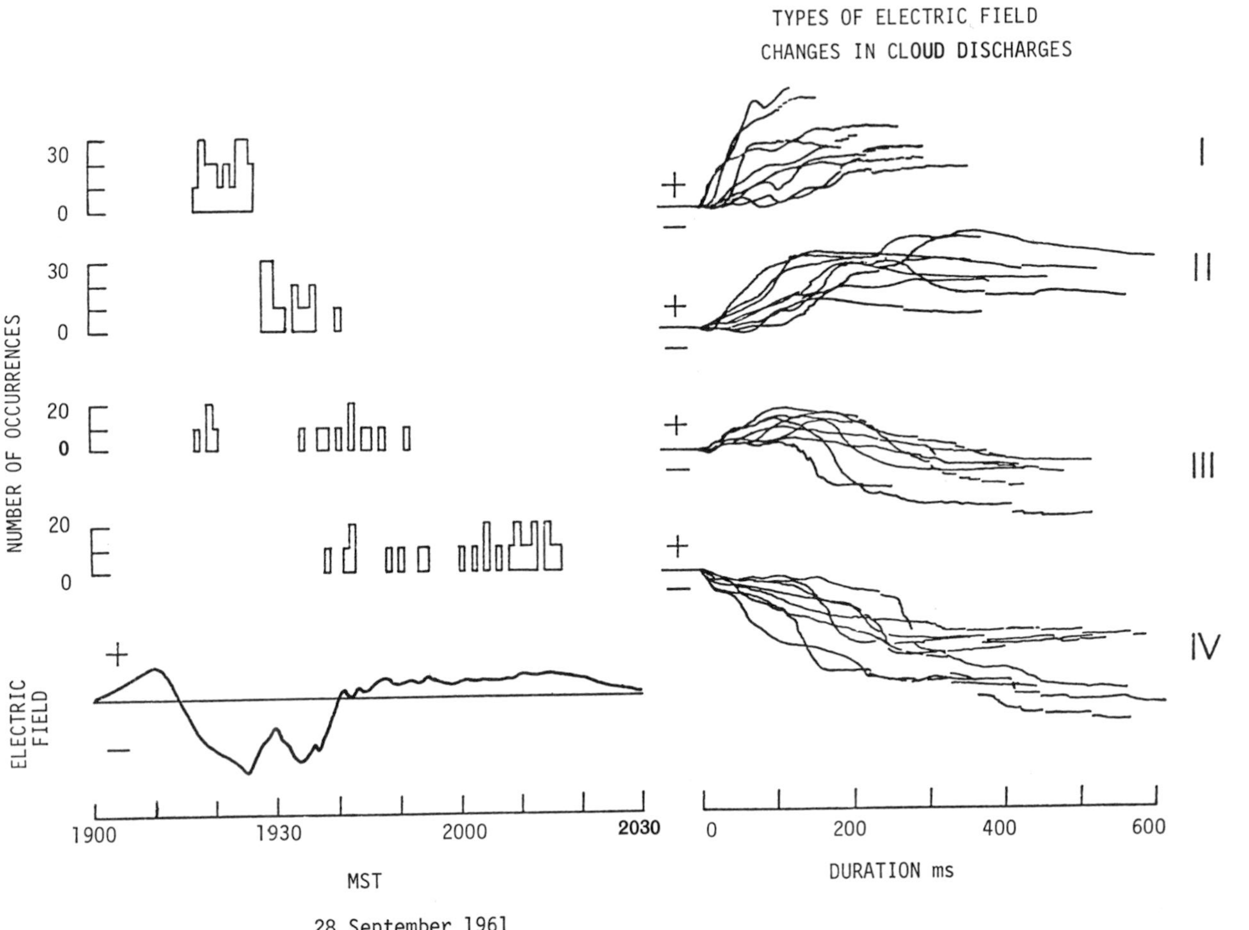

Fig. 3. Types of cloud discharge field-change and their occurrence throughout various states of the storm shown in Fig. 2. The electric field curve is also included for reference to show the time of occurrence of each type (from

motion of charge in the cloud. Interpreting the data obtained, Pierce (1955b) showed that the slow positive field-change could be produced by a downward moving leader from the negatively charged centre, and the complex field-change by the combination of two leaders from the negatively charged region, one proceeding upward and the other proceeding downward. A similar approach was considered by Takagi (1961). Smith (1957), using his Florida two-station measurements, discussed variations of field-change with distance for the cases of positive charge lowered, negative charge raised, negative charge lowered and positive charge raised. Ogawa and Brook (1964) considered a unidirectional streamer to be associated with both the initial and the very active parts of the discharge, the total duration of those parts being approximately one-half of the total discharge duration. For simplicity a vertical channel along which the streamer distributed was assumed and because of corona losses and channel capacitance, a charge of uniform density. The streamer field-changes were calculated on these assumptions. A similar set of calculations was later published by Khastgir and Saha (1972). The following more general treatment of a developing streamer is based upon Huzita and Ogawa (1976b).

Suppose that a streamer starts effectively from the positive space charges at a height H_1 and proceeds toward the negative charge centre at a height h_1 along a line making an angle θ with the vertical. The streamer produces an extended line charge which constitutes a charge distribution having cylindrical symmetry. Figure 4 is a schematic diagram showing models of positive and

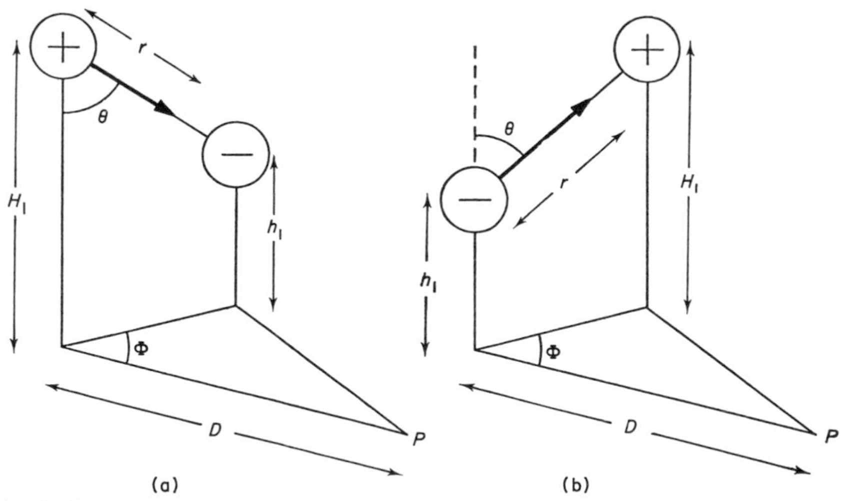

Fig. 4. Geometry for tilted models of cloud discharges. The positive and negative streamers proceed at an angle θ from the vertical line. D is the distance from the point directly underneath the streamer-source charge region to the observing point P.

negative developing line charges of length r and charge density ρ per unit length above a conducting flat earth. The electric field at an observing station P due to the streamer charge and its electrical image is given by

$$E_1 = \frac{2\rho}{4\pi\varepsilon_0}\int_0^r \frac{(H_1 - r\cdot\cos\theta)\,\mathrm{d}r}{[(H_1 - r\cdot\cos\theta)^2 + (r\cdot\sin\theta\sin\phi)^2 + (D - r\cdot\sin\theta\cdot\cos\phi)^2]^{\frac{3}{2}}}, \quad (11)$$

where ρ is assumed constant, and ϕ is the angle which the plane containing the streamer makes with the line connecting the projection of the source charge onto the surface and the observer. Integration of Equation 11 after some substitutions results in

$$E_1 = \frac{2\rho a}{4\pi\varepsilon_0 l}\left(\frac{\left(\frac{H_1}{a} - \frac{m}{l}\right)r + \frac{n}{l} - \frac{mH_1}{la}}{\frac{ln - m^2}{l^2}(lr^2 - 2mr + n)^{\frac{1}{2}}} - \frac{\frac{n}{l} - \frac{mH_1}{la}}{\frac{ln - m^2}{l^2}n^{\frac{1}{2}}}\right), \quad (12)$$

where $a = \cos\theta$, $b = \sin\theta\cdot\sin\phi$, $c = \sin\theta\cdot\cos\phi$, $l = a^2 + b^2 + c^2$, $m = aH_1 + cD$ and $n = H_1^2 + D^2$. For the special case of a vertically moving streamer ($\theta = 0$),

$$E_{10} = \frac{2\rho}{4\pi\varepsilon_0}\left(\frac{1}{[D^2 + (H_1 - r)^2]^{\frac{1}{2}}} - \frac{1}{(D^2 + H_1^2)^{\frac{1}{2}}}\right). \quad (13)$$

If the line charge is negative, then the sign of the electric field should be reversed.

As the positive streamer extends, the increasing charge on its length results in a decrease of charge in the source volume. The field-change at the ground due to this decrease in the source charge is given by

$$E_2 = -\frac{2\rho(L - r)\,H_1}{4\pi\varepsilon_0(D^2 + H_1^2)^{\frac{3}{2}}} \quad (14)$$

where L is the maximum extension of the streamer.

The total field resulting from the streamer extension and the source-charge decrease can be written as

$$E = E_1 + E_2. \quad (15)$$

We now consider the simple cases of a positively charged streamer moving vertically downward from a spherically symmetric positively charged volume, and of a negatively charged streamer moving vertically upward from a negatively charged volume. We consider only the most common case in which the positive charge is uppermost. In this case, the effective upper end of the streamer is at H_1 and the effective lower end at h_1. A constant streamer velocity, v, is assumed so that $r = vt$.

Figure 5 shows the calculated field variation with time, as a function of distance from observer to flash, for the vertical streamer case, in which

$H_1 = 7{\cdot}5$ km and $h_1 = 2{\cdot}5$ km. The initial and very active portion (the first half of the field-change curves in Fig. 3) should be compared with the curves of Fig. 5. It is apparent that the data presented in Fig. 3 favour the model of a descending positive streamer.

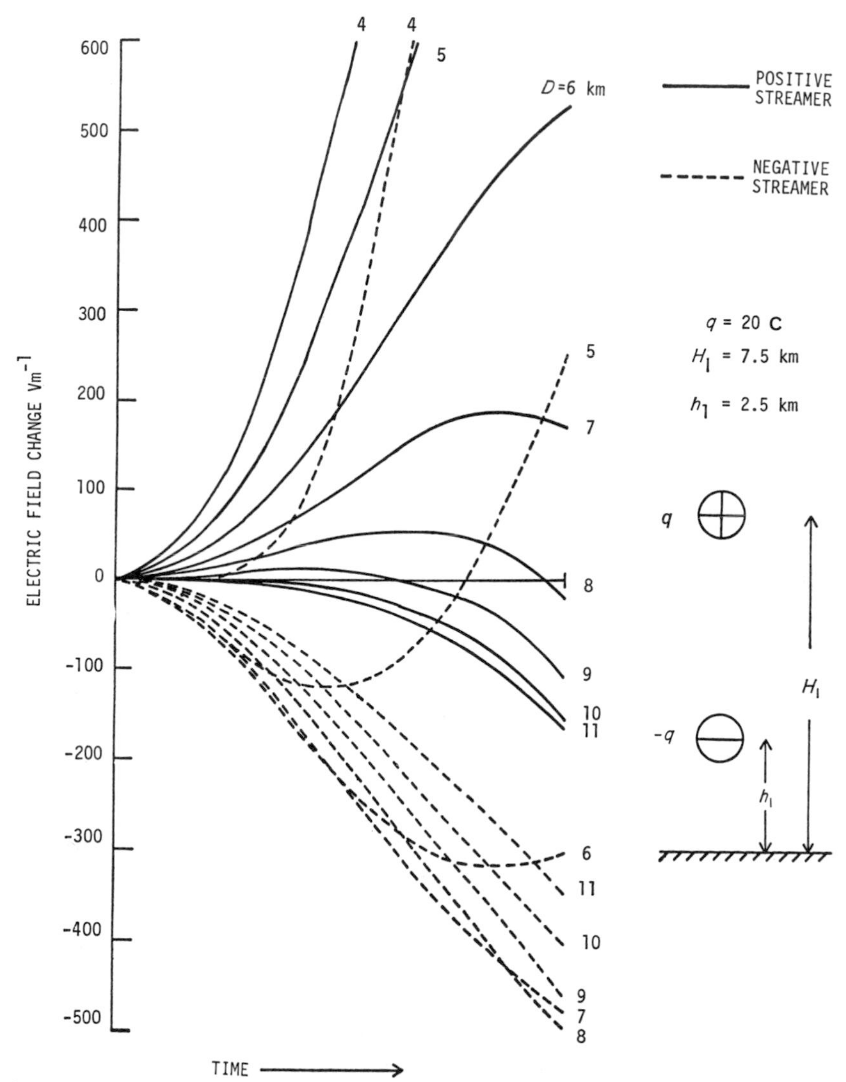

Fig. 5. Calculated variation with time in the electric field accompanying the vertical development of a descending positive streamer (solid curves) and an ascending negative streamer (dashed curves) as a function of distance from observer to flash (from Ogawa and Brook, 1964).

Kitagawa and Brook (1960) have shown that there is a very real difference in the electrical nature of the field-change between the stepped leader of the cloud-to-ground discharge and the breakdown streamer of the intra-cloud discharge. We believe that the observed distinction is a consequence of the difference in the nature of a negative and a positive streamer; the positive streamer produces a field which accelerates electrons into the tip, while the negative streamer disperses the electrons ahead of it (see Chapter 5.5.4).

The average neutralized charge in the cloud discharges studied by Ogawa and Brook (1964) was ≈ 30 C. Using the average time duration of 250 ms, the average current for the initial streamer can be calculated to be 120 A. If the streamer length is taken to be the total discharge length, i.e. ≈ 2 km, then the streamer speed can be calculated to be 8×10^3 m s^{-1}, a value similar to that reported by Takagi (1961), who concluded from a statistical treatment of the data that the main process in a cloud discharge involves a branched positive streamer which proceeds downward toward the negatively charged region with a velocity of about 10^4 m s^{-1}.

Ishikawa (1961) photographed three types of different-velocity streamer with a high-speed lightning-flash camera. The velocity values are $(3 \text{ to } 4) \times 10^6$ m s^{-1}, $(1 \text{ to } 2) \times 10^5$ m s^{-1} and about 5×10^4 m s^{-1}. Ishikawa called these the fast streamer, the medium velocity streamer and the slow streamer, respectively. The slow streamer corresponds to the present initial streamer; Ishikawa's velocity measurements are about six times larger than the present rough estimate. The fast streamer of Ishikawa (1961) corresponds to the K change which will be described in Section 5.

Ishikawa (1961) analysed statistically a large number of electrostatic pulses, observed during 1956 to 1959 with an intermediate antenna of time constant 3 ms, and found that the most often observed type of discharge in a thundercloud was the combination of a "positive complex streamer" and "negative local streamers" in a cloud of positive polarity. This description may be interpreted, in terms of the present terminology, as that a positive initial streamer followed by negative K changes in a cloud of positive polarity is the most often observed cloud discharge sequence.

Mackerras (1968) estimated approximate current measurements using "ramps of greatest slope" in the electric field. Differentiating Equation (9) with respect to time we get

$$2I(H_1 - h_1) = -4\pi\varepsilon_0 \Delta \dot{E}_1 D^3 \text{ A m} \tag{16}$$

where $I = \dot{q}$ and $\dot{E}_1 \equiv (\mathrm{d}E_1/\mathrm{d}t)$. A median value of 10^3 A km was found for the right side of Equation (16). Assuming $2(H_1 - h_1) = 10$ km, Mackerras (1968) estimates the median maximum current as 100 A, and suggests that similar currents occur in non-impulsive discharge components in all types of flash, such as in the long continuing currents of ground flashes.

The cloud-discharge field-change due to a tilted streamer of both positive and negative sign will be discussed in Section 4.4.

4.3 Multistation Measurements

Smith (1957) made simultaneous measurements of slow electric field-changes at two stations 13·2 km apart near Orlando in central Florida in 1955. He discussed the shapes of the field-changes, interpreting them in terms of moving charges in cloud. During the investigation, a total of 693 slow field-changes were recorded by the two stations. Of these, 69% were simple field-changes, the remaining 31% were complex, having at least one maximum or minimum during the field-change. Examples of slow field-changes photographed from the original records obtained by Smith (1957) are shown in Fig. 6. A comparison of the two-station records (1 and 2) shows that the two field-changes measured at stations 13·2 km apart are seldom similar either in type or in the sign of the net field change, or both. The degree of "decorrelation" between records of the same event taken 13 km apart should be staggering to the uninitiated!

Smith (1957) analysed field-change data from 54 intra-cloud discharges. From these data and a knowledge of which station was closer to a given discharge, the sign of the equivalent charge and its direction of motion as the discharge developed were determined. Thirty-nine out of 54 field-changes were of the positive dipole type, and the remaining 15 were of the negative dipole type. Negative charge was raised in 30 out of 39 positive dipole discharges, and positive charge was lowered in 9 cases. In 15 negative dipole cases, upward moving positive charge occurred 9 times and downward moving negative charge 6 times. Smith (1957) concluded from these data that the majority (83%) of slow field-changes can be explained satisfactorily by the uniform movement of a single charge within the cloud and, in the most frequent case, the cloud discharge involved the raising of negative charge. This conclusion is at odds with the results of Takagi (1961) and Ogawa and Brook (1964).

4.4 Tilted Streamer

To date it has not been possible to reconcile the above results except by assuming that in some storms the effective movement of charge in a cloud flash is negative charge raised while in others it is positive charge lowered,

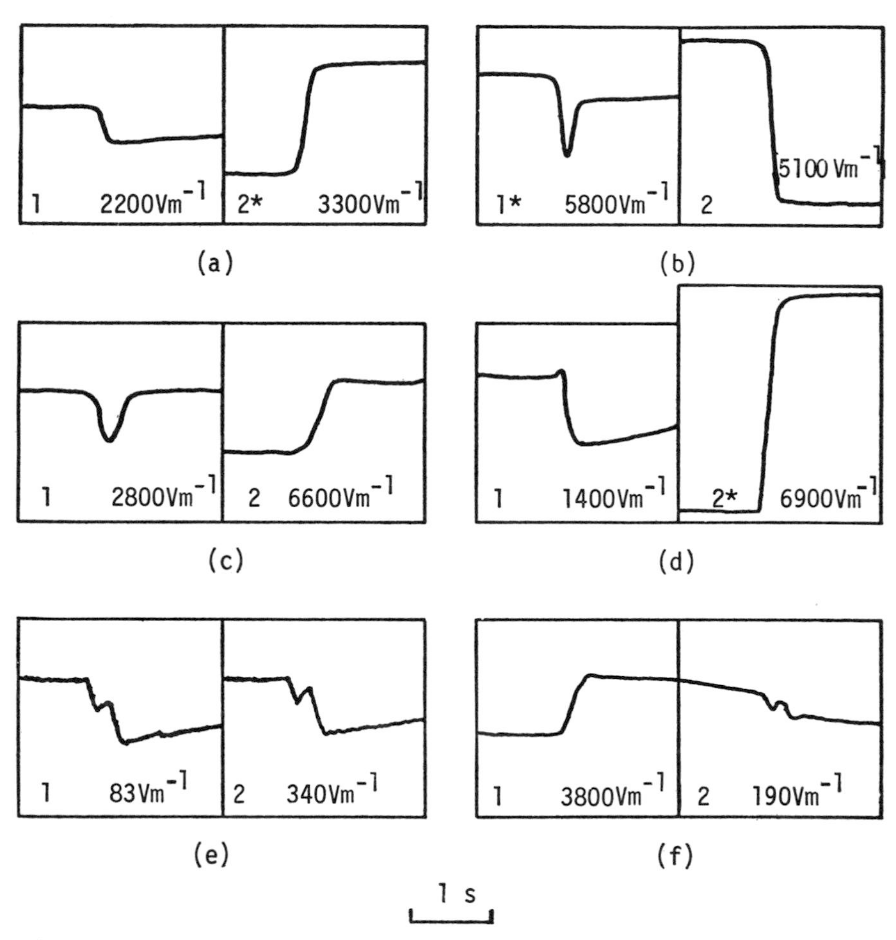

Fig. 6. Slow electric field-changes recorded simultaneously at two stations, 13·2 km apart in central Florida during July–August, 1955. The amplitude of each field-change is given. The station known to be closer to the storm is indicated by an asterisk (from Smith, 1957).

assuming a vertical motion. We now consider the effect of a tilted streamer and, in the extreme, that of a horizontally moving streamer.

It is of great importance in analysing the cloud discharge field-change data to know whether the dipole is vertical or inclined. As has been described earlier, Workman *et al.* (1942) reported that the cloud discharge channel is more horizontal than vertical, while Reynolds and Neill (1955) and Takagi (1961) found greater vertical than horizontal components. Whether a cloud discharge is vertical or not may be determined by careful examination of field-change patterns observed simultaneously at a large number of stations. For a more refined examination of observed field-change data, slow electric field-changes due to cloud discharges of the tilted dipole type may be calculated from Equation (15).

The results of such a calculation, taken from Huzita and Ogawa (1976b), are shown for positive streamers in Fig. 7 and for the negative streamers in Fig. 8. The positive streamer starts from $H_1 = 7{\cdot}5$ km and the negative

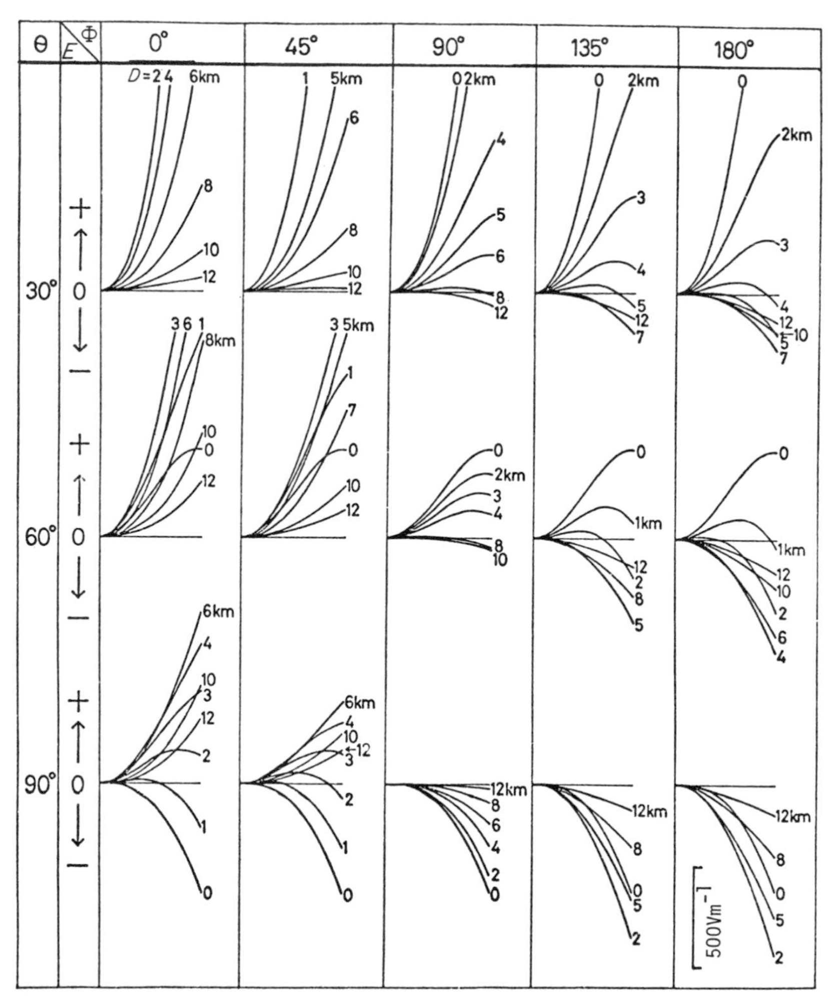

Fig. 7. Calculated curves of electric field-change for the tilted positive streamer during its extension for the geometry shown in Fig. 4 (from Huzita and Ogawa, 1976b).

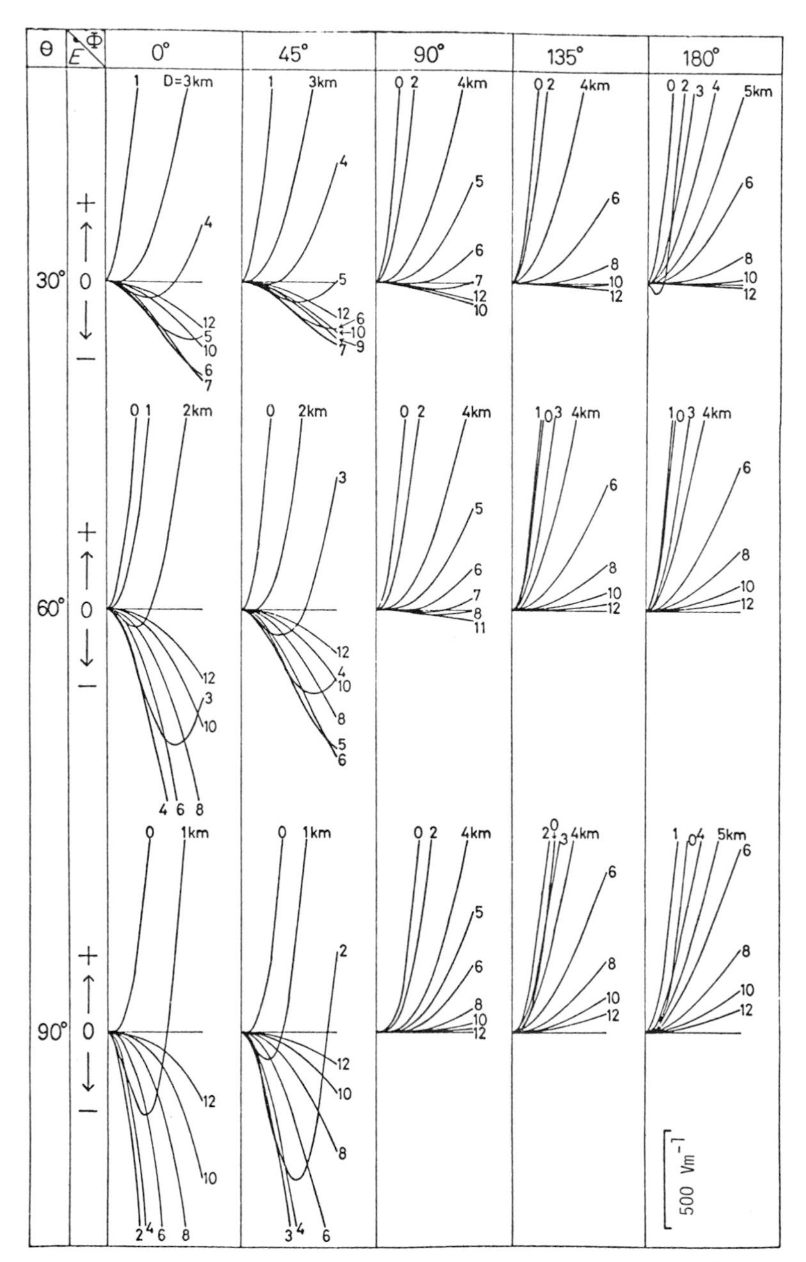

Fig. 8. Calculated curves of electric field-change for the tilted negative streamer during its extension for the geometry shown in Fig. 4 (from Huzita and Ogawa, 1976b).

streamer from $h_1 = 3{\cdot}5$ km. The total streamer length is $L = 5$ km and the line charge density is $q = 4$ C km^{-1}. Inclination of the streamer from the vertical, θ, is taken for three angles, 30°, 60° and 90°. The angle, ϕ (see Fig. 4), is taken for five values: 0°, 45°, 90°, 135° and 180°. There are in total 15 cases for the combinations of θ and ϕ values. In each case, electric field-change curves are drawn as a function of the distance, D, to the observer.

In the case of the positive streamers, for constant θ, the electric field-change becomes more and more negative as ϕ varies from 0° to 180°, i.e. as the streamer approaches, or moves away from, the observer. Similar results can be seen, for constant ϕ, as θ varies from 30° to 90°, i.e. from vertical to horizontal.

In the case of the negative streamer, for constant θ, the electric field-change becomes more and more positive, as ϕ varies from 0° to 180°. It can be seen in Fig. 8 that there is a distinct difference between the cases of $\phi = 0°$ and 45°, and $\phi = 90°$, 135° and 180°; in the former cases remarkably large negative field-changes can be seen, while in the latter cases no such negative changes appear. This result is of course a consequence of the low height of the negative streamer source charge.

Aina (1972) published calculations for a horizontal extension of the discharge within a cloud, and also for the combination of a vertical channel within the cloud and a horizontal channel underneath the cloud base to explain the results of observations in Ibadan.

An interesting variation, in which the source charge remains constant while the streamer develops has been calculated by Khastgir and Saha (1972). This model would seem to allow for both positive and negative streamers. Some evidence for source charge replenishment by streamers from adjacent volumes is to be found in the work of Proctor (1974b).

With regard to the differences between the work of Smith (1957) and Takagi (1961) and Ogawa and Brook (1964), noted in Section 4.3, the tilted streamer model does indicate possible ambiguities in interpretation when the number of observing stations is small and assumptions regarding channel inclination have to be made, but it is not convincing. Little more need be said on this point except to admit that streamer propagation, positive and negative, needs further study with relation to discharge initiation.

5. *K* Changes

5.1 Observation

Kitagawa and Kobayashi (1958) made measurements of electric field-changes using both an electrostatic fluxmeter and an oscilloscope during the thunderstorm seasons of 1955 and 1956. They found small and rapid changes occurring in the junction or *J* process of the ground discharge. These small changes were called *K* changes* and are characterized by a duration of less than 1 ms, an amplitude of a fraction of that of the return stroke (Chapter 5.2.1), and a recurrence period of about 10 ms. Further, Kitagawa and Kobayashi (1958) made simultaneous measurements, with a photomultiplier, of luminosity accompanying the field-changes between adjacent return strokes, and found that only the *K* changes radiated light during the *J* process. *K* changes are considered to result when an advancing streamer reaches a densely charged region, of opposite sign, of the order of several 100 m in diameter, corresponding to the subcell structure of the thunderstorm cloud.

For cloud flashes, Kitagawa and Kobayashi (1958) also found the same kind of small, rapid field-changes with accompanying pulses of luminosity. The character of those field-changes was very similar to the *K* changes in the ground discharges, and they were also so designated. Accordingly, the discharge process of the cloud discharge was assumed to be similar to that of the *J* process in the ground discharge.

Kitagawa and Brook (1960) compared *K* changes in cloud and ground flashes in detail and again showed that the *K* field-changes in the later part of the cloud discharge are similar in appearance to the *K* changes which appear during the *J* process in the ground discharge. Usually, no field-change is noticeable in the intervals between the *K* changes; the net field-change during the final part of the discharge is effectively the sum of the individual *K* changes. The slope of the final or *J*-type part is also determined by the polarity of the *K* changes. It is significant that *K* changes do not occur in the initial and very active parts; i.e. they are not a part of the initial breakdown streamer process.

Ogawa and Brook (1964) observed *K* changes in the final stage and in the *J*-type part of the cloud discharge throughout the storm described earlier. The magnitude and the polarity of the largest of the observed *K* changes of each cloud flash are plotted in Fig. 9 as a function of time. A single cloud discharge may produce as many as 20 detectable *K* changes, the most frequent number being 6 per flash. These smaller *K* change values will fall on a line under the largest one, hence, for clarity in Fig. 9, only the largest *K* change in each flash is plotted.

* We always suspected that the designation *K*, which Kitagawa explained as short for the German "Kleine Veränderungen", was really short for the authors Kitagawa and Kobayashi, just as Khastgir and Saha (1972) refer to "the Malan or *M* component". In any event, the above reasons are as good if not better than those which can be mustered to justify the circus of alphabetic jargon presently used to characterize lightning events.

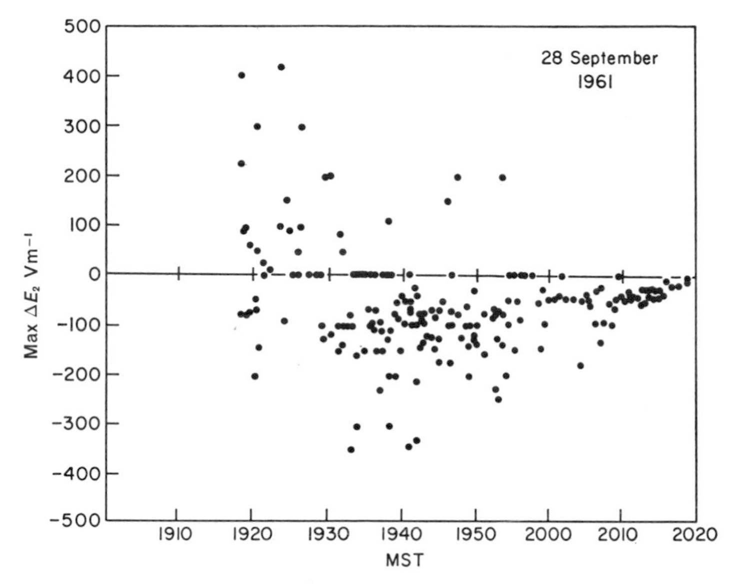

Fig. 9. The maximum *K* changes, ΔE_2, plotted versus time as they occurred in the individual flashes in the storm shown in Fig. 2. The weaker *K* changes (on the average there are about six per flash) have been omitted for clarity (from Ogawa and Brook, 1964).

5.2 Characteristics of the *K* Changes

At intermediate distances, a *K* change often shows an initial negative component, although the net electric field change remains negative. If we compare the shape of the *K* change with the calculated curves for the breakdown streamer discussed in Section 4.2, we must interpret the *K* change as a recoil or return stroke of opposite polarity. If, for example, we assume that the breakdown streamer is a positive streamer with a descending vertical component, the *K* change must then be a negative return stroke which travels backward from the tip of the channel already formed by the preceding streamer process. This interpretation was confirmed by the photographic analyses described in the next section.

In Fig. 9 the maximum *K* change magnitudes are plotted as they occurred throughout the storm of Fig. 2. The sign of the field-change produced by the *K* change is seen to have changed from positive to negative somewhat earlier in time (and hence also distance) than did both the electric field *E* and the field-change ΔE_1. This reversal occurred at approximately 19.32 MST, when the storm was about 6·5 km from the station.

Analysing ΔE_2 and ΔE_1 in a manner similar to the analysis of E and ΔE_1 again suggests a similar constraint upon the effective end-points of the K streamer. We conclude that the K streamer is probably shorter than the total discharge streamer. A calculation, using Equation (6), shows that the upper end of the K change discharge is at $H_2 = 5{\cdot}3$ km assuming a starting point for the K change recoil streamer at $h_2 = 4{\cdot}0$ km, i.e. at the end-point of the initial streamer. We can calculate the charge involved in the K change by using Equation (4) and the measured field-change value of $\Delta E_2 = 10$ V m^{-1} at a distance of 8 km. The result is a charge $q_2 = 1{\cdot}4$ C, and the moment change $\Delta M_2 = 3{\cdot}6$ C km. With this charge magnitude, noting that the duration of a K change is of the order of 1 ms, it is possible to estimate, very roughly, the current involved in a K change. Thus we arrive at an average current, associated with the K change, of about 1,400 A.

The K change channel length can be estimated as $H_2 - h_2 = 1{\cdot}3$ km in length, which corresponds to about 65% of the total cloud discharge channel length. Using the K change duration of the order of 1 ms, we arrive at a velocity of approximately $1{\cdot}3 \times 10^6$ m s^{-1}, which is near the most frequently measured value of the dart leader velocity in strokes to ground (see Chapter 5.5.4). This estimate is in fair agreement with the values given by Ishikawa (1961) and Takagi (1961) who estimated the velocities of $(3 \text{ to } 4) \times 10^6$ m s^{-1} and of the order of 10^6 m s^{-1} respectively.

Ishikawa (1961) estimated mean values of the charge dissipated in cloud flashes and gave a value of 32 C on the basis of 16 events, 0·47 C for a local discharge (K change) and 7·5 C for the summed value of 16 local discharges occurring successively in a discharge.

6. Photographic Analyses

Sourdillon (1952) studied a series of air-discharges with a Boys' camera and pointed out, for the first time, the repeating nature of some cloud discharge channels. In a similar type of study, Ogawa and Brook (1964) made a detailed analysis of a cloud discharge which exhibited several long streamer paths outside the cloud. The streamers were photographed with a slowly moving camera (image speed on film of 5 cm s^{-1}). We present the following concise summary of their results.

(a) The developing streamers exhibit continuing luminosity with some superimposed fluctuations. They are definitely not of the stepped leader type, but are more optically comparable to the continuing current luminosity in ground strokes with superimposed fluctuations due to M components.

(b) As the streamer progresses, branches are formed, each branch in turn propagates as a continuing current (luminosity) streamer.

(c) After about 0·25 s, the luminosity disappears for from 0·01 to 0·1 s followed by a series of short duration, bright events separated in time by from 3 to 30 ms. These bright events are associated with the *K* changes.

(d) The bright *K* change events can be seen to propagate back along the previously formed channel, the luminosity decreasing as it does so. This change in luminosity is especially noticeable as the event reaches a branch point much in the same manner as the cloud-to-ground return stroke travels up the trunk from ground and out the branches (Schonland, 1956). The *K* changes are considered to be recoil streamers, the cloud flash analogue of the return stroke.

(e) Existence of the recoil streamer implies that the initial continuing current streamer of, for example, positive polarity has reached a region of concentrated opposite (negative) charge; the *K* change then follows as a neutralizing event signalling the final stage of the cloud discharge.

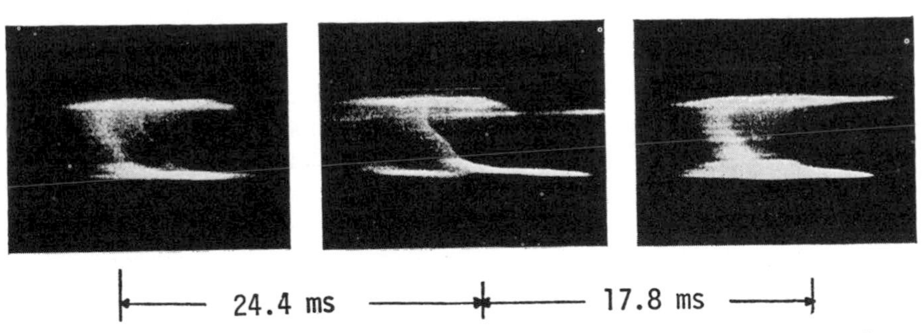

Fig. 10. Moving-camera photograph of a repetitive cloud flash exiting and re-entering the cloud base. Time between the events is shown. Duration of the "strokes" is ≈ 1 ms.

Although there have been several other attempts at analysing cloud-flash luminosity, the above analysis confirmed, for the first time, that the initial and very active portions of the cloud flash correspond to the propagation of a more or less continuing current streamer, while the final or *J*-type portion corresponds to the charge-neutralizing events. Confirmed in the photographs is the lack of significant activity (luminosity) between the *K* changes, strengthening the conclusions of Kitagawa and Brook (1960) that the final portion of the cloud discharge is essentially a field change equal to the sum of the individual *K* changes.

As an example of a series of three *K* change events reminiscent of a return stroke, we refer to the photograph shown in Fig. 10. Here we show a "repetitive cloud stroke" taken with a high-speed moving film camera in Socorro. In each of the events, the two horizontal luminous streaks are due to cloud illumination in the region where the streamer exited and then re-entered the cloud. The duration of each event is about 1 ms (the film has been cut and shortened to save space). The times between these *K* change events are shown to be 24·4 and 17·8 ms. No significant luminosity was registered in the interval between the events. A number of "repeating elements" are also shown by Sourdillon (1952).

7. Radiation from Cloud Flashes

There is a considerable body of literature relating to lightning-generated radio noise (see Chapter 10), covering a large number of aspects which do not directly bear on the subject presently under discussion. We have therefore chosen to discuss only a few works dealing with radiation fields from cloud flashes; they were chosen because they help clarify the relationship between radiation pulses and electrostatic field-changes.

Lightning channels are usually several kilometres in length, and, during impulsive current events, may be regarded as good antennae for radiating electromagnetic energy. The study of the radiation fields of lightning discharges over a wide range of frequencies constitutes a valuable complementary source of information on, among other things, the discharge process, the attenuation of radio waves in a neutral and ionized atmosphere, and the investigation of plasma parameters in space.

Malan (1958) studied the relationship between electrostatic field-change events and radiation at frequencies ranging from 3 kHz to 12 MHz. His measurements showed that the radiation from cloud flashes closely resembled those of the *J* and *F* processes of ground discharges. Here again was good evidence that the *K* change process is common to cloud and ground discharges.

Malan found that, from 3 to 10 kHz, there were usually only a few small radiation pulses, not necessarily associated with the largest *K* field-changes. This behaviour is not surprising if we consider that the orientation of the channel with respect to the observer will determine, in part, the strength of the measured radiation. At higher frequencies, up to 2 MHz, more and more radiation pulses appear, but those associated with the *K* changes remain the largest. From 4 to 12 MHz the radiation becomes essentially continuous and the *K* pulses can no longer be distinguished.

Malan (1958) also made radiation amplitude measurements of discharges to ground from the same storm. These records were used to form ratios of return stroke to cloud stroke amplitudes as a function of frequency. Table II lists his results. Note the steady decrease in relative return-stroke amplitude as the frequency is increased.

Table II
Comparison between radiation from ground and cloud discharges

Frequency	Amplitude ratio (return stroke/cloud stroke)
3 kHz	20/1 to 40/1
6 kHz	10/1 to 20/1
10 kHz	10/1
20 kHz	5/1
30 kHz	2/1 to 3/1
50 to 100 kHz	1/1 to 1·5/1
1·5 to 12 MHz	1/1

Above 50 kHz one can no longer neglect the contribution of cloud flashes to lightning-generated radio noise in the atmosphere. Tepley (1961) further confirmed the importance of some cloud-discharge events, previously neglected, by noting that

> "in the generation of negative 'slow tails' in sferics only the intra-cloud stroke (specifically, the element of the discharge referred to as the '*K* change') is likely to be statistically important".

Brook and Kitagawa (1964) also observed the radiation field as well as the electrostatic field at 420 and 850 MHz from lightning occurring 10 to 30 km from the observatory. In this microwave frequency range the cloud discharge is as strong or even stronger a source than the ground discharge. The initial portion of the cloud discharge produced continuous radiation as well as strong pulse radiation. In the very active portion there appear large and frequent radiation pulses. In the *J*-type portion *K* changes emit radiation pulses of the same intensity as the *K* change in the *J* process of the ground discharge.

By far, the most detailed and significant work relating to the location of streamers in cloud has recently been reported by Proctor (1974a). He investigated VHF radio pulses from cloud flashes by means of a five-station hyperbolic receiving system (Proctor, 1971). This system operates at 253 MHz and is capable of resolving 25, 25 and 150 m in rms errors in the three Cartesian coordinates to locate each radio pulse source. Thus VHF radio pictures of lightning were obtained by locating the sources of a large number of pulses from the same discharge. The fixes obtained in this way were used to find the approximate paths of lightning streamers. Data concerning positions, extents, directions of streamers, source size and step length were obtained. Along with the radiation, Proctor measured electric field, field-change and radar precipitation-echo location.

Radiation from cloud discharges was classified by Proctor into two distinct types by the rates at which pulses were emitted during their initial and very active phases, i.e. low pulse rates of $\approx 2\times10^{3}\ s^{-1}$ and high pulse rates of 3×10^{4} to over $5\times10^{5}\ s^{-1}$. The low pulse-rate type of cloud flash was suitable for study with this hyperbolic system because pulses corresponding to the same event could be identified easily on the recorded responses of the spaced receivers. The source of most of the pulses could be found readily, although the analysis was most laborious. Of the pulses 25% were suitable for determining their source sizes. Sources were located near the tip of the advancing streamer and 182 source extents of 9 flashes were found to be 347 m on the average with standard deviation of 291 m and rms value of 453 m. Sources were found to form (or propagate) at high apparent speed. The mean of the speeds obtained by dividing each source length by the duration of the pulse it radiated was $2{\cdot}89\times10^{8}\ m\ s^{-1}$. A significant number formed at speeds exceeding light velocity showing that the measured velocity was a phase velocity. The average duration of the pulses was $1{\cdot}0\ \mu s$.

An example of a picture of one of these flashes is shown in vertical cross-section in Fig. 11. This flash had a single main channel without large branches. It developed in an upward direction. Points in the main channel are labelled A to M. Other streamers developed below A without producing large field-changes. The symbols relate to time, and change irregularly at intervals that are approximately 10 ms.

Proctor (1974b) estimates, in one case, the magnitudes of charge, charge line-density and current in the initial and very active portion by using field-change recordings and the known path. The magnitude of the charge for the flash shown in Fig. 11 was about 10 C along the path length of about 10 km so that the line density was estimated as $1\ C\ km^{-1}$. The base current ranged from -100 to $+400$ A. The negative sign of the current indicates a reversed current direction.

In his detailed analysis, Proctor (1974b) found that in the final stages, cloud flashes radiate long pulse trains,* up to several hundred microseconds in length, at intervals from five to several hundred milliseconds. They almost invariably accompany K changes which are delayed from 20 to 70 μs after the start of the noise. He found that they were emitted during two kinds of events: (a) a recoil streamer returning from the channel tip, and (b) a

* Adding to the already-thick alphabet soup, he apologetically calls these continuous trains "Q noise".

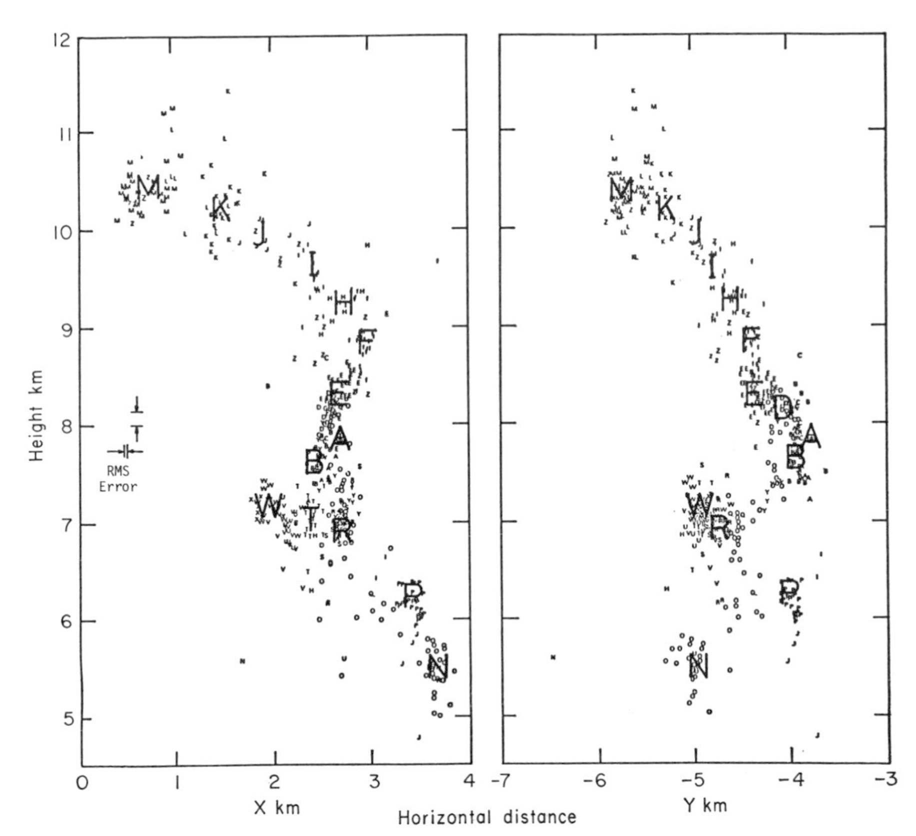

Fig. 11. Elevation plots of VHF radio pictures of lightning which were obtained by locating the sources of pulses with a hyperbolic direction-finding system using five spaced receivers. These are projections onto the *XZ* and *YZ* planes. Symbols refer to the time of pulse emission alphabetically. The first source to emit is marked *A*. Heights are 1,430 m above MSL. This plot includes all but three of the pulses emitted by the main channel. The error bar shown on the *XZ* elevation applied to the *YZ* elevation also (from Proctor, 1964a).

shorter streamer moving towards the starting point of the flash from the adjoining space but *not* by way of the existing channel. In his analysis Proctor found that since the sign of the *K* change depends on the direction and position of the streamer in relation to the field meter, the indication is that the *K* streamer is a positive one.

These results are especially interesting because (a) they are the most precise measurements to date, and (b) they are consistent with a model in which negative charge is raised in the cloud. To de-emphasize this last point, however, we note that in about two out of a total of nine cloud flashes, positive charge appeared to have been lowered in the cloud.

What does seem important now, in the light of Proctor's work, is to establish whether the initial continuing current streamer, inferred by Ogawa and Brook (1964) to be a positive streamer, can also be a continuing-current *negative* streamer.

Apparently related to Proctor's (1974a) observed high pulse repetition frequency noise are the observations by Krider *et al.* (1975) on bursts of uniform pulses produced by cloud discharges in Florida and Arizona. The half-width of the large-amplitude pulses was typically 0·75 μs, with time intervals between them of about 5 μs. Krider *et al.* suggested that

> "the source of these pulses is an intra-cloud dart-stepped leader process similar to that which has been photographed in discharges to ground. Unfortunately, because the technique of study involves a triggered oscilloscope recording system, the relationship of the radiation pulses to other cloud flash events was lost."

8. Summary and Conclusions

In Table III we have collected a summary of numerical values of cloud flash parameters.

There are not many published studies which bear directly on the mechanism of the cloud discharge, but, in perspective, a reasonable description of the major aspects of the discharge has emerged. The work of Pierce (1955a, b), Smith (1957), Kitagawa and Brook (1960), Takagi (1961) and Ogawa and Brook (1964) provided much insight into the details and fine structure of the discharge; while the studies of Workman and Holzer (1942), Reynolds and Neill (1955) and Tamura *et al.* (1958) provided information regarding charge distribution in cloud through the use of multiple-station measurements.

With all due regard for the excellent work of the investigators mentioned in this chapter, it appears to the authors that we are now for the first time in many years in a position to mount an effective attack on the nature and mechanism of the cloud discharge. The work of Proctor (1974a, b) has demonstrated a technique which, although not simple, needs to be implemented and used in combination with other methods recently demonstrated. The work of Krehbiel *et al.* (1974), following up on the original multiple-station technique of Workman and Holzer (1942), has shown that the effective origins of return stroke charges in clouds can be located with accuracies ranging from 50 to ≈ 300 m. The work of Nakano (1973) and of Teer and Few (1974) demonstrated the value of acoustic methods in locating lightning channels. And the fast-scanning "noise" radar of Brook and Krehbiel (1975), which covers an entire hemisphere of sky in 15 s, can now provide the precipitation-echo information on a time scale suitable for correlation with electrical events.

Table III
Numerical data for cloud discharge

	Total discharge	Initial streamer	*K* change	References
Charge	32			Workman and Holzer (1942)
(C)	21			Reynolds and Neill (1955)
	90			Hatakeyama (1958)
	32		0·47	Ishikawa (1961)
	15			Wang (1963b)
	10			Proctor (1974b)
	30		1·4	Ogawa and Brook (1964)
	33 ± 27			Average value
Moment	70			Workman and Holzer (1942)
(C km)	21			Reynolds and Neill (1955)
	81			Pierce (1955a)
	20			Tamura *et al.* (1958)
	200			Wang (1963b)
	100			Mackerras (1968)
	120		3·6	Ogawa and Brook (1964)
	87 ± 62			Average value
Height	5·8 to 4·7			Workman and Holzer (1942)
(km)	5·8 to 5·2			Workman *et al.* (1942)
	5·5 to 5·1			Reynolds and Neill (1955)
	10 to 6			Hatakeyama (1958)
	6			Tamura *et al.* (1958)
	(7–11) to (3–6)			Takagi (1961)
	8·2 to 4·8			Wang (1963b)
	5			Teer and Few (1974)
	6·0 to 4·0		5·3 to 4·0	Ogawa and Brook (1964)
Duration			<1	Kitagawa and Brook (1960)
(ms)	500	250		Ogawa and Brook (1964)
Speed		5×10^4	$(3 \text{ to } 4) \times 10^6$	Ishikawa (1961)
($m\ s^{-1}$)		10^4	10^6	Takagi (1961)
		8×10^3	$1{\cdot}3 \times 10^6$	Ogawa and Brook (1964)
Current		100 (max)		Mackerras (1968)
(A)		120	1,400	Ogawa and Brook (1964)

What appears to be needed, in addition to "average" values which seem to preoccupy most investigators, is to achieve a balance of the electrical budget of a storm throughout its life cycle. We must also be able to balance the electrical budget involving currents and charges for the anomalously large storms, not only for the average ones. The reader will justifiably wonder about the utility of a concept of "average charge height", for example, where the average height is given as 6 km but ranges from 1 to 12 km. Yet values of charge height, current, charge, etc. are presently given in this way.

It is obvious that discharges in clouds occur at all heights, and that currents and charges range over at least one to two orders of magnitudes. Accepting these facts, we ought to proceed to put the discharge *back* into the cloud environment to try to understand the conditions which lead to the various impulsive events, and to determine the parameters which make the currents small, or large, or negative, or positive, etc. A concerted effort, using complementary techniques simultaneously, is the obvious approach. It appears to these authors that in a very few years a far better understanding of electrical processes and their relationships to the cloud environment will be forthcoming.

References

Aina, I. (1972). Lightning discharge studies in a tropical area. III. The profile of the electrostatic field changes due to non-ground discharges. *J. Geomagn. Geoelect.* **24**, 369–380.

Brook, M. and Kitagawa, N. (1964). Radiation from lightning discharges in the frequency range 400 to 1,000 Mc/s. *J. geophys. Res.* **69**, 2431–2434.

Brook, M. and Krehbiel, P. (1975). A fast scanning meteorological radar. Proc. 16th Radar Met. Conf. Houston, Texas.

Gish, O. H. and Wait, G. R. (1950). Thunderstorms and the earth's general electrification. *J. geophys. Res.* **55**, 473–484.

Hatakeyama, H. (1936). An investigation of lightning discharge with the magnetograph. *Geophys. Mag.* **10**, 309–319.

Hatakeyama, H. (1949). The electric charge neutralized by the lightning discharge. *J. Geomagn. Geoelect.* **1**, 4–6.

Hatakeyama, H. (1958). The distribution of the sudden change of electric field on the earth's surface due to lightning discharge, *in* "Recent Advances in Atmospheric Electricity" (L. G. Smith, Ed.), pp. 289–298. Pergamon Press, London.

Hewitt, F. J. (1957). Radar echos from inter-stroke processes in lightning. *Proc. phys. Soc. Lond.* B **70**, 961–979.

Holmes, C. R., Brook, M., Krehbiel, P. and McCrory, R. A. (1971). On the power spectrum and mechanism of thunder. *J. geophys. Res.* **76**, 2106–2115.

Huzita, A. and Ogawa, T. (1976a). Charge distribution in the average thunderstorm cloud. *J. met. Soc. Japan* **54**, 285–288.
Huzita, A. and Ogawa, T. (1976b). Electric field changes due to tilted streamers in the cloud discharge. *J. met. Soc. Japan* **54**, 289–293.
Ishikawa, H. (1961). Nature of lightning discharges as origins of atmospherics. *Proc. Res. Inst. Atmos. Nagoya Univ.* **8** A, 1–274.
Jacobson, E. A. and Krider, E. P. (1976). Electrostatic field changes produced by Florida lightning. *J. atmos. Sci.* **33**, 103–119.
Khastgir, S. R. and Saha, S. K. (1972). On intra-cloud discharges and their accompanying electric field changes. *J. atmos. terr. Phys.* **34**, 115–126.
Kitagawa, N. and Brook, M. (1960). A comparison of intracloud and cloud to ground lightning discharges. *J. geophys. Res.* **65**, 1189–1201.
Kitagawa, N. and Kobayashi, M. (1958). Field changes and variations of luminosity due to lightning flashes, *in* "Recent Advances in Atmospheric Electricity" (L. G. Smith, Ed.), pp. 485–501. Pergamon Press, London.
Krehbiel, P., McCrory, R. A. and Brook, M. (1974). The determination of lightning charge location from multistation electrostatic field change measurements. Proc. Conf. on Cloud Phys., Tucson, Arizona, Oct. 21–24.
Krider, E. P., Radda, G. J. and Noggle, R. C. (1975). Regular radiation field pulses produced by intra-cloud lightning discharges. *J. geophys. Res.* **80**, 3801–3804.
Mackerras, D. (1968). A comparison of discharge processes in cloud and ground lightning flashes. *J. geophys. Res.* **73**, 1175–1183.
Malan, D. J. (1955). La distribution verticale de la charge négative orageuse. *Annls Geophys.* **11**, 420–426.
Malan, D. J. (1956). Visible electrical discharges inside thunderclouds. *Geofis. pura Appl.* **34**, 221–236.
Malan, D. J. (1958). Radiation from lightning discharges and its relation to the discharge process, *in* "Recent Advances in Atmospheric Electricity" (L. G. Smith, Ed.), pp. 557–563. Pergamon Press, London.
Malan, D. J. (1963). "Physics of Lightning," pp. 1–176, The English Universities Press, London.
Malan, D. J. and Schonland, B. F. J. (1953). Charge distribution and electrical processes deduced from lightning discharges, *in* "Thunderstorm Electricity" (H. R. Byers, Ed.), pp. 238–250. University of Chicago Press, Chicago.
Meese, A. D. and Evans, W. H. (1962). Charge transfer in the lightning stroke as determined by the magnetograph. *J. Franklin Inst.* **273**, 375–382.
Nakano, M. (1973). Lightning channel determined by thunder. *Proc. Res. Inst. Atmos. Nagoya Univ.* **20**, 1–7.
Nelson, L. N. (1968). Magnetographic measurements of charge transfer in the lightning flash. *J. geophys. Res.* **73**, 5967–5972.
Ogawa, T. and Brook, M. (1964). The mechanism of the intracloud lightning discharge. *J. geophys. Res.* **69**, 5141–5150.
Ogawa, T. and Brook, M. (1969). Charge distribution in thunderstorm clouds. *Q. Jl R. met. Soc.* **95**, 513–525.
Pierce, E. T. (1955a). Electrostatic field changes due to lightning discharges. *Q. Jl R. met. Soc.* **81**, 211–228.
Pierce, E. T. (1955b). The development of lightning discharges. *Q. Jl R. met. Soc.* **81**, 229–240.

Proctor, D. E. (1971). A hyperbolic system for obtaining VHF radio pictures of lightning. *J. geophys. Res.* **76**, 1478–1489.

Proctor, D. E. (1974a). Sources of cloud-flash sferics. CSIR Special Report No. TEL 118, Pretoria, South Africa.

Proctor, D. E. (1974b). VHF radio pictures of lightning. CSIR Special Report No. TEL 120, Pretoria, South Africa.

Reynolds, S. E. and Neill, H. W. (1955). The distribution and discharge of thunderstorm charge-centers. *J. Met.* **12**, 1–12.

Schonland, B. F. J. (1953). "Atmospheric Electricity", pp. 1–95. Methuen, London.

Schonland, B. F. J. (1956). The lightning discharge, *in* "Encyclopaedia of Physics" (S. Flügge, Ed.) Vol. 22, pp. 576–628. Springer-Verlag, Berlin.

Simpson, G. C. and Robinson, G. D. (1941). The distribution of electricity in thunderclouds, II. *Proc. R. Soc.* A **177**, 281–329.

Smith, L. G. (1957). Intracloud lightning discharges. *Q. Jl R. met. Soc.* **83**, 103–111.

Sourdillon, M. (1952). Étude à la chambre de Boys de "l'éclair dans l'air" et du "coup de foudre à cime horizontale". *Annls Geophys.* **8**, 349–364.

Takagi, M. (1961). The mechanism of discharges in a thundercloud. *Proc. Res. Inst. Atmos. Nagoya Univ.* **8** B, 1–106.

Takeuti, T. (1965). Studies on thunderstorm electricity. I. Cloud discharge. *J. Geomagn. Geoelect.* **17**, 59–68.

Tamura, Y. (1949). On the distribution of electricity in thunderclouds. *J. Geomagn. Geoelect.* **1**, 22–25.

Tamura, Y., Ogawa, T. and Okawati, A. (1958). The electrical structure of thunderstorms. *J. Geomagn. Geoelect.* **10**, 20–27.

Teer, T. L. and Few, A. A. (1974). Horizontal lightning. *J. geophys. Res.* **79**, 3436–3441.

Tepley, L. R. (1961). Sferics from intracloud lightning strokes. *J. geophys. Res.* **66**, 111–123.

Uman, M. A. (1969). "Lightning," pp. 1–264, McGraw-Hill Book Company, New York.

Wang, C. P. (1963a). Lightning discharges in the tropics. 1. Whole discharges. *J. geophys. Res.* **68**, 1943–1949.

Wang, C. P. (1963b). Lightning discharges in the tropics. 2. Component ground strokes and cloud dart streamer discharges. *J. geophys. Res.* **68**, 1951–1958.

Wilson, C. T. R. (1920). Investigations on lightning discharges and on the electric field of thunderstorms. *Phil. Trans. R. Soc.* A **221**, 73–115.

Winn, W. P. and Byerly, L. G. III (1975). Electric field growth in thunderclouds. *Q. Jl R. met. Soc.* **101**, 979–994.

Winn, W. P. and Moore, C. B. (1971). Electric field measurements in thunderclouds using instrumented rockets. *J. geophys. Res.* **76**, 5003–5017.

Workman, E. J. and Holzer, R. E. (1942). "A preliminary investigation of the electrical structure of thunderstorms." Tech. Notes Natn. Advis. Comm. Aeronaut. No. 850.

Workman, E. J., Holzer, R. E. and Pelsor, G. T. (1942). "The electrical structure of thunderstorms." Tech. Notes Natn. Advis. Comm. Aeronaut. No. 864.

Wormell, T. W. (1939). The effects of thunderstorms and lightning discharges on the earth's electric field. *Phil. Trans. R. Soc.* A **238**, 249–303.

7. The Long Spark

T. E. ALLIBONE

University of Leeds, England

1. Introduction

I reviewed the subject "Electrical Breakdown at High Voltages" in the Baird Memorial Lecture at Strathclyde University in 1967 (Allibone, 1967); even at that time the subject was an old one but one of absorbing interest containing many mysteries. Eight years have elapsed since then and the number of additional scientific contributions devoted to its study is extremely large; many mysteries have been unravelled and many of the physical processes involved are now understood; we can almost calculate the electrical breakdown potential of large gaps in air.

From an engineering point of view the subject has become of even greater importance due to the anticipated growth of extra-high-voltage transmission systems, indeed the work of the past decade on the long spark shows us that there may be an upper limit of transmission voltage set solely by the sparkover characteristics of the very large protective gaps which have to be used on the system.

This review is strictly practical and greatly restricted, dealing only with breakdown in air at atmospheric pressure and mainly between electrodes spaced so far apart that the electric field between them is far from uniform, and the metal of the electrode and the vapour from it can play no significant part in the breakdown process.

Interest in the subject stems from two sources; the physicist is attracted by the apparent similarity between lightning and the long spark, the engineer is concerned with the insulation of the transmission system to withstand the operating voltage and the electrical transients which flow due to lightning and to switching operations, so a study of the long spark comes close to the study of lightning itself.

The long spark was first studied with alternating potentials to provide data for transmission-line construction and the effects of atmospheric temperature, pressure, humidity and rain were well understood by the late 1920s, but it was not until unipolar potentials were produced by the impulse voltage generator that the effects of polarity were appreciated. Some of these are profound and are not yet fully measured nor understood. The earliest records of transients on transmission lines due to lightning showed (see Chapter 9.1.2.2) that the voltage on the line rose to a peak value in times of the order of a few microseconds and that the transient lasted for many microseconds and so it came about that in the 1930s so-called "standard" impulse voltage wave-shapes were adopted with which to test apparatus for use in high-voltage transmission systems.

The first successful photograph of a lightning flash revealing the details of its progress to earth was achieved in 1933–34 in South Africa by Schonland and Collens. As described in Chapter 5, the flash was seen to be composed of several separate strokes and each stroke was found to be composite in character, consisting of a leader stroke blazing an ionized trail from cloud to ground, succeeded immediately by a return stroke from ground to cloud, having great luminosity; this return stroke could therefore at once be regarded as the carrier of the large current of thousands of amperes which engineers had been measuring on transmission towers struck by lightning (see Chapter 9.3).

The dual nature of the stroke was at once shown by Allibone and Schonland (1934) to characterize the million-volt spark generated in the laboratory, the time between the start of the leader and the occurrence of the main stroke was identified with the "time-lag of sparkover" which had been the subject of much study. The authors decided to use the same terminology as had been used to describe the main features of the lightning stroke; thus began the physical study of the long spark in air, sparks of the order of one metre in length by Allibone (1938) and by Allibone and Meek (1937, 1938).

Since the war the planning of much higher transmission-line voltages has necessitated the building of impulse generators of up to 10 MV and the physical study of the long spark has been greatly extended. In addition, switching surges generated on transmission lines have assumed greater importance and the sparkover characteristics of air insulation between a variety of electrodes have had to be re-examined with a variety of voltage wave-shapes, thus adding to the complexity of the study of the long spark both as an engineering necessity and as a physical exercise. The physical understanding of the influence of wave-form on spark development has already been beneficial to the engineer and may soon have advanced sufficiently far to reduce the amount of testing of transmission-line insulation and to enable improvements to be made to the geometric forms of such insulation.

2. Sparkover under Unipolar Lightning Impulse Voltages and Switching Impulse Voltages

2.1 General Description of Breakdown Characteristics

As the breakdown* process of air in a long gap develops in times of the order of microseconds it is greatly influenced by the rate of application and removal of the applied voltage and to a smaller extent by some details of the whole voltage/time characteristic of the applied stress. Precise determination of the wave-shape was not possible until the high-voltage cathode-ray oscillograph and potential-divider had been developed, roughly by 1930–1932; thereafter most engineering investigations were done with impulse waves of almost double-exponential form, rising to a crest value in one microsecond (a time referred to as T_1 or as T_{cr}) and then falling to half the crest value in times of 5, or 50 or 500 μs, referred to as T_2, care being taken to include sufficient resistance and capacitance in the circuit to prevent high-frequency oscillations from occurring on the rising wave-front and crest of the wave. It was important to remove these oscillations because at that time crest voltages were invariably measured with the sphere spark-gap and it was known that this could respond to waves of very short duration and might thus spark over at the crest of an oscillation whereas the sparkover of point gaps, for example, was relatively insensitive to minor disturbances on the voltage wave. The sphere-gap no longer plays the dominant role in voltage measurement, but that the subject of voltage generation of "pure" wave-shapes is still very important is shown by a thorough study of impulse-voltage generating circuits, particularly for the Ultra High Voltage (UHV) systems now being planned, see Blasius *et al.* (1973).

It should be noted that although the term "impulse voltage" has frequently been applied to all short-duration surges generated in the laboratory, the modern tendency is to restrict it to surges having short wave-fronts, of the order of a few microseconds, indeed to call them lightning impulse voltages

* "Sparkover" and "breakdown" are used by different workers to describe the same phenomenon; the writer prefers to use the sparkover of a gap between electrodes recognizing that it is the air which breaks down but consistency throughout has not been achieved.

and to use the words "switching impulse voltages" to describe surges having long wave-fronts of the order of 10 to 1,000 μs. BS 923, 1972 gives full details of the definitions and measurement of T_1, T_2, T_{cr}, Time-lag T_c or T_{chop}, for different wave shapes (I.E.C. Publication 52, 1960).

2.2 Sparkover of the Rod/Plane, Rod/Rod and Sphere/Plane Gaps, Under Lightning Impulse Voltages

A comprehensive investigation of the lightning impulse sparkover of a wide range of such gaps up to 1 MV done as soon as the oscillograph was available may be cited as indicative of our knowledge in the early 1930s (Allibone *et al.*, 1934). Briefly the extreme examples of sparkover were encountered with the most divergent form of electric field, that of the point/plane gap, the sparkover voltage of which rose almost linearly with spacing, the mean gradient being 520 kV m^{-1} for the positive point/plane gap and 1,000 kV m^{-1} for the negative, both gaps being tested with a 1/500 μs wave; slightly higher gradients were found as T_2 was reduced. For other, less divergent electric fields, such as occur in the sphere/plane and the point/point gaps, sparkover occurred at intermediate gradients, exact values depending on the duration of the wave, and negative sparkover gradients were always found to be higher than positive; all the work was done with $T_1 = 1$ to 1·5 μs, later called the "standard impulse wave-front". The time-lag of breakdown increased with gap spacing, and on negative polarity was shorter than on positive.

The sparkover voltage of such gaps could not be defined precisely, there was a range of voltages over which the probability of sparkover varied from zero to unity; in early days this voltage range might have been due to the imprecision of generation of a succession of impulses but we now know that it is an inherent feature of the breakdown process though its cause has not yet been determined; it occurs with the most carefully regulated impulse voltage generator and can be as much as $\pm 10\%$ of the voltage for which the fractional probability of sparkover is 50%, usually referred to as V_{50}. There have been many suggested ways of determining the probability function of the sparkover voltage but no one way has been proved to be the best, as recently confirmed by the Les Renardières Group (1974, p. 62). Usually as many as 20 impulses of the (supposedly) same voltage are applied at each of a succession of voltages differing by 1 or 2% and the results, plotted on probability paper, usually lie on a straight line implying a Gaussian probability distribution but there have been many exceptions to this. Almost all investigators have found that if a complete test is repeated, even on the same day, a slightly different ogive (here meaning the curve relating the percentage sparkover to the applied voltage) will result and the cause of this is not known. More serious than this however is the fact that determinations of the sparkover probability of a gap of, apparently, precisely defined geometric form done in different laboratories the world over differ significantly from one another: four international comparisons have been organized between 1937 and 1964 (Allibone, 1937; Berger, 1956; Baatz, 1962; Carrara, 1964) and in the latest (1964) comparison in which a rod/plane and a rod/rod were tested up to 8 MV in 15 laboratories, the values of V_{50} differed from the mean of all the determinations by $\pm 6\%$ for positive 1/50 μs waves, and by $\pm 10\%$ for negative waves, an overall performance no better than the comparison done in 1937. The explanation of this spread may be, in part, the uncertainty of the correction factor to be applied for atmospheric humidity, in part the different field configurations, in part wave-shapes, atmospheric ionization, voltage measurement and variability of voltage generation. Humidity correction factors have been widely studied and international agreement has now been reached for standard impulse waves. The mean values of sparkover derived from these 15 laboratories are shown in Fig. 1; values were corrected to standard air density and to the agreed humidity of 11 g m^{-3} water vapour.

Reference is made in Section 2.7 to the effect of atmospheric ionization on breakdown, but that this is a significant cause of the wide distribution given above, has recently been shown by Allibone and Dring (1974b) in the study of a 60 cm rod/plane gap: the ogive was constructed from 2,000 applied impulses and then the gap was irradiated with an extremely strong source of γ-rays (100 mCi) and a second 2,000-impulse test showed that the ogive was displaced in the direction of lower voltage by amounts ranging from 0·5 to 10%, so the day-to-day variations of residual atmospheric ionization can cause a marked difference in the sparkover voltage of a gap. It was established in that investigation that the value of V_{50} for the irradiated gap was always the same whenever the gap was tested over a two-year period, provided the correction factors for air density and humidity were applied.

The most recent determination of the value of V_{50} for a rod/plane gap up to a spacing of 7 m using a 1/44 μs wave of positive polarity (Bahder *et al.*, 1974) confirms the work reported by Carrara (1964) using a 1·2/50 μs wave on gaps up to 8 m; the value rises linearly with spacing, the mean gradient being 547 kV m^{-1}. On negative polarity Carrara reported a fall in gradient from 1,000 kV m^{-1} for a 1 m gap to 700 kV m^{-1} for a 6 m gap and the most recent measurement (Gruber *et al.*, 1975) using a wave of approximately 3/1,000 μs shape yielded a gradient of 680 kV m^{-1} for a 5 m gap (by extrapolation, probably 750 kV m^{-1} for a 1·5/1,000 μs wave).

For the rod/rod gap, V_{50} rises linearly with spacing up to 7·5 m (Carrara, 1964); the mean gradients for the two polarities fall between those of the rod/plane gap but their exact values depend upon the height of the gap above ground, and also on whether the gap is mounted horizontally or vertically; $+V_{50}$ increases, and $-V_{50}$ decreases as the rod/rod gap is raised further from the earthed plane of the laboratory, but the two values do not coincide even when the tip of the lower electrode is raised above the ground by a distance of four times the gap spacing.

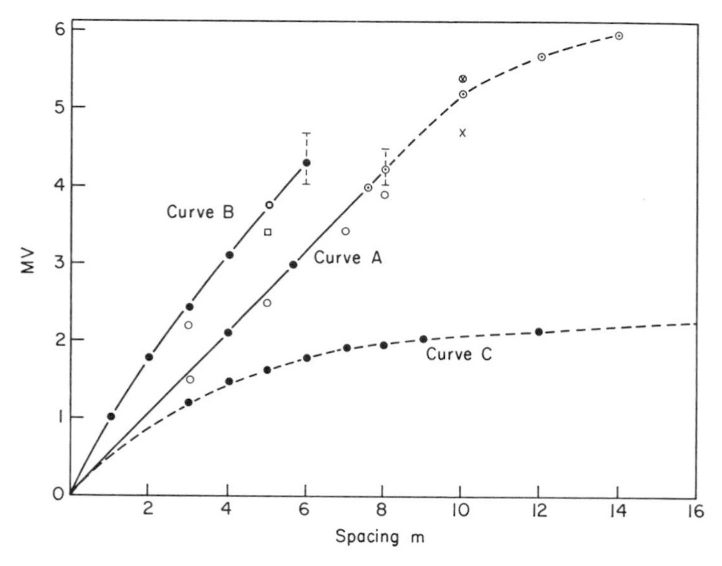

Fig. 1. Rod/plane gaps. Lightning impulse voltage sparkover/spacing calibrations, positive and negative polarities; 1/50 μs waves; also 50 Hz sparkover calibration. *Curve A* (+1/50 μs); ● Carrara (1964), 0 to 8 m; corrected for RAD and humidity; ⫶ limits of the mean values obtained in the 15 cooperating laboratories, values confirmed by Bahder *et al.* (1974); ⊙ values theoretically derived by Bahder *et al.*, from 8 to 14 m but not confirmed by experiment; ○ Udo (1963); × Les Renardières Group (1974), for 10 m gap with 7/3,500 μs wave; ⊗ Les Renardières Group (1974), probable value for T_{cr} = 1·5 μs. *Curve B* (−1/50 μs): ● Carrara (1964), 0 to 6 m; corrected for RAD and humidity; ⫶ limits of mean values obtained by cooperating laboratories; □ Gruber *et al.* (1975), for 5 m gap, 3/1,000 μs wave; **O** Gruber *et al.* (1975), probable value for a 1·5/1,000 μs wave. *Curve C* (50 Hz. *crest values*); Ryan and Powell (1972) up to 5 m; Stekolnikov *et al.* (1962) up to 9 m; Aleksandrov *et al.* (1962) up to 16 m.

Although the sparkover voltages for rod/plane gaps quoted above are rising about linearly with gap spacing from 2 to 6 or 8 m Bahder *et al.* (1974) have forecast, from a study of the growth of the discharge, that the sparkover voltage of the rod/plane gap for positive impulses will not continue to rise linearly, and that for a 14 m gap the average gradient will have fallen to

430 kV m^{-1}, see Fig. 1: this forecast is discussed in Section 4.2.1, but at the time of writing there is no experimental support for this prediction and from a curve given by Les Renardières Group (1974) relating V_{50} to T_{cr} for a 10 m rod/plane gap, the value of V_{50} for a $T_{cr} = 1{\cdot}5$ μs wave would appear to be over 5·5 MV (the lowest value of T_{cr} used in this work was 7 μs so the extrapolation of the curve to give V_{50} at $T_{cr} = 1{\cdot}5$ μs cannot be made accurately).

There is no agreement between different workers as to the magnitude of the standard deviation (σ) of test results from the value V_{50}, but from the very large number of tests reported by Carrara (1964) for large gaps and by Allibone and Dring (1974a) for smaller gaps it appears that σ is smaller, 1 to 5% for point/plane gaps under positive polarity, and larger, 4 to 8% for these gaps under negative polarity; for rod/rod gaps the values of σ lie between these extremes: σ in general falls as gaps are increased and intense irradiation of gaps up to 1 m reduces the values of σ for all gaps.

The shape of the small electrode, be it conical or a rod of square cross-section, has little effect on the sparkover voltage of large gaps.

The effect of increasing the radius of curvature of the high-voltage electrode is to raise the sparkover voltage of the gap; the amount by which it exceeds that of the rod/plane gap increases with the radius of curvature for a given spacing, and the sparkover of a sphere/plane gap is higher than that of the rod/plane gap until the spacing exceeds eight times the sphere radius. The effects are similar for impulses of both polarities.

2.3 Sparkover under Switching Impulses

2.3.1 Double-exponential and Linearly Rising Switching Impulses

As the wave-front of the applied impulse is increased the sparkover voltage of almost all gaps falls and then rises again.

In 1960 Bazelyan *et al.* first observed that as the wave-front was increased from 1 to 500 μs, with impulses having T_2 values of several milliseconds, the positive sparkover voltage of a 1 m rod/plane gap fell to a minimum value of about 25% below the value for a 1/1,000 μs wave and at this value the average time-to-chop was 100 μs; as the wave-front duration was further increased the sparkover voltage continued to rise. As the gap was increased the minimum value of the sparkover voltage occurred after longer time-lags, being some 200 μs for a 3·0 m gap; see Fig. 2. Three years later Stekolnikov and Shkilev (1963c) and Udo (1963) showed that a similar fall, followed by a rise, occurs in the sparkover voltage of a rod/plane gap on negative polarity as

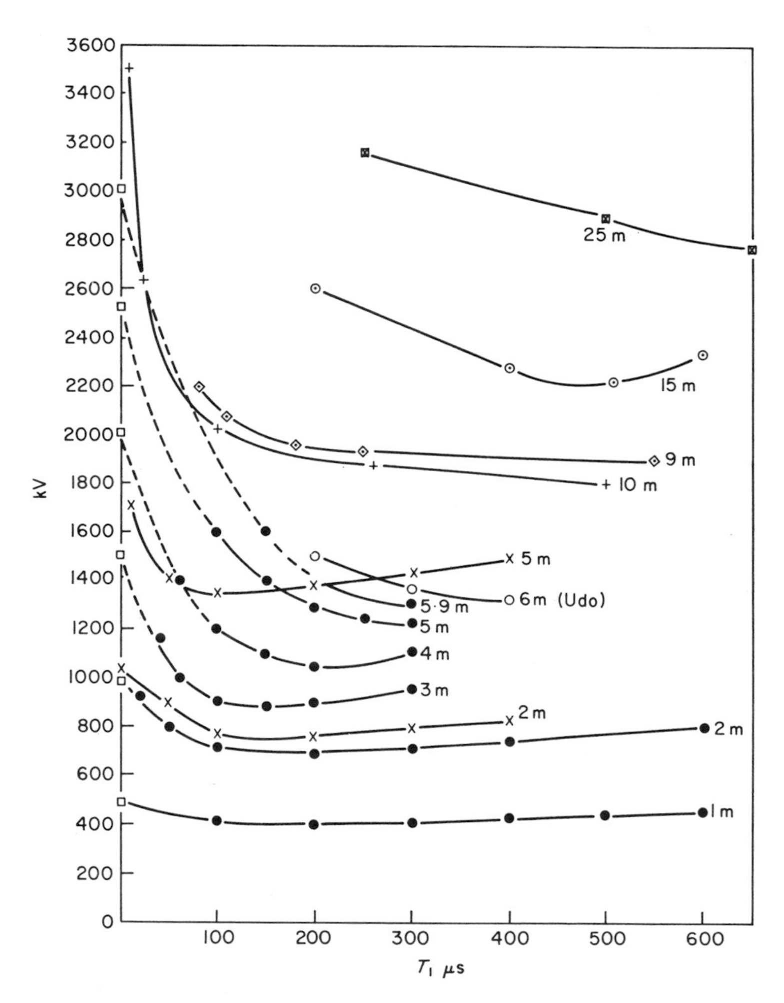

Fig. 2. Rod/plane gaps, variation of sparkover voltage with wave-front time T_1. Positive polarity switching impulse voltages; □ lightning impulse voltages, 1/1,000 μs waves taken from Fig. 1; ● Bazelyan *et al.* (1960), Stekolnikov *et al.* (1962), 1 to 5·9 m; × Hughes and Roberts (1965), 2 and 5 m; ○ Udo (1963), 6 m; + Les Renardières Group (1974), 10 m; ◈ Watanabe (1967), 9 m; ⊙ Kachler *et al.* (1971), 15 m; ⊠ Barnes and Winters (1971, 25 m.

the value of T_1 is increased; see Fig. 3, but the minimum occurs with smaller values of T_1 than on positive polarity for the same gap spacing. As these slower rates of rise of voltage more closely resemble the disturbance created by switching operations on transmission lines they have received a very great amount of attention as they probably set the minimum clearances to ground on extra-high-voltage transmission systems. Whilst the different wave-forms of switching surges on lines are numerous, three classes of switching impulse wave-forms have been used in the laboratory, a double-exponential form like the lightning impulses but in general having a high T_2 value, 1 to 10 ms, a semi-sinusoidal wave-form as delivered by a transformer, and a mixture of these two having an alternating wave superimposed on a double-exponential wave. The wave-forms which have been recognized as "standard" internationally have $T_1/T_2 = 60/2{,}500$, 125/2,500 and 375/1,500 but these have not yet been generally followed, so the complexity of test results is extremely great. As far as the writer has been able to ascertain there is reasonable agreement between results of tests in different laboratories in spite of these complexities, but much more work of a comparative nature remains to be done and no international comparisons of results have yet been organized.

In studying papers on breakdown under switching impulses there are two further complexities. If the oscillogram shows that sparkover of a gap occurred on the rising wave-front, some authors quote the actual voltage at which sparkover took place, others quote the "prospective voltage" V_{cr}, that is, the crest value of the applied wave when, on 50% of the applications of voltage, sparkover did not occur, and sometimes the language used is imprecise. Secondly, in describing the fall and subsequent rise of the sparkover voltage as the wave-front of the applied stress is increased some authors quote the duration of the wave-front (T_1, or more generally T_{cr} as defined in the Standards literature BS 923, 1972), whilst others quote the average value of the oscillographically observed time-to-chop T_c, which might or might not coincide with T_{cr}. In the Russian work quoted above actual sparkover voltages V_{chop}, and times to sparkover T_{chop}, were given. In general it can be said that when the impulse has the wave-form which causes sparkover at the lowest voltage (that is, at the trough of what is now called "the U-curve") sparkover takes place on the crest of the wave; if lower values of T_1 are used sparkover occurs on the wave-tail, and conversely if high values of T_1 are used sparkover occurs mainly on the rising wave-front. There are, however, big variations of time-lag from the mean value.

The work on positive polarity switching impulses was extended two years later by Stekolnikov *et al.* (1962) to gaps of 5·9 m, see Fig. 2. The minimum sparkover voltage for this 5·9 m gap fell 50% below the impulse sparkover

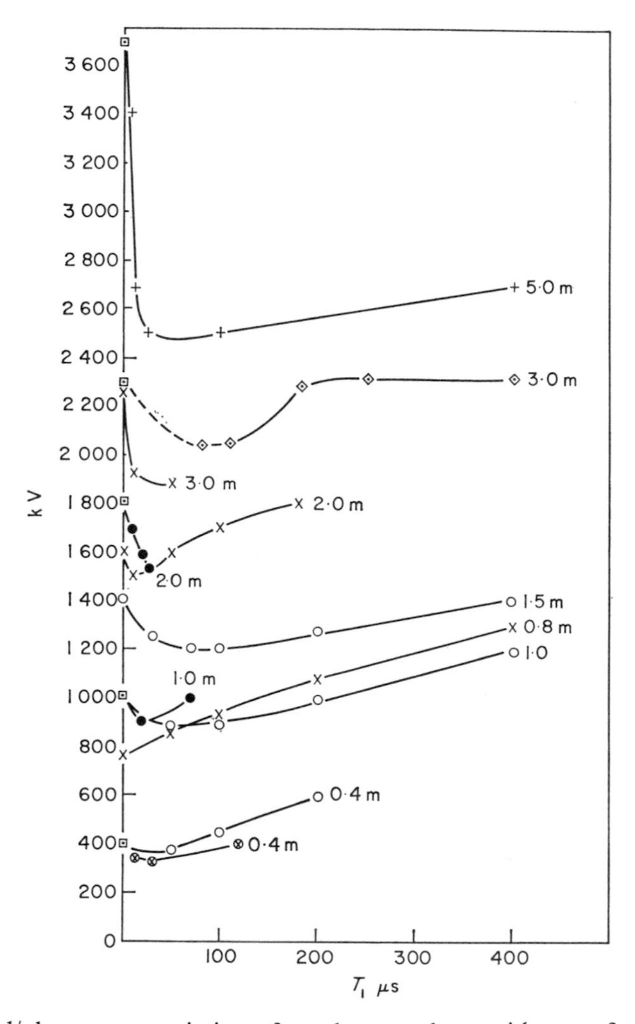

Fig. 3. Rod/plane gaps, variation of sparkover voltage with wave-front time T_1, negative polarity switching impulse voltages; ⊡ lightning impulse voltages, 1/1,000 μs waves taken from Fig. 1; ; ● Stekolnikov and Shkilev (1962), 1 and 2 m gaps; × Hughes and Roberts (1965), 0·8 to 3 m gaps; ○ Udo (1963), 0·5 to 1·5 m gaps; + Gruber *et al.* (1975), 5 m gap; ◈ Watanabe (1967), 3 m gap; ⊗ Allibone and Dring (1974c), 0·4 m gap.

voltage (to an average gradient of 230 kV m^{-1}) and it occurred when the time-to-chop was over 300 μs; this sparkover voltage is 25% lower than the power-frequency sparkover voltage of the gap sparkover, which occurs on the positive half-cycle. A corresponding fall in the sparkover voltage of rod/rod gaps was also found; for a 3 m gap it was 35% below the impulse sparkover voltage, the average gradient being 370 kV m^{-1} and 25% below the power-frequency sparkover voltage.

These results were almost immediately confirmed by Udo (1963) for gaps up to 6 m, see Fig. 2 for his rod/plane sparkover values; they agree very closely with those already reported. Some results by Hughes and Roberts (1965) show less agreement; only two sets of values are given in Fig. 2, for 2 and 5 m. Whereas the earlier work had shown that the trough of the *U*-curve shifted to longer times as the gap widened, Hughes and Roberts found that the trough occurred at 100 μs for all gaps from 0·8 to 5·0 m. On negative polarity, Fig. 3, they found no trough for 0·8 and 1·2 gaps but for 2 and 3 m gaps the results confirm the earlier work. Larger gaps have been studied recently; Watanabe (1967) has determined the trough of the *U*-curve for a 9 m rod/plane gap on positive polarity and a 3 m gap on negative, the Les Renardières Group (1974) has determined values for a 5 and 10 m gap, Gruber *et al.* (1975) for a 5 m gap on negative polarity, Kachler *et al.* (1971) for a 15 m rod/plane gap on positive polarity and, finally, Barnes and Winters (1971) report sparkover values for a 25 m gap with positive waves up to 1,000 μs wave-front but even with this the trough of the curve has not been reached. These values are all shown in Figs 2 and 3.

Some sparkover voltages for rod/plane gaps are given in Fig. 4 for positive and negative switching impulses. Many of the results correspond to the trough of the *U*-curve for any particular gap but in some cases the results for several gaps are available for only one stated value of T_1, Watanabe's (1967) up to 13 m for $T_1 = 180$ μs and Barnes and Winters (1971) up to 29 m for $T_1 = 250$ μs. On both polarities the average sparkover gradient is falling as the gap increases. On positive polarity it is 150 kV m^{-1} for a 15 m gap and for a 25 m gap it is 110 kV m^{-1} with a wave having $T_1 = 1{,}100$ μs but at the trough of the *U*-curve for this gap the gradient might be only 100 kV m^{-1}. On negative polarity the lowest gradient so far reported (Gruber *et al.*, 1975) is 470 kV m^{-1} for a 7 m rod/plane gap; although the curve is beginning to show a possible saturation effect, note that the gradient is still more than twice the positive polarity gradient for the same spacing.

The rod/rod calibration has been extended to 15 m by Kachler *et al.* (1971) for different heights of the grounded point above the ground plane; all values lie between the rod/plane positive and negative calibrations and the lowest gradient for the longest gap is 170 kV m^{-1}.

Harada *et al.* (1973) compared the behaviour of gaps when subjected to switching impulses of double-exponential form and to linearly rising impulses of the form used originally in the Russian (Bazelyan *et al.*, 1960) work and they corrected all results for atmospheric conditions so that they could be accurately compared. For the rod/plane and rod/rod gap the sparkover values are about 5% higher when the linearly rising surges are applied and the value of T_{cr} for the minimum value of V_{50} is 30 to 100% greater; both these findings are to be expected as the breakdown process depends very much on the rate of change of voltage near the crest value of the surge.

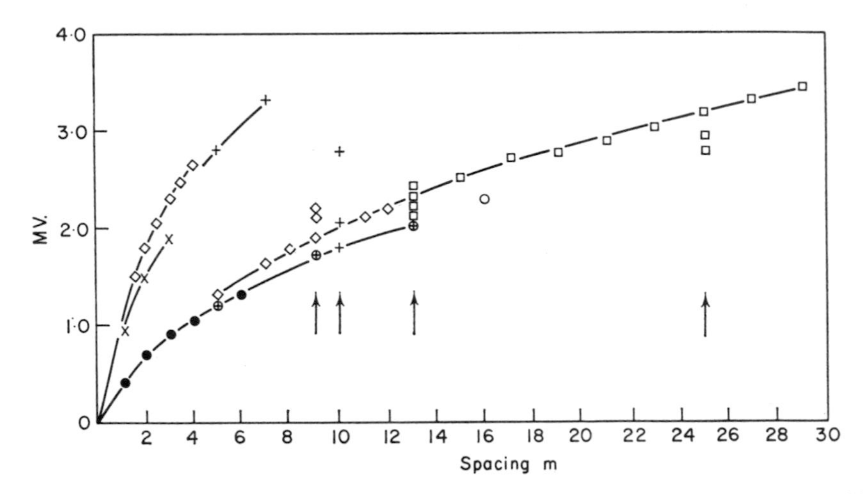

Fig. 4. Rod/plane gaps, sparkover voltages of rod/plane gaps stressed with switching impulse voltages of both polarities (sparkover voltages at the trough of the *U*-curves unless stated otherwise); ● Bazelyan *et al.* (1960), positive only; ● Stekolnikov *et al.* (1972), positive only; × Hughes and Roberts (1965), negative only; + Les Renardières Group (1974), at 10 m with $T_1 = 500$, 2,200 and 22 μs in ascending positions; ◇ Watanabe (1967), $T_1 = 180$ μs, also at 9 m with $T_1 = 80$, 180 and 550 μs; □ Barnes and Winters (1971), at 13 m with $T_1 = 500$, 1,100, 250 and 1,700 μs in ascending positions, at 25 m with $T_1 = 1{,}100$, 500 and 250 μs in ascending positions, at all other spacings, 9 to 29 m $T_1/T_2 = 250/5{,}000$; ⊕ Volkava and Chernyshov (1967); + Gruber *et al.* (1975), negative only, $T_1 = 300$ μs, 5 and 7 m only; ◇ Watanabe (1967), negative to 4 m, $T_1 = 180$ μs; ○ 50 Hz at 16 m.

The standard deviation σ from the value of V_{50} is generally greater for switching impulse voltages than for short wave-front impulse voltages and for negative polarity than for positive polarity discharges, but published results vary greatly. Watanabe (1967) gives values of σ for some rod/plane gaps ranging from 1 to 13 m and for surges of both polarities but no trend can be seen in these and he considered that the range was "due mainly to the variation of the surrounding atmospheric conditions". Harada *et al.* (1973), using only surges of positive polarity, quote larger values of σ for rod/rod gaps (5 to 6%) than for rod/plane gaps (3 to 4%) and show that for a linearly rising wave-fronted surge σ increases as the rate of rise falls. The Les Renardières Group (1974) gives values of σ for the positive polarity sparkover voltage, at the trough of the *U*-curve, for rod/plane gaps ranging from 1 to 13 m, σ rises from 2% for a 1 m gap to 5% for a 10 m gap and then diminishes; for negative polarity switching impulses applied to a 5 m rod/plane the determination of σ "appears difficult to determine, values between 4% and 13% have been encountered": this same remark about the large values of σ for negative polarity switching impulses appears in Allibone and Dring (1974a).

2.3.2 "Non-standard" Switching Impulses

Little work has so far been done with these switching impulses but three examples will suffice to indicate general results. Harada *et al.* (1973) applied double-exponential waves on which a periodic oscillatory component was superimposed; the periodic time T of this component was varied from 50 to 500 μs, and all results were corrected to standard atmospheric conditions so that they could be compared with the results—already referred to in Section 2.3.1—obtained with the two other forms of switching impulses. The familiar *U*-curve was obtained as the time to the first crest was lengthened and the minimum value of V_{50} at the trough of the curve was almost the same as for the double-exponential switching impulse but it occurred at a higher value of T_{cr}. Caldwell and Darveniza (1973) tested 56 cm rod/rod and rod/plane gaps with two forms of non-standard waves, (a) double-exponential waves chopped by a 30% drop in voltage after a few microseconds, and (b) bi-polar oscillatory waves of a few microseconds periodic time oscillating about zero voltage. With the partially chopped waves of positive polarity the sparkover voltages were always 10 to 20% greater than the corresponding values for double-exponential waves of any values of T_1/T_2; with the oscillating waves sparkover voltages were 25 to 50% greater than for the double-exponential wave. On negative polarity the partially chopped waves gave sparkover voltages almost the same as for the double-exponential waves and, again, the oscillatory waves gave breakdown voltages 10 to 25% higher.

Allibone and Dring (1974c) and Allibone *et al.* (1975), tested rod/rod and rod/plane gaps up to 60 cm with damped oscillatory waves superimposed on the double-exponential wave having a value of $T_2 = 1{,}000$ μs and the periodic times of the oscillatory component ranged from 14 to 84 μs. They found that sparkover frequently occurred on the second or subsequent oscillation, not only at the crest of these but also at almost any point on the wave where the voltage was only 30 to 80% of the first crest voltage. However, the voltage V_{50} of both types of gap was seldom more than 15% in excess of the values for the double-exponential wave having the same value of T_{cr} and frequently the difference was negligible; this general summary applied to positive and negative waves.

2.4 Breakdown of a Window Gap Surrounding a Transmission Line

In some designs of transmission-line towers each phase conductor passes through a gap between steel spars of the structure and this is generally called a window gap. Work on the breakdown of window gaps has been confined mostly to positive polarity switching impulses since these constitute the most important group when selecting the design of a window. Attempts have been made to derive a "gap factor" for any window design based on a study of the field distribution between conductor and window and ground; see Paris (1967) and Paris and Cortina (1968). The gap factor is defined as the sparkover voltage for this window gap divided by that of a point/plane gap of the same size, and it has been shown that this factor is almost independent of the wave-shape of the applied impulse. The conductor/plane strength is some 15 to 20% higher than that of the rod/plane gap of this same spacing because of its extended length and a conductor to window more resembles a small sphere/point gap than a point/plane gap: this is fairly obvious when it is remembered that the sparkover of a point/point gap increases—as already noted—as the grounded point is mounted further from the grounded plane. As sparkover of the small sphere/point gap changes in almost the same way as does that of a rod/plane gap when the wave-front is lengthened it follows that the gap factor, K, does not alter much with wave-shape of the applied impulse. K does not change much with the spacing d, provided that all the surrounding geometrical parameters change in proportion as d is changed: hence a determination of the electric field in an electrolytic tank would appear to be a good way of determining K for a given transmission-line diameter and window design and thus avoid a huge amount of testing, see Schneider and Weck (1974). Encouraging progress on this aspect of surge work is reported for spacings in the range 3 to 8 m. It may be noted that the values of V_{50} given by Paris and Cortina (1968) for gaps between a large variety of electrode shapes, conductors, cross-arms, window structures, etc. fall between those of the positive and negative rod/plane gaps, at least up to 4 m thus confirming that these gaps present the extreme forms of divergent electric field, so gap factors can range from 1·0 to approximately 2·5.

Tests on window gaps of up to 12 m are reported by Kachler *et al.* (1971); the breakdown voltage on positive polarity of a 12 m conductor/window gap with a 350/2,300 μs wave is a few per cent higher than for the rod/plane gap.

The most recent work by Menemenlis and Harbec (1974) on window structures tested with a range of switching impulses of positive polarity to derive the shapes of the *U*-curves confirms that for tower structure/conductor gaps ranging from 4 to 9 m the values of V_{50} at the troughs of the *U*-curves all lie above that of the rod/plane gap.

2.5 Sparkover between two Phase Conductors Separately Stressed

Switching operations in three-phase systems can result in non-symmetric surges travelling on the lines and the complexities of the stress pattern are legion. It must suffice here to mention just one example taken from a recent CIGRE Task Force report (1973). Two switching impulses of different wave-shapes, of opposite polarity and of equal magnitude, were generated simultaneously and were applied to two electrodes—two rods mounted horizontally in the example here cited—and tests were made with gaps from 2 to 12 m. Provided the values of T_{cr} for the negative waves were always longer than those for the positive waves, the total sparkover voltages measured were independent of the precise values of T_{cr} for the negative wave. When, however, the T_{cr} of the negative wave was smaller than that of the positive wave, a higher total sparkover voltage resulted, whatever the gap spacing. It is not possible to make a strict comparison of the results with calibration curves of rod/rod gaps when one gap was grounded because no curves exist for the same wave-shapes and values are very dependent on the height of the horizontal gap above the ground; by comparison with results from Barnes and Winters (1971) it would appear that both curves for the double-impulse tests lie above that for the rod/grounded rod. The general conclusion from the double impulse test is that the discharge from the positive rod is dominant unless that from the negative rod precedes it in time; this matter is further discussed in Section 4.3.

2.6 Effect of Air Density and Humidity

The effect of air density was thoroughly examined a long time ago using alternating voltages and the sparkover of many gaps was found to vary directly with air density over the relatively small range encountered in the laboratory and this linear relationship has been accepted for lightning and switching impulse sparkover voltages of gaps up to 1 or 2 m. Recently tests have been made at high altitudes; Phillips *et al.* (1967) tested at 10,500 ft in Colorado, and Harada *et al.* (1970a) at 6,100 ft in Japan. Both groups found that for lightning impulse tests of positive polarity, where flashover varies linearly with spacing, flashover varied in proportion to the relative air density, but for impulse tests of negative polarity and for switching impulses of both polarities where, with large gaps, the flashover voltage does not increase linearly with spacing but begins to show a saturation effect, the flashover voltage of a given gap varies in proportion to δ^n where n varies between 0·7 and 1·0. Harada *et al.* showed that for these large gaps the correction factor for air density should be applied to the gap spacing instead of to the flashover voltage; thus if a calibration curve of sparkover be taken for a range of gaps when the relative air density (RAD) is δ, then the sparkover voltage of a gap D at standard air density will be the sparkover voltage read off this curve for a gap D/δ.

The effect of humidity is more complex. For gaps of small to medium length calibration curves have been recorded for lightning impulses and for switching impulses having a large range of wave-shapes, but agreement between different laboratories is not very good. An overall view for these gaps is that (a) for positive impulse voltages (T_1 ranging from 1 to 3 μs) the sparkover voltage increases by 1% to 1·3% for each increase of 1 g of water vapour per m^3 of air for both rod/plane and rod/rod gaps, and then as T_1 is further increased, this correction factor rises to 1·7% for long switching impulses; (b) for negative impulse voltages humidity exerts almost no effect on sparkover voltage when T_1 is small, but the humidity coefficient rises to 1·0% for long switching impulses. For large gaps Harada *et al.* (1971) reported that the coefficient remained constant as the gap increased for those situations when the sparkover voltage/spacing graph is a straight line, but when the graph becomes curved as spacing increases—implying a saturation effect already mentioned—then the humidity correction coefficient falls as the gap is increased. From a study of these and other results Kachler *et al.* (1971) considered that the humidity correction factor should be applied to the spacing instead of to the sparkover voltage and that the RAD correction factor should be similarly treated. There the matter must rest till new information is available. Prabhaker *et al.* (1971) found that at very high humidities the sparkover voltage became independent of further increase of humidity, but doubt has been shed on this result by Allibone and Dring (1974d). The Japanese work (Harada *et al.*, 1971) shows no abrupt change in the sparkover/humidity curve at very high humidities.

Under rain-test conditions the positive switching impulse sparkover voltage is almost the same (Paris, 1967) as in the absence of rain, but the negative rod/plane breakdown falls markedly—by 15% in this reference, indicating probably that the "plane" has become transformed into a "pointed plane" by long vertical water splashes. In contrast with this conclusion, for positive impulses, Barnes and Winters (1971) happened to be testing 21 m and 23 m rod/plane gaps with positive switching impulses when rain fell and they found the sparkover voltage was reduced by 10%: this solitary result should be borne in mind when considering the lightning discharge.

2.7 Time-lags, Standard Deviations and Withstand Voltages

The time-lag of breakdown is a very variable characteristic and it is almost impossible even to generalize. With "standard impulse voltages", i.e. 1·5/50 μs waves, time-lags may range from 5 to 20 μs on positive polarity for V_{50} for rod/plane gaps, and from 2 to 8 μs on negative polarity, though it will be recalled that $-V_{50}$ is roughly twice $+V_{50}$. For rod/rod gaps intermediate values are found. All time-lags diminish as higher voltages are applied to the same gap spacing, and, by contrast, near the withstand voltage some long time-lags are encountered. This is particularly noticeable when the wave-tail T_2 is increased; with $T_2 = 1{,}000$ μs time-lags of 100 μs are not infrequent for small gaps and it is of interest to note that intense irradiation of gaps with γ-rays (Allibone and Dring, 1974a) practically eliminates these.

As T_1 is increased, the sparkover voltage of a gap falls and then rises—the U-curve already referred to in Section 2.3.1. At the trough of the U-curve sparkover occurs most frequently at the crest voltage; beyond the trough sparkover occurs most frequently on the rising wave-front, but the spread of values of the time to chop is very large until the fractional probability of sparkover reaches 100%. When overvoltages of 10% or above are applied, gaps break down with marked consistency of time-lag.

Reference has already been made to the standard deviation σ of sparkover voltages about the value V_{50}; an applied voltage of $V_{50} - 3\sigma$ almost invariably fails to cause sparkover and has been taken to represent the withstand voltage of a gap. Carrara (1964) reported values of σ (based on approximately 300 impulses per test in each of the cooperating laboratories) ranging from 0·8 to 2% for rod/plane gaps from 1 to 8 m, using standard 1·5/50 μs impulse waves of positive polarity, and from 1 to 5% for 1 to 6 m gaps on negative polarity; almost exactly the same range was obtained from tests on rod/rod gaps. With switching impulses of 180/200 μs, Watanabe (1967) reported values of σ (based on 60 to 100 impulses per test) ranging from 3·5 to 6·5% for rod/plane and rod/rod gaps of 4 to 13 m tested on positive polarity, and 1·5 to 5 m tested on negative polarity, and values of σ within this range have been reported by other investigators. It is surprising therefore to note that Menemenlis and McGillis (1974) obtained values for σ of 7% for a 3 m rod/plane gap and 9% for a 5·7 m tower window gap (almost a rod/rod gap) for impulses of short wave-front; both these gaps tested with switching impulses at voltages V_{50} corresponding to the trough of the *U*-curves yielded a value for σ of 4%. On the basis of these values the authors considered that the withstand voltages $V_{50} - 3\sigma$ are only slightly affected by the time-to-crest T_{cr} of the applied impulse, the high values of σ which they reported for impulse tests with low values of T_1 are quite exceptional. At lower voltages Allibone and Dring (1975) report that, on the basis of very long tests of 2,000 impulses per test, the values of σ for short and long wave-fronts were almost identical so that the withstand voltages fell into a *U*-curve just like the values of V_{50} for the same gap.

3. Sparkover under Alternating and Constant Voltage

It is of interest to note here the power-frequency alternating-voltage sparkover characteristics of long rod/plane gaps. Up to 5 m, figures given by Ryan and Powell (1972), and up to 9 m, given by Stekolnikov *et al.* (1962), are higher than the corresponding sparkover voltages shown in Fig. 4 for switching impulses; up to 9 m the switching impulse figures are almost the troughs of the *U*-curves for each gap. But for very long gaps (Aleksandrov *et al.*, 1962) up to 16 m, the power-frequency sparkover voltages are lower than the switching impulse figures in Fig. 4; however these are almost certainly not the lowest trough voltages for these extra long gaps, as may be judged by reference to the three values of sparkover of a 25 m gap quoted by Barnes and Winters (1971).

From the two curves for the sparkover of very long gaps under switching impulse voltages and power-frequency voltages it is easy to see how abnormally long discharges arise from time to time.

Under constant voltages figures for the sparkover of rod/plane and rod/rod gaps up to 3 MV are available, see Fig. 5. Knudsen and Iliceto (1970) report that for the rod/plane gap sparkover rises linearly with spacing at a rate of 480 kV m^{-1} on positive polarity; on negative polarity the gradient is 1,400 kV m^{-1} over the first 50 cm and falls to an average of 1,050 kV m^{-1} over a gap of 150 cm. Thus on positive polarity the d.c. sparkover characteristic is almost identical to that of the 1/50 μs impulse sparkover whilst on negative polarity it is higher than that of the negative 1/50 μs impulse sparkover. These values virtually extend the earlier figures of Udo and Watanabe (1968). For vertical rod/rod gaps where the earthed rod stands high above the earthed plane—at least twice the gap spacing—the d.c. sparkover on both polarities rises linearly with spacing up to 3 m at a mean gradient of

570 kV m^{-1}; this value is a little higher than that reported by Colombo and Mosca (1972) for a 3 m gap standing 6 m above the ground. In all tests moderately high resistances have been used in series, 100 kΩ to 2·5 MΩ and though Colombo and Mosca report that the resistance did not affect the results, supporting information is not available. Knudsen and Iliceto report that the corona consisted of a background of 1 mA with spikes superimposed at frequency of 4 to 10 kHz, when the polarity was positive; on negative polarity the corona was a smooth current with small ripples of 20 to 50 kHz.

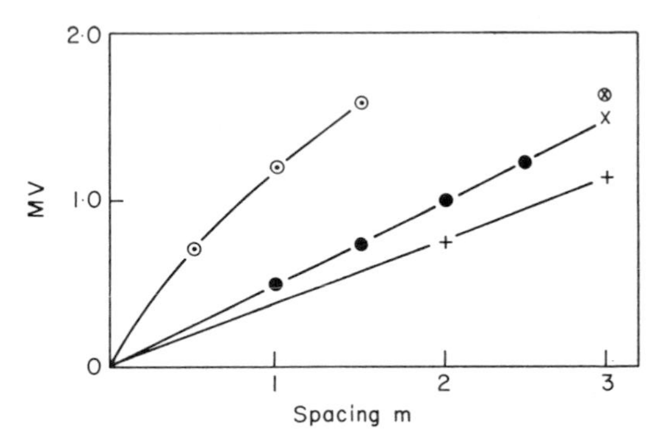

Fig. 5. Rod/plane and rod/rod gaps stressed with constant and 50 Hz alternating voltages; ● positive polarity rod/plane; ⊙ negative polarity rod/plane (Knudsen and Iliceto (1970); × positive polarity rod/rod at 3 m only; ⊗ negative polarity rod/rod at 3 m only (Colombo and Mosca, 1972), lower rod raised 6 m above earth; + 50 Hz rod/plane.

4. Physical Mechanism of Spark Formation

4.1 The Leader Stroke/Main Stroke Mechanism

4.1.1 General Description

As already mentioned, it was discovered by Allibone and Schonland (1934) that the million-volt spark from an impulse generator was preceded by a leader stroke from a rod to plane; it travelled at approximately 5×10^6 cm s^{-1} so that it reached the ground in 10 μs, roughly the time-lag of breakdown of the 2 m gap, using a 1/580 μs wave. This was followed by a return stroke, so called by analogy with the lightning stroke (see Chapter 5.5.6), extending from ground back to the high-voltage electrode, but the time-resolution of the camera was not sufficient to determine its velocity. Allibone (1937) and

Allibone and Meek (1937, 1938) made a comprehensive study of the leader stroke/main stroke phenomenon, using a quartz lens to record as much as possible of the ultraviolet light on a moving photographic film. They discovered that any section of the leader stroke as it progressed from a point to a plane had a high luminosity for less than 0·5 μs, and that the length of this luminous section could not be much more than 1 cm. As the leader traversed the upper quarter of the gap, its velocity diminished, from 2×10^6 cm s^{-1} to $0{\cdot}5 \times 10^6$ cm s^{-1} and then the leader ceased to grow and the gap failed to break down; or, alternatively, its velocity increased for a short time and again diminished, and then it increased to 3×10^6 cm s^{-1} and traversed the remainder of the gap. At each of the sudden increases in velocity the path of the leader back to the electrode was reilluminated very fast and in this sense the hesitant progress of the leader nearly resembled the stepped leader-stroke to the first stroke of a lightning flash (Schonland and Collens, 1934). To reduce halation from the return stroke of the spark high resistances of 10 to 1,000 kΩ were incorporated in the lead to the rod, and thus the wave-fronts of the impulses were of the order of 10 to 200 μs; the mean velocity of the leader fell as the resistance was increased, i.e. as T_1 was increased and the number of reilluminations and steps of the leader increased. At pressures of less than one atmosphere (Allibone and Meek, 1938) these effects were greatly enhanced. It was found that if an earthed rod projected upwards from the ground plane the downward positive leader was met by a short ascending negative leader travelling at a lower speed. When the high-voltage electrode was of negative polarity a negative leader stroke descended at $1{\cdot}0 \times 10^6$ cm s^{-1} and was *always* met by an ascending positive leader, indeed if the cathode was a large sphere, the positive leader traversed the whole gap. As shown in Chapter 5.4.2, McEachron (1939) first described a positive leader stroke ascending from the top of the Empire State Building to the thundercloud above. Oscillograms recorded currents in the million-volt leader stroke of the order of 1 A, and it was noted that current was flowing all the time the leader progressed although the luminosity between the leader and the main stroke had fallen to a very low value. In the same years Stekolnikov and Belyakov (1937, 1938) began a similar study but without an ultraviolet lens; they too used a high resistance to avoid halation and studied gaps of 2 m. They measured the velocity of the return stroke as 4×10^8 cm s^{-1} and noted that this value was influenced by circuit resistance. No further work was done until the war was over.

In what follows precedence will be given to describing the discharge across a point/plane gap when lightning and switching impulses of either polarity are applied to it; brief reference will be made to the effect of increasing the size of the point, thereby reducing the stress concentration at the electrode. The phenomena at a point/point gap are a complex mixture of the two basic rod/plane phenomena and they have not been studied deeply.

The nomenclature used here in describing the development of the spark is fairly widely accepted but there are many variants. The first manifestation, corona, or impulse corona, is filamentary in structure, very similar to the famous eighteenth-century Lichtenberg figures, indeed they resemble closely the two patterns, positive and negative (Chapter 5.5.4). Sometimes the corona is succeeded by a second and third burst of corona; T_i is used to denote the time of inception of these. When the leader stroke (sometimes still called a streamer) develops, bursts of leader corona appear at its tip.

4.1.2 The Initial "Impulse Corona" Discharge

When a long point/plane gap is stressed with a positive or negative polarity impulse having a wave-front T_1 of 0·2 to 1·0 μs duration and having an amplitude near to the breakdown voltage of the gap a large burst of corona discharge extends from the point over a considerable fraction of the gap, indeed the dark-accustomed eye can see some of the shafts of the corona extending almost over the whole of a 1 m gap. Komelkov (1947) was the first to show that the corona developed at very great speed, about 10^8 cm s^{-1}; he used a new technique of photographing the discharge through slits placed at right angles to the axis of the discharge, and light then fell upon a drum camera rotating on a line parallel to the slits. In this way the discharge from a fast-fronted wave could be recorded as it reached various distances from the anode. The corona travelled over most of the gap as the impulse rose towards crest value and it was followed by the leader stroke travelling at 10^6 cm s^{-1}; from time to time a new burst of corona discharge emerged from the tip of the leader until at last filaments of corona reached the plane; shortly after that the main stroke developed. Work of a similar kind was reported by Saxe and Meek (1948, 1955) who recorded the light passing through the slits with a photomultiplier and established that the corona lasted only a small fraction of a microsecond and that its development was not influenced by circuit resistances of 10 to 40 kΩ.

Park and Cones (1956), using an inverted geometry, i.e. plane/point, established that in a 20 cm gap the impulse corona developed at 10^8 cm s^{-1} and extended over the whole gap, but with longer gaps, 150 to 300 cm, Brago and Stekolnikov (1958) showed that the impulse corona did not extend over the whole gap unless a 20% overvoltage was applied. The corona current measured by Park and Cones at the anode rose to a crest value of many amperes in 0·3 μs and declined to half-value in a further 0·3 μs, during which brief time several microcoulombs of charge had been injected into the gap. The peak current was proportional to the actual voltage on the wave-front at which the burst of corona occurred and the length of the corona filaments was proportional to this voltage.

The influence of the initial burst of corona on the subsequent behaviour of the gap was first noted by Akopian *et al.* (1954), who showed that if a gap was first stressed with the minimum impulse sparkover voltage and then on top of this a pulse of 50% higher voltage lasting for 1 μs was imposed, the gap failed to break down. Waters and Jones (1959a) observed that there were long "dead-times" between the corona development and the breakdown of the gap; by imposing a 15% voltage step at various times after the corona they proved that the strength of the gap had been increased by the corona and that it was continuously changing. They concluded that the space charge had interrupted the further development of breakdown for varying periods of time and that the dispersion in the sparkover voltage of a gap was probably caused by variations in the amount of space charge created by the corona. They measured (1961) corona currents rising to 50 A in 0·3 μs in a 2 m gap stressed with 0·5/2,000 μs waves and photographed corona streamers 130 cm long carrying a total charge of 12 μC. There were considerable pulse-to-pulse variations which they (like Park and Cones) ascribed (1964) to the actual voltage on the wave-front at which the corona had developed but the mean space charge increased with the square of the applied voltage. They reported that, on negative polarity, the extent of the impulse corona was smaller than that on positive polarity for the same voltage, the individual corona streamers were fewer and of longer radius and some streamers exhibited branching backwards towards the cathode as well as forward—these were referred to as "retrograde streamers". Using for the first time the electron-optical image converter (E.O.C.) which gave immensely greater time resolution, Stekolnikov (1961) (see Fig. 6a) showed that the streamers of the impulse corona began with a speed of 5×10^9 cm s^{-1} and then their speed fell to 2×10^8 cm s^{-1} for most of their length. Baldo and Gallimberti (1970), using waves having $T_1 = 0{\cdot}9$ μs and various values of T_2 ranging from 23 to 1,500 μs, showed that almost the same charge was injected at the same voltage irrespective of wave-tail duration, and confirmed the square-law dependency of Q on V_{cr}, see Fig. 7.

The effect of the impulse corona on the gradient at the electrode was first examined by Bazelyan (1964) who used a segmented 12·5 cm diameter, spherical grounded electrode 100 cm below a plane cathode to which an impulse having $T_1 = 7$ μs was applied: as the corona developed at the crest of the wave the anode gradient fell from 45 kV cm^{-1} to 30 kV cm^{-1}. From the field measurements he deduced that the corona space charge lay at the

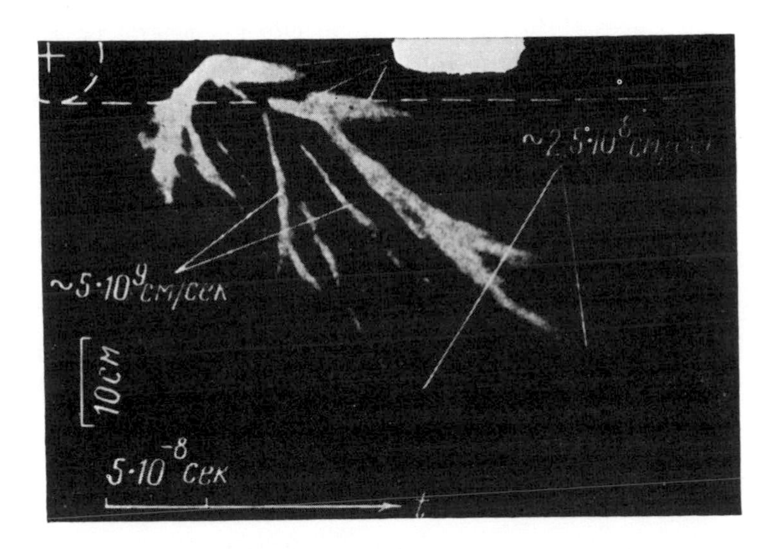

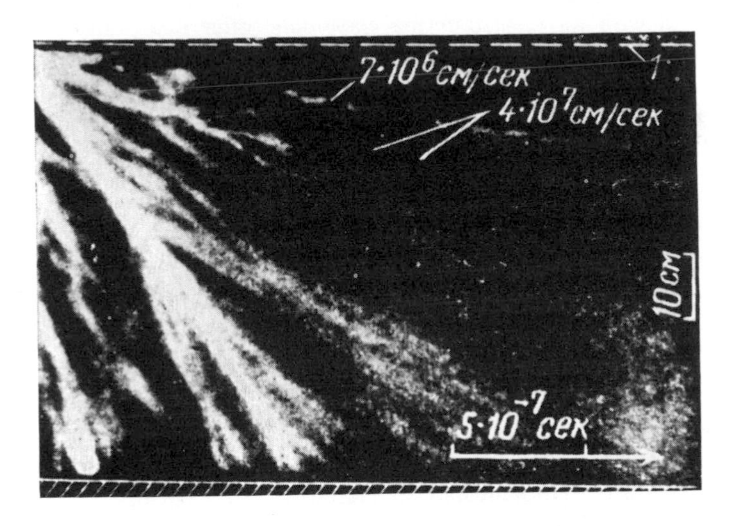

Fig. 6. The first electron-optical converter (image intensifier) camera records of corona (after Stekolnikov, 1961). (a) Impulse corona; (b) leader corona.

ends of the branches of the impulse corona—thus confirming a hypothesis of Park and Cones (1956)—leaving behind either a neutral or a weakly charged medium. Meek and Collins (1965) and Collins and Meek (1965) successfully measured the gradient at the rod and at the plane using small

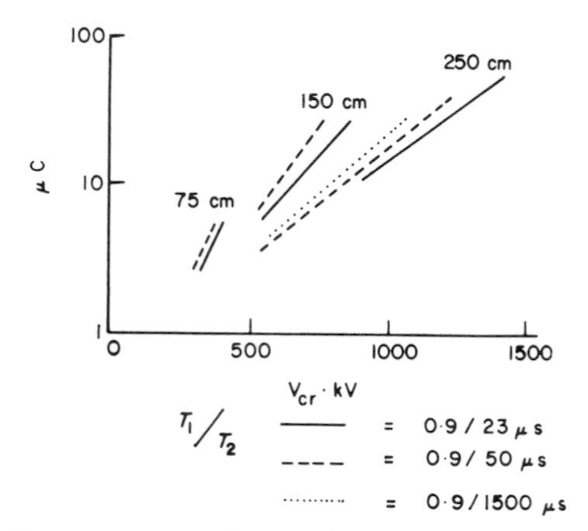

Fig. 7. Variation of corona space charge with applied voltage for three rod/plane gaps stressed with waves of differing wave-tails (after Baldo and Gallimberti, 1970).

embedded probes in a 15 cm gap stressed with a 0·4/30 μs wave; see Fig. 8. At the rod the gradient fell from 40 to 10 kV cm^{-1}, at the plane it rose from the geometric value of 2 kV cm^{-1} to over 4 kV cm^{-1} as the impulse corona spread into the gap; as the voltage was increased some corona streamers crossed the whole gap and thus neutralized some of the charge induced in the plane. The authors concluded that the space charge lay in a hemispherical volume radiating from the anode tip. The technique has been widely used since then and has contributed greatly to the further understanding of the discharge. Bazelyan (1966) applied it to a 200 cm gap and correlated the charge injected, Q, with the field measurement; the larger the corona, the larger the fall in the anode gradient; sometimes a second corona discharge occurred (he was using a wave having $T_1 = 11$ μs) but the following discharges did not start till the gradient had recovered to 28 kV cm^{-1}.

Waters *et al.* (1968) added a further technique, an electro-static flux meter located at various positions in the earthed plane electrode; with its aid the spatial variation of induced charge over the surface of the plane could be determined, so that, together with a knowledge of the net space charge in the gap, the spatial distribution of this charge could be deduced: the measurements were correlated with excellent time-resolved photographs of the corona. They too found that the charge increased as the corona inception voltage increased, that the gradient at the plane rose from the geometric value of 2 kV cm^{-1} to as high as 10 kV cm^{-1} and that it only fell to half-value in very many milliseconds as the charge slowly disappeared. The location of the space charge (Waters *et al.*, 1970), is shown in Fig. 9: the highest

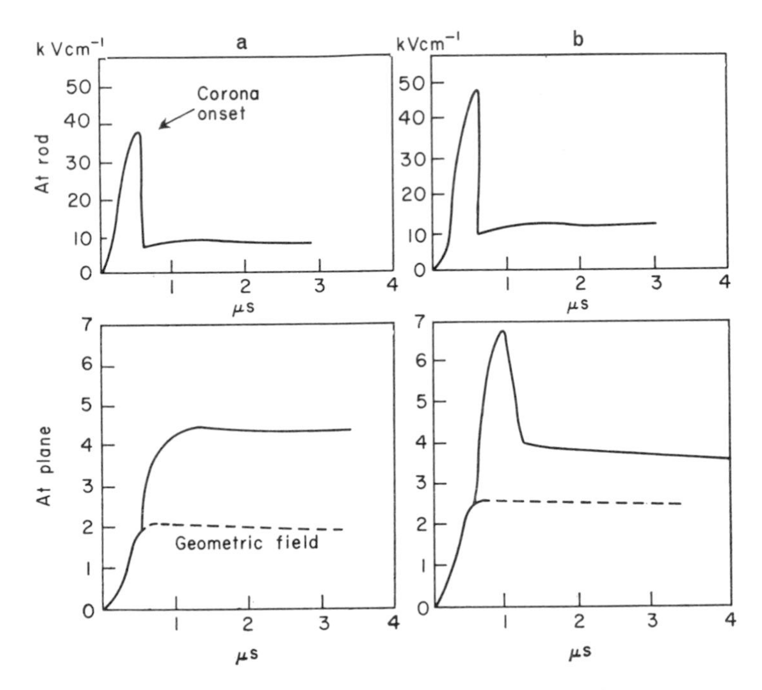

Fig. 8. Field variations at anode rod and cathode plane when corona traverses part of the gap—curve a and when some corona streamers reach the plane—curve b. The "geometric field" is the field at the plane in the absence of corona (after Collins and Meek, 1965).

concentration of positive ions occurred in a dish-shaped volume located roughly at the extremities of the main shafts of the corona, thus confirming Bazelyan's (1964) deduction. At higher voltages when filaments of corona reach the plane large numbers of electrons are released from the plane and many are trapped in the mid-gap region as negative ions, a reversal of the field at the plane occurs and thus all fields in the gap are altered, paving the way for the post-corona developments.

The initiation of corona appeared to be due to dissociation of negative ions in the vicinity of the anode. Waters *et al.* (1965) and Boylett *et al.* (1966)

applied steady drift fields to the anode; when these were such as to repel negative charges the onset of corona was delayed, and with an attracting field, time-lags were reduced.

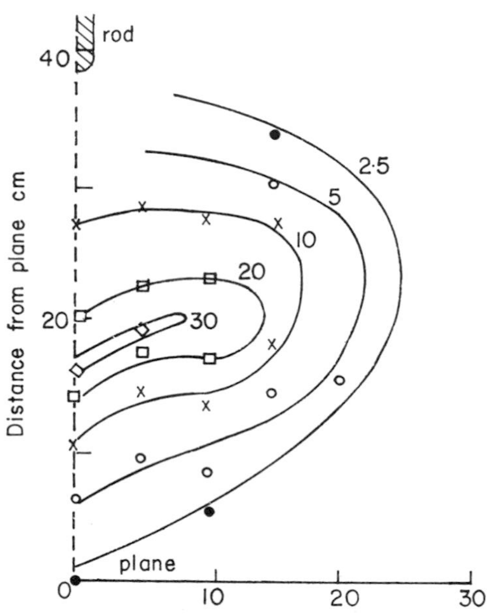

Fig. 9. Contour representation of positive space-charge density in a 40 cm rod/plane gap when corona has not traversed the whole gap. Densities are given in μC m^{-3} (after Waters *et al.*, 1970).

4.1.3 Corona formed by Switching Impulses

In the first account of breakdown with oblique-sloping voltage waves, Bazelyan *et al.* (1960) ascribed the lower sparkover voltage at the trough of the *U*-curve to the discovery that less corona space charge was injected into the gap at this voltage than when the wave-fronts were shorter, or longer, but no quantitative evidence was submitted in support of this. Stekolnikov *et al.* (1962) showed oscillograms of the integrated current in a 100 cm gap stressed with waves of long front duration; the charge injected rose in small quanta of about 0·05 μC to a total of about 2 μC, which is roughly one third of the charge deposited by a 1/1,000 μs wave causing breakdown of the same gap. After the first impulse corona, further corona flares occurred, coinciding in time with the jumps recorded oscillographically. Stassinopoulos and Meek (1967) observed that the field at the plane, when the gap was stressed with a 240/1,200 μs wave, likewise rose in small steps after a small initial rise due to a small impulse corona; these they ascribed to the growth of the leader stroke whether it crossed the whole gap or not.

As these subsequent bursts of corona which occur on the rising wave-front of switching impulses are closely associated with the leader stroke they will be considered later in Section 4.2.2.

4.2 The Positive Leader Stroke

4.2.1 Lightning Impulse Voltages

The optical study of the leader stroke preceding the breakdown of large gaps stressed with impulses rising to maximum in a microsecond has been assisted by the development of the electon-optical converter camera (E.O.C.). With ordinary cameras having a time-resolution as high as 0·5 mm μs^{-1}, halation from the main stroke generally obscures details of the leader stroke and sometimes of the impulse corona. To circumvent this, higher circuit resistance may be used—but then the impulse ceases to have a 1 μs front—or the voltage can be removed from the gap by a chopping circuit triggered to break down as the leader approaches the opposite electrode (Waters and Jones, 1962). This, however, introduces the complication of an enormous increase in light due to the discharge of the space charge back into the chopping circuit and the original pattern of leader and leader corona may be obscured. There is yet a third well-known technique for studying the leader, namely to apply lightning impulses of short duration, T_2, ranging from, say, 5 to 40 μs; this provides all the required evidence of impulse corona and the start of the leader stroke, but the latter becomes slowly arrested in mid-gap as the anode voltage declines; halation is non-existent.

After the impulse corona has developed there is sometimes a period of time during which apparently no current flows into the gap and no discharge can be recorded photographically. The first records of this inactive period have already been mentioned (Waters and Jones, 1959a). When a 0·2/1,000 μs wave was applied to a 76 cm gap, breakdown took place either in a 10 to 20 μs period, or after a long intervening "dead-time" of 100 to 200 μs. Either the rotating camera showed the immediate development of a leader, or there was a long hiatus in which no light was recorded—except occasionally a second burst of corona appeared—and then ultimately a leader stroke developed and the main stroke followed. In both cases the leader velocity was 1 to 2 cm μs^{-1}.

In a second paper by Waters and Jones (1959b) there may be seen a leader starting immediately after the impulse corona, travelling at $4 \cdot 10^6$ cm μs^{-1} in steps of 2 cm and from each step a shower of filamentary discharge—

generally called leader corona—is seen extending towards the plane, and it does so at a speed high compared with that of the leader. As the leader advanced its speed fell and then, having crossed 15% of the gap, it began to accelerate and had reached 2×10^7 cm s^{-1} and was carrying a current of over 100 A when the chopping gap operated. Calculation indicated that the gradient in the leader was over 2 kV cm^{-1}. The leader corona was further studied (Waters and Jones, 1961); its extent diminished as the leader progressed, and ceased when the leader ceased to develop; but if, instead, the leader accelerated, the leader corona rapidly developed to the cathode plane. Each of the steps of the leader was "connected back" to the anode by a reillumination specially noticeable when the leader channel happened to deviate by a sufficiently large angle from the general direction of progression. Its diameter was 1·5 mm over the first part of the gap when it was carrying a current of $\sim$1 A. It is interesting to note that, apart from the actual time-scale, the details of growth are similar to those reported by Allibone and Meek (1938) when larger values of T_1 were used, 10 to 100 μs.

Gallimberti and Stassinopoulos (1971), using a 2/2,200 μs wave, have further elucidated the development of the leader after one of the long "dead-times" of 100 μs; during that dead-time no current was recorded, then a current consisting of superimposed pulses amounting to a few amperes flowed, and either this fell to zero, or it increased rapidly and breakdown followed in a few microseconds, thus supporting oscillographically the time-resolved photographs (Waters and Jones, 1959a, b) which were rendered suspect by the chopping technique.

Details of the growth of the stepped leader were exhibited on a far larger scale by Stekolnikov (1961) in the first records taken with the electron-optical camera, see Fig. 6b. Out of the impulse corona from a 12·5 cm sphere the leader stroke developed at 7×10^6 cm s^{-1} accompanied by showers of leader corona travelling towards the cathode at 4×10^7 cm s^{-1}, a shower developing at every 2 to 3 cm step of the leader's progress. When the leader had crossed 60 cm of a 400 cm rod/plane gap corona from its tip reached the plane. As Stekolnikov argues, if the leader has a high conductivity, the 340 cm of unbridged gap will be stressed with a sufficiently high overvoltage for the leader corona to bridge it. Indeed, in an extended account by Waters (1962) of the Waters and Jones (1961) experiments a calculation shows that after the leader-current had fallen to 0·1 A coinciding with minimum leader velocity of 10^6 cm s^{-1}, the current and velocity then increased and the gradient across the unbridged part of the 200 cm gap rose from the average of 5·5 kV cm^{-1} to 9 kV cm^{-1} before the leader was arrested by the chopping gap.

The advance of the leader as revealed by the third technique, that of using waves of short duration, has been studied by Gallimberti and Rea (1970,

1973), by Aked and McAllister (1972) and by Bahder *et al.* (1974). The first used a 0·9/23 μs wave; the leader can be followed across part of a 150 cm rod/plane gap and then from its tip numerous very strong streamers reached the plane; the stationary photograph strongly resembles the time-resolved records of Waters and Jones (1961). Oscillograms show the leader current (after the impulse corona) falls and then rises again to a few amperes for 10 μs, but then the leader ceases to grow because the anode voltage has fallen. The rapid irregularities of the anode current are indicative of the repeated reilluminations of the leader-track already mentioned. Similar information is provided by the second group of workers using 1/16 μs and 1/39 μs waves. The third group photographed 3 to 7 m rod/plane gaps stressed with 1·0/44 μs waves at V_{50}: the discharges which failed to proceed to breakdown exhibited leader strokes bridging 22% of a 3 m gap and 17% of a 7 m gap. By extrapolation the authors forecast that for gaps larger than 8·5 m the breakdown voltage will not continue to rise linearly with spacing—as already mentioned in Section 2.2—the length of the leader at the moment the streamers of the leader-corona reach the plane is increasing (by a somewhat uncertain extrapolation) so rapidly that breakdown of the gap will require only a modest increase in voltage (see Fig. 1). If this proves to be the case it will be an interesting example to the practical use of leader study.

4.2.2 *Switching Impulse Voltages*

Apart from the accidental use of impulses having T_1 from 10 to 100 μs by Allibone and Meek (1938) and Stekolnikov and Belyakov (1938) due solely to the inclusion of a high resistance in circuit, no study of the leader stroke was made using these voltages until Stekolnikov and Shkilev (1963a) applied the E.O.C. to the study of a 2 m rod/plane gap; as their results have well stood the test of time and substantially describe the leader process as we know it today they are illustrated in Fig. 10 for five values of wave-front times. When T_1 is small one or more bursts of impulse corona occur in the first few microseconds and then the leader stroke starts from the anode and travels at 5×10^6 cm s^{-1}, accompanied by a continuous flow of leader corona, the boundary of which advances to the plane at 4×10^7 cm s^{-1}; when the first streamers of the impulse corona reach the plane the leader has advanced 20 to 30 cm. When $T_1 = 120$ μs, leader velocity falls to less than 10^6 cm s^{-1} and the velocity of the boundary corona to 2×10^6 cm s^{-1}, but in 100 μs, when the corona reaches the plane, the leader is 70 cm long. When $T_1 = 400$ μs, the leader appears to advance in discrete steps and the leader corona in well-defined bushels, velocities have fallen, at 2×10^5 cm s^{-1} and at 400 μs the

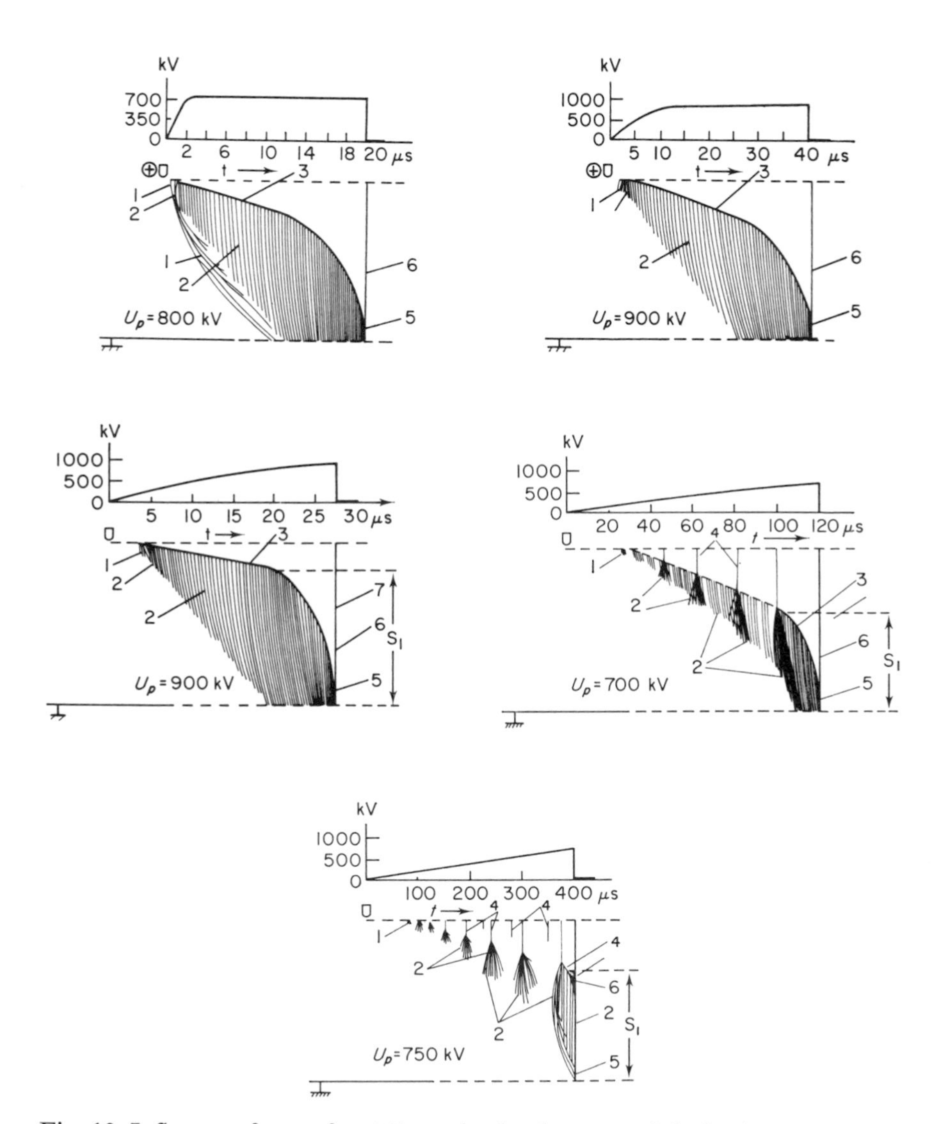

Fig. 10. Influence of wave-front T_1 on the development of the leader stroke (after Aleksandrov *et al.*, 1968). 1. Impulse corona. 2. Leader corona. 3. Leader stroke. 4. Reillumination of leader channels. 5. Junction of downward-directed and upward-directed leaders. 6. Main stroke.

leader has only advanced 50 cm. Thus the critical length of the leader is longest for those impulses which break the gap down at the lowest voltage —for 2 m, $T_1 = 120$ μs.

Bazelyan (1966) correlated the leader advance with the field strength at the anode surface: the leader started to propagate as soon as the gradient reached 28 kV cm^{-1}, irrespective of the value of T_1. The total charge carried into the gap by the leader was far larger than that deposited by the impulse corona, but for a given spacing, the impulse voltage corresponding to the trough of the U-curve produced the smallest charge injection. Stassinopoulos (1968) measured the field at both electrodes for a wave having $T_1 = 200$ μs; he reported an anode gradient of only half the above value, and also found the field at the plane rising in steps for 100 μs as the leader developed. He confirmed (1969) the observation by Stekolnikov and Shkilev (1963a) that in a 200 cm rod/plane gap for which U_{min} occurs when $T_{cr} = 125$ μs the leader bridged 25% of the gap, and with higher values of T_{cr} the leader length diminished.

Aleksandrov *et al.* (1968, 1969) reported leader development over a 15 m rod/plane gap stressed with a voltage rising to maximum in 4300 μs; after the initial impulse corona, bursts of leader corona occurred at intervals of the order of 200 μs, and at each burst, the path back to the anode was reilluminated, the effective velocity was 10^5 cm s^{-1}. After the leader had covered 20% of the gap its further progress was continuous at 10^6 cm s^{-1}, the boundary of the leader corona advancing to the plane faster than this; indeed all the details of this 15 m discharge agree closely with those in Fig. 10. They showed (1970) that if the high-voltage electrode were more extensive, such as a large shield or ring of the kind used on transmission lines to reduce localized stress, the leader inception voltage is increased so that when it starts to develop an overvoltage is applied to it and it then progresses continuously instead of intermittently. Gänger and Maier (1972) recorded intermittent steps with extremely strong reilluminations of the leader channel back to the anode in 8 m gaps, similar to those in Fig. 10: they were associated with extremely large leader coronas contributing 10 to 50 μC each and in the whole leader 300 μC were injected.

The start of the "final jump" has been the subject of investigations ever since 1938. Stassinopoulos and Esposti (1972) applied ramp voltages to a 150 cm rod/plane gap; three rates of rise of voltage were chosen, and the length of the leader was measured up to the point when it started to accelerate for the final jump, the critical length. They confirmed the fact, already established for switching impulses of the double-exponential wave-form, that the longest leader occurred for the minimum ramp voltage to break down the gap, so the final jump, plotted against time to breakdown, displayed a U-curve.

At Les Renardières a concerted attack has been made on the mechanism of breakdown of rod/plane gaps of 5 to 13·5 m using many techniques to study various aspects simultaneously (Les Renardières Group, 1972, 1974). Positive switching impulse voltages of three wave-shapes 90/3,500 μs, 220/3,500 μs and 450/3,500 μs were used in the 1972 tests, and three others, 22/1,500, 500/10,000 and 2,200/12,000 μs were used in the 1974 tests. Cameras record the discharge from the two directions at right angles so that the actual length of the leader can be derived. Large numbers of results have been recorded from which analyses of statistical significance have been made. Nothing has been discovered contrary to anything that has already been reported here; many of the observations for large gaps are identical to those already noted for small gaps. Several shapes of anode were used—cone, hemispherically ended rod and sphere; the following observations relate to the conically ended electrode except where the 1·5 m diameter sphere is specially mentioned.

The effect of the inception voltage on the magnitude of the impulse corona was examined for the 10 m gap; the corona causes the fall in the gradient at the anode and as noted in earlier work, the charge and the fall in gradient are smallest for the wave-shape which causes breakdown of the 10 m gap at the lowest voltage, the trough of the *U*-curve. But the size of the impulse corona does not determine whether the gap will break down or not: this is new information based on a large statistical analysis.

After the impulse corona there is a dead-time for the 10 m gap with all wave-shapes, lasting from 10 to 20 μs, the length depending on the magnitude of the impulse corona. In the dead-time no light was recorded on the photo-multiplier, hence the alternative name, "dark period".

The dead-time is terminated by the start of the leader stroke. For the shortest wave-front, $T_1 = 22$ μs, a second burst of corona occurs—as indeed it does always with waves $T_1 = 1$ μs; this is followed at once by the leader stroke which then advances continuously for low values of T_1, but for $T_1 = 2{,}200$ μs the leader advances in discrete steps. In all cases it carries a current of 0·1 to 1 A depending on the applied voltage. The growth patterns are shown in Fig. 11 for the 10 m gap and at V_{50} the mean total charge injected with each wave is marked by *a*; it will be noticed that this charge is about twice as high for $T_1 = 22$ μs as for the longer wave-fronts. The maximum field at the plane due to the space charge (separated from the geometric field) for a given gap is related to the total charge injected so it rises with increasing leader length and, for a given voltage, is lowest for the wave-front for which the breakdown voltage is the minimum.

The reillumination of the leader channel has been well coordinated with the field-change measurements; an example is given in Fig. 12 for a 10 m gap stressed with a wave for which $T_1 = 500$ μs. A reillumination occurred

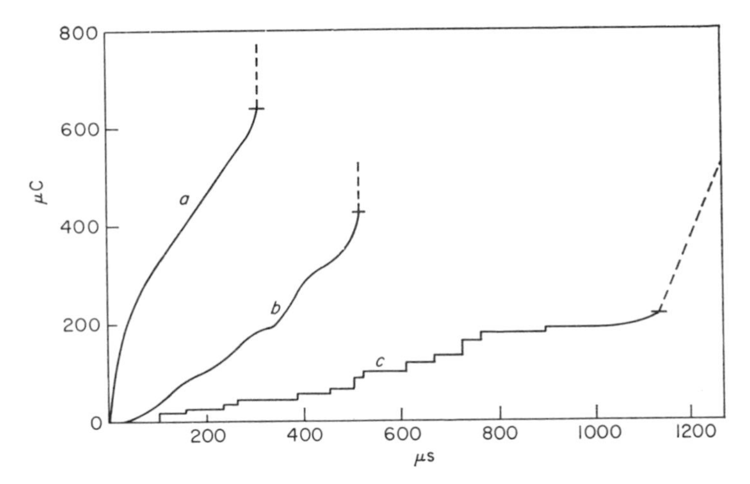

Fig. 11. Charge injected into a 10 m rod/plane gap stressed with V_{50}. The line — marks the start of the "final jump": Curve (a) $T_1 = 22$ μs, (b) = 500 μs, (c) = 2,200 μs.

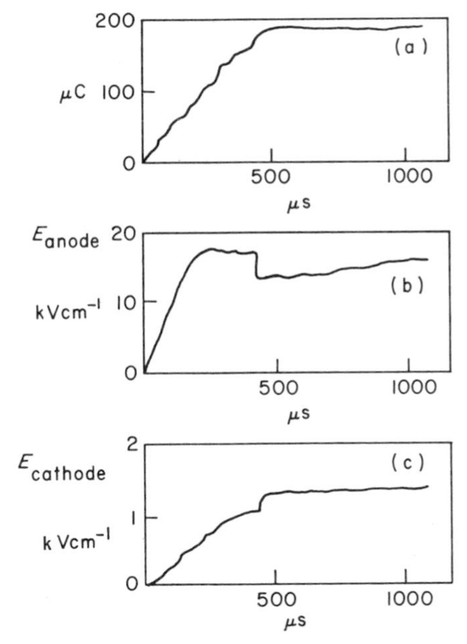

Fig. 12. Effect of reillumination of a leader channel at 380 μs on (a) total charge injected into the gap; (b) the anode gradient; (c) the cathode gradient; for a 10 m rod/plane gap with $T_1/T_2 = 500/10{,}000$ μs wave.

after 380 μs; the leader was 4 m long, the anode gradient fell 2 kV cm^{-1}, the cathode gradient rose 0·3 kV cm^{-1}, the injected charge jumped 40 μC and, from this, calculation shows that the new charge is located at the leader tip—indeed exactly where Bazelyan (1964) showed it to be! Further calculations prove that it is the conductivity of the leader that has changed when this reillumination occurs.

The tortuosity of long sparks is well known to all workers in the field, and the two cameras at right angles have shown the relationship between true leader stroke length, l, and its length projected on the z-axis, l_z. Typically a 5 m gap length is bridged by a 7·5 m leader. For voltages below the breakdown voltage the spread in values of l is longer than the spread when leaders cause breakdown, i.e. those which pursue a moderately straight path will be the more successful in reaching the cathode. The time to breakdown or chop, T_{chop}, is made up of three components, the time of leader inception, t_1, the time of leader propagation, t_2, being the time from t_1 to the moment t_f, when the leader suddenly accelerates and makes the final jump, and the duration of this jump, $T_{chop} - t_f = t_3$. Now t_1 ranges from, say, 15 to 25 μs, t_3 has been found for a 10 m gap to range from 14 to 17 μs; so as T_{chop} can vary over a very wide range, 600 to 100 μs if overvoltages are applied, the variation must mainly occur in values of t_2, i.e. in the amount of tortuosity; it is a well-observed fact that when overvoltages are applied, sparks are very straight.

Reference has already been made to the use of the electrostatic flux meter to locate space charge deposited in a 0·4 m gap (Waters *et al.*, 1968, 1970). Measurements on the 10 m gap showed that the space charge was indeed mainly located at the end of the leader.

At the highest voltage no corona streamers reached the plane, whereas in 0·4 m gaps they did without causing breakdown, so the observation on the long gap fully supports the observation of Brago and Stekolnikov (1958) to which reference has been made. From these measurements two very important gradients may be calculated, see Fig. 13 curve (a); the average value in leaders of different (projected) lengths, l_z, taken from individual shots (but probably also indicative of the temporal change in gradient during the leader's progress), and curve (b) the average value across the unbridged portion of the gap across which the final jump will occur at the point where the curves meet.

The fluxmeter measurements also show the area over which the dish-shaped positive space charge is distributed. For the 5 m gap it is a dish of 4 m diameter; for the 13·5 m gap it is 10 m diameter, larger than one would be led to expect from a study of the photographs. Several seconds elapse before the field created by this charge falls to insignificant values—certainly repetitive impulse voltage testing at 30 s intervals is unlikely to show a drift in the sparkover probability due to accumulation of space charge.

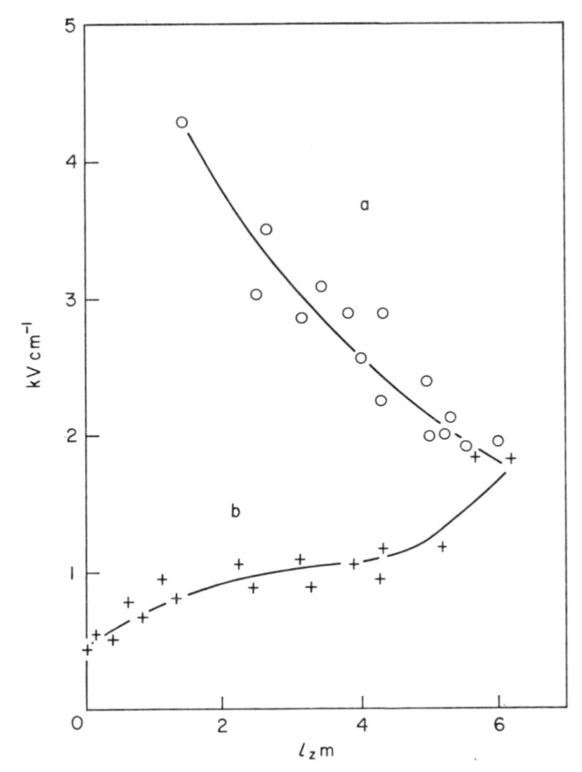

Fig. 13. Average values of gradients for different lengths of leader strokes; (a) in leaders of different (projected) lengths l_z; (b) in the unbridged portion of the gap.

As already stated, a criterion for the start of the final jump is that streamers (or a sufficient number of streamers) of the leader corona must first have reached the cathode plane and thus have liberated a large electron current from the cathode. At that moment T_f, the leader current, luminosity and velocity start of increase sharply and the field at the plane rises. The rapidly accelerating leader has been traced now right to the plane, travelling at over 10^8 cm s^{-1} before the main stroke occurs. The current in the leader in its final advance is primarily set by circuit parameters, oscillograms of breakdown frequently show a gradual fall of voltage before the sudden collapse.

The actual determination of the length l_{zf} at the moment when the final jump begins presents difficulty because of the great halation from the main stroke which follows a few microseconds later. The technique adopted was to measure l_z on image converter photographs which just did not include the main stroke, to note the time of shutter closure and T_f from the oscillograms, and then extrapolate to l_{zf}.

A very important feature of the final jump has been revealed in these coordinated studies though the phenomenon had been noted before. The average projected leader length l_z when V_{50} is applied and the gap does not break down is given in Fig. 14, curves (a) and (b); note that for a 10 m gap

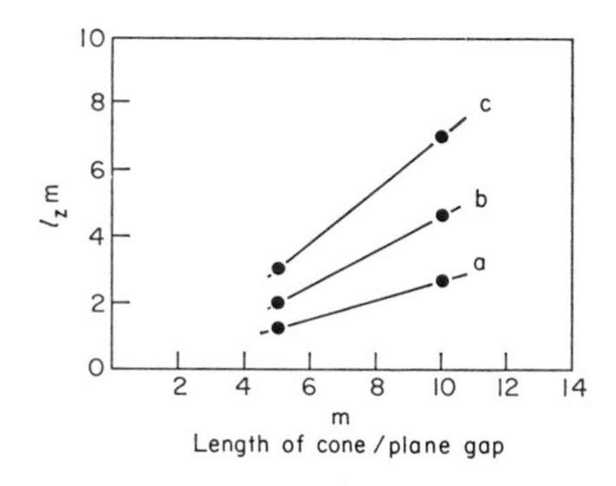

Fig. 14. Measured (projected) lengths l_z of leader strokes in rod (cone)/plane gaps at the voltage V_{50}: (a) mean values of l_z when no sparkover occurs; (b) maximum values of l_z when no sparkover occurs; (c) lengths of leader l_z before the start of the final jump.

the mean is only 3 m and the maximum (b) recorded length was only 4·7 m. But when breakdown occurs the leader has travelled the distance shown in curve (c), l_{zf} before the start of the final jump; for the 10 m gap it has travelled a mean of 7 m. It would therefore appear that the decision to break down or not to break down was taken well before the moment T_f. What then is happening between the moment when the leader, having reached the length $l_{\text{withstand}}$, either ceases to propagate or continues and the moment T_f when its velocity suddenly increases for the final jump? At present no theory gives a firm answer to this question but a possible explanation lies in the change, with time, of the gradient at any one point in the channel. Two phenomena operate: (a) the channel expands at 3×10^4 cm s^{-1} and in doing so its temperature falls, and (b) the channel is constrained by the force of its own self-magnetic field—the pinch-effect force—which has little influence when the density is high but which becomes significant as the channel expands. The net result of these is that at any position in the channel, the radius, and the gradient, follow similar increases and then decrease as shown in Fig. 15. So the change in leader growth from "withstand" to "go" may be due to chance variation in the age of the leader—the difference between the maximum and mean lengths of withstand leaders is shown in the curves (a) to (b) in Fig. 14, it is quite large. The longer, older, channel will be advancing as the

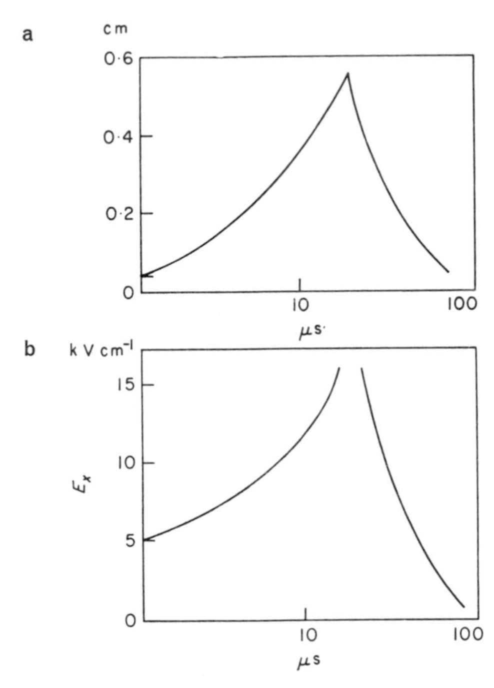

Fig. 15. Change with time, at any point x along a leader stroke, of (a) the radius of the channel, (b) the gradient in the channel.

gradient is falling, for the younger channel the gradient at any point over its length will be increasing. The time-dependency of the gradient may also be the cause of the *U*-curve; the long propagation time, and thus the low gradient along the leader, results in a high gradient of 5 kV cm^{-1} being left across the unbridged section of the gap. Waters (1974) and Klewe *et al.* (1974) report calculations of the temporal growth of the leader over a 10 m rod/plane gap, see Fig. 16; the main gradient over its whole length falls as the leader grows, but its length then remains nearly constant for 100 μs before advancing to the final jump.

Although theory gives a reasonably satisfactory estimate for the sparkover voltage of gaps up to 25 m the estimate for lightning ranges from 3·4 to 32 MV, the higher value being favoured (see Chapter 9.5.3). But all the work reported so far has been on the positive spark, and lightning is essentially a negative discharge—and is d.c., not impulsive!

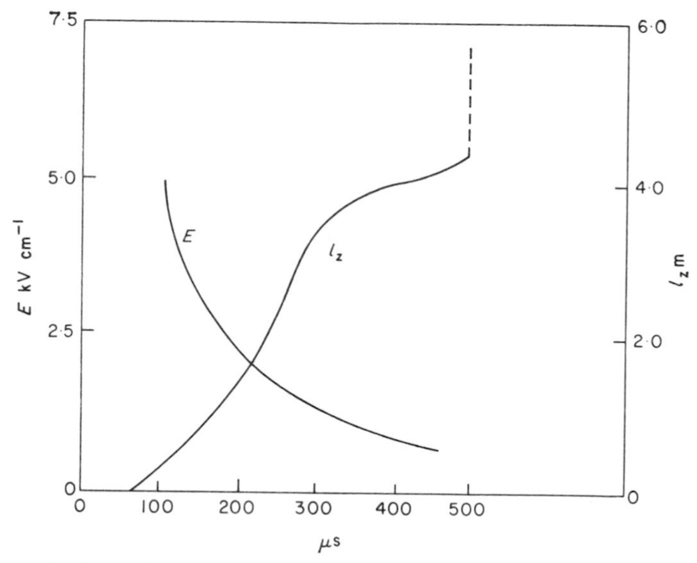

Fig. 16. Fall of gradient E, and growth of the leader stroke (projected) length l_z in a 10 m rod (cone)/plane gap. T_1 = 500 μs and sparkover at the wave crest.

As stated earlier, the characteristics of the final jump are mostly controlled by circuit parameters. There is general agreement that the average gradient across the gap from leader tip to plane is 5 kV cm^{-1} when the last jump starts, and that it proceeds in a very direct line to the plane. Dale and Aked (1974), Garcia and Hutzler (1974), Suzuki and Miyake (1975) and Baldo *et al.* (1975) report that after the streamers have reached the plane they are reilluminated, the head of reillumination proceeding from the plane back to the rapidly descending leader, both velocities being well over 10^8 cm s^{-1}, but fine details of these events are at present on the limit of observation. Circuit parameters likewise control the main stroke.

If this same phenomenon is true of the last jump of the lightning leader stroke any oscillogram taken of the rise of current in a transmission tower or tower of a TV aerial struck by lightning would show a very rapid rise in a few microseconds from the few amperes associated with the rising positive leader to the many thousands of amperes of the main stroke (see Chapter 9.4.1.1), remembering, of course, that the lightning stroke is predominantly of negative polarity.

One matter of interest to students of the lightning discharge has been considered by Dawson *et al.* (1968) and by Uman *et al.* (1968); both groups were working on the same 3·3 MV discharge over a 4 m gap (and therefore there were overvoltages on both polarities). The light output was studied with calibrated photodiodes, and the channel diameters were derived from spectra. Of the total estimated input energy only a small fraction is retained as ionization and thermal energy of the gas, a smaller fraction is radiated in the whole spectral range from ultraviolet to infrared, the balance, probably 90%, appears as shock-wave energy in the 1,000 Hz region.

4.3 The Rod/Plane Discharge of Negative Polarity

Far less attention has been given to the long gap stressed with negative polarity; no theories have been propounded for the observed facts and only some of the differences are worthy of a brief note.

Allibone and Meek (1938) listed the important differences for rod/plane gaps up to 150 cm stressed with waves having T_1 ranging from 10 to 150 μs;

(a) the negative corona is different in character—just as the Lichtenberg figures differ;
(b) the leader develops at first in discontinuous fashion with time intervals of about 5 to 10 μs between the separate advances, until one of them then travels downwards in continuous manner;
(c) the leader is always met by a positive leader ascending from the plane in typical continuous fashion at 3×10^6 cm s^{-1}; the height of the junction point, expressed as a percentage of the gap, decreases as the gap is increased, being 30% for a 150 cm gap;
(d) some distance from the cathode there are filamentary discharges branching in both directions, back towards the cathode as well as onwards; they were named "mid-gap streamers", and they may also be seen in Lichtenberg figures;
(e) there is an apparent mingling of the downward and upward developing shafts of leader corona over quite long distances (but only one camera was available so one discharge might have been "behind" the other);
(f) finally, all these phenomena were most vividly displayed when the pressure was reduced to a half-atmosphere.

Stekolnikov and Shkilev (1962, 1963, 1963b, c) used the EOC to study the negative discharge in a 3 m gap using a 1·5/1,000 μs wave. They too observed the enhanced step-like character of the leader corona, the very luminous tuft of discharge which terminated every corona burst, and they proved that reillumination of the leader corona went backwards to the leader at 10^8 cm s^{-1} —this had not been observed for positive leader corona. Details are best presented pictorially, see Fig. 17. When one of the leader corona shafts reached the plane the positive leader grew upwards. Waters and Jones (1964), using a fast wave, 0·2/2,000 or 0·5/2,000 μs, described the fine filamentary corona radiating from the rod; through these were a few long shafts of corona which branched forwards, but also branched back to the cathode. From the corona, the leader developed—there was no dead-time and its speed was 3×10^7 cm s^{-1}. They too found that when a shaft of leader corona reached the anode the positive leader started to develop.

The fact that the upward leader did not develop before streamers reached the plane was further proved by Gallimberti and Rea (1969) who measured the field at the plane in a 150 cm gap stressed with 1/50 μs waves. The field rose from the "geometric" value of 2 kV cm^{-1} to 11 kV cm^{-1} when a voltage of 90% V_{50} was applied to the cathode; the rise took place as current increments occurred in the negative leader and no corona reached the plane at that voltage.

Switching-impulse voltages were applied to gaps up to 3 m by Stekolnikov and Shkilev (1963a), T_1 ranging from 10 to 30 μs: they noted that no considerable change took place as T_1 was increased, their records are shown in Fig. 18 diagrammatically. Negative leader velocity fell to $0{\cdot}3 \times 10^6$ cm s^{-1} and with long waves, as reported by Allibone and Meek (1938), the leader advances were discrete, separated by many microseconds. The leader corona was still accompanied by a reillumination of the channel back to the cathode at $1{\cdot}3 \times 10^8$ cm s^{-1} and when a streamer of the leader corona reached the anode "light flashed up the whole gap at more than 10^8 cm s^{-1}" before the positive leader channel grew to meet the negative.

Levitov *et al.* (1972) compared the development of the two leaders using waves of $T_1 = 1{\cdot}5$, 20 and 120 μs, $T_2 = 3{,}500$ μs, applied to a rod/rod–plane gap, $d = 5{\cdot}4$ m, the earthed rod being a short projection rising out of the plane. As T_1 increased, the negative leader covered an increasingly large fraction of the gap, and in all cases, when the leaders met, a reillumination of them both extended from the junction point to the cathode and anode. Unfortunately the EOC was cut off just before the meeting point so there was no confirmation of the "ball of light" at that point which Allibone and Meek (1938) reported lasted for 700 μs.*

* The writer has often wondered whether this ball of light lasting an extremely long time could be a manifestation of "ball lightning" (see Chapter 12), but he has never had an opportunity of pursuing the research necessary.

Suzuki (1975) extended the above range of values of T_1 up to 470 μs when studying gaps of 1·0 and 1·5 m between rod and plane: for all values of T_1 from 33 μs to 470 μs the discharge from the cathode was intermittent in form, as in Fig. 18(c), the main discharges numbering 3 to 5 whatever the wave-front, and he therefore likened them to the lightning leader stroke. Each carried a current of 5 to 50 A, the current increasing as the instantaneous voltage on the cathode increased up to the breakdown value, and the last covered one-third of the gap before the transition to breakdown began.

Finally, Gruber *et al.* (1975) have reported briefly on tests at Les Renardières on gaps from 1·5 to 7 m, using waves having T_1 ranging from 3 to 900 μs. The leader development is exactly as drawn by Stekolnikov *et al.* (1962), see Figs 17 and 18, and the growth of reillumination in both directions, mid-gap streamers, was confirmed. With extremely long wave-fronts the discontinuous nature of the leader was greatly enhanced, intervals of 80 μs elapsing between reilluminations of the leader channel. The growth of the continuous phase of leader from the last of the discrete shafts was reported and when one or more streamers reached the plane from a height of 2·8 m for a 4 m gap, the final jump began; positive leaders grew up from the plane and junction occurred at 25% of the 4 m gap from the plane.

4.4 The Sphere/Plane Discharge

Norinder and Salka (1951) studied extensively the corona patterns which developed from a small sphere/plane gap. Obviously the inception voltage for corona increased as the sphere diameter increased and all discharges of both polarities developed from the sphere along very straight paths for 1 to 10 cm before they started to branch; thereafter they resembled the corona patterns from a conical electrode. Time-resolved photographs (Allibone and Meek, 1938; Stekolnikov, 1961, see Fig. 6) showed that these straight shafts developed at speeds in excess of 10^8 cm s^{-1}, but the further progress in the form of a leader stroke was indistinguishable from that from a rod electrode.

5. Future Problems

One of the main outstanding problems from both an engineering and a physical point of view is the full reconciliation of the positive with the negative polarity discharge. The different appearance was shown 200 years ago, yet no satisfactory explanation for this has been given and the lightning strokes of different polarities manifest these differences too.

a

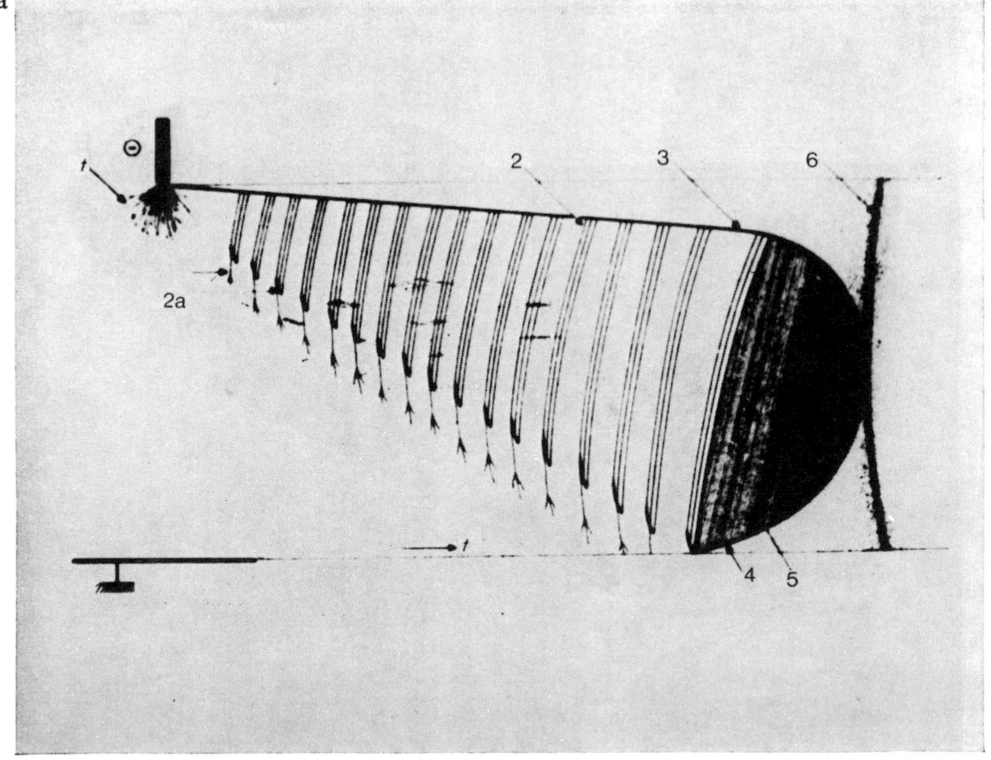

b

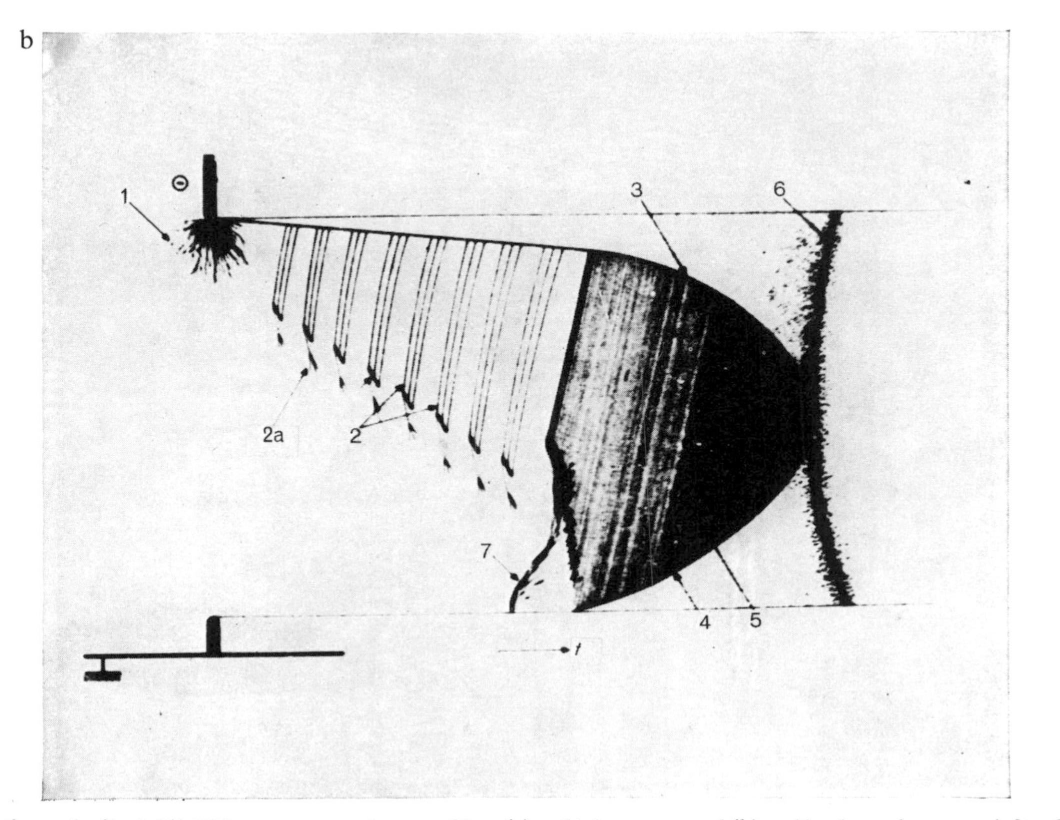

Fig. 17. Negative polarity 1·5/1,000 μs wave sparkover of 3 m (a) rod/plane gap and (b) rod/rod on plane gap (after Stekolnikov and Shkilev, 1962, 1963c). 1. Impulse corona. 2. Reilluminated leader corona. 2a. Leader corona. 3. Leader stroke. 4. Positive upward developing leader. 5. Intermingled corona. 6. Main stroke. 7. Rod/rod on plane gap; positive impulse corona.

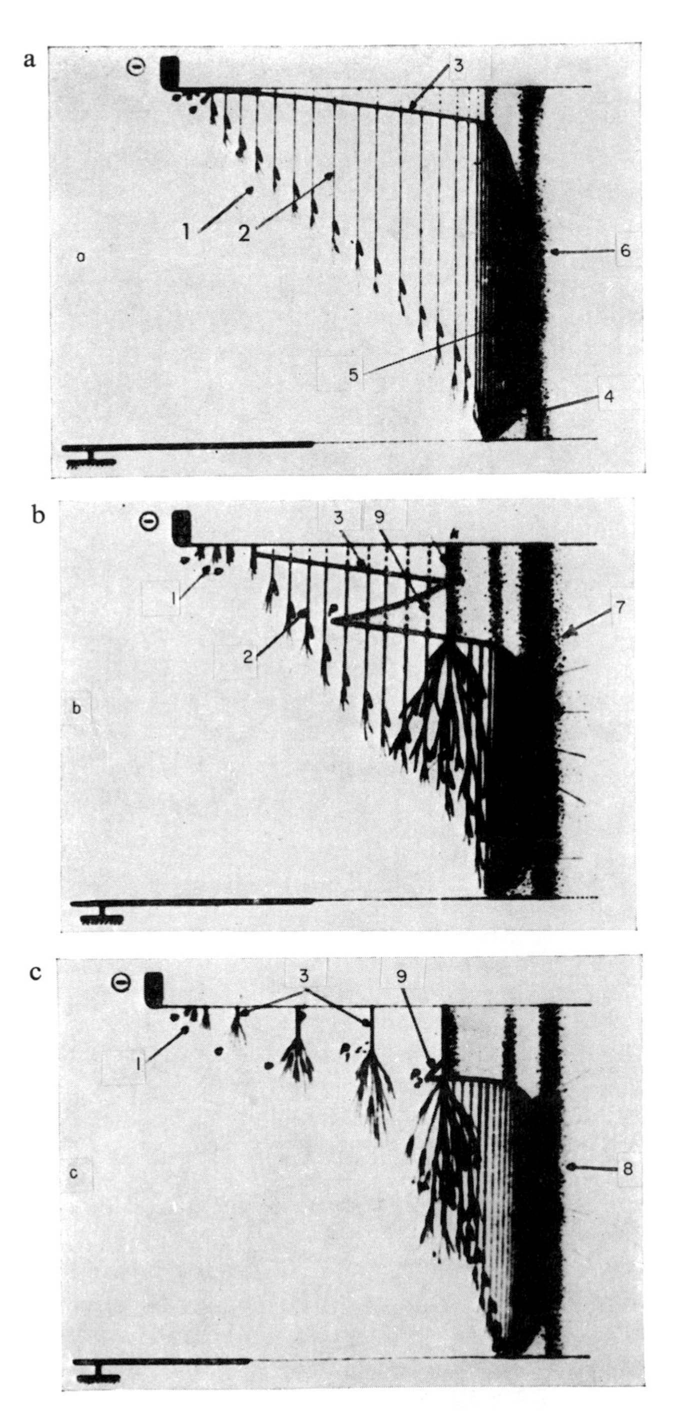

Fig. 18. Negative polarity switching impulse sparkover of a 3 m rod/plane gap (after Stekolnikov and Shkilev, 1963a). (a) 250 kV μs^{-1}; (b) 100 kV μs^{-1}; (c) less than 40 kV μs^{-1}. 1. Leader corona. 2. Reilluminated corona stems. 3. Leader stroke. 4. Upward developing positive leader. 5. Intermingled negative and positive leader coronas. 6, 7, 8. Main stroke. 9. Retrograde discharges.

The reasons for the difference in the mean sparkover gradients between the two polarities need to be further elucidated; when higher test voltages become available does this difference still persist, and in the case of the negative polarity spark does the sparkover voltage/spacing curve bend over to indicate an asymptotic limit?

The student of the long spark would like to see the lightning stroke photographed with all the techniques available for the study of the laboratory spark. In particular, he would like to know more about the corona at the ends of the steps of the stepped leader stroke; very little has been added to knowledge of this since before the Second World War. How near is the pattern of growth to that of the 10 million volt discharge which is being so well studied in the laboratory?

The breakdown of large gaps under d.c. voltages might add significantly to knowledge of lightning leader stroke progress. From an engineering point of view the part played by the upward-developing positive leader stroke from a high point to meet the descending negative leader stroke needs much further study with the highest voltages available; more might then be learnt about the "attractive power" of transmission towers and lines for descending lightning leader strokes and more about the shielding angles of earthed conductors.

Enough information has been given in Section 2 to show that although agreement on switching impulse sparkover data is fair, it is not good; not good enough to form the basis for any accepted tests for electrical apparatus, and careful comparative studies should be organized with a view to reducing the spread of test results.

Finally, the study of the physics of the discharge will present challenges for many years but for this the very high voltages are not necessary; their use has been very valuable but the discharge does not differ in character for being very long, indeed the value of the 10 MV studies lies in proving that the discharge has not changed its nature by growing in length.

References

Aked, A. and McAllister, I. W. (1972). The effect of impulse voltage wavetail duration on the pre-breakdown phenomena in long positive point/plane gaps. Int. Conf. HV Discharges, Munich, pp. 260–264.

Akopian, A. A., Larianov, V. P. and Torosian, A. S. (1954). On impulse discharge voltages across high voltage insulation as related to the shape of the voltage wave. Conf. int. grands Res. Elect. Report No. 411.

Aleksandrov, G. N., Kizevetter, V. E., Rudakova, V. M. and Tushnov, A. N. (1962). The discharge voltages of long air gaps and strings of insulators in the presence of a.c. *Electrichestvo* **5**, 27–32.

Aleksandrov, G. N., Redkov, V. P., Gorin, B. N., Stekolnikov, I. S. and Shkilev, A. V. (1968). Peculiarities of the electric breakdown of extremely long airgaps. *Dokl. Akad. Nauk SSSR* **183**, 1048.

Aleksandrov, G. N., Redkov, V. P., Gorin, B. N., Stekolnikov, I. S. and Shkilev, A. V. (1969). Investigation of the development of spark discharges in long airgaps. Ninth Int. Conf. Phen. Ionized Gases, Bucharest, p. 281.

Aleksandrov, G. N., Redkov, V. P., Gorin, B. N., Stekolnikov, I. S. and Shkilev, A. V. (1970). Peculiarities of spark discharge in long airgaps of high voltage structures. Instn Elect. Engrs Gas Discharge Conf. pp. 309–312.

Allibone, T. E. (1937). International comparison of impulse-voltage tests. *J. Instn elect. Engrs* **81**, 741–750.

Allibone, T. E. (1938). The mechanism of the long spark. *J. Instn elect. Engrs* **82**, 513–521.

Allibone, T. E. (1967). "Electrical Breakdown at High Voltages", pp. 1–12. University of Strathclyde Press, Glasgow.

Allibone, T. E. and Dring, D. (1974a). Factors influencing sparkover of large gaps in air when stressed with impulse voltages. Inst. elect. electron. Engrs Winter Meeting, January, Paper C47 036–0.

Allibone, T. E. and Dring, D. (1974b). The variation of ion density in a high voltage laboratory during impulse voltage testing. *Proc. Instn elect. Engrs* **121**, 401–402.

Allibone, T. E. and Dring, D. (1974c). Sparkover of a rod/plane gap with non-standard impulse voltages; effect of added radiation. Instn Elect Engrs Colloquium on Protection of Electrical Systems, May, Paper 6.

Allibone, T. E. and Dring, D. (1974d). Effect of humidity on sparkover of air gaps under impulse voltages. *Proc. Instn elect. Engrs* **121**, 221–222.

Allibone, T. E. and Dring, D. (1975). Influence of the wavefront of impulse voltages on the sparkover of rod gaps and insulators. *Proc. Instn elect. Engrs* **122**, 235–238.

Allibone, T. E., Dring, D. and Dragan, G. (1975). Sparkover of rod gaps stressed with impulse voltages of non-standard waveshapes. *Proc. Instn elect. Engrs* **122**, 581–584.

Allibone, T. E., Hawley, W. G. and Perry, F. R. (1934). Cathode-ray oscillographic studies of surge phenomena. *J. Instn elect. Engrs* **75**, 670–688.

Allibone, T. E. and Meek, J. M. (1937). Development of the spark discharge. *Nature* **140**, 804.

Allibone, T. E. and Meek, J. M. (1938). The development of the spark discharge. *Proc. R. Soc.* A **166**, 97–126, **169**, 246–268.

Allibone, T. E. and Schonland, B. F. J. (1934). Development of the spark discharge. *Nature* **134**, 735–736.

Baatz, H. (1962). Comparative tests with impulse voltages on rod gaps. Conf. int. grands Res. Elect. Report No. 325.

Bahder, G., Garcia, F. G., Barnes, H. C. and McElroy, A. J. (1974). Nonlinearity in impulse breakdown of very large air gaps. *Trans Inst. elect. electron. Engrs* **93**, 960–964.

Baldo, G. and Gallimberti, I. (1970). Charge and field measurements in rod/plane gaps under impulse voltages. Instn Elect. Engrs Gas Discharge Conf. pp. 576–579.

Baldo, G., Gallimberti, I., Garcia, H. N., Hutzler, B., Jouaire, J. and Simon, M. F. (1975). Breakdown phenomena of long gaps under switching impulse conditions, influence of distance and voltage level. *Trans. Inst. elect. electron. Engrs* **94**, 1131–1140.

Barnes, H. C. and Winters, D. E. (1971). UHV transmission design requirements—switching surge flashover characteristics of extra long air gaps. *Trans Inst. elect. electron. Engrs* **90**, 1579–1588.

Bazelyan, E. M. (1964). The measurement of the space charge in the initial stages of a positive long spark. *Zh. tekh. Fiz.* **34**, 474–483.

Bazelyan, E. M. (1966). Effect of corona charge on the formation of a long positive spark under the action of a voltage pulse. *Zh. tekh. Fiz.* **36**, 365–373.

Bazelyan, E. M., Brago, E. N. and Stekolnikov, I. S. (1960). The large reduction in mean breakdown gradients in long discharge gaps with an oblique-sloping voltage wave. *Dokl. Akad. Nauk SSSR* **133**, 550–553.

Berger, K. (1956). Comparative tests on spark gaps. Conf. int. grands. Res. Elect. Report No. 326.

Blasius, P., Schneider, K. H., Vogel, O. and Weck, K. H. (1973). Generation of lightning impulses in UHV-test circuits. *Electra* **27**, 35–63.

Boylett, F. D. A., Edwards, H. G. J. and Williams, B. G. (1966). Impulse corona discharge. Seventh Int. Conf. Phen. Ionized Gases, pp. 647–650.

Brago, E. N. and Stekolnikov, I. S. (1958). *Izv. Akad. Nauk SSSR* **11**, 50.

Caldwell, R. O. and Darveniza, M. (1973). Experimental and analytical studies of the effect of non-standard wave shapes on the impulse strength of external insulation. *Trans. Inst. elect. electron. Engrs* **92**, 1420–1428.

Carrara, G. (1964). Investigation on impulse sparkover characteristics of long rod/rod and rod/plane gaps. Conf. int. grands Res. Elect. Report No. 328.

Collins, M. M. C. and Meek, J. M. (1965). Measurement of field charges preceding impulse breakdown of rod/plane gaps. Seventh Int. Conf. Phen. Ionized Gases, Belgrade, pp. 581–585.

Colombo, A. and Mosca, W. (1972). Performance of sphere and rod/rod gaps under high direct voltages. *Trans. Inst. elect. electron. Engrs* **91**, 501–509.

Conf. int. grand Res. Elect. Task Force Committee 33.03.03 (1973). Switching impulse test procedure for phase-to-phase air insulation. *Electra* **30**, 55.

Dale, S. J. and Aked, A. (1974). Effect of gap configuration on the breakdown probability and on the discharge development in long point/plane gaps. Inst. Elect. Engrs Gas. Discharge Conf. pp. 192–196.

Dawson, G. A., Richards, C. N., Krider, E. P. and Uman, M. A. (1968). Acoustic output of a long spark. *J. geophys. Res.* **73**, 815–816.

Gallimberti, I. and Rea, M. (1969). Field current measurements during negative rod/plane discharges in long air gaps. Ninth Int. Conf. Phen. Ionized Gases, p. 170.

Gallimberti, I. and Rea, M. (1970). Leader development in positive rod/plane discharges. Instn. Elect. Engrs Gas Discharge Conf. pp. 298–302.

Gallimberti, I. and Rea, M. (1973). Development of spark discharges in long rod/plane gaps under positive impulse voltages. *Alta Freq.* 264–275.

Gallimberti, I. and Stassinopoulos, C. A. (1971). Types of breakdown in positive rod/plane gaps. Tenth Int. Conf. Phen. Ionized Gases, p. 170.

Gänger, B. E. and Maier, E. G. (1972). Studies of spark formation at high switching voltages of positive polarity. *Trans. Inst. elect. electron. Engrs* **91**, 2427–2436.

Garcia, H. N. and Hutzler, B. (1974). Electrical breakdown in long air gaps. Instn Elect. Engrs Gas Discharge Conf. pp. 206–210.

Gruber, G., Hutzler, B., Jouaire, J. and Riu, J. P. (1975). Long sparks of negative polarity. High Voltage Conference, Zurich.

Harada, T., Aoshima, Y., Ishida, Y., Ichihara, Y., Anjo, K. and Nimura, N. (1970). Influence of air density on flashover voltages of air gaps and insulators. *Trans. Inst. elect. electron. Engrs* **89**, 1192–1202.

Harada, T., Aihara, Y. and Aoshima, Y. (1971). Influence of humidity on lightning and switching impulse flashover voltages. *Trans. Inst. elect. electron. Engrs* **89**, 1433–1442.

Harada, T., Aihara, Y. and Aoshima, Y. (1973). Influence of switching impulse wave shape in flashover voltages of air gaps. *Trans. Inst. elect. electron. Engrs* **92**, 1085–1091.

Hughes, R. C. and Roberts, W. J. (1965). Application of flashover characteristics of air gaps to insulation co-ordination. *Proc. Instn elect. Engrs* **112**, 198–202.

Kachler, A. J., La Forest, J. J. and Zaffanella, L. E. (1971). Switching surge flashover of UHV transmission line insulation. *Trans. Inst. elect. electron. Engrs* **90**, 1604–1611.

Klewe, R. C., Waters, R. T. and Jones, B. (1974). Review of models of breakdown Inst. Elect. Electron Engrs Summer Conference, pp. 29–40.

Knudsen, N. and Iliceto, F. (1970). Flashover tests on large air gaps with d.c. voltage and with switching surges superimposed on d.c. voltage. *Trans. Inst. elect. electron. Engrs* **89**, 781–787.

Komelkov, V. S. (1947). Structure and parameters of a leader discharge. *Bull. Acad. Sci. USSR* **8**, 955–966, *Dokl. Akad. Nauk SSSR* **58**, 57.

Les Renardières Group (1972). Research on long airgap discharges at Les Renardières. *Electra* **23**, 53–157.

Les Renardiéres Group (1974). Research on long airgap discharges at Les Renardiéres. *Electra* **35**, 49–156.

Levitov, V. I., Bazelyan, E. M., Pulavskaja, I. G. and Volkova, O. V. (1972). The peculiarities of discharges developed in long gaps with negative voltage polarity. Int. Conf. HV Discharges, Munich, pp. 314–318.

McEachron, K. B. (1939). Lightning to the Empire State Building. *J. Franklin Inst.* **227**, 149–217.

Meek, J. M. and Collins, M. M. C. (1965). Measurement of electric fields at electrode surfaces. *Electronics Letters* **1**, 110.

Menemenlis, C. and Harbec, G. (1974). Switching impulse breakdown of EI–IV transmission towers. *Trans. Inst. elect. electron. Engrs* **93**, 255–261.

Menemenlis, C. and McGillis, D. (1974). Switching impulse breakdown of airgaps with application to the design of EHV/UHV external insulation. Conf. int. grand. Res. Elect. Report No. 33.08.

Norinder, H. and Salka, O. (1951). Mechanism of positive spark discharge with long gaps in air at atmospheric pressure. *Ark. Fys* 3, **19**, 347.

Paris, L. (1967). Influence of airgap characteristics on line-to-ground switching surge strength *Trans. Inst. elect. electron. Engrs* **86**, 936–946.

Paris, L. and Cortina, R. (1968). Switching and lightning impulse discharge characteristics on large airgaps and long insulator strings. *Trans. Inst. elect. electron. Engrs* **87**, 947–955.

Park, J. H. and Cones, H. N. (1956). Surge voltage breakdown of air in a non-uniform field. *J. Natn. Bur. Stand. USA* **56**, 201–224.

Phillips, T. A., Robertson, L. M., Rohlfs, A. F. and Thompson, R. L. (1967). Influence of air density on electrical strength of line insulation *Trans. Inst. elect. electron. Engrs* **86**, 948–961.

Prabhakar, B. R., Nandagopal, M. R. and Gopalakrishna, H. V. (1971). Effect of humidity and temperature on impulse flashover of airgaps. *Proc. Instn elect. Engrs* **118**, 823–824.
Ryan, H. M. and Powell, C. W. (1972). 50 Hz breakdown characteristics of long airgaps. Instn Elect. Engrs Int. Conf. Gas Discharges, pp. 30–32.
Saxe, R. F. and Meek, J. M. (1948). Development of spark discharges. *Nature* **162**, 263.
Saxe, R. F. and Meek, J. M. (1955). The initiation mechanism of long sparks in point/plane gaps. Instn Elect. Engrs Monograph 124.
Schneider, K. H. and Weck, K. H. (1974). Parameters influencing the gap-factor. *Electra* **35**, 25–45.
Schonland, B. F. J. and Collens, H. (1933). Progressive lightning. *Nature* **132**, 407.
Schonland, B. F. J. and Collens, H. (1934). Progressive lightning. *Proc. R. Soc.* **143**, 654–674.
Stassinopoulos, C. A. (1968). Simultaneous electric field measurements at both electrodes of a rod/plane gap. *Proc. Instn elect. Engrs* **115**, 1225–1226.
Stassinopoulos, C. A. (1969). The influence of waveform on impulse breakdown of positive rod/plane gaps. Ninth Int. Conf. Phen. Ionized Gases, Bucharest, p. 261.
Stassinopoulos, C. A. and Esposti, G. D. (1972). Behaviour of rod/plane gaps under positive ramp voltages. Second Instn Elect. Engrs Gas Discharge Conf. pp. 27–29.
Stassinopoulos, C. A. and Meek, J. M. (1967). The influence of space charges on spark growth in rod/plane gaps. Eighth Int. Conf. Phen. Ionized Gases, Vienna, p. 206.
Stekolnikov, I. S. (1961). New information on initial stages of a spark. *Dokl. Akad Nauk SSSR* **141**, 1076–1077.
Stekolnikov, I. S. and Belyakov, A. P. (1937). Investigation of a spark discharge. *Elektrichestvo* **8**, 49–50.
Stekolnikov, I. S. and Belyakov, A. P. (1938). Experimental study of spark discharge. *J. exp. theor. Phys.* **8**, 444.
Stekolnikov, I. S. and Shkilev, A. V. (1962). New information on the development of a negative spark and its comparison with lightning. *Dokl. Akad. Nauk SSSR* **145**, 782–785.
Stekolnikov, I. S. and Shkilev, A. V. (1963). *Trans. Sov. Phys. -Dokl.* **7**, 712–716.
Stekolnikov, I. S. and Shkilev, A. V. (1963a). Development of a long positive spark in the case of an exponential voltage wave front. *Dokl. Akad. Nauk SSSR* **151**, 837–840.
Stekolnikov, I. S. and Shkilev, A. V. (1963b). Investigation of the mechanism of the negative spark. *Dokl. Akad. Nauk SSSR* **151**, 1085–1088.
Stekolnikov, I. S. and Shkilev, A. V. (1963c). The development of a long spark and lightning. Int. Gas Discharge Conf. Montreux, pp. 466–481.
Stekolnikov, I. S., Brago, E. N. and Bazelyan, E. M. (1962). Decrease of discharge potentials over wide gaps by use of oblique waves. *Zh. Tekh. Fiz.* **32**, 993–1000, Leatherhead Conference, pp. 139–146.
Suzuki, T. (1975). Breakdown process in rod to plane gaps with negative switching impulses. *Trans. Inst. elect. electron. Engrs* **94**, 1381–1389.
Suzuki, T. and Miyake, K. (1975). Breakdown process of long airgaps with positive switching impulses. *Trans. Inst. elect. electron. Engrs* **94**, 1021–1033.
Udo, T. (1963). Sparkover characteristics of large gap spaces and long insulation strings. *Trans. Inst. elect. electron. Engrs* **83**, 471–490.

Udo, T. and Watanabe, Y. (1968). D.C. high voltage sparkover characteristics of gaps and insulator strings. *Trans. Inst. elect. electron. Engrs* **87**, 266.
Uman, M. A., Orville, R. E. and Sletten, A. M. (1968). Four-meter sparks in air. *J. appl. Phys.* **39**, 5162.
Volkava, O. V. and Chernyshov, V. I. (1967). Electrical characteristics of 500 kV protective gaps. *Elektrichestvo* **3**, 19–21.
Watanabe, Y. (1967). Switching surge flashover characteristics of extremely long airgaps. *Trans. Inst. elect. electron. Engrs* **86**, 933–936.
Waters, R. T. (1962). Streak photography and other studies of the long spark in air, *in* "Gas Discharges and the Electricity Supply Industry", pp. 38–53, Butterworth, London.
Waters, R. T. (1974). Electric gradient in the leader channel of the long spark. Instn Elect. Engrs Gas Discharge Conf. pp. 182–186.
Waters, R. T. and Jones, R. E. (1959a). Space charge phenomena in impulse sparkover. Fourth Int. Conf. Phen. Ionized Gases, pp. 413–417.
Waters, R. T. and Jones, R. E. (1959b). Time-resolved photography of arrested discharges. Fourth Int. Conf. Phen. Ionized Gases, pp. 408–411.
Waters, R. T. and Jones, R. E. (1961). High voltage impulse breakdown in rod/plane gaps in air. Fifth Int. Conf. Phen. Ionized Gases, Munich, pp. 992–1002.
Waters, R. T. and Jones, R. E. (1962). New methods of producing high-voltage square waveforms and their application to spark photography. *Proc. Instn elect. Engrs* **109** A, 144–150.
Waters, R. T. and Jones, R. E. (1964). The impulse breakdown voltage and time-lag characteristics of long gaps in air. (I) The positive discharge. (II) The negative discharge. *Phil. Trans. R. Soc.* A **256**, 185–234.
Waters, R. T., Jones, R. E. and Bulcock, C. I. (1965). Influence of atmospheric ions on impulse corona. *Proc. Instn. elect. Engrs* **112**, 1431–1438.
Waters, R. T., Rickard, T. E. S. and Stark, W. B. (1968). Electric field and current density in the impulse corona discharge in a rod/plane gap. *Proc. R. Soc.* A **304**, 187–210.
Waters, R. T., Rickard, T. E. S. and Stark, W. B. (1970). The structure of the impulse corona in a rod/plane gap. *Proc. R. Soc.* A **315**, 1–25.

8. Lightning Spectroscopy

R. E. ORVILLE

National Center for Atmospheric Research, Boulder, Colorado, U.S.A.

1. History

The physical processes occurring in the lightning discharge have been an object of study for just over a century. Of the present techniques for studying lightning, which include measuring the current, the electric and magnetic fields, the acoustic signal, the light emissions and the spectral emissions from the channel, it was the observation of the latter which provided the first clues to the physical characteristics of the lightning flash. Among the earliest observers were Herschel (1868), Gibbons (1871), Holden (1872) and Clark (1874). It was Herschel who first identified a nitrogen line as being the brightest in the visible spectrum and who noted that the relative intensities of lines change from spectrum to spectrum. Furthermore, he noted a variation between the line and continuum spectrum and recorded a sharp line in the red region which in later years would be identified as *H*-alpha. Gibbons made similar observations and suggested prophetically that this area of research would be a "promising field of investigation". Holden was one of the first to record his visual observations by noting the relative positions of seven lines in the spectrum which today can be identified as six lines due to singly ionized nitrogen atoms and one line from hydrogen, *H*-alpha. Further observations of the bright continuum were reported by Clark.

It appears that Schuster (1880) made the first systematic identifications of the lines in the spectrum of lightning. He studied the region 5,000 to 5,800 Å using a direct vision spectroscope with a slit moveable by means of a micrometer screw. Wavelengths could then be determined by leaving the instrument set for night-time observations and in the morning observing the Fraunhofer lines in the vicinity of the previous night's settings. In this way, singly ionized nitrogen lines (NII) were identified at 5,002 and 5,681 Å. Many bands were also observed, but Schuster correctly suggested that these were probably numerous lines which could not be resolved.

During the next half century, investigators constructed a table of spectral lines produced by the lightning discharge. Using an 8 inch telescope equipped with an objective prism, J. H. Freese (Pickering, 1901) recorded a spectrum on film showing nearly 30 bright lines, including several from the Balmer series of hydrogen. Accurate wave-length identifications, however, were difficult to obtain using an objective prism with its inherent nonlinear dispersion. This problem was corrected by Slipher (1917) who made a significant contribution by obtaining the first photographic spectra of lightning using a slit spectrometer. Lines of nitrogen and oxygen were identified in the region between 3,830 and 5,000 Å with good accuracy and comparison was made with a laboratory spark spectrum in air. The wave-length identifications were subsequently extended by Dufay (1926) to the ultraviolet region where ozone limited the observations to 2,860 Å and into the red region to the limit of detection at 6,550 Å.

Additional spectroscopic studies of lightning over the next few decades confirmed and added lines to those already attributed to oxygen and nitrogen and extended the species identifications to minor constituents in the atmosphere. Israel and Wurm's (1941) work is representative of the thorough studies which characterized this period of research. A few years later, Dufay (1947) reported an OH band at 3,068 Å, an NH band at 3,362 Å and a possible identification of NO. Hu (1960) confirmed the presence of OH and NO in both lightning and laboratory spark studies.

The first studies of the infrared spectrum of lightning were reported by Jose (1950) in the region 7,400 to 8,800 Å. Lines due principally to neutral nitrogen and oxygen atoms (NI, OI) were observed. Petrie and Small (1951) extended the infrared observations to 9,100 Å using an auroral slit spectrograph which recorded one spectrum from many flashes in one hour. They identified OI, NI, AI and CI* and concluded that all radiation in the 7,100 to 9,100 Å region is from neutral atomic species. The following year, Knuckles and Swenson (1952) filled in the red region of the spectrum with a detailed study between 6,200 and 7,100 Å. Within this narrow region they reported the emissions from hydrogen, CI, NI, NII, OI, OII, AI, N_2(2p) and N_2^+ and possibly several weak bands due to OH and NH.

It is interesting to note that from the earliest spectroscopic studies of lightning in the 1870s to those in the 1950s, the research was almost exclusively devoted to identifying the emission features of the spectrum. Dufay (1949) and Israel and Fries (1956) were the first to consider the spectrum of lightning

* The latter species identified as CI at 9,078 Å was not found by Wallace (1964) and was reported by him to be NI at 9,061 Å.

as a source of quantitative information about physical conditions in and around the discharge channel. The latter especially noted the necessity of obtaining the spectra of individual strokes in a flash. However, it was not until Salanave (1961) independently implemented photography and quantitative analysis of slitless spectra of lightning strokes that a real change in the direction of optical studies of lightning occurred. The remaining part of this chapter is due in large part to his initiative and interest which stimulated over a decade of research in lightning spectroscopy.

2. Experimental Techniques and Qualitative Results

The principal techniques for observing the light emissions from the lightning channel have utilized slit spectrometers, slitless spectrometers, and recently photoelectric systems.

2.1 Slit Spectrometers

The most common spectrometer involves the use of an entrance slit, a collimating mirror, a diffraction grating, a focusing mirror, and a detector which in lightning studies is usually photographic film. Each spectral line is in effect an image of the slit. The advantages of this spectrometer are that wave-lengths can be accurately determined and good wave-length resolution can be obtained. The best example of these data has been published by Wallace (1964), who obtained spectra primarily of lightning reflections from clouds.

Using a Meinel 9 inch f/0·8 spectrograph, Wallace (1964) carried out a detailed examination of lightning spectra in the region 3,150 to 9,800 Å. He identified a number of molecular species and attributed them to N_2 second positive and N_2^+ first negative bands, CN violet bands and NH bands. Neutral atomic species were observed and identified as NI, OI, H, AI and CI. The NI multiplets 20, 21 and 31 (Moore, 1959) were considerably broader than the instrument profile and Wallace concluded that they were pressure broadened. The OI (10) multiplet also appeared diffused and was probably pressure broadened since its upper level is only 0·85 eV below the ionization potential. The only hydrogen lines observed were H-alpha and H-beta in the Balmer series. Numerous singly ionized lines were identified, but all were attributed to previously reported NII emissions. No OII emissions were observed. Absorption features were apparent and could be associated with the usual water vapour and molecular oxygen features.

Unfortunately, the disadvantages of the slit spectrometer in general and in the above data in particular are that many discharges must be observed before the light produces an acceptable recorded image. The result is that any physical differences between lightning flashes are obscured during the integration time. Consequently, no conclusions can be made concerning physical differences between flashes or differences between the luminous components of the lightning flash. Fortunately, this problem can be largely eliminated by adapting a slitless spectrometer to the study of lightning.

2.2 Slitless Spectrometers

When a light source is narrow, as in the case of lightning, the slit can be eliminated and the source serves effectively as the slit. The result is that now the spectral lines are images of the lightning channel and their shape duplicates that of the channel. The slitless spectrometer produces wavelength dispersion of the image of the channel. The characteristics of the slitless spectrometer include (a) the effective slit width is the width of the luminous source, which provides a measurement of the source width for close flashes and determines the spectrometer resolution for distant flashes, (b) one exposure is sufficient for a recorded image and (c) the channel properties can be studied as a function of height. Since 1960, slitless spectrometers have been used extensively in lightning spectroscopy. Their use can basically be divided into two applications, (a) spectrometers which record the spectrum of a lightning flash and thus *time integrate* the emissions from the components in one flash and (b) spectrometers which have the capability to isolate a narrow section of the channel and move the recording film at a speed to *time resolve* the luminous components in one flash.

2.2.1. Time-integrated Data

Salanave (1961) and Salanave *et al.* (1962) revived an interest in lightning spectroscopy with the publication of the most detailed spectra in the visible region followed by similar accomplishments in the ultraviolet (Salanave, 1964) and in the infrared regions (Salanave, 1966). The best examples of these data are shown in Fig. 1 and cover the region from 3,000 to 8,700 Å. The spectrum throughout this region consists of atomic and singly ionized emission lines and molecular bands in emission as well as in absorption against a weak continuous spectrum.

The UV slitless spectrum in Fig. 1a is dominated by singly ionized nitrogen and oxygen lines, the latter being notably absent in the slit spectra obtained by Wallace (1964). Salanave (1964) first reported the prominent lines due to ionized nitrogen and oxygen at 3,325–3,329 Å and 3,437 Å (NII) and 3,727 and 3,749 Å (OII) which had not previously been detected in slit spectra. Four unidentified features at 3,150, 3,550, 3,840, and 3,883 Å, and a fifth at 3,914 Å (unmarked) are near regions where emissions are sometimes strong in slit spectra and attributed to N_2, 3,159 Å, N_2^+, 3,548 and 3,914 Å and CN, 3,851 and 3,883 Å. These emissions are generally believed to occur in the continuing current luminosity and are usually weak or absent in slitless spectra (Orville and Salanave, 1970), but generally prominent in slit spectra (Wallace, 1964). The horizontal streaks of continuum emission are observed to become very faint in the 2,850 Å region and this is due to absorption by the Hartly absorption band of ozone. A similar u.v. spectrum of a flash which occurred only 0·85 km from a spectrograph enabled Orville (1967) to estimate that a few tenths of a millimetre (0·01 to 0·05 cm) of ozone existed between the spectrograph and the flash. This is significant in that it is one order of magnitude greater than the clear air value and may have existed as a mantle around the lightning channel or have been evenly distributed in the intervening path. One previous analysis for ozone using lightning spectra was published by Dufay (1949). He used a slit spectrograph, obtained a u.v. spectrum from several flashes, and calculated the ozone necessary to explain the sharp cut-off in the 3,020 to 3,060 Å region in his spectrum. The resulting reduced ozone thickness of 0·3 cm, however, cannot be directly compared to Orville's (1967) value since the distances to Dufay's (1949) flashes are unknown. It would appear nevertheless that the thunderstorm can be a significant source of ozone.

Many of the first slitless spectra published by Salanave were obtained with an aerial camera equipped with a transmission diffraction grating in front of an f/6·0 Aero-Tessar lens of 61 cm focal length. A representative spectrum is reproduced in Fig. 1(b) and is one of over two hundred spectra. This slitless spectrum in the 3,900 to 6,900 Å region was obtained with a dispersion of 25 Å mm^{-1} and a resolution ($\lambda/\Delta\lambda$) of 1,500 at 5,000 Å. Over 160 emission lines have been reported (Salanave *et al.*, 1962) and most of them can be attributed to neutral and singly ionized nitrogen and oxygen. Two lines are attributed to hydrogen, H-beta, a broad and diffuse line at 4,861 Å and H-alpha, the most intense line in this spectrum, located at 6,563 Å. A faint O_2 (1–0) absorption band is observed at 6,867 Å and no doubly ionized lines are reliably identified. Densitometer traces of this type of spectra show that

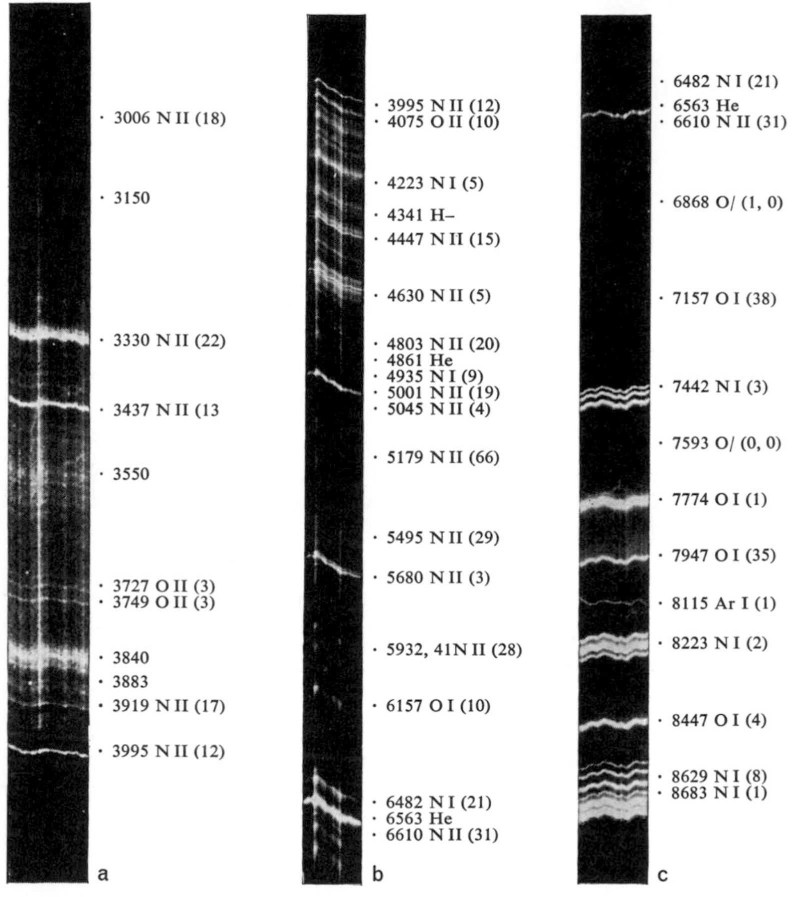

Fig. 1a. The ultraviolet spectrum of lightning obtained on Kodak Plus X film with a transmission grating and fore-prism, both of fused quartz. A Schott UG-5 filter cuts off the unwanted visible spectrum; ozone around the channel and/or in the intervening air path cuts off the atmospheric transmission somewhat below 3,006 Å, as shown by the fading part of the figure (sharp horizontal streak in upper part of picture is a scratch on the film) (Orville and Salanave, 1970).
Fig. 1b. The visible spectrum of lightning, photographed on Tri-X Aerecon film with a slitless spectrograph. The wave-lengths, in ångströms, are only approximate and are merely intended, along with the emitting atom or ion, to identify the source or the principal emitted in cases of a blend. H-beta and H-gamma are weak and greatly broadened by the Stark effect (Orville and Salanave, 1970).
Fig. 1c. The infrared spectrum of lightning obtained on Kodak High-speed infrared film and using the same spectrograph as for Fig. 1b. The Aero Tessar lens is designed to perform especially well in the red to infrared region, hence the excellent focus over the entire spectrum of H-alpha to nearly 8,700 Å (Orville and Salanave, 1970).

neutral emissions are more intense higher in the channel and singly ionized emissions are less intense, a variation of relative line intensity which was first observed by Fox (1903).

Infrared slitless spectra of lightning were reported by Salanave in 1966 and one of the best examples is reproduced in Fig. 1(c). The spectrum is dominated by line emissions from neutral atomic species superimposed upon a continuum which apparently extends without attenuation throughout the recorded infrared region. The same spectrograph used to record Fig. 1(b) was used except that Kodak high-speed infrared film was substituted. Figure 1(c) shows the spectrum from a 350 m vertical section of a lightning channel. Several O_2 absorption bands are apparent, the Fraunhofer A band at 7,593 Å and the B band at 6,868 Å. It has been suggested by Salanave (1966) that the O_2 absorption features could be used to estimate the distance to lightning flashes within 10 km but no quantitative results have been published to date. All of the lines identified in the infrared slitless spectra (Salanave, 1966) agree with the lines identified by Wallace (1964) in his slit spectra and no new lines were observed in the slitless spectra.

It is interesting to note the differences first reported by Meinel and Salanave (1964) in the N_2^+ emission in lightning. The strength of these emissions is distinctly different when observed with a slitless spectrograph as compared with a slit spectrograph. Meinel and Salanave conclude that the wide variations reported in the literature are a function of the method of observation and suggest that the molecular emissions recorded in slit spectra occur at some distance from the visible lightning stroke, probably in terminal streamers in the cloud. The occasional recording of molecular emissions in slitless spectra, particularly N_2, probably arises from the continuing current luminosity in cloud-to-ground strikes. It is worth noting that in over a hundred years of lightning spectroscopy, and particularly in the last fifteen years of increased activity, only *one* spectrum of an intra-cloud flash has been obtained, and that was of poor quality (Larsen, 1905).

A survey of the spectral emissions in the visible region (e.g. Fig. 1b) led Salanave and Brook (1965) to suggest and publish the results of the use of a narrow-band interference filter centred at H-alpha to record lightning flashes in daylight. The system works in part because the H-alpha line is the location of the intense absorption (Fraunhofer C) line in the solar spectrum. Krider (1966a) furthered this technique by pointing out the advantages of Kodak Linagraph Shellburst film with its excellent red sensitivity. Later Krider and Marcek (1972) noted that a Kodak 92 filter used in conjunction with Kodak Shellburst film was sufficient to restrict the spectral sensitivity to a narrow region near the H-alpha line in the lightning spectrum. This latter system is desirable because no interference filter is used with its inherent off-axis shift of transmission properties and is therefore suitable for wide-angle applications.

The spectra in Fig. 1 are representative of records which integrate all the emissions from a lightning stroke or flash over the duration of the emissions. It is interesting for qualitative and certainly for quantitative purposes to time-resolve the emissions in the discharge on a time scale which will reveal significant temporal changes in the spectrum of one flash.

2.2.2 *Time-resolved Data*

The first attempt to time-resolve the luminous events comprising the lightning flash was apparently made by Israel and Fries (1956). They used a slitless spectrograph with a slotted disc moving rapidly in front of fixed film. Observations were not successful in Germany, however, and so the instrument was loaned to Salanave in Arizona who obtained data the first time the spectrograph was used (Salanave, 1965). Unfortunately, the instrument is designed so that time and position on the channel are both variable and consequently it is difficult to interpret the results. With a fixed slot and moving film, only time is variable and interpretation of the results is much less ambiguous. Furthermore, the time resolution was only 25 μs which is almost an order of magnitude too slow. Fortunately, the development of high-speed cameras in the United States provided instruments with sufficient time resolution and light gain to allow their use as slitless spectrographs.

The first photographic records of spectral emissions with good time resolution (2 to 5 μs) and good wave-length resolution (10 Å) were published by Orville (1966a, 1968a). The data were obtained by converting a Beckman and Whitley high-speed streaking camera to a slitless spectrograph and adding a Bausch and Lomb replica transmission grating in front of the camera's objective lens. The grating was blazed for 5,500 Å, had 600 lines/mm and produced a dispersion of 72 Å mm^{-1} in the focal plane. A 200 mm objective lens focused the return stroke on a 0·5 mm horizontal slit. Thus a 10 m section of the lightning channel was effectively isolated at a distance of 4 km. Data obtained with this system are presented in Figs 2, 3 and 4 and a quantitative analysis of these data is presented in Section 3.3.

Time-resolved spectral emissions from all components of the lightning flash have been recorded. Figure 2 shows a return stroke (Orville, 1968a), Fig. 3 shows the stepped leader preceding a first return stroke (Orville, 1968b), and Fig. 4 shows the dart leader preceding a subsequent return stroke (Orville, 1975). It is apparent in Figs 2 to 4 that all recorded emissions in the return stroke are due to neutral hydrogen or to neutral or singly ionized atoms of nitrogen and oxygen. No molecular or doubly ionized emissions have been identified in over 100 spectra of which Figs 2 to 4 are three examples. The return stroke spectra show that the luminosity rises from zero to peak in less than 10 μs and that emission lines due to neutral species persist for over 100 μs

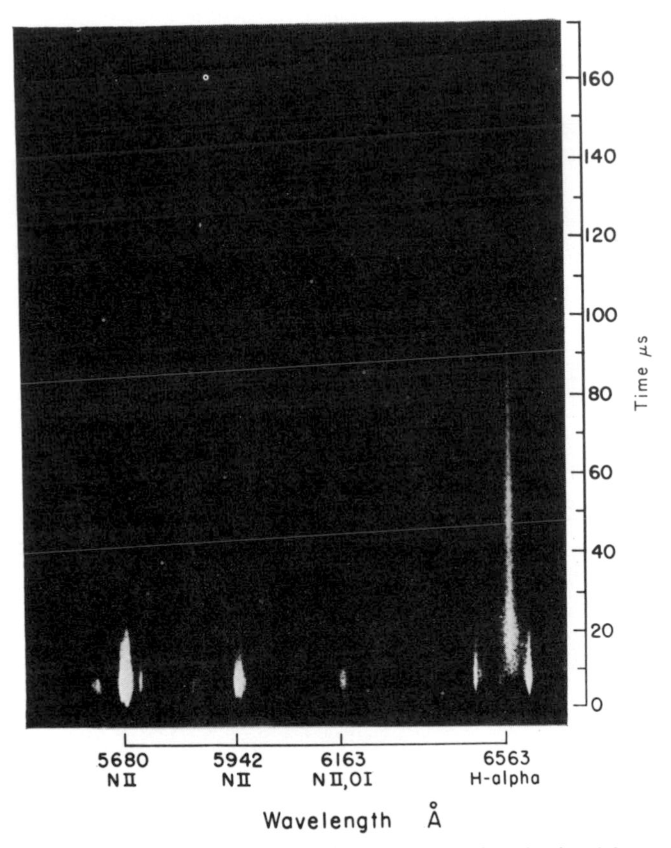

Fig. 2. Time-resolved slitless spectrum of a return stroke obtained from a narrow vertical section of the lightning channel with 5 μs resolution. The NII emissions have been analysed for temperature and the H-alpha emission for electron density (Orville, 1968a).

in Figs 2 and 4 and for more than 800 μs in Fig. 3. The recorded time sequence of emissions in the return stroke is (a) line radiation from singly ionized atoms, (b) continuum radiation and (c) line radiation from neutral atoms. It should be pointed out that this sequence is consistent with the effective excitation potential of the background continuum lying between that of the ions and neutrals and hence this emission probably results from radiative recombination or radiative attachment (Uman, 1966).

The presence of NII radiation at 5,942 Å in Fig. 2 is interesting in view of its absence in Wallace's (1964) slit spectra. This may be the result of the short emission time and a water vapour absorption band in this region

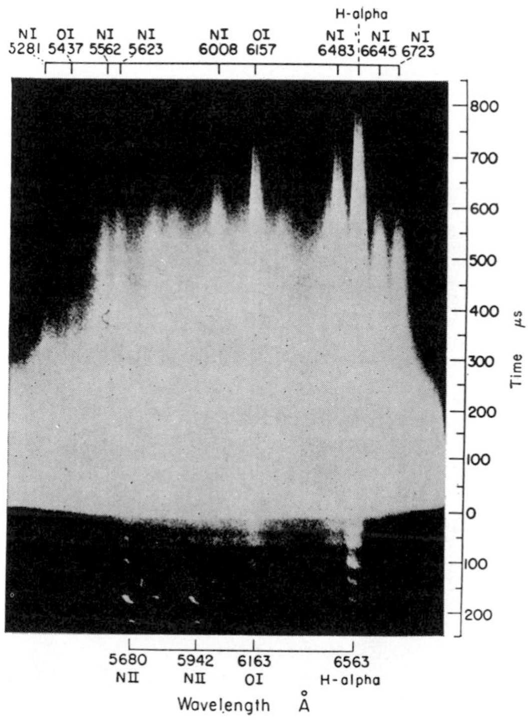

Fig. 3. Time-resolved slitless spectrum of a return stroke showing the stepped leader emissions from a vertical section of the channel which precede the first stroke in a flash. The time resolution is approximately 20 μs. The bright step at 180 μs before the return stroke has been analysed for electron temperature from the NII emissions at 5,680 and 5,942 Å (Orville, 1968b).

which prevents any significant exposure being recorded in the slit spectra of Wallace. The variation in spectral regions covered by the slitless spectra in Figs 2 to 4 is because the fortuitous positioning of the lightning channel relative to the slitless spectrographs determines the wave-length range recorded.

The chance positioning of a lightning stroke within 0·5 to 1·0 km of the spectrograph and the intermittent emissions of the stepped leader produced the spectrum shown in Fig. 3. Emissions were recorded from a 2 m vertical section of the channel approximately 100 m above the ground. Distinct NII emissions occur in the steps at intervals of 31 to 42 μs and are superimposed upon a modulating continuum and neutral emissions whose intensity increases with time. It is interesting that the characteristic stepping in the leader is only apparent in the singly ionized radiation. This leads to the obvious suggestion that if Schonland and his co-workers had used film sensitive to H-alpha emissions, the first Boys camera photographs of lightning would have revealed a "pulsating leader" instead of the "stepped leader" which precedes each first negative return stroke (Chapter 5.5.4).

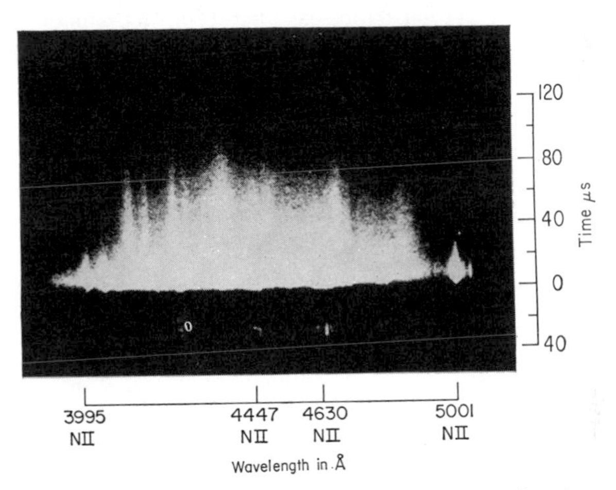

Fig. 4. Time-resolved slitless spectrum similar to Figs 2 and 3, but now the dart leader spectral emissions are shown preceding a subsequent flash. The time resolution is approximately 10 μs (Orville, 1975).

The spectrum of a subsequent stroke from a different flash is shown in Fig. 4 and provides a direct comparison of the dart leader spectrum to the stepped leader spectrum. Singly ionized emissions from a 13 m section of the channel dominate the spectrum and there is no evidence of emissions from H-beta at 4,861 Å except in the subsequent return stroke. The spectral emissions decrease in intensity after the passage of the dart, but are continuously recorded until the return stroke occurs. A quantitative analysis of these data is presented in Section 3.3.3. Film, of course, is only one medium for recording the spectral emissions and spectrometers equipped with photoelectric detectors have provided valuable corroborative data for the photographic studies.

2.3 Spectrometers using Photoelectric Detectors

The first successful attempt to time-resolve the spectral features of individual return strokes in cloud-to-ground lightning discharges was actually performed a year before the successful experiments reported in the previous section. Krider (1965) used narrow bandpass interference filters to isolate spectral regions of interest and in particular studied four emission features monitored simultaneously from first return strokes. He used photomultiplier tubes with oscillographic recording and time-resolved the emissions with 5 μs resolution. Krider studied the NII multiplets at 3,995 and 4,041 Å and the continuum and molecular emissions at 3,650 and 3,914 Å, respectively. An H-beta filter was used to study the HI emissions because H-alpha was close to the photomultiplier threshold. Krider's results included the observation of different zero-to-peak intensity rise times for different emission features. The NII lines reach intensity maxima before the continuum, which in turn is at a maximum before the H-beta line. The effect of the continuum radiation was found to be small, particularly during the early stages of a stroke.

A year later, Krider (1966b) extended his studies to the first photometric analysis of lightning with correlated still photographs of the discharge channels. Krider used a calibrated silicon diode and an oscilloscope; the combination acting as a fast response lightning photometer covering the visible and infrared portions of the spectrum from 0·4 to 1·1 μm. The photodiode and associated circuitry had a response time of less than 1 μs. Since a vertical section of the channel was not isolated, Krider observed signals rising to peak intensity in 40 to 100 μs which was just the time for the geometrical growth of the channel. Krider estimated the energy deposition in the channel to be 10^7 W m^{-1}.

At the time of Krider's studies in Arizona, similar experiments were being performed in Sweden. Scuka (1969) and Lundquist and Scuka (1970) developed an electronic optical system for lightning research which uses interference filters to isolate regions centred at 3,905 and 6,555 Å with a 15 Å width at 6 dB attenuation. With 2 μs resolution, they clearly see a faster rise time in the blue as compared to the red region in 150 oscillograms obtained from 18 thunderstorms in 1966. Simultaneously, measurements were made of the electrostatic field and the electrostatic field changes. There is an indication that the red pulses appear about 15 μs before the violet pulses and the electric field changes. No explanation is given, but it may be that the emissions in the red region (H-alpha plus continuum radiation) are produced by leaders preceding the return stroke.

One of the more thorough studies of the lightning discharge with photodetectors was conducted by Barasch (1970). He measured peak spectral intensities emitted by first and subsequent return strokes, stepped and dart leaders, cloud-to-air flashes, intra-cloud pulses, and *M*-components of return strokes in five narrow regions of the visible and near visible spectrum. All data were obtained on a millisecond scale and the detectors were wide-field narrow-bandwidth radiometers which viewed the entire channel so that signals were a combination of geometrical growth of the channel and real spectral changes within the channel. Barasch observed that the spectral intensities in averaged return stroke emissions relative to 3,914 Å (N_2^+ plus continuum) are 0·95 at 4,140 ° (continuum), 2·1 at 6,563 (H-alpha plus continuum), 4·8 at 8,220 Å (NI (2) plus continuum) and 0·7 at 8,900 Å (continuum). It was found that return strokes vary in intensity by a factor of 1,000 from storm to storm and by a factor of 50 within one flash. This result is probably a function of the varying peak currents in strokes (Chapter 9.3) and stroke duration (Chapter 9.4) which could not be resolved by the low time resolution of the system (1 ms). Similar spectral characteristics were observed for leaders and return strokes although the time resolution was inadequate to resolve the pulsating spectral characteristics shown in Fig. 3. A curious result is the lack of any strong N_2^+ (3,914 Å) radiation in Barasch's data which is present in some slit spectrograms in the literature.

The highest time resolution reported to date has been published by Krider (1974) who recorded, with 0·2 μs resolution, relative light intensities of five leader steps immediately preceding a return stroke in a cloud-to-ground flash. He used a silicon photodiode with a circular 50° field of view which was filtered to have a spectral response which was relatively flat over the visible and near infrared from 0·4 to 1·0 μm. Krider observed that a typical step rises from zero to peak in about 1 μs, falls to half peak in 1 to 2 μs, and then decays slowly until the next step occurs. No attempt has been made yet to use interference filters and isolate spectral emissions from the leader with this same high time resolution.

2.4 Calibration Techniques

The use of photodetectors avoids calibration problems and provides data whose relative intensities can readily be obtained. Film, on the other hand, has a detector response or film density which is a function of wave-length, the time of exposure, humidity and temperature at the time of exposure, and the time, type and temperature of the development of the film. Furthermore, the film sensitivity is a nonlinear function of wavelength. It should be clear that the measurement of spectral line intensities by integrating the area under lines and above the continuum, and the measurement of line profiles requires a careful calibration procedure to obtain meaningful relative intensities. Details describing the calibration process involved are given by Prueitt (1963) and by Orville (1966b, 1968a).

3. Quantitative Analyses

3.1 Introductory Remarks

The first analyses of the lightning spectrum beyond line identifications were performed by Dufay and Dufay (1949) when they reported from the analysis of the Stark broadening of H-beta that the fraction of ionized atoms and molecules was of the order of 5×10^{-4}. A decade later, the first estimates of the lightning temperature were published by Wallace (1960) who suggested, on the basis of the CN band spectra in flash integrated spectra, that the rotational temperature was in the range of 6,000 to 30,000 K. A year later, Zhivlyuk and Mandel'shtam (1961) observed the broadening of H-alpha and H-beta lines in data obtained with slit spectrographs and estimated that the electron density in lightning flashes was greater than or equal to 10^{17} cm^{-3}, a value suggesting that the earlier work of Dufay and Dufay was considerably in error. Zhivlyuk and Mandel'shtam further presented a calculation showing that a 10 cm diameter lightning channel is optically thick in the centre of the 3,995 Å NII line for reasonable temperatures and densities. They therefore assumed that within $\pm 0{\cdot}5$ Å from the centres of three NII lines and one OII line that the emitted radiation is blackbody. Consequently, they compared the intensities at the centres of these lines with the Planck radiation law to deduce an approximate temperature of 20,000 K. All values, however, were derived from slit spectra composed of the integrations of emissions from many flashes. Consequently, the properties of individual return strokes were not first deduced until Salanave (1961) obtained spectra of individual return strokes. The analysis of these data and subsequent high-speed time-resolved spectra (Figs 2, 3, 4) requires first a consideration of theory applicable to the lightning flash.

3.2 Theory

A quantitative analysis of time-resolved lightning spectra is based upon the assumptions that (a) the lightning channel is optically thin to the spectral line of interest, (b) the region of the channel cross-section from which the radiation of interest is observed must to some reasonable approximation be at uniform temperature and (c) the discrete atomic energy levels from which the transitions leading to the measured NII line emission occur must be populated according to Boltzmann statistics (Uman, 1969). The temperature determined under these assumptions is then called the electron temperature. If the more restrictive assumption is used, namely that (c′) the lightning channel is in local thermodynamic equilibrium (LTE), then the temperature determined by the analysis of singly ionized radiation is a temperature in the thermodynamic sense. We will first consider the assumption (a) of optical thinness.

We assume that the energy levels from which the NII radiation originates are populated according to Boltzmann statistics, then

$$N_n = \frac{Ng_n}{B(T)} \exp(-\varepsilon_n/kT), \tag{1}$$

where N_n is the number density of atoms in energy level n, N is the total number density of atoms, ε_n is the excitation energy and g_n the statistical weight of level n, k is Boltzmann's constant, T is the absolute temperature and $B(T)$ is the partition function

$$B(T) = \sum_j g_j \exp(-\varepsilon_j/kT), \tag{2}$$

where the summation is over all energy levels. From Equation (1) the ratio of the number density for two levels, n and m, is

$$\frac{N_n}{N_m} = \frac{g_n}{g_m} \exp[(\varepsilon_m - \varepsilon_n)/kT]. \tag{3}$$

If the gas is optically thin, then we can write the intensity of an emission line due to a transition from level n to level r as

$$I_{nr} = CN_n A_{nr} k\nu_{nr}, \tag{4}$$

where C is a geometrical factor, A_{nr} is the Einstein transition probability, and ν_{nr} is the frequency of the emitted photon. If we assume that LTE exists, then from Equation (1) we can write

$$I_{nr} = \frac{CNg_n A_{nr} k\nu_{nr}}{B(T)} \exp(-\varepsilon_n/kT). \tag{5}$$

A similar expression can be written for transition from level m to level p and then from Equation (5) the ratio of the intensity of two emission lines from the same species is

$$\frac{I_{nr}}{I_{mp}} = \frac{g_n A_{nr} \nu_{nr}}{g_m A_{mp} \nu_{mp}} \exp[(\varepsilon_m - \varepsilon_n)/kT]. \tag{6}$$

If the two lines are within the same multiplet then the upper energy levels and the line frequencies are approximately the same so that

$$\frac{I_{nr}}{I_{mp}} = \frac{g_n A_{nr}}{g_m A_{mp}} \tag{7}$$

or

$$\frac{I_{nr}}{I_{mp}} = \frac{g_n f_{nr}}{g_m f_{mp}} \tag{8}$$

since f is proportional to $A\lambda^2$. Note that the emission oscillator strength, f_{nr}, is related to the absorption oscillator strength, f_{rn}, by

$$f_{nr} = -\frac{g_r f_{rn}}{g_n}. \tag{9}$$

It is the absorption oscillator strength which is usually listed in tables, e.g. Griem (1964, p. 363).

Uman and Orville (1965) have used Equation (8) to check the opacity in flash-resolved, time-integrated strokes by comparing measured intensity ratios on the left-hand side of Equation (8) with theoretically calculated gf ratios by Griem (1964) and experimental laboratory measurements of the gf ratios by Mastrup and Wiese (1958). Despite experimental intensity ratio errors of 10 to 30% with the lightning data there is generally good agreement which suggests that the lightning channel is optically thin. However, Hill (1972) points out that the Uman and Orville (1965) values are consistently lower by $24 \pm 8\%$ than the laboratory values observed by Mastrup and Wiese for an optically thin source. Hill further claims that there is large absorption of H-alpha from the heated air continuum in a lightning channel, particularly during the first few microseconds of the return stroke. Hill's results appear to be highly model dependent, but they nevertheless indicate that the lightning channel may not be optically thin during the first few microseconds of the discharge. Orville (1966b) analysed one time-resolved stroke spectrum for opacity and found deviations from optical thinness in the first few microseconds of the stroke. Clearly, the assumption of optical thinness in the first few microseconds of a lightning return stroke has not yet been proved. Additional data with greater wave-length resolution are required to answer this important question.

Now consider the second assumption (b) that the region of the channel cross-section from which the radiation of interest is observed must to some reasonable approximation be at uniform temperature. This assumption to a great extent is a function of the opacity of the channel. Uman (1969) has theoretically shown that the volume of the lightning channel radiating visible

NII lines must have a diameter of approximately 1 mm or less if the temperature is 30,000 K and 1 cm or less if the temperature is 20,000 K for the volume to be optically thin. If the channel is optically thin, then a temperature profile which is not flat will quickly become and remain flat due to optically thin radiation loss. This could occur even if the channel is optically thick in the ultraviolet and at some line centres. It appears that assumption (b) of a flat temperature profile is probably true in the stages after the first few microseconds because the channel becomes optically thin and the temperature profile becomes and remains flat.

The third assumption (c) requires that the atomic energy levels involved in the transitions leading to visible NII line radiation are populated according to Boltzmann statistics, a condition which requires that the collisional rate processes from a level exceed the radiative decay rate from the same level. It should be noted that the collisional rate processes increase as the spacing between energy levels decreases, that is with increasing energy levels, and that the radiative decay rates decrease with increasing energy levels (Griem, 1964). For electron densities of 10^{17} cm^{-3} and higher, Uman (1969) has shown that the collisional transition rate from the lowest energy level used in lightning temperature measurements, the $3s\,^3P^\circ$ level at 18·4 eV above the ground state in the NII ion, is in excess of 10^{10} s^{-1} and exceeds the radiative decay rate. Consequently, over a hundred collision-induced transitions occur in 0·01 μs. Therefore if the channel changes its properties on a time scale longer than 0·01 μs, a Boltzmann distribution will exist above the lowest $n = 3$ level as a function of time (Uman, 1969, p. 954).

Frequently, a more restrictive assumption replaces (c), namely that to a reasonable approximation LTE exists in the return-stroke channel. This is probably true when the channel is near atmospheric pressure because the effect of the electric field in separating particle temperature in the return-stroke channel should not be significant. On the other hand, in the first few microseconds the channel pressure is high (Chapter 11.3.2), the electric field is also probably high and we would not expect LTE to exist in the channel.

Under the preceding assumptions we can now obtain an expression for the temperature in the lightning channel. From Equation (6), we write

$$T = (\varepsilon_m - \varepsilon_n)/k \Big/ \ln\left(\frac{I_{nr}\,g_m\,A_{mp}\,\nu_{mp}}{I_{mp}\,g_n\,A_{nr}\,\nu_{nr}}\right). \tag{10}$$

Under assumption (c), this is the electron temperature and under assumption (c′), this is the true thermodynamic temperature.

Prueitt (1963) has used Equation (10) to calculate temperatures from flash-resolved stroke-integrated slitless spectra obtained by Salanave. The relative intensities of NII multiplets centred at 3,995, 4,041, 4,433, 4,630 and

5,679 Å were measured for five different strokes. Two to four positions were densitometered along three of the channels to give nine temperature values which range from 24,200 to 28,400 K with standard deviations which range from 400 to 1,000 K. Prueitt concluded that the temperatures differ from stroke to stroke and also vary at different positions on the stroke.

No multiplets other than NII have been used for temperature calculations. Hill (1977) has recently rewritten Equation (10) for the intensity ratio of the H-alpha/H-beta lines to obtain

$$T = \frac{0{\cdot}66}{k} \Big/ \ln\left(\frac{0{\cdot}46\, I_\alpha}{I_\beta}\right) \tag{11}$$

but no reliable data have yet been analysed to determine how temperatures determined from neutral lines compare with temperatures determined from NII lines.

In addition to the temperature, the electron density in the return stroke can be determined by noting the broadened image of the H-alpha line which is predominantly the result of the Stark effect. Fortunately, the electron density determination is independent of the assumption of LTE. Griem (1964) has tabulated the Stark profiles, shifts and widths of many lines including those of hydrogen. Consequently a measurement of the Stark-broadened H-alpha line is sufficient to determine the electron density. Uman and Orville (1964) report electron densities between $1 \times 10^{17}\,\text{cm}^{-3}$ and $5 \times 10^{17}\,\text{cm}^{-3}$ for three strokes which were studied by Prueitt (1963). These data, we recall, were stroke-integrated slitless spectra so that the values represent an average for the duration that the spectral emissions were recorded. Furthermore, these results assumed that (a) the H-alpha line is predominantly radiated while the stroke has one value of electron density, (b) the electron, ion and neutral hydrogen densities across the stroke are approximately constant, (c) all effects which may broaden the H-alpha line are negligible in comparison with the Stark effect, and (d) there is little self-absorption of H-alpha within the stroke. The extent to which these assumptions are valid can be partially determined by turning to the analysis of high-speed time-resolved data, which display the emissions of the visible components of lightning flashes on the order of microseconds (Figs 2, 3, 4).

3.3 Results of Stroke-resolved Spectroscopy

All quantitative analyses discussed to this point have been the result of emissions integrated over the duration of the stroke, which also include any emissions from the leaders preceding the strokes. We will now consider the physical characteristics derived from the high-speed time-resolved spectra of return strokes, a stepped leader and several dart leaders.

3.3.1 *Return Strokes*

Ten return-stroke spectra, of which Fig. 2 is one example, have been analysed for temperature as a function of time using Equation (10) (Orville, 1968a). Eight spectra were obtained with 5 μs resolution (e.g. Fig. 2) and two were obtained with 2 μs resolution. The peak temperatures range from 20,000 to 36,000 K with peak values in five of the ten spectra in a narrow interval from 28,000 to 31,000 K. Errors are reported to be of the order of 10 to 25%. One example of a temperature versus time curve is shown in Fig. 5.

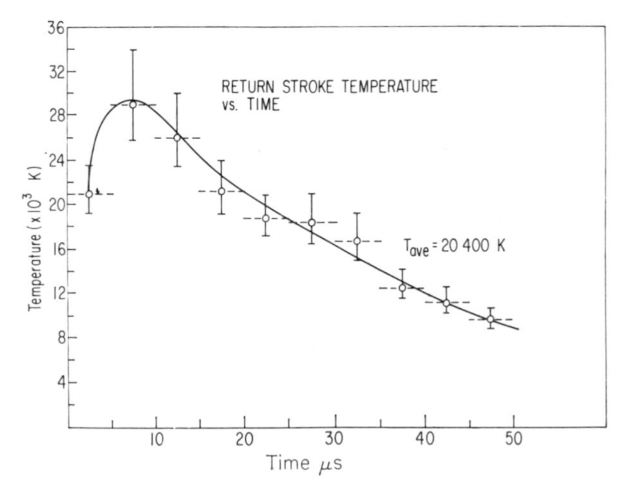

Fig. 5. Return-stroke temperature as a function of time obtained from the analysis of NII spectral emissions. The dashed lines show the time intervals over which the data points are averaged (5 μs resolution). The vertical bars are estimated systematic errors (Orville, 1968a).

The vertical bars represent systematic errors and the horizontal dashed lines represent the time intervals over which the temperatures are averaged. The temperature rise in the first 5 μs does not appear in most of the data and may be spurious due to the uncertainties in opacity, but the peak temperature occurring in the first 5 to 10 μs appears to be typical. The average temperature of 20,400 K was obtained by integrating the emissions for 50 μs and taking the ratio to obtain a temperature which can be compared to Prueitt's (1963) values. All peak temperatures occurred within the first 10 μs or less and decayed uniformly with time. In general, the NII line intensities are observed when the temperatures are 15,000 K or higher. Hill (1977) has questioned the last three temperature values reported in Fig. 5 and reproduced from Orville (1968a) on the grounds that significant NII radiation should not be expected at these temperatures at atmospheric pressure. Hill's suggestion is probably correct and it is likely that the errors in the intensity ratios of weak NII multiplets are greater beyond 35 μs than originally estimated. These three temperature values, it would appear, may have errors of the order of 50%.

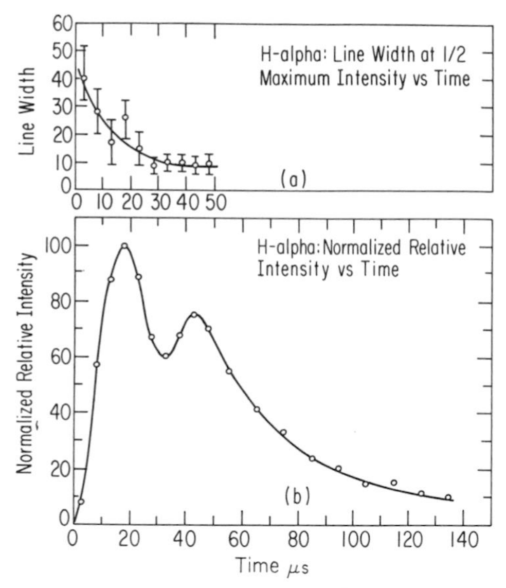

Fig. 6. Analysis of the H-alpha emissions in Fig. 2 yields (a) the full line width at half the peak intensity and (b) the relative intensity as a function of time (Orville, 1968a).

The return-stroke spectrum reproduced in Fig. 2 has been analysed for the H-alpha full-line width at half the maximum intensity (FWHM) as a function of time (Fig. 6a). Furthermore the H-alpha normalized relative intensity has been determined by measuring the area under the line profile and above the continuum every 10 μs (Fig. 6b). The surge in the relative intensity in the 40 to 50 μs region may be the result of a momentary current increase resulting from a branch above the isolated section of the channel providing an additional source of charge (see Chapter 9.4.1.1) and consequently increasing the H-alpha emissions. The line width measurements in Fig. 6(a) have been used to determine the electron density as a function of time and the results are plotted in Fig. 7. The values begin near 10^{18} cm^{-3} and decay to a nearly constant value of about 10^{17} cm^{-3} after 15 μs. It should be noted that most of the H-alpha radiation appears after 15 μs so that the time-resolved measurement is in good agreement with the time-averaged result of 2×10^{17} cm^{-3} (Uman and Orville, 1964). If H-alpha is very broad or heavily absorbed in the first 5 μs, then electron densities in excess of 10^{18} cm^{-3} might be present at these times and not be detected (Krider, 1973). Unfortunately, only one H-alpha spectrum has been analysed for the electron density as a function of time and no other lightning spectra have been obtained with sufficient wavelength resolution to study any line other than H-alpha.

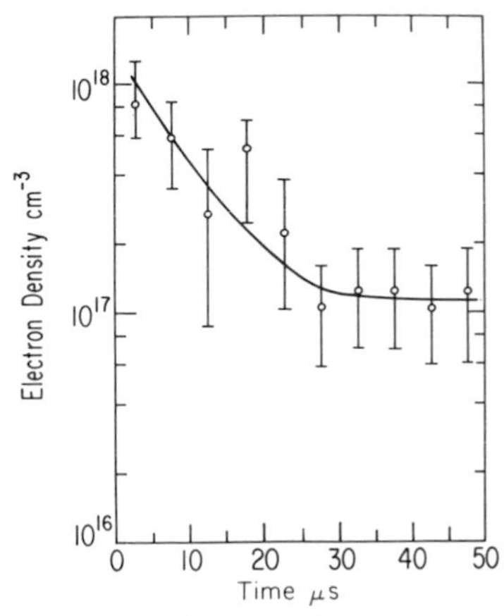

Fig. 7. Return-stroke electron density as a function of time computed from the H-alpha line widths in Fig. 6a. Each data point is averaged over 5 μs with the vertical bars showing estimates of the systematic errors. The electron density from 50 to 150 μs was less than $1{\cdot}5 \times 10^{17}$ cm^{-3} (Orville, 1968a).

Once the temperature and the electron density have been determined as a function of time, it is possible to use the tables of Gilmore (1967) and determine other physical characteristics of the lightning channel. Under the assumption of LTE, which is probably valid after the first few microseconds in the return-stroke channel, the channel pressure, relative mass density, per cent ionization and the relative populations of NI, NII and NIII versus time are shown for a typical return stroke in Figs 8 to 11 and will now be discussed.

The pressure–time curve shown in Fig. 8 indicates an average pressure of 8 atm in the return-stroke channel in the first 5 μs, decreasing to atmospheric pressure in about 20 μs. It is important to realize that the peak pressure is a poorly determined quantity which reflects the large errors in the electron density measurements (Fig. 7). On the other hand, if a pressure of 8 atm is correct, based on a correctly determined electron density, it is still possible that much higher pressures exist in the first microsecond or so and are not detected. High temperatures and electron densities would produce a broad H-alpha Stark width that would blend into the continuum and remain undetected. Consequently, electron densities calculated from the H-alpha line width (FWHM) in the first 5 μs are weighted toward densities existing in the latter part of the 5 μs exposure time; the pressure is then weighted in a similar way towards lower pressures. Therefore, the "peak pressure" of 8 atm can only be accepted as an indication of the high pressures existing within the channel in the first 5 μs.

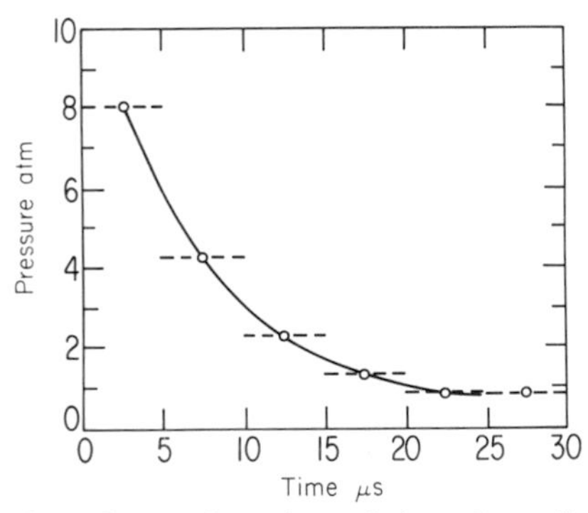

Fig. 8. The pressure plotted as a function of time for a lightning return stroke. The horizontal dashed lines in this figure and in subsequent figures (9–11) indicate the time intervals over which the data points are averaged (Orville, 1968a).

The relative mass density (ρ/ρ_0) as a function of time is shown in Fig. 9, where ρ_0 is the mass density of air and ρ is the mass density in the return-stroke channel. The rapidly expanding channel, characterized by high temperatures, electron density and pressure, has a mass density that decreases in the 0 to 25μs period. The minimum relative mass density is 3×10^{-3} and this is attained in approximately 20 to 25 μs. At 25 μs, atmospheric pressure has been attained in the channel and further cooling of the channel produces an increase in the relative mass density.

The relative per cent ionization as a function of time is shown in Fig. 10. A value of 100% ionization means that in a given volume the number of electrons equals the number of molecules, atoms and ions. Molecular species, however, are effectively non-existent for the temperatures and electron densities existing in the 0 to 30 μs period of the return stroke. The physical implication of a channel characterized by 100% ionization, as is clearly shown in Fig. 10, is that the dominant species is singly ionized.

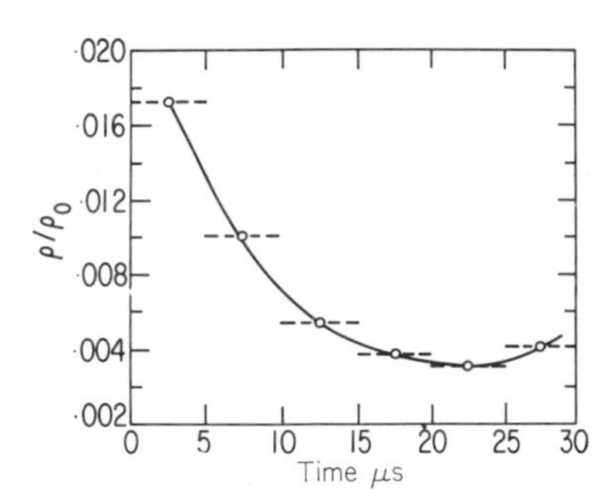

Fig. 9. Relative mass density plotted as a function of time for a lightning return stroke (Orville, 1968a).

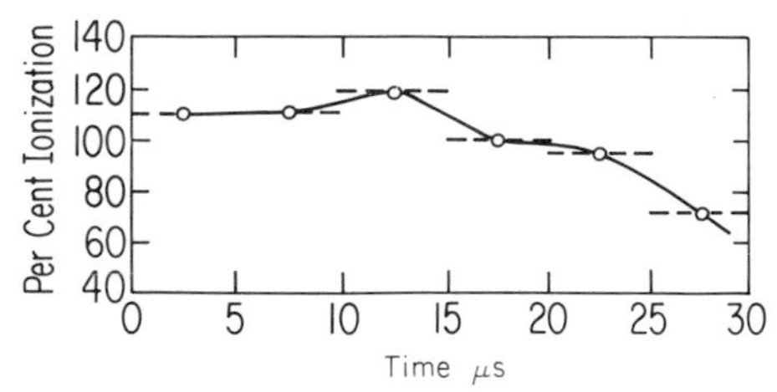

Fig. 10. The per cent ionization plotted as a function of time for a lightning return stroke. A gas composed entirely of singly ionized species (e.g. NII) would be represented by 100% ionization (Orville, 1968a).

A representative variation of species concentrations is presented in Fig. 11. Gilmore's (1967) tables for the equilibrium composition of air contain the concentrations of 38 species, but it is sufficient to plot the variation for three species of the principal constituent of air. Neutral nitrogen atoms (NI), singly ionized nitrogen atoms (NII) and doubly ionized nitrogen atoms (NIII) have concentrations which vary relative to each other in precisely the same way as similar species of oxygen. In the 0 to 20 μs period all three species concentrations are decreasing indicating that the pressure reduction during channel expansion dominates the concentrations. At 20 μs the channel has reached the ambient pressure (Fig. 8). A decreasing temperature beyond 20 μs (Fig. 5) now dominates the species concentration and consequently recombination accounts for the increasing NI at the expense of NII and NIII. These results are in qualitative agreement with the time-resolved spectral features presented in Figs 2 to 4. For example, the NII emissions in the return stroke dominate the optical spectrum in the early phase of the return stroke and the NI emissions (and H-alpha) appear at later times. The presence of NIII ions has not been detected because the channel temperature is apparently insufficient to excite the NIII ions to the energy levels from which optical emissions occur (30 to 40 eV). The NII and OII ions radiate in the visible region from energy levels 20 to 30 eV above their ground states. If doubly ionized species are eventually detected in the lightning return stroke, it is clear that the emissions will occur in the very early phase of the stroke.

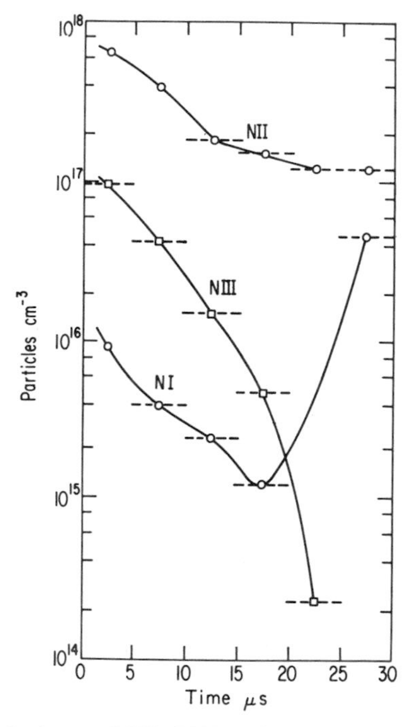

Fig. 11. Relative populations of NI, NII and NIII plotted as a function of time for a lightning return stroke (Orville, 1968a).

A word of caution should be expressed with respect to the physical characteristics of the return-stroke channel presented in Figs 5 to 11. These results are based only on spectrometer records which are the result of bright strokes and these discharges may not be representative of "typical" lightning return strokes.

3.3.2 Stepped Leader

The one stepped-leader spectrum recorded to date and reproduced in Fig. 3 has been analysed for temperature using the ratio of NII multiplets in the first step (Orville, 1968b). A value of 30,000 K with an uncertainty of +5,000 K and −10,000 K has been obtained. Since the luminous emissions occur in less than 2 μs (Orville, 1968b; Krider, 1974) it is doubtful if LTE exists and the reported value is probably the electron temperature. The pulsating nature of the H-alpha emissions in the leader spectrum has not been quantitatively analysed, but its modulated appearance for many microseconds in a section of the channel after the passage of a step suggests that the emissions occur at atmospheric pressure and are characteristic of a channel at approximately 15,000 K (Maecker, 1951, p. 314). Because the channel is oriented at an angle from the vertical in the spectrum (Fig. 3), it is not possible to determine the profile of the H-alpha line, consequently an electron density estimate cannot be determined from these data.

3.3.3 Dart Leader

The dart-leader spectrum reproduced in Fig. 4 is one of five which have been recorded in one flash and one of three which have been analysed for temperature (Orville, 1975). Again, only NII multiplets were measured and in particular the relative intensities of the 4,433, 4,447 and 4,630 Å multiplets were used. Dart temperatures of the order of 20,000 K ± 10% have been calculated.

The dart-leader slitless spectrum (Fig. 4) covers a wave-length section which would include the H-beta emissions (4,861 Å) if they were recorded but excludes the H-alpha emissions (6,563 Å). The absence of any neutral emissions, particularly H-beta in the dart-leader spectrum, is not in agreement with Barasch (1970), who used narrow-passband interference filters in conjunction with wide-field photoelectric detectors. Barasch reported neutral plus continuum emissions in the dart leader at 6,563 and 8,220 Å. Based on the similarities of the dart-leader spectrum to the return-stroke spectrum, it is reasonable to predict that the neutral emissions in the dart spectrum will be photographically recorded when the red and infrared regions are observed.

4. Future Experiments

The emphasis in future lightning experiments will be on correlating observations such as those reported in this chapter, with data on the electrical current and electric field changes. Thus, the thermodynamic properties should be correlated with the electrical properties such as current and charge transferred. Other useful experiments not involving correlations can also be performed. For example, the spectral recording with high time resolution should be extended as far as possible into the infrared and into the ultraviolet. Greater wavelength dispersion is desired for more line profile studies and higher time resolution is required to resolve the nature of the temperature rise, if any, in the first few microseconds of the return stroke.

It is interesting to note that all the discussion and indeed lightning spectroscopic research for the past century has been exclusively concerned with the study of cloud-to-ground lightning flashes. The majority of lightning occurs within clouds and to date good spectral recordings of these discharges have not been obtained. The possible significance of lightning in producing important, but minor constituents in the atmosphere such as O_3 and NO_x, can only be evaluated when we have spectral records and know the physical characteristics of lightning flashes within clouds.

References

Barasch, G. E. (1970). Spectral intensities emitted by lightning discharges. *J. geophys. Res.* **75**, 1049–1057.

Clark, J. W. (1874). Observations on the spectrum of sheet lightning. *Chem. News, Lond.* p. 28, July 17.

Dufay, M. (1926). Spectres des éclairs. *C. r. hebd. Séanc. Acad. Sci., Paris* **182**, 1331–1333.

Dufay, M. (1947). Sur le spectra des éclairs dans les régions violette et ultraviolette. *C. r. hebd. Séanc. Acad. Sci., Paris* **225**, 1079–1080.

Dufay, M. (1949). Recherches sur les spectres des éclairs, deuxième partie: Étude du spectre dans les régions violette et ultraviolette. *Annls Géophys.* **5**, 255–263.

Dufay, M. and Dufay, J. (1949). Spectra des éclairs photographies au prism—objective. *C. r. hebd. Séanc. Acad. Sci., Paris* **229**, 838–841.

Fox, P. (1903). The spectrum of lightning. *Astrophys. J.* **14**, 294–296.

Gibbons, J. (1871). Spectrum of lightning. *Chem. News, Lond.* p. 96, Aug. 25.

Gilmore, F. R. (1967). "Thermal Radiation Phenomena," Vol. I. "The Equilibrium Thermodynamic Properties of High Temperature Air." (Lockheed Missiles and Space Co., Palo Alto, California, May 1967); Defense Documentation Center AD-654054, unclassified.

Griem, H. R. (1964). "Plasma Spectroscopy" 580 pp. McGraw-Hill, New York.

Herschel, J. (1868). On the lightning spectrum. *Proc. R. Soc.* **15**, 61–62.

Hill, R. D. (1972). Optical absorption in the lightning channel. *J. geophys. Res.* **77**, 2642–2647.

Hill, R. D. (1977). Anomalous behaviour of H lines in lightning spectra. *in* "Electrical Processes in Atmospheres", (H. Dolezalek and R. Reiter, Eds). Steinkopff, Darmstadt. To be published.

Holden, E. S. (1872). Spectrum of lightning. *Am. J. Sci. and Arts.* **4**, 474–475.
Hu, R. (1960). The lightning spectra in the visible and ultra-violet regions with grating spectrograph. *Sci. Rec. Peking* **4**, 380–383.
Israel, H. and Fries, G. (1956). Ein Gerät zur spektroskopischen Analyse verschiedener Blitzphasen. *Optik* **13**, 365–368.
Israel, H. and Wurm, K. (1941). Das Blitzspektrum. *Naturwissenschaften* **52**, 778–779.
Jose, P. D. (1950). The infra-red spectrum of lightning. *J. geophys. Res.* **55**, 39–41.
Knuckles, C. F. and Swenson, J. W. (1952). The spectrum of lightning in the region λ6,159–λ7,157. *Annls Géophys.* **8**, 333–334.
Krider, E. P. (1965). Time-resolved spectral emissions from individual return strokes in lightning discharges. *J. geophys. Res.* **70**, 2459–2460.
Krider, E. P. (1966a). Comment on paper by Leon E. Salanave and Marx Brook, "Lightning photography and counting in daylight, using H_α emissions." *J. geophys. Res.* **71**, 675.
Krider, E. P. (1966b). Some photoelectric observations of lightning. *J. geophys. Res.* **71**, 3095–3098.
Krider, E. P. (1973). Lightning spectroscopy. *Nucl. Instrum. Meth.* **110**, 411–419.
Krider, E. P. (1974). The relative light intensity produced by a lightning stepped leader. *J. geophys. Res.* **79**, 4542–4544.
Krider, E. P. and Marcek, G. (1972). A simplified technique for the photography of lightning in daylight. *J. geophys. Res.* **77**, 6017–6020.
Larsen, A. (1905). Photographing lightning with a moving camera. *Smithson. Inst. Rept.* **60**, Part I, 119–127.
Lundquist, S. and Scuka, V. (1970). Some time correlated measurements of optical and electromagnetic radiation from lightning flashes. *Ark. Geofys.* **5**, 585–593.
Maecker, H. (1951). Der elektrische Lichtbogen. *Ergebn. exakt. Naturw.* **25**, 293–358.
Mastrup, F. and Wiese, W. (1958). Experimentelle Bestimmung der Oszillatorenstärken einiger NII und OII Linien. *Z. Astrophys.* **44**, 259–279.
Meinel, A. B. and Salanave, L. E. (1964). N_2^+ emission in lightning. *J. atmos. Sci.* **21**, 157–160.
Moore, C. E. (1959). A multiplet table of astrophysical interest—Parts I and II. National Bureau of Standards Technical Note 36. Available from U.S. Dept. of Commerce, Office of Technical Services, Washington, D.C.
Orville, R. E. (1966a). High-speed time-resolved slitless spectrum of a lightning stroke. *Science, N.Y.* **151**, 451–452.
Orville, R. E. (1966b). A Spectral Study of Lightning Strokes. Ph.D. Dissertation, University of Arizona. Available from University Microfilms, Inc. Ann Arbor, Michigan, No. 67–147.
Orville, R. E. (1967). Ozone production during thunderstorms measured by the absorption of ultraviolet radiation from lightning. *J. geophys. Res.* **72**, 3557–3561.
Orville, R. E. (1968a). A high-speed time-resolved spectroscopic study of the lightning return stroke, Parts 1, 2, 3. *J. atmos. Sci.* **25**, 827–856.
Orville, R. E. (1968b). Spectrum of the lightning stepped leader. *J. geophys. Res.* **73**, 6999–7008.
Orville, R. E. (1975). Spectrum of the lightning dart leader. *J. atmos. Sci.* **32**, 1829–1837.

Orville, R. E. and Salanave, L. E. (1970). Lightning spectroscopy—photographic techniques. *Appl. Optics* **9**, 1775–1781.
Petrie, W. and Small, R. (1951). The near infrared spectrum of lightning. *Phys. Rev.* **84**, 1263–1264.
Pickering, E. C. (1901). Spectrum of lightning. *Astrophys. J.* **14**, 367–369.
Prueitt, M. L. (1963). The excitation temperature of lightning. *J. geophys. Res.* **68**, 803–811.
Salanave, L. E. (1961). The optical spectrum of lightning. *Science, N.Y.* **134**, 1395–1399.
Salanave, L. E. (1964). "The Optical Spectrum of Lightning." Advances in Geophysics, Vol. 10, pp. 83–98. Academic Press, New York.
Salanave, L. E. (1965). The photographic spectrum of lightning; determinations of channel temperature from slitless spectra, *in* "Problems of Atmospheric and Space Electricity," (S. C. Coroniti, Ed.) pp. 371–383. Elsevier, Amsterdam.
Salanave, L. E. (1966). The Infrared Spectrum of Lightning. Inst. Elect. Electron. Engrs, Region Six Conference Record, Tucson, Arizona.
Salanave, L. E. and Brook, M. (1965). Lightning photography and counting in daylight, using H_α emission. *J. geophys. Res.* **70**, 1285–1289.
Salanave, L. E., Orville, R. E. and Richards, C. N. (1962). Slitless spectra of lightning in the region from 3,850 to 6,900 ångströms. *J. geophys. Res.* **67**, 1877–1884.
Schuster, A. (1880). On spectra of lightning. *Proc. phys. Soc.* **3**, 46–52.
Scuka, V. (1969). Electronic optical system for lightning research. *Ark. Geofys.* **5**, 569–584.
Slipher, V. M. (1917). The spectrum of lightning. *Bull. Lowell Obs.* No. 79, Vol. III, No. 4.
Uman, M. A. (1966). Quantitative lightning spectroscopy. *Inst. elect. electron. Engrs Spectrum* **3**, 102–110.
Uman, M. A. (1969). Determination of lightning temperature. *J. geophys. Res.* **74**, 949–957.
Uman, M. A. and Orville, R. E. (1964). Electron density measurement in lightning from stark-broadening of H-alpha. *J. geophys. Res.* **69**, 5151–5154.
Uman, M. A. and Orville, R. E. (1965). The opacity of lightning. *J. geophys. Res.* **70**, 5491–5497.
Wallace, L. (1960). Note on the spectrum of lightning in the region 3670 to 4280 Å. *J. geophys. Res.* **65**, 1211–1214.
Wallace, L. (1964). The spectrum of lightning. *Astrophys. J.* **139**, 994–998.
Zhivlyuk, Yu. N. and Mandel'shtam, S. L. (1961). On the temperature of lightning and the force of thunder. *Soviet Phys. JETP* **13**, 338–340.

9. Lightning Currents and Related Parameters

R. H. GOLDE

*London, England**

1. Historical Survey

1.1 General

In many respects the lightning current is the most important single parameter of the lightning discharge. With a knowledge of the wave-shape and amplitude of the current the electrical problems of protection against lightning can be solved. Similarly, from a knowledge of the lightning current the physicist can derive information concerning the charge, energy, electric moment and other related parameters as is shown in Section 5 of this chapter.

It must be emphasized that the information needed by the physicist is much less complete than that required by the protection engineer. The engineer is concerned with the lightning current at the point of strike and most of that information is now available. In contrast, the physicist is interested in the characteristics of the lightning channel along its entire length and little is known about the variation of the current over this distance. Thus, unless otherwise stated, this chapter is restricted to a description of the characteristics of the lightning current at the point of strike.

Many previous surveys of lightning currents are confined to flashes to transmission lines, structures of usual height and open ground. With the steadily increasing number of tall television masts and other buildings of exceptional height the characteristics of lightning discharges which are initiated by upward leaders are becoming of major practical importance. Both types of discharge are therefore considered.

Misconceptions occasionally persist long after improved information has been available. The lightning current provides an interesting example of this

* Formerly, Electrical Research Association, Leatherhead, Surrey.

type. It may have been the flickering appearance of a multiple lightning discharge which first gave rise to the idea that the lightning current was oscillatory. Arguing from experiments with Leyden jars, Sir Oliver Lodge (1892) convinced himself that lightning was an oscillatory phenomenon with a frequency of about 1 MHz.

Later it became fashionable to calculate this frequency by visualizing a capacitor, formed by a charged cloud and the earth, discharging to earth through the channel which was represented by its internal resistance and inductance. As late as 1924 this circuit was regarded as sufficiently valid for Creighton to suggest that the current amplitude could reach 1,450 kA and its rate of rise 400 kA μs^{-1}! The same treatment was still retained in a distinguished textbook published 25 years later (Biermanns, 1949).

In a paper which even today would be remarkable for its wide scope Humphreys argued, in 1918, that the lightning current was more likely to be aperiodic than oscillatory and this was explained by the suggestion that the internal resistance of the lightning channel exceeded the critical value for damping. However, with the numerical values adopted by Peek (1924) a cloud was assumed to be discharged in one single stroke and this led to the conclusion that the lightning current would last no more than a few microseconds. On this basis Fortescue (1930) pointed out that, if a charge of 20 C, a figure which had been established by Wilson (1920), was discharged in 2 μs, this would require a current amplitude of 10,000 kA, "obviously an impossible value".

The argument about the oscillatory or aperiodic nature of the lightning current lasted several decades (Bermbach, 1908; Emde, 1910; Peek, 1924; Rump, 1926; Binder, 1928).

Today, the current flowing to earth in a lightning stroke is known to be invariably unidirectional (see also Chapter 5.5.6) although, in rare cases, one component stroke in a multiple lightning flash may involve positive polarity in a sequence of negative component strokes.

1.2 Early Estimates

1.2.1 Current Amplitude

The earliest serious estimate of lightning current amplitudes seems to have been made by Kohlrausch (1888) who found that the fusing of a metal conductor had required a current of 30 kA lasting about 2 ms. Forty years later this and two other similar cases were recalculated (Binder, 1928) and led to the conclusion that values of between $6{\cdot}6\times10^4$ and 10^5, for the action integral $\int i^2\,dt$, had been involved. The thermal effects of lightning currents on metal electrodes were used repeatedly in later years to estimate the charges involved in lightning strikes and these aspects are discussed in Section 5.1.

Humphreys (1918) calculated the forces which caused two hollow copper conductors to be crushed and he deduced currents of 19 and 47 kA respectively. The information then available on the fusing, crushing and pitting of conductors was critically examined and compared with laboratory results by Bellaschi (1935). He concluded that lightning currents of 100 kA or more, lasting perhaps 100 μs had to be postulated to explain some field observations. A more recent case of the crushing of parallel conductors is mentioned in Section 5.1.1.

The most important method of measuring current amplitudes was developed by Pockels (1897) who found that the remanent magnetism of basalt rock struck by lightning depended exclusively on the maximum value of the magnetizing field intensity, even if this lasted for only a short time. He fashioned pieces of basalt measuring $4 \times 2 \times 1{\cdot}5$ cm^3 and had them installed on lightning conductors on tall exposed buildings. The first results, obtained on an Italian observatory (Pockels, 1901), produced values of about 11 and 20 kA. Pockels commented that these estimates may have been too low since the basalt pieces could have lost some of their residual magnetism due to shaking in transport. This feature was rediscovered several decades later after the development of the magnetic link in which natural basalt was replaced by a bundle of steel strips of high retentivity.

The magnetic link has become the principal tool for measuring the amplitudes of lightning currents and the results are discussed in Section 3.

1.2.2 Current Wave-shape

A current flowing through a straight conductor sets up a magnetic field and, if the current changes with time, the resulting field change can be recorded in a suitably constructed loop aerial. Following an earlier attempt by de Blois (1914), Norinder (1935) utilized this fact to record oscillographically the field changes produced by earth flashes in the vicintiy of a lightning observatory in Sweden.

In order to translate the electromagnetic field records into current waveshapes Norinder determined the distance of the flash and its angle with the plane of the aerial by means of triangulation. In addition he assumed that the current along the entire lightning channel was the same at any instant, an assumption which, as already mentioned in Section 1.1, cannot be upheld. While the extensive results of this investigation are mainly of historical interest, Norinder's basic idea has been applied again in recent years and this aspect is discussed more fully in Section 4.1.6.

The development of the high-speed cathode-ray oscillograph (c.r.o.) opened the way to direct information on the time variation of lightning currents. This method was first used by Norinder (1925) to record lightning overvoltages on electrical transmission lines and he soon established the fact that lightning surges were unidirectional and that apparent polarity reversals were caused by reflections at the line terminations. This use of the c.r.o. spread quickly to Switzerland (Berger, 1929, 1943) and the U.S.A. (for the summary of results see Wagner and McCann, 1941).

The derivation of the wave-shapes of lightning currents from c.r.o. records of lightning surges on lines is subject to several important limitations. On the electrical supply lines existing some 30 years ago, all but the weakest direct lightning strikes caused flashover of the insulation, so that frequently no more than part of the wave-front was recorded. Surges of high amplitude on conductors are subject to considerable attenuation and distortion, which is greatest for chopped surges, and in most cases the distance of the point struck from the point of recording is unknown. Finally, the surge impedance of a conductor does not remain constant with changing surge-voltage amplitude. It may be for these reasons that few systematic attempts have been made to deduce statistical conclusions from these recordings on the characteristics of lightning currents. The most comprehensive survey of 123 clear oscillograms recorded in different countries seems to be due to Neuhaus and Strigel (1935) and Müller-Hillebrand (1936) who produced cumulative frequency distributions for the front length and time to half value of the current.

The early oscillographic recording of lightning currents was also subject to certain limitations. Apart from the need for a distortionless current shunt, the high cost of the equipment could be justified only at positions which could be expected to receive a large number of direct lightning strikes. This recording technique was therefore applied, in the first place, to captive balloons (Stekolnikov and Valeev, 1937; Davis and Standring, 1947) and to tall structures, particularly in the U.S.A. The results of these investigations are discussed later.

2. Time Resolution of Lightning Current

The mechanism and the temporal development of the normal earth flash, initiated by a downward leader stroke, are described in Chapter 5. In the upper part of Fig. 1 this development is illustrated schematically, as recorded by a camera sweeping from left to right. During the progress of the first leader, displacement currents flow into the earth. These lead to conductive currents when the corona sheath surrounding, and preceding, the leader channel reaches the ground. The current then increases rapidly when the filamentary discharges, described in Chapter 7.4.1.2, are concentrated into one, or several, plasma channels. This is the manner in which, in my view, contact is established between the downward moving leader and ground or with a short upward growing streamer or leader (see Chapter 17.2.1). It is at, or shortly after, this instant that the lightning current, as visualized by myself and Wagner (1967), begins to flow. It will continue to flow until the charge tapped in the original volume of cloud has been neutralized and this is indicated schematically in the lower part of Fig. 1.

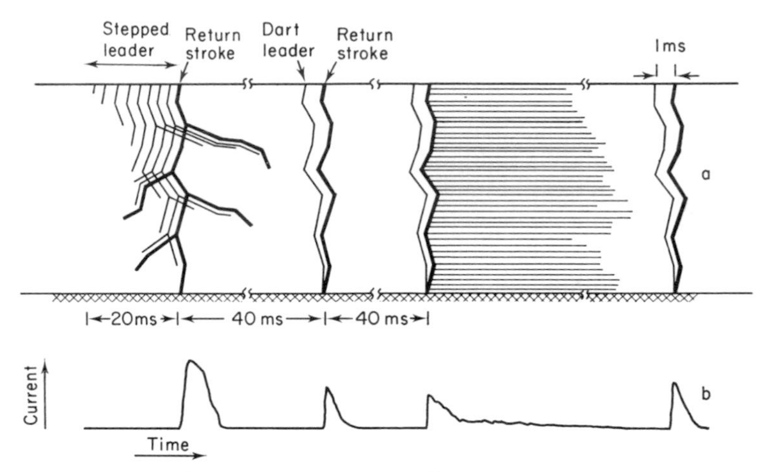

Fig. 1. Temporal development of downward lightning flash as recorded by camera moving from left to right (a) and associated current (b) (not to scale).

In the case of a multiple flash, further current bursts will follow when a subsequent dart leader is followed by a return stroke. Whether any current

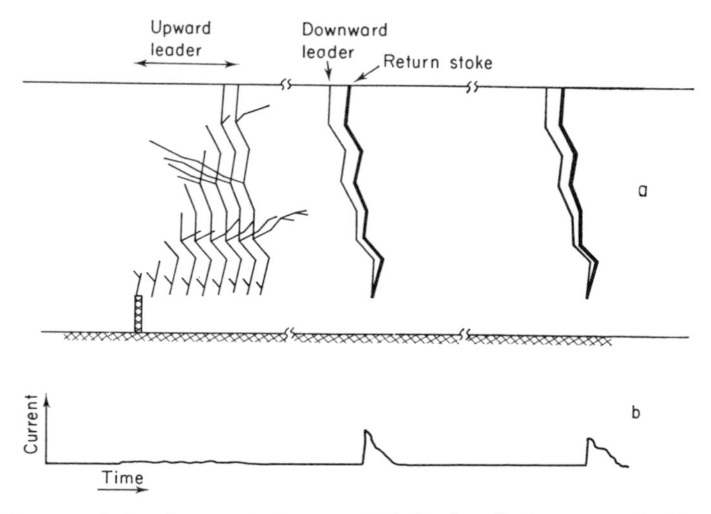

Fig. 2. Temporal development of upward lightning flash as recorded by camera moving from left to right (a) and associated current (b) (not to scale).

flows in the intervals between component strokes is a question which is discussed in Section 4.1.4.

The development of an upward discharge is shown in the upper part of Fig. 2. As shown in Chapter 5.4.3.3, upward leaders from towers under the influence of a normal negative or a rare positive charge centre show considerable differences. However, when these leaders are followed by return strokes—and this chapter is primarily concerned with the currents flowing during the return-stroke process—these two types of discharge produce similar impulse currents. Thus the prolonged streamer current from the tower may be followed by a current pulse. If this initial stroke is followed by subsequent strokes the normal process of a multiple downward flash ensues, as indicated schematically in the lower part of Fig. 2.

3. Current Amplitude

3.1 Introductory Remarks

In any analysis of lightning-current amplitudes account must be taken of the different nature of flashes initiated respectively by downward or upward leaders and of the effect of multi-stroke flashes. Lightning strikes to ground or to structures of low height are almost invariably initiated by downward leaders. As the height of a structure increases above, say, 100 m an increasing proportion of strikes are initiated by upward leaders (Golde, 1973, Fig. 22). A differentiation must therefore be made between these two types of discharge.

The majority of lightning flashes are negative and multiple. Complete information on current amplitudes in individual strokes is available from current recordings by means of oscillographs or fulchronographs. Magnetic links, on the other hand, merely record the maximum current amplitude in any multiple flash. It is now established that the crest value of the first component stroke in a multiple flash is nearly always higher than those of subsequent strokes (see Section 3.2.2).

It has occasionally been argued that the frequency distribution of lightning currents varies with such factors as the height of the structure, altitude of the terrain and soil resistivity. These questions are examined in Sections 3.2.1.2 and 3.2.1.3.

The amplitudes of lightning currents have been derived from fusing or crushing of conductors and from burn marks on, or holes in, metal surfaces although these latter cases are normally evaluated in terms of electric charge and are therefore discussed in Section 5.1. A summary of results from thermal effects is given in Bellaschi (1935) but, in view of the small number of examples, these are not further utilized in this survey.

3.2 Downward Discharge

3.2.1 *First Strokes*

3.2.1.1 *General survey.* By far the largest number of measurements of lightning currents is due to magnetic links on high-voltage transmission lines. The literature on this subject is so extensive that reference must suffice to surveys undertaken by Lewis (1950), Uman (1969), Szpor (1969) and Cianos and Pierce (1972). These measurements are subject to several restrictions or errors the most important of which are the following:

(a) great care must be exercised in the determination of the remanent magnetism of links (Ouyang, 1966);
(b) the minimum distance between a magnetic link and the magnetic centre of a piece of angle iron (tower leg) is such that currents below several kiloamperes cannot be detected; this lower limit is not invariably specified. On the other hand, links can be saturated by intense lightning currents;
(c) the current discharged through a lattice tower is subdivided in a complex manner between the four corner legs and cross-members (Anderson and Hagenguth, 1958) and currents can be induced in the lattice structure by earth-wire currents (Golde 1946);
(d) due to a phase shift between earth-wire and tower currents, algebraical addition of these currents leads to an overestimate of lightning currents (Golde, 1946; Schlomann *et al.*, 1957);
(e) magnetic links can be affected by power-frequency fault currents flowing through earth wires and towers;
(f) during transport the remanent magnetism of magnetic links can be affected by adjacent magnetized links and by shaking.

To overcome at least some of the most serious of these uncertainties, magnetic links were installed later on rods erected on top of transmission towers and on lightning conductors of tall chimneys so as to ensure a unique discharge path. The results from some of these installations have been critically analysed by Popolansky (1972). Confining the analysis to currents exceeding 2 kA, he finds that the resulting cumulative frequency distribution curve is best described by a log-normal law. These results are further discussed below.

Two different sets of data are available from the high-speed c.r.o. records obtained on Mount San Salvatore (Berger *et al.*, 1975) and, within the last few years, on Italian research stations (Garbagnati *et al.*, 1974). From these records the crest currents of all component strokes in a multiple flash initiated by a downward leader can be read off directly and separately. The resulting cumulative frequency distributions for 101 negative and 26 positive first strokes from San Salvatore are plotted in Fig. 3, together with their respective

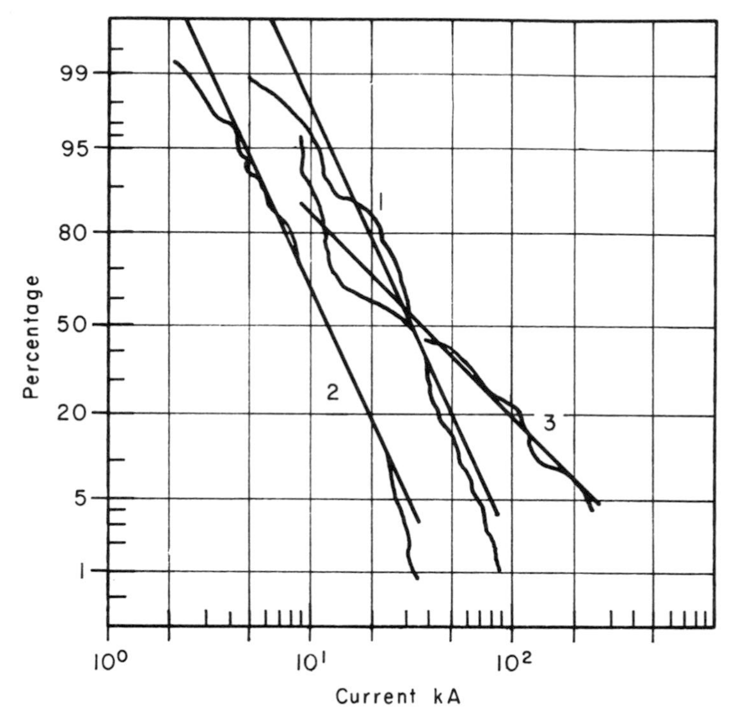

Fig. 3. Cumulative frequency distribution of lightning current amplitudes (Berger *et al.*, 1975). 1. Negative first strokes. 2. Negative subsequent strokes. 3. Positive strokes.

lines of best fit. Positive flashes to the towers on San Salvatore which are initiated by downward leaders consist, with one exception, of single strokes (Berger, 1967). The median value of negative first strokes is 30 kA as compared with 35 kA for the less frequent positive strokes. However, as seen from Table I, while 5% of all negative strokes exceed 80 kA, the corresponding value for positive strokes is 250 kA. Positive strokes are thus much more intense than negative strokes (see also Section 3.2.1.4). The Italian stations gave a median of 28·4 kA from 27 first strokes, most of them of negative polarity.

The data collected by Popolansky (1972) which are mentioned above include not only results obtained on tall chimneys and lightning rods but also the negative and positive first strokes recorded by Berger, totalling 624 registrations. The resulting cumulative frequency distribution curve is reproduced in Fig. 4, together with the line of best fit and the 95% confidence limits. This curve produces a median value of 28 kA. Berger *et al.* (1975) conclude that the median values obtained respectively on Mount San Salvatore and on tall chimneys are so close that the lightning strikes to the

Table I

Lightning parameters—downward flashes (Berger *et al.*, 1975)

Number	Parameter	Unit	Percentage exceeding tabulated value		
			95%	50%	5%
	Current amplitude (crest) exceeding 2 kA				
101	Negative first strokes and flashes	kA	14	30	80
135	Negative subsequent strokes	kA	4·6	12	30
26	Positive flashes	kA	4·6	35	250
	Charge				
93	Negative first strokes	C	1·1	5·2	24
122	Negative subsequent strokes	C	0·2	1·4	11
94	Negative flashes	C	1·3	7·5	40
26	Positive flashes	C	20	80	350
	Impulse charge				
90	Negative first strokes	C	1·1	4·5	20
117	Negative subsequent strokes	C	0·22	0·95	4·0
25	Positive flashes	C	2·0	16	150
	Time to crest				
89	Negative first strokes	μs	1·8	5·5	18
118	Negative subsequent strokes	μs	0·22	1·1	4·5
19	Positive flashes	μs	3·5	22	200
	Maximum d*i*/d*t*				
92	Negative first strokes	kA μs^{-1}	5·5	12	32
122	Negative subsequent strokes	kA μs^{-1}	12	40	120
21	Positive flashes	kA μs^{-1}	0·2	2·4	32
	Time to half-value				
90	Negative first strokes	μs	30	75	200
115	Negative subsequent strokes	μs	6·5	32	140
16	Positive flashes	μs	25	230	2,000
	Action integral				
91	Negative first strokes and flashes	A^2 s	$6{\cdot}0 \times 10^3$	$5{\cdot}5 \times 10^4$	$5{\cdot}5 \times 10^5$
88	Negative subsequent strokes	A^2 s	$5{\cdot}5 \times 10^2$	$6{\cdot}0 \times 10^3$	$5{\cdot}2 \times 10^4$
26	Positive flashes	A^2 s	$2{\cdot}5 \times 10^4$	$6{\cdot}5 \times 10^5$	$1{\cdot}5 \times 10^7$

television towers on top of the steep Swiss mountain may be deemed to be comparable with those striking tall structures in open country. While this conclusion appears justified it must be pointed out that the slopes of the curves of best fit in Figs 3 and 4 do not fully coincide.

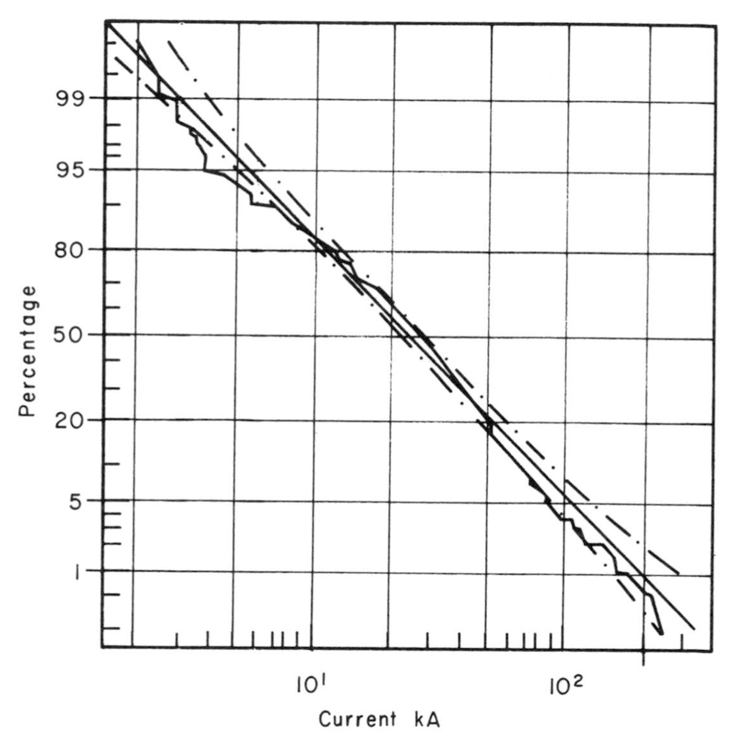

Fig. 4. Cumulative frequency distribution of lightning current amplitudes (Popolansky, 1972).

The currents examined in this section so far constitute the highest amplitudes discharged in single and multiple lightning flashes. They thus represent the data on which calculations concerning lightning protection have to be based. In recent years it has become customary to plot these cumulative frequency distributions of lightning currents on a logarithmic scale, to replace the actual curve by a line of best fit and to express this by a mathematical probability function (Whitehead, 1969; Popolansky, 1972). This method is of great benefit for statistical assessments of the effects of lightning currents but, in my view, there is no physical reason why the frequency distribution of the lightning-current amplitude should follow a log-normal distribution over its entire range. This conclusion is supported by a thorough statistical analysis undertaken by Kröninger (private communication). Care should therefore be exercised before accepting the low (or high) probability values. So far as lightning currents of less than 2 kA are concerned attention may, for instance, be drawn to results obtained on Mount San Salvatore with the Dudell oscillograph (Berger, 1955) which is capable of registering currents down to between 30 and 100 A.

Two different distribution curves which have recently been reported (Uman *et al.*, 1973b) and which are derived from magnetic field measurements are discussed in Section 4.1.6. Both distributions comprise all stroke currents in multiple flashes. The first curve refers to flashes within 10 km distance, is based on a lower limit of practically zero and produces a median value of about 12 kA. The second curve covers distances beyond 10 km, has a lower limit of about 15 kA and gives a median value of about 34 kA. These results cannot therefore be directly compared with the curves in Figs 3 and 4 which are confined to the highest currents in multiple flashes. On the other hand, it may be noted that the foregoing median value of 12 kA is the same as that recorded on San Salvatore for subsequent strokes (Section 3.2.2) which is based on a lower recording limit of 2 kA.

3.2.1.2 *Effect of structure height.* The results discussed so far refer to lightning strikes to structures varying in height between 22 and 140 m (the two towers on Mount San Salvatore are 70 and 90 m high). Thus the question arises as to whether the cumulative frequency distribution curves given in Figs 3 and 4 apply equally to strikes to open country or to very high structures.

The former problem was first examined by Golde (1954). As shown in Chapter 17.3 it is now generally accepted that the distance over which a structure attracts a downward leader is a function of the charge on the leader which, in turn, is related to the amplitude of the current in the return stroke (see Section 6). It would thus follow that lightning strikes to open ground should have a greater proportion of lower currents than those in Figs 3 and 4 while taller structures might be expected to be subjected to a higher number of more intense currents. A predicted cumulative frequency distribution curve for flashes to open ground is given in the reference quoted.

This theoretical approach was further developed by Schwab (1965), Sargent (1972) and Horváth (1973). They derived frequency distributions of currents liable to be discharged into structures varying in height from 100 to 500 ft. Good agreement was shown between the theoretically predicted curve and Berger's curve plotted in Fig. 3. In all these data, currents produced by flashes which were initiated by upward leaders are, of course, excluded.

In order to test the foregoing theoretical results against observed current amplitudes, Popolansky (1974) investigated a total of 209 flashes to the lightning conductors of chimneys. He subdivided the chimneys into four groups of height. For a total of 169 negative flashes the median current appeared to decrease with increasing height while, for 40 positive flashes, the opposite tendency prevailed. While these results contradict the theory, Popolansky admits that the size of the sample can hardly be accepted as adequate.

No use has been made in Section 3.2.1.1 of the many thousands of results obtained on transmission lines for the reasons stated. The great majority of these lines were erected on towers, the heights of which infrequently exceeded 30 m. The median current values deduced from these investigations (Lewis and Foust, 1940; Gross and Lippert, 1942; Baatz, 1951) are of the order of 20 kA, notably lower than the 30 kA derived in Section 3.2.1.1. Many of these measurements were made on rods on transmission towers or by adding the currents in earth wires due to strikes to the span. In my view (Golde, 1946) these latter results are more likely to constitute overestimates than underestimates of the correct current values. If this is so, these results could be adduced to support the theoretical conclusion that the probability of flashes discharging high current amplitudes increases with increasing structure height. It is evident that more work is required to answer this question convincingly.

3.2.1.3 *Effect of topographical features*. The top of Mount San Salvatore is 914 m above sea level (see Chapter 5.4.3). It was seen in Section 3.2.1.1 that the current-amplitude distributions obtained on this mountain and on structures in the plains of Europe are comparable. Up to about 1,000 m, altitude appears therefore to have no noticeable effect on lightning current amplitudes. On the other hand, Schwab (1965) argues that, in mountainous regions, the height of thunderclouds above ground is lower than over plains. In consequence, the critical breakdown strength of air is reached before large charges are built up and very high current amplitudes are therefore less likely to occur over mountains than over plains.

The amplitudes of 164 lightning currents on an 88 kV line in Rhodesia were examined by Anderson and Jenner (1954) and later compared by Anderson (1971) with the results obtained on San Salvatore. He concludes that the cumulative frequency-distribution curves coincide up to 30 kA but that the Rhodesian data indicate more flashes of higher crest value. It is not clear whether this difference is due to the high altitude of the Rhodesian plateau (1,000 to 1,600 m above sea level) or to the nature of tropical thunderstorms.

Magnetic-link measurements on transmission lines in the U.S.S.R. (Alizade *et al.*, 1968) are claimed to indicate that the mean values of current amplitudes decrease with increasing altitude. This conclusion is derived from results on 330 kV lines at 250 m above sea level, 220 kV lines at 550 m and 110 kV lines at 1,140 m. In view of the differences in the heights of the transmission towers involved this claim can only be accepted with reservations.

The same conclusion was reached earlier from results on a single 100 kV transmission line in Colorado (Robertson *et al.*, 1942) which rises from an altitude of 1,800 to 4,100 m, passing three times over the Continental Divide. It may be noted that, at the higher altitudes, 36% of all flashes discharged positive currents as against only 5% at low altitudes.

In complete contrast, current measurements on a 50 kV line crossing the Andes Mountains (Foust *et al.*, 1953) indicate an increase in current amplitudes at high altitudes. The magnetic links were installed on rods on top of the transmission towers and on separate diverter rods acting as lightning protection, the line rising from an altitude of 650 to 4,300 m.

In an attempt to explain the differences between the results obtained in the last two investigations it was suggested (Foust *et al*, 1953) that tower-footing resistances exerted a critical influence on lightning current amplitudes. The possible effect of the earthing resistance at the point struck was first examined theoretically by Ollendorf (1932). By applying Maxwell's field equations to the lightning channel and the return path of the current through earth and the atmosphere he found that the earthing resistance not only greatly affected the amplitude of the current in the return stroke but also the actual formation of the leader channel and its branching. The physical picture of the lightning discharge visualized in this approach can no longer be upheld and the results cannot therefore be accepted as valid.

A basically similar concept was advanced by Stekolnikov (1935) who suggested that the resistivity of the soil would affect both the striking point and the amplitude of the return stroke. Burgsdorf (1941) and Stekolnikov and Lamdon (1942) claimed to have proved the latter conclusion in magnetic-link measurements on transmission towers and unearthed wood poles in the Caucasus. They stated that the predicted effect increased with increasing amplitude of the lightning current. Thus 1% of lightning currents were found to exceed 100 to 200 kA in the plains of the U.S.S.R. but no more than 70 kA in the Caucasian mountain areas with high soil resistivity.

The characteristics of altitude and soil resistivity are clearly interrelated but, in view of the contradictory results published so far, their effects on the amplitudes of lightning currents remain unresolved.

3.2.1.4 *Effect of polarity*. It is generally accepted that the overwhelming proportion of lightning flashes at low altitudes has negative polarity and this may be the reason why some of the most numerous observations by means of magnetic links on transmission lines are not subdivided according to polarity. There is, however, strong evidence that the relative frequency of positive flashes increases with increasing altitude (Robertson *et al.*, 1942; Berger *et al.*, 1975).

The inherent inaccuracy in the interpretation of magnetic-link measurements on transmission towers and earth wires mentioned in Section 3.2.1.1 fortunately does not affect a direct comparison with respect to polarity. The most comprehensive information of this type is due to Baatz (1951) whose results show a decrease in the frequency of positive flashes over a range of 10 to 40 kA above which this ratio remains constant up to about 100 kA. Flashes above 100 kA, however, comprise 4 out of a total of 124 of positive polarity but only 22 out of 859 of negative polarity. It would thus appear that the proportion of positive flashes is highest at both the lower and upper limits of current amplitude. Similar conclusions are drawn by Davis and Standring (1947) from measurements on tethered balloons and by Popolansky (1960) who finds a particular increase in the proportion of positive flashes at amplitudes below about 5 kA.

The results from Mount San Salvatore (Berger *et al.*, 1975) show that 101 negative first strokes produce a median value of 30 kA while 26 positive flashes give 35 kA (see Table I). In fact, all currents exceeding 100 kA were of positive polarity in the period 1963–71 (Berger, 1972) while only one out of 81 negative flashes discharged a current of 105 kA in the period 1955 to 1963. A similar conclusion concerning exceptionally severe positive discharges is reached from the observations on some transmission lines on the North American continent (Bell, 1940; Foust *et al.*, 1953).

The relatively great severity of the rare positive lightning flashes can thus be accepted with confidence. Their relative frequency at the lower end of the current range, on the other hand, may possibly be explained by the occurrence of small pockets of positive charge in the lower regions of a thundercloud (Simpson and Scrase, 1937). Alternatively, it is suggested that some of these records are actually not due to lightning strikes to the structure on which the magnetic links are installed but to the redistribution of bound positive charges following a negative lightning flash to earth in close proximity. I have recorded such positive currents on adjacent chimneys, one of which discharged a high negative current while magnetic links on the other indicated a positive current of several kiloamperes (see also McCann and Harder, 1948, and Hyltén-Cavallius and Strömberg, 1959).

3.2.2 Subsequent Strokes

Magnetic links merely record the highest current amplitude discharged in a multiple lightning flash. In the fulchronograph (see Chapter 5.1.2) an indication is obtained of the entire current flow throughout the duration of a flash. Such records, obtained from strikes to tall structures (McCann, 1944; Shaw, 1970), show that the highest current is usually discharged in the first component stroke and that the median value of all stroke components is less than half that of the first strokes. Nevertheless, very occasionally, this general rule is broken as evidenced by a lightning strike to a 192 m high mining stack in Montana, U.S.A. (Wagner *et al.*, 1941), which involved a flash of 22 component strokes, the first of which carried 8 kA while six subsequent strokes discharged currents between 8·2 and 12·5 kA.

The general rule for the relative amplitudes of lightning currents was also deduced from the magnetic field records of Norinder and Knudsen (1961). Brook *et al.* (1962) report that the most frequent value of charge brought to earth by 24 first strokes lies between 3 and 4 C while this value is reduced to 0·5 to 1 C in subsequent strokes. Even though the time to half-value in first strokes is, on average, about twice that for subsequent strokes, the foregoing results again confirm that first strokes are usually the most intense.

The most comprehensive information on current amplitudes in subsequent strokes of multiple flashes is available from Mount San Salvatore (Berger *et al.*, 1975). Curve 2 in Fig. 3 shows the resulting cumulative frequency distribution. From Table I it appears that the median current of 135 subsequent strokes is 12 kA as compared with 30 kA for 101 first strokes. These results refer to negative strokes since no multiple positive flashes were recorded in this investigation.

In the Italian investigation 52 subsequent strokes gave a median value of 10·2 kA (Garbagnati *et al.*, 1974). The fact that the ratio of current amplitudes in first over subsequent strokes is higher in the results of Brook *et al.* (1962) than those obtained in the European Alps is likely to be due to the much larger number of subsequent strokes recorded in the former investigation, and this difference, in turn, may be explained by the different nature of thunderstorms in New Mexico and in the European Alps.

3.3 Upward Discharge

The mechanism of upward growing leader strokes was first investigated by McEachron (1941) and his co-workers on the Empire State Building in New York which rises to 380 m above street level. The current registered at the top of the building normally starts with a continuous negative discharge of several hundred amperes lasting for nearly a second and this may be followed by a sudden negative current pulse of the type described in the preceding section. The current flow in such a flash is shown schematically in the lower part of Fig. 2. This type of discharge was further studied and elaborated by Berger on Mount San Salvatore (Chapter 5.4.3).

The initial continuous currents from the Empire State Building, 86% of which were negative throughout, exhibited a maximum value of the order of 1,450 A although most of them amounted to less than 400 A (Hagenguth and Anderson, 1952). About half of all 135 strikes were not followed by any impulsive current peak; of the 84 impulsive currents which were recorded about 50% exceeded 10 kA but these comprised strokes initiated by either upward or downward leaders. Upward discharges from a tower on Mount San Salvatore begin with a current of the order of 100 A lasting up to a few tenths of a second. When these continuous currents are followed by current pulses with normal downward leader and return strokes, the current amplitudes are substantially lower than those indicated in Fig. 3 and do not exceed about 12 kA (Berger, 1967).

The nature of upward strokes initiated by rockets was first studied by Newman *et al.* (1967) (see Chapter 21.3) who recorded 17 flashes, most of them multiple, with a maximum of 41 kA. In a similar investigation in a field laboratory at St Privat, France (Gary *et al.*, 1975), the highest amplitude recorded amounted to 19 kA (private communication). When a lightning discharge is initiated artificially as in these investigations the charges in the cloud may conceivably be lower than those associated with normal downward strokes and the resulting current amplitudes may, consequently, also, on an average, be lower.

4. Current Wave-shape

4.1 Downward Discharge

4.1.1 First Negative Strokes

The large number of current oscillograms obtained on Mount San Salvatore (Berger, 1972) and in the Italian investigation (Garbagnati *et al.*, 1974) exhibit notable differences between the shapes of first and subsequent strokes and, amongst the former, between negative and positive strokes. Negative first strokes which comprise the largest amount of data will be discussed first.

It must be understood from the start that no two current oscillograms are identical. However, a consistent similarity in the different wave-shapes was found to exist in the steeply rising current front and the individual oscillograms were therefore aligned at the point where the current amplitude attains 50% of its crest value. A mean wave-shape was thus computed (Kröninger, 1974) and this is reproduced in Fig. 5. The first part of the current shown in curve B is based on 88 records while the longer wave-shape, curve A, is based on only 10 records and this difference explains the apparent ripple at the point of transition.

By no means all current oscillograms show the comparatively smooth shape of Fig. 5. Many exhibit a pronounced double hump near the crest (Berger, 1967, 1972) and less pronounced humps are evident on current tails (Fig. 8, Chapter 5). The sudden increases in current are due to the effect of strong branches from the main lightning channel as demonstrated by the sudden increase in luminosity of a lightning channel at the instant when the return stroke reaches a branching point (Malan and Collens, 1937). The correlation between luminosity and current intensity has been examined by Flowers (1944) and Mackerras (1969).

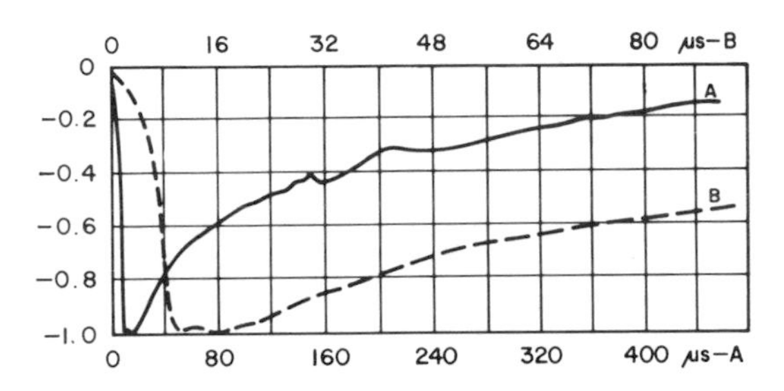

Fig. 5. Mean current wave-shape of negative stroke (Berger *et al.*, 1975). A. Full recorded length. B. Expanded front.

It has been suggested at various times that these wave-shapes are strongly affected by the height of the metal towers on top of a high mountain. Furthermore, information has been published during the last few years purporting to indicate that the current wave-shapes arising from lightning strikes to open ground differ from the foregoing results. These aspects are discussed in Section 4.1.6. In the present context the wave-shapes illustrated in Fig. 5 will now be compared with results from other investigations which, although not sufficiently numerous to be examined on a statistical basis, are pertinent in a discussion of the wave-shape of the current in the first component stroke.

As can be seen from Fig. 5 the front of the normal negative current wave in a strike to a tower on San Salvatore has a concave shape. The median value of its time to crest is 5·5 μs (Table I). In the Italian research stations (Garbagnati *et al.*, 1974) this value amounts to 7 μs. Some oscillographic current records obtained on the Empire State Building (Hagenguth and Anderson, 1952), on the Cathedral of Learning in Pittsburgh (McCann, 1944) and on several other tall structures (McCann and Harder, 1948; Wagner and Hileman, 1958) show similar concave fronts and times to crest within the range indicated in Table I. One of the first clear current records, obtained by Stekolnikov and Valeev (1937) on a balloon cable, exhibits a double-exponential wave-shape with a wave-front of about 8 μs.

Transmission lines constitute much lower structures than those considered so far and, although, for the reasons stated in Section 1.2.2, lightning-surge oscillograms often cannot be interpreted in terms of lightning currents, some relevant records are available. The clearest oscillograms have been collected together by Wagner and McCann (1941). The most famous of these is the oscillogram obtained at a distance of only 40 m from the point at which flashover occurred on a 220 kV line (Bell and Price, 1930). This oscillogram which is reproduced in Fig. 6 shows the difficulty, in those early days, of

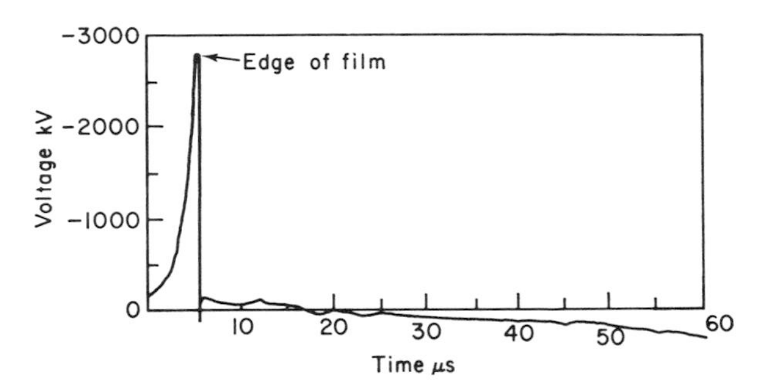

Fig. 6. Oscillogram of lightning surge voltage on a 220 kV line (Bell and Price, 1930).

ensuring tripping of the oscillograph at the first instant of current rise. Furthermore, flashover intervened before the crest of the current (voltage)

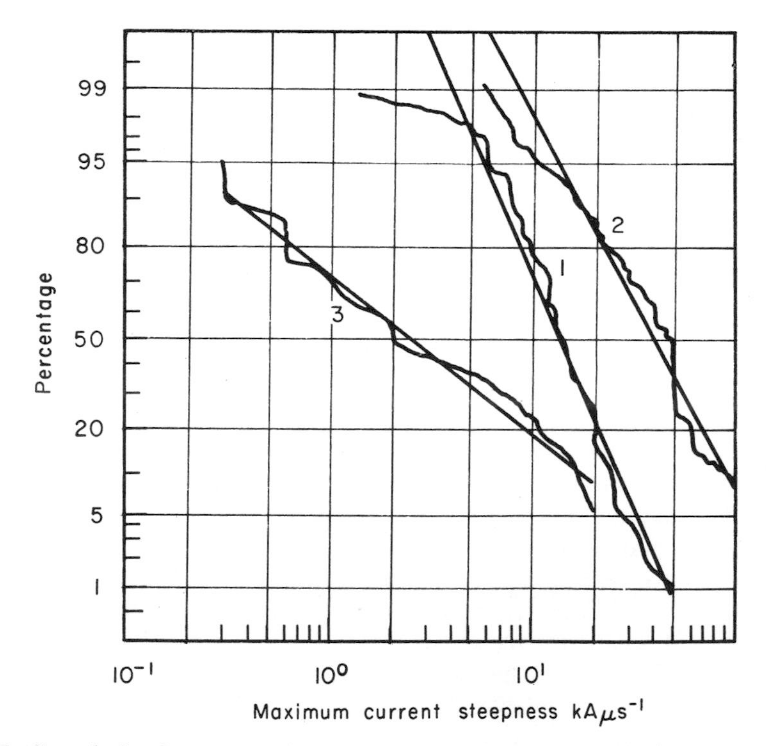

Fig. 7. Cumulative frequency distribution of maximum rates of rise of lightning currents (Berger *et al.*, 1975). 1. Negative first strokes. 2. Negative subsequent strokes. 3. Positive strokes.

was reached. Nevertheless, this oscillogram convincingly supports the view that the concave current wave front of Fig. 5 is not confined to objects of exceptional height. Concave fronts have also been recorded (Hagenguth *et al.*, 1952) in the currents in long electric sparks and these have since been frequently confirmed.

The rate of rise of current is a figure which is of particular importance to the problem of lightning protection. A comparison of published values is made more difficult by differing definitions adopted by different authors and different responses of some of the early recording instruments. For many practical problems the maximum rate of rise is required and this is seen from Fig. 5 to occur immediately before the crest value of the current amplitude is reached. Curve 1 in Fig. 7 shows the cumulative frequency distribution of di_{max}/dt with a median value of 12 kA μs^{-1} (Table I). The corresponding value from 19 Italian currents is 9·8 kA μs^{-1} (Garbagnati *et al.*, 1974).

It is customary to describe the tail of the impulsive part of the lightning current by the time from the start of the current wave to the instant when its amplitude has fallen to one-half of its crest value. From 90 San Salvatore

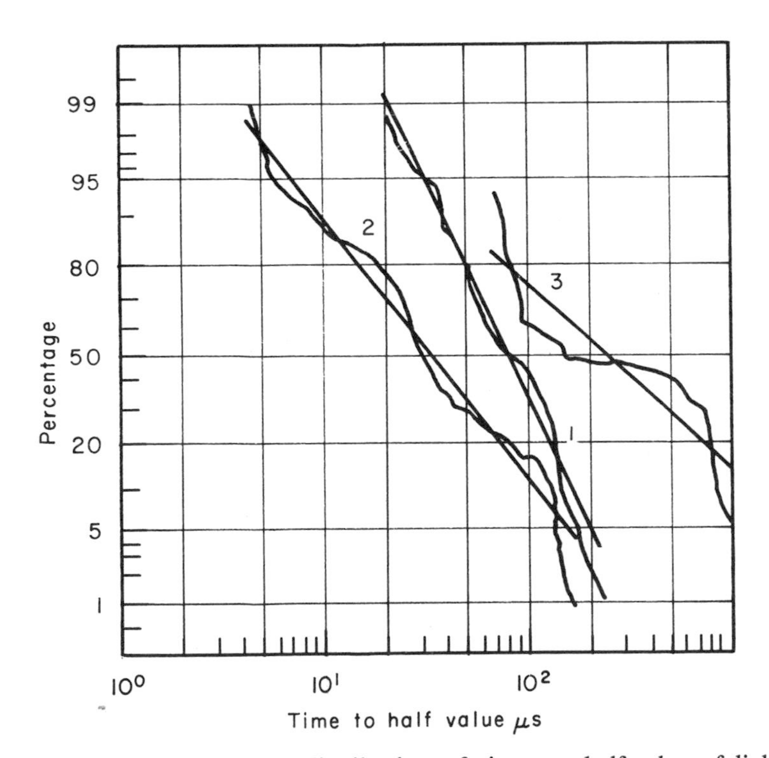

Fig. 8. Cumulative frequency distribution of times to half-value of lightning currents (Berger *et al.*, 1975). 1. Negative first strokes. 2. Negative subsequent strokes. 3. Positive strokes.

records which are plotted in Fig. 8 this "time to half-value" is found to have a median value of 75 μs, see also Table I. In this case the result from 19 records obtained in Italy is notably shorter, namely 44 μs (Garbagnati *et al.*, 1974). No strictly comparable data are available from other sources since the durations of first and subsequent strokes are not differentiated.

According to the results obtained by Hagenguth and Anderson (1952), the impulsive current component is, occasionally, followed by what these authors call an "intermediate current". This may reach 2,500 A and may last for one or a few milliseconds. Such long current tails had already been deduced from records of electrostatic field changes by Appleton and Chapman (1937). These intermediate currents may, or may not, be followed by continuing currents, as discussed in Section 5.1.1.

4.1.2 Positive Strokes

Lightning flashes of positive polarity usually consist of a single stroke as mentioned before. Furthermore, their wave-shapes, illustrated in Fig. 9 and

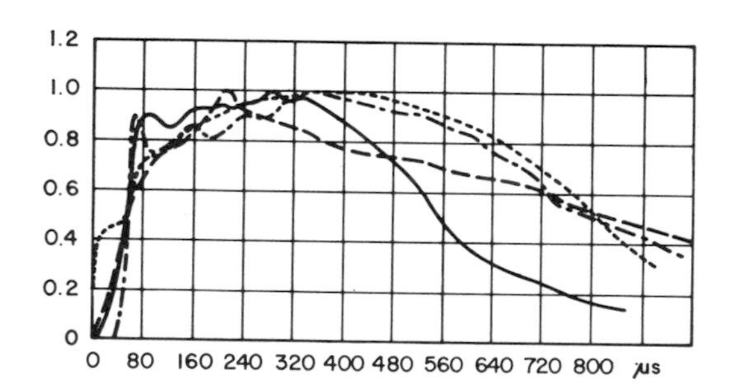

Fig. 9. Typical wave-shapes of positive currents (Berger *et al.*, 1975).

selected from 21 records, give a median front time of 22 μs (Table I). The maximum rate of rise has a median of 2·4 kA μs^{-1} and a time to half-value of 230 μs (Table I). Positive strokes, as recorded on Mount San Salvatore, are usually associated with long upward (connecting) leaders from the towers (see Chapter 5.4.3.2) and this is believed to be the reason that they have notably longer fronts and tails than negative first strokes. Combined with their somewhat higher average crest values, positive strokes are therefore much more intense than negative first strokes, an observation which is confirmed by their respective charges which are discussed in Section 5.1.

Polarity reversals in rare multiple flashes have occasionally been observed (McCann, 1944; Hagenguth and Anderson, 1952; Berger, 1972). The wave-shapes of the individual stroke currents do not appear to differ significantly from those of multiple flashes discharging one polarity only.

4.1.3 Subsequent Strokes

From 76 subsequent negative strokes recorded oscillographically on San Salvatore the mean current wave-shape reproduced in Fig. 10 was constructed (Berger *et al.*, 1975). As compared with the first strokes, subsequent strokes have a much shorter front with a median of 1·1 μs (Table I). In these circumstances it is understandable that their maximum rates of rise are notably higher than those recorded for first strokes, with a median of 40 kA μs^{-1} and, as shown in Table I and curve 2 in Fig. 7, a maximum value exceeding 120 kA μs^{-1}. The wave-tails are slightly shorter than those of first strokes with a median of 32 μs. Comparable results from Italy (Garbagnati *et al.*, 1974) give 1·3 μs for the median front duration of 48 strokes and a median rate of rise of 19·8 kA μs^{-1}, obtained from 37 records.

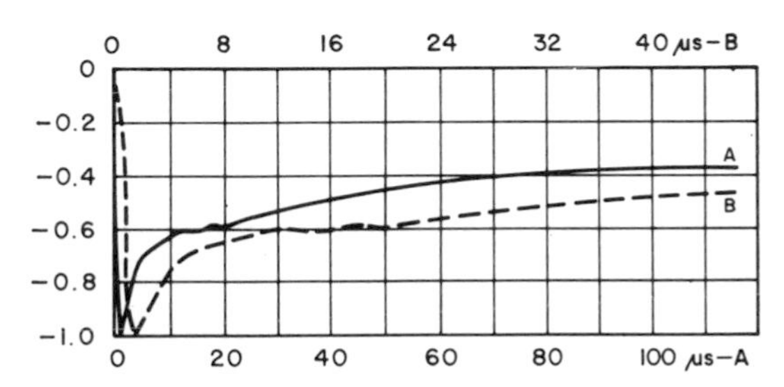

Fig. 10. Mean current wave-shape of negative subsequent stroke (Berger *et al.*, 1975). A. Full recorded length. B. Expanded front.

Following the impulsive component, so-called "continuing currents" may be discharged in some subsequent strokes. These currents have been particularly investigated in flashes to open ground in New Mexico (Williams and Brook, 1963; Kitagawa, 1965) and are discussed further in Section 5.1.1. They last at least 40 ms and carry currents between a few tens to over 500 A, with an average of about 180 A.

4.1.4 General Survey

As mentioned in Section 1.2.2, Norinder (1935) recorded the magnetic field changes produced by ground flashes with a view to deriving the wave-shape of the current in the return stroke. This method was, in recent years, further developed by Uman *et al.* (1973a). Their field records and the wave-shapes of the currents derived from them exhibit a sharp initial peak which has never been recorded in the current oscillograms obtained in any of the structures discussed in the preceding sections. Furthermore, both first and subsequent currents produced very similar rise times with mean values of 3·7 μs for

436 first strokes and 3·0 μs for 480 subsequent strokes (Fisher and Uman, 1972). These data thus contradict the results discussed so far. Some of the concepts underlying this approach are examined in Section 4.1.6.

In all other investigations the characteristics of first and subsequent strokes have been lumped together. Maximum rates of rise have been measured with the aid of loop antennae in combination either with magnetic links (Komelkov, 1941; Stekolnikov, 1941), with klydonographs (Szpor, 1969) or with explosive gaps (Schlomann *et al.*, 1957; Hyltén-Cavallius and Strömberg, 1959). The results of these investigations, which have been summarized by Szpor (1969), give values which are within the range obtained on San Salvatore, but these methods of registration give no indication of the particular point on the current wave-front at which the instrument registered.

Current oscillograms obtained on tall structures in the U.S.A. have been analysed in terms of "effective" current rise, defined as the rate of rise between the 10 and 90% values on the current front (McCann, 1944; Hagenguth and Anderson, 1952). These gave effective rates of current rise up to about 45 kA μs^{-1}, a value which seems to be comparable with the highest rate of rise in first strokes derived from records on San Salvatore (Berger, private communication).

Times to half-value were measured in lightning flashes to chimneys (Hyltén-Cavallius and Strömberg, 1959). They give values from 20 to 200 μs. From 57 fulchronograms recorded on several tall structures in the U.S.A. (Wagner *et al.*, 1941) values between 25 and 100 μs with an average value of 50 μs were obtained. Oscillograms of 82 currents discharged into the Empire State Building gave similar values with a maximum of 120 μs (Hagenguth and Anderson, 1952). However, both investigations also showed that the initial current peaks lasting several tens of microseconds were usually followed by currents of the order of one or two kiloamperes lasting several hundred microseconds, to be followed in turn by continuing currents* of a few hundred amperes lasting several milliseconds. On Mount San Salvatore the occurrence of continuing currents was observed in about one-third of all strokes (Berger, 1967).

The character of these continuing currents has been examined in greater detail in lightning flashes to open ground in New Mexico. They are defined as currents which last not less than 40 ms. Durations of up to 300 ms have been observed and their maximum amplitude is given as 590 A (Krehbiel *et al.*, 1974). As pointed out by Brook (private communication), the current values are subject to a variation of at least a factor of 2, not only due to

* These *continuing currents* which constitute extended tails of high current pulses must be clearly distinguished from currents, termed *continuous currents*, which flow during the progression of upward lightning strokes from tall structures as described in Section 3.3.

measurement uncertainties but also to variations with the type of storm, namely daytime storms over a mountain or night storms over the plains.

Early investigations on the Empire State Building led to the conclusion that a small continuous current flow is maintained between cloud and ground throughout the duration of a multiple flash (McEachron, 1941). A physical argument for the need of such a current was advanced (Bruce and Golde, 1941; Kitagawa *et al.*, 1962). A continuous current flow was also observed in a flash to the Anaconda Copper Mining smoke stack (McCann, 1944), but in 12 other flashes to tall structures examined by McCann no such continuous current flow was detected and it was concluded that this was not a necessary condition for flashes to lower objects. Berger (1967) expressed the view that if such currents exist, their amplitudes must be below 1 A.

4.1.5 Analytical Representation of the Lightning Current

For many calculations concerning the effects of a lightning flash to earth it is clearly helpful to have a simple mathematical expression describing the wave-shape of a typical lightning current. Independently and simultaneously Stekolnikov (1941) and Bruce and Golde (1941) suggested a double-exponential expression of the form:

$$I = I_0[\exp(-at) - \exp(-bt)]. \tag{1}$$

Numerical values for the constants a and b were derived from three properties of the lightning discharge in accordance with the knowledge then available. These are the charge density along the leader channel, the velocity of the return stroke and the rate of recombination of the charges on the leader during the return-stroke process.

Equation (1) which is often referred to in the literature as the "Bruce–Golde equation" is clearly imperfect. In particular, as pointed out by Dennis and Pierce (1964) the inherent assumption that the current amplitude is constant at any instant along the return-stroke channel is wrong, at least during the time while the tip of the return stroke travels upwards. Barlow *et al.* (1954) found that better agreement with radio-noise disturbances was obtained by adding a third term of the form:

$$I = I_0[a \exp(-\alpha t) - b \exp(-\beta t) + c \exp(-\gamma t)]. \tag{2}$$

For further discussion of current wave-shape and radio noise see Chapter 10.3.2.1.

Müller-Hillebrand (1962), being concerned with the magnetic fields produced by lightning strokes, extended Equation (1) by adding terms to represent the sudden changes in current amplitude when a branching point from the main channel is reached. Equation (1) was also further developed, to take account of the long current tail, by Hepburn (1957). Several Indian research workers, most recently Gosh and Khastgir (1972), have also contributed to this aspect. The most comprehensive analytical expressions are due to Cianos and Pierce (1972), who have critically surveyed the information and have developed a statistical model of the current discharge for engineering usage.

It is evident that the constants in any analytical expression for the current will have to be adjusted in accordance with a standard current shape which may ultimately be accepted. More importantly, if the concave current front illustrated in Fig. 5 is regarded as representative of a first stroke, this may be described, as suggested by Whitehead *et al.* (1973), by a general expression of the form:

$$I = I_0[\exp(\alpha t/T - 1)]/\exp(\alpha - 1). \tag{3}$$

The basic concept of any expression such as Equation (1) has been rejected by Uman *et al.* (1973a) since it does not explain the magnetic field changes observed by them. They have therefore suggested replacing it by a transmission-line model. In this they follow an approach originally elaborated by Wagner and Hileman (1961) and examined in Chapter 5.5.6. This aspect is discussed in the section which follows.

4.1.6 Critical Review

It has been shown that the extensive results obtained by Berger on Mount San Salvatore and later by Garbagnati *et al.* (1974) on Italian mountain tops exhibit distinct differences in the wave-shapes of the currents in first and subsequent strokes. It has furthermore been shown that these results are supported, in principle at least, by observations on other tall structures and on high-voltage transmission lines. Objections have, however, been raised against accepting these results as valid for lightning flashes to open ground (Griscom *et al.*, 1965).

Cianos and Pierce (1972) suspect that the slow current rise in the first stroke as pictured in Fig. 5 is influenced by the development of upward leaders; they argue that the effective height of the tower is increased to almost 300 m by the configuration of the mountain (Pierce, 1972). This argument overlooks, in the first place, that the concave wave-front depicted in Fig. 5 applies to flashes initiated by downward, and not upward, leaders and, in the second place, that upward streamers, although of considerably shorter length, have also been observed from flat open ground (Golde, 1967, and Chapter 17.2.1).

Many years ago in a discussion I myself advanced the view that electrostatic coupling between a downcoming leader channel and a tall tower would result in a notably higher charge density near the lower tip of the leader than in a leader over open ground. The high concentration of charge near the leader tip was then thought to favour a short wave-front. The same view was expressed by Wagner and McCann in the discussion of one of McEachron's papers (1941).

Because of some recent work which is discussed later and because I have since changed my views it is pertinent to examine the features of a first lightning leader just before contact is established with the object to be struck.

First lightning strokes are almost invariably branched. It appears legitimate to assume that the charge density distribution in the ground and on the leader and its branches is then determined by the electrostatic coupling between individual sections of the leader and ground, namely by the capacitances C_1, C_2 and C_3 in Fig. 11. Before this stage is reached, point-discharge currents flow upwards from the tower, their amplitude increasing with increasing field strength between the tip of the leader and the ground. As illustrated in Fig. 4, Chapter 17, this field gradient increases very rapidly as the leader approaches the tower or ground so that the point-discharge current becomes concentrated in one or several upward streamers. The current through the tower grows correspondingly rapidly until the leader channel and one of the upward streamers meet and it is suggested in Section 2 that it is at this instant or very shortly afterwards that the current reaches its crest value. This argument seems to be supported by the abrupt change in slope in almost all of Berger's current oscillograms immediately after the crest value has been reached. It is interesting to add that early laboratory tests with sparks have shown (Allibone and Meek, 1938) that maximum luminosity frequently occurs at the point where downward and upward leaders meet so that this would be the point at which the current in the "return spark" reaches its maximum.

As the tip of the return current moves upwards along the leader channel its amplitude is largely determined by the electrostatic coupling between the leader channel and the nearest branch, as indicated by the capacitances C_4 in Fig. 11. If the branch is long, the capacitance C_4 is high and if the branch is fairly low, this may result in a second current peak which, on occasions, may exceed the first crest as indicated in Fig. 8, Chapter 5.

The physical picture outlined above is believed to be in line with ideas advanced by McEachron and McMorris (1936), and Wagner (1967). Furthermore it would seem to lend support to the development of a concave wave-front. In addition this argument is strongly supported by present knowledge of the switching-impulse breakdown of long gaps (Chapter 7.4). The amplitude and duration of the current tail are likely to be affected by the number and spatial distribution of side branches and, more generally, by the rate of recombination of the charges on the leader channel by the return stroke (Bruce and Golde, 1941). There is also reason to believe that it will be influenced by soil resistivity. Thus it has been suggested (McEachron, 1941; Wagner *et al.*, 1941) that continuing strokes are most likely to develop in areas of high soil resistivity.

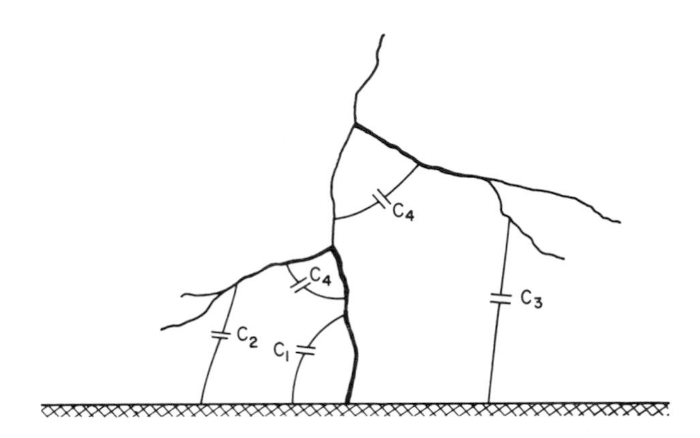

Fig. 11. Capacitive coupling between branched lightning channel and ground.

Subsequent strokes differ basically from first strokes. Their leaders do not progress into virgin air but are normally guided by residual conductivity along the initial discharge channel: they are hardly ever branched, and they do not normally give rise to upward streamers. For all these reasons the front in the subsequent return stroke may be expected to be shorter and the general shape of the current tail more regular and smoother than in a first stroke as indeed confirmed by Berger (1967).

Very different current wave-shapes from those shown in Fig. 5 have been derived in two investigations. The first of these is explained by the so-called "prestrike theory" of Griscom *et al.* (1965). This was discussed and rejected by Wagner (1967) and no further consideration appears to be warranted.

The second much more important investigation is due to Uman *et al.* (1973). They have assembled a large body of records of field-changes produced by flashes to open ground in Florida at distances between 0·5 and 100 km. Some 500 first strokes produced a mean rise time of 3·7 μs while a similar number of subsequent strokes gave a mean value of 3·0 μs (Fisher and Uman, 1972). They thus concluded that the rise times of field-changes in first and subsequent strokes were the same.

In order to translate the records of field-changes into current wave-shapes these authors then developed a "transmission-line or propagating current-pulse model" which is based on several assumptions (Uman *et al.*, 1973b). These include the concept that the peak current does not vary throughout the length of the lightning channel and that the return stroke starts at ground level. Both these assumptions are in contradiction to the picture of the first return stroke outlined above. As concerns the first aspect, Schonland (1956) specifically states that the luminosity of a first return stroke "decreases abruptly and markedly in intensity after each branching point" and he adds

> "the observations as a whole indicate clearly that the return streamer is engaged in removing charge which has been brought into the air by the leader and is *held chiefly on its branches*".

As to the second aspect, it is suggested that part of the positive charge by which the negative charge near the tip of the leader channel is neutralized is concentrated on the upward streamer. Thus, if the upward streamer is long, i.e. when the stroke is intense, the crest value of the current in the return stroke begins to flow upwards from a point at a substantial height above ground level. In contrast, because of the absence of branches in subsequent strokes, the reduction in amplitude of the return current along the discharge channel would be expected to be much less than for a first return stroke and subsequent currents would indeed start from ground level. The first conclusion is supported by Schonland's (1956) observation that the luminosity of subsequent return strokes "tends to decrease as it travels upwards but the effect is not very marked".

The foregoing considerations seem to invalidate the basic assumptions underlying the evaluation of current wave-shapes by Uman *et al.* However, in view of the importance of definitely establishing the current wave-shapes in strikes to open ground it is hoped that the interpretation of field-change records will be further refined. In this connection, it may be fruitful to reconsider the approach initiated by Wagner (1960) who showed that the principal characteristics of field-changes recorded by Uman and his colleagues can be explained by accepting a convex current front of about 10 μs duration.

4.2 Upward Discharge

Of 135 flashes to the Empire State Building about 50% have no high current peak and consist entirely of a continuous-current discharge as illustrated in the first part of Fig. 2 (Hagenguth and Anderson, 1952). Only three similar cases were observed in flashes to the Cathedral of Learning (McCann, 1944). The duration of current flow may be as long as 1·6 s but some 80% of all discharges did not last longer than 400 ms.

In the course of nine years Berger (1967) recorded 243 negative flashes initiated by positive upward leaders, 157 of which were not followed by any high-current impulses. As in the American results, the continuous current discharge was of an irregular nature. Current amplitudes usually reached several hundred amperes and lasted for several hundred milliseconds.

The great majority of upward leaders carry positive charges, that is they develop under the influence of the predominantly negative charges in the lower regions of the thundercloud. When these are followed by high-current strokes, the shapes of these strokes resemble those of subsequent strokes in the normal negative earth flashes.

In only five cases did upward discharge start from a negatively charged tower (Berger and Vogelsanger, 1966). These were followed by intense positive strokes discharging currents ranging from 22 to 106 kA. It appears that, in these cases, the negative upward discharge meets a downward leader which progresses from one of the high positive charge centres in the upper region of the cloud. Berger calls these discharges "connecting streamers" and they are described in Chapter 5.3.2. Such leaders are confined to mountains or very tall earthed objects and frequently involve the so-called giant flashes of positive polarity to which reference has been made in Section 4.1.2. The highest current recorded on San Salvatore amounted to 270 kA (Berger, 1972).

5. Related Parameters

5.1 Charge

5.1.1 Charges in Flashes

Some of the investigations to determine the charges involved in lightning flashes were performed on tall structures where the direction of progression of the leader was either unknown or unspecified. For these reasons the two types of discharge must, initially, be considered together.

The first estimate of the total charges involved in flashes to open ground is due to Wilson (1920), who used a special capillary electrometer to measure the charge which flowed through an earthed conductor which was alternately exposed to, and shielded from, the local electrostatic field. The electric field change due to a lightning flash within a distance of a few kilometres is proportional to the electric moment (see Chapter 10.2) of the discharge. Wilson assumed that the charge centre in the cloud was at a height of 2 km and derived a figure of 20 C for the mean charge dissipated in an earth flash. This value was confirmed by more extensive investigations (Wormell, 1939) which also showed that it applied to both positive and negative flashes.

Wilson's method was used by Wang (1963) in Singapore. From over 3,000 records he deduced an average charge of 25 C and suggested that this value, higher than that found in England, could be due to the nature of tropical thunderstorms. From his recordings of electromagnetic field-changes Norinder (1952) concluded that the most frequent value of charge was between 10 and 15 C.

Very much higher values, up to 1,065 C, were recorded in two investigations (Meese and Evans, 1962; Nelson, 1968) in which the magnetic field-changes caused by earth flashes were recorded by magnetometers. The recording technique was criticized by Cianos and Pierce (1972) whose comments also apply to measurements reported by Hatakeyama (1958).

The foregoing estimates refer to the total charge dissipated in an earth flash. These include the substantial charges which are neutralized during the development of the initial leader channel and its extensive corona envelope (Bruce and Golde, 1941).

The charges conveyed from the lightning channel to earth have been determined by integrating oscillographic current records. The currents recorded by McCann (1944) were obtained on structures of such height that they undoubtedly included a small number of upward flashes. The values derived from 37 flashes ranged up to about 100 C with a most frequent value of between 5 and 10 C.

A different method of estimating the charges in lightning flashes was used by McEachron and Hagenguth (1942). From laboratory tests with heavy impulse currents they derived a relation between the size of the hole burnt in a thin metal sheet and the charge in the test arc. They then examined the sizes of holes burnt in the course of six years in a thin copper sphere on top of a 290-m high radio tower. A total of 150 holes gave an average value for the charge of 15 C with a maximum of 240 C. As in the foregoing cases examined by McCann, it is unknown how many of these holes were due to flashes with upward as opposed to downward leaders, but a normal downward flash was undoubtedly involved in a large hole burnt in a metal-roofed rural home which, analysed by the same method, suggested a charge of 210 C.

Hill (1963) reverted to the thermal effects of the lightning channel to examine, in strictly physical terms, the microscopic structure of small pips of molten copper (fulgamites) produced by lightning. For this purpose small copper caps were placed over the ends of four lightning rods on top of a 330-m high radio tower in Illinois. During one storm these caps were subjected to 11 flashes, after which they were removed for analysis. The charges derived ranged from 0·015 C to about 14 C, with a mean value of about 2·5 C.

A particularly interesting case involving the crushing of two parallel earth conductors of a television mast was reported by Meister (1973). Careful calculation, supported by laboratory tests, led to the conclusion that a positive giant stroke discharged a current of 300 kA with 500 μs time to half-value and a charge of 215 C.

The great majority of flashes to the Empire State Building are initiated by upward leaders. Of 73 flashes which could be analysed, 50% involved charges in excess of 19 C, with a maximum value of 164 C (Hagenguth and Anderson, 1952). A large part of this charge was contributed by continuous currents.

The current records of downward flashes to the towers on Mount San Salvatore were integrated to determine the associated charges (Berger *et al.*, 1975). A total of 94 negative flashes gave a median value of 7·5 C. The corresponding values for 26 positive flashes had a median of 80 C. These results which are plotted in the form of cumulative frequency-distribution curves in Fig. 12 and listed in Table I confirm the greater severity of positive

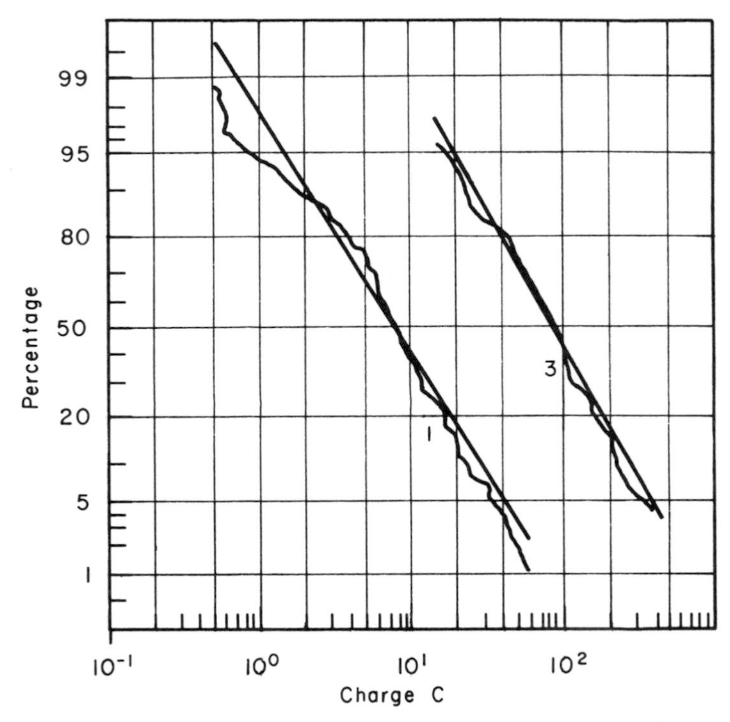

Fig. 12. Cumulative frequency distribution of charges in lightning flashes (Berger *et al.*, 1975). 1. Negative flashes. 3. Positive flashes.

flashes to which attention has been drawn in Section 3.2.1.4. From 32 flashes recorded in Italy (Garbagnati *et al.*, 1974) a median value of 3·2 C was deduced.

Flashes with upward leaders from one of the towers on San Salvatore involved negative charges up to 200 C while those in positive flashes reached 300 C (Berger, 1967). However, for both polarities charges exceeding 100 C were comparatively infrequent. Contributions to these values by continuous currents were substantial only in negative flashes.

By far the most detailed investigation of the charges in flashes to open ground has been reported from New Mexico (Williams and Brook, 1963; Kitagawa, 1965). The physical interpretation of simultaneous photographic and field records is discussed in Chapter 5 and the nomenclature used in that account is followed here. About one-half of all multiple flashes is found to be of the usual type but the other half is characterized by continuing discharges which, as mentioned in Section 4.1.3, last at least several tens of milliseconds and have amplitudes up to a maximum of 590 A. These continuing discharges do not usually follow either single strokes or the first stroke of a multiple flash (Kitagawa *et al.*, 1962). The average charges in flashes without continuing currents amount to 20 C whereas those in hybrid flashes, that is with continuing currents, are 34 C (Kitagawa, 1965). Continuing currents have also been observed in flashes initiated by upward leaders and these are discussed in Chapter 5.4.3.2.

From the foregoing survey it appears that the charges involved in earth flashes and reported by different authors vary over a wide range. Early derivations from measurements of the electric moment (see Section 5.2) suffer from lack of knowledge of the heights of the charge centres tapped in individual strokes of a multiple flash. Flashes initiated by upward leaders are likely to be associated with greater charges than those in downward discharges because of the charges required in building up the corona sheath in the downward leader, as mentioned earlier in this section. Furthermore, flashes without continuing currents carry significantly smaller charges than flashes with such currents. With these points in mind, the apparent differences in the results reported by different authors are less difficult to explain.

5.1.2 Charges in Strokes

The charges dissipated in component strokes to the masts on San Salvatore (Berger *et al.*, 1975) were analysed by the authors under the heading "impulse charge". This term was intended to include the rapidly changing part of the current wave and was determined by inspection. It thus definitely excludes contributions by continuing currents exceeding 2 ms in duration. From 90 first negative strokes a median charge of 4·5 C was deduced and 117 negative subsequent strokes gave a median of 0·95 C. The rarer 25 single positive strokes had a median of 16 C (see Table I and curve 3 in Fig. 13).

No positive flashes were recorded in the New Mexico investigations (Kitagawa, 1965). The most frequent value of charge involved in first negative strokes is given as 3 to 4 C with a maximum of about 20 C, that associated with subsequent strokes as 0·5 to 1 C with a maximum of about 15 C.

Although the results from San Salvatore are expressed by their median values and those from New Mexico in terms of most frequent occurrence, the respective frequency distributions indicate that the charges associated with individual strokes are of the same order of magnitude in both localities. On the other hand, the charges involved in flashes differ considerably as shown in the preceding section, with a median value of 7·5 C from San Salvatore (omitting positive strokes which were absent in New Mexico) and a most frequent value of 20 C without continuing strokes and 30 C with long-duration currents in New Mexico. Apart from the greater frequency of continuing strokes in New Mexico, the foregoing differences seem to be mainly due to the relative frequency of occurrence of multiple flashes. Thus only 24% of all flashes on San Salvatore are multiple (Berger, 1972) as against 86% in New Mexico (Kitagawa, 1965).

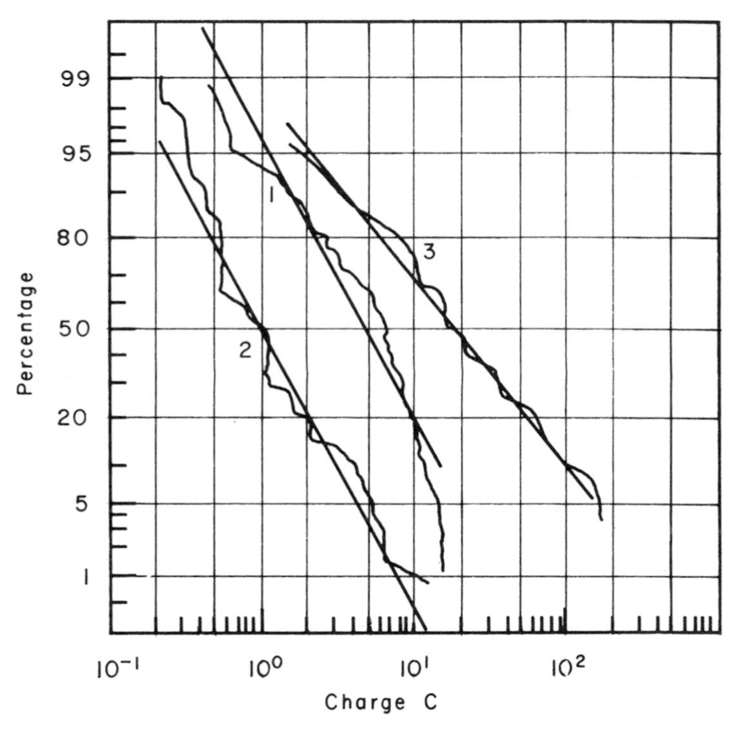

Fig. 13. Cumulative frequency distribution of impulse charges in lightning strokes (Berger *et al.*, 1975). 1. Negative first strokes. 2. Negative subsequent strokes. 3. Positive strokes.

5.2 Electric Moment

Although, at first sight, the electric moment is not directly related to the lightning current it forms the basis of some of the estimates of the charges dissipated in lightning flashes, as shown in Section 5.1. It therefore warrants consideration in the context of this survey. The electric moment of a lightning flash is given by

$$M = 2QH,$$

where Q is the charge involved in the flash, this charge being assumed to be uniformly distributed in a sphere with its centre at a height H above ground, while an identical charge of opposite polarity is assumed to be concentrated at a depth H vertically below the cloud charge.

Wilson's (1920) observations of 179 earth flashes produced a mean value, for both negative and positive flashes, of about 100 C km. In the extension of this investigation by Wormell (1939) it was found that the most frequent moment of negative flashes was about 110 C km and that of positive flashes 220 C km. The values were found to vary over a considerable range with a maximum of 1,200 C km.

This investigation was further extended and refined by Pierce (1955, 1956) who examined the contribution to the total electric moment by the component strokes of multiple flashes. Negative flashes produced an average value of 115 C km while the rare single positive strokes gave 110 C km.

From over 3,000 flashes recorded in Singapore, Wang (1963) derived an average moment of 210 C km. In South Africa, Schonland and Craib (1927) obtained a mean value of 94 C km from 82 discharges, while a later analysis by Barnard (1951) of 10 discharges produced a value of 182 C km.

The electric moments destroyed in earth flashes in New Mexico were examined in detail by Brook *et al.* (1962). The average value for discrete flashes, that is flashes without continuing current flow, was found to be 151 C km as against 346 C km for hybrid flashes, thus clearly showing the large contribution by continuing discharges. Combining these types gave an average value for all negative flashes of 249 C km, a figure which is notably higher than the average value of 150 C km determined by Pierce in England. This may be taken to indicate a genuinely greater severity of lightning flashes in the desert country of New Mexico as compared with temperate regions.

5.3 Action Integral and Energy

When a unidirectional current of constant amplitude i flows through a metallic conductor for a time t, the temperature rise is proportional to

$$\int i^2 \, \mathrm{d}t.$$

This is called the "action integral". Cianos and Pierce (1972) determined its numerical value separately for the impulsive component of the current and for its continuing tail.

Berger *et al.* (1975) derived numerical values of the action integral from the oscillographic records on Mount Salvatore. The results are given in Table I and the cumulative distribution curves reproduced in Fig. 14. From 91 negative first strokes they obtained a median value of $5{\cdot}5 \times 10^4$ A^2 s. Corresponding values for 88 negative subsequent strokes were $6{\cdot}0 \times 10^3$ A^2 s, while the value for 26 positive flashes amounted to $6{\cdot}5 \times 10^5$ A^2 s. These figures once again indicate the greater severity of the rare positive discharges. Garbagnati *et al.* (1974) derived a median value of $2{\cdot}9 \times 10^4$ A^2 s from 26 first strokes.

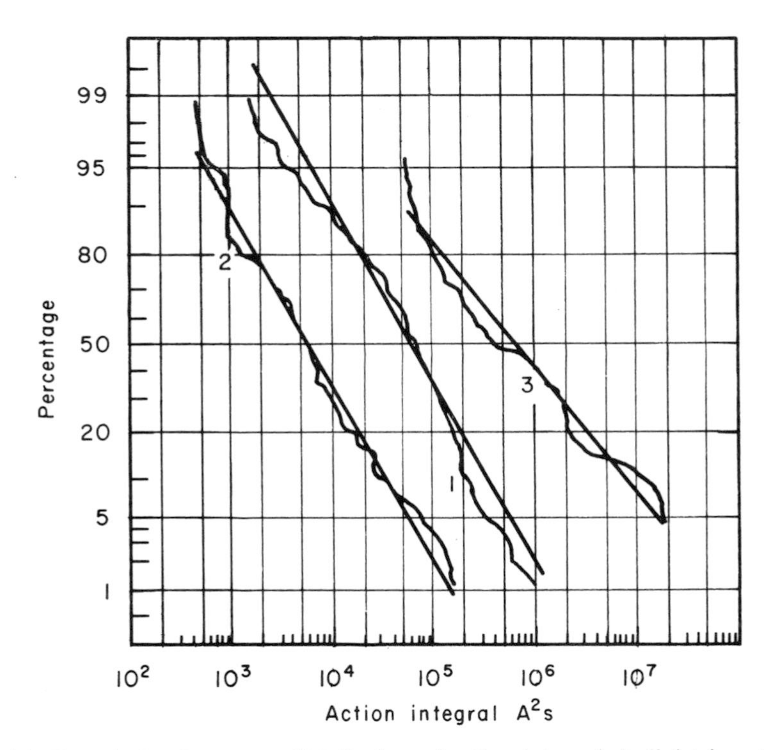

Fig. 14. Cumulative frequency distribution of action integrals in lightning strokes (Berger *et al.*, 1975). 1. Negative first strokes. 2. Negative subsequent strokes. 3. Positive strokes.

The potential of a thundercloud cannot be measured by present-day methods. The best estimates (Bruce, 1944; Wagner, 1967) suggest that it lies between 10^7 and 10^8 V. From Fig. 4, Chapter 17, it can be concluded that, when the tip of a leader channel has approached ground to within about 200 m, its potential is 2×10^7 V, a figure which agrees closely with the $1{\cdot}8 \times 10^7$ V derived in Chapter 5.5.10. Adding to this value the likely potential drop along the leader, Berger obtains a cloud potential of between 3·6 and $4{\cdot}8 \times 10^7$ V. Assuming an average potential of 4×10^7 V and an average charge of 10 C, the average energy in a lightning flash may thus be taken to be of the order of 2×10^8 J (55 kW h or 2×10^{15} erg) as compared with Wilson's (1920) classical estimate of 10^{17} erg. Flashes containing continuing currents might be expected to give values somewhat exceeding the foregoing estimate.

6. Correlation of Parameters

With a view to providing a model of the lightning current wave-shape Cianos and Pierce (1972) examined the likelihood of an interrelation between the various parameters and the entire discharge. However, the question of specific correlations was first raised when early observational data about lightning currents became available.

The first question examined was that of a correlation between the rate of rise of the current and its crest value. McCann (1944) and Alizade *et al.* (1968) found that the rate of rise tended to increase with increasing amplitude, while Komelkov (1941) and Hyltén-Cavallius and Strömberg (1959) could not detect any connection between these factors. The types of measurement from which these contradictory conclusions were reached are discussed in Section 4.1.4 where it is shown that these authors adopted different definitions of rate of rise of current and that the number of observations in any single survey was not large. Popolansky (1970) subjected 157 current records obtained on chimneys to a statistical test and found no evidence of a positive correlation.

The oscillographic current records obtained on Mount San Salvatore were subjected to statistical correlation tests. Berger's (1972) first examination of the regression lines associating a variety of parameters showed that a strong positive correlation existed between the crest value of the current and both the impulse charge (as defined in Section 5.1.2) and the action integral if the contribution by continuing current flow was excluded.

The results of the complete analysis of the data from San Salvatore were reported by Kröninger (1974). Coefficients of correlation were computed between a variety of parameters and assessed by comparing the derived value with the critical coefficient for a 1% level of significance. Correlation coefficients were determined for negative and positive downward flashes as well as for negative and positive first component strokes and negative subsequent strokes and, within these categories, for peak current, front duration, maximum rate of rise, impulse charge, action integral, total stroke charge, flash duration and total flash charge.

Amongst the large number of combinations of these factors a strong positive correlation was found to exist between the peak value of the current in the first negative stroke, its impulse charge and its action integral and also between the total charge in a flash and its duration. Somewhat lesser, though still significant, correlations were found between total stroke charge and total charge in a multiple flash and between the total charge in the first stroke and its peak current.

In subsequent negative strokes the highest coefficients were detected between the impulse charge, the action integral and the peak current. Still significant is also the correlation between total stroke current and action integral.

Positive strokes were shown in Section 4.1.2 to exhibit rather variable wave-shapes and it is therefore not surprising to find that strong correlations exist merely between total stroke duration and charge and between impulse charge and action integral. However, the correlation between impulse charge and peak current, although outside the 1% level of significance, can still be accepted as reasonably established.

7. Outlook

The critical reader may have noted that, in contrast to many other reviews, the results of different investigations of the same parameter are not presented in tabular form nor in summarizing graphs. The reason for this deliberate omission is the fact that many results are not strictly comparable and that so many footnotes would have been required as to make such presentations more confusing than illuminating. Nevertheless, every investigation adds to the sum total of present knowledge of the lightning current.

The frequency distribution of lightning current amplitudes in earth flashes is reasonably well established. Over most of its range it can be said to follow a log-normal distribution but the frequency of occurrence of exceptionally high and low crest values remains to be established by greatly increased numbers of observation. Further study is also required of current amplitudes in tropical storms and in flashes to very tall structures. The most vital gap in present knowledge concerns the spatial and temporal variations of the current along the entire length of the discharge channel from ground to cloud.

In my view, the wave-shapes of first and subsequent earth strokes can also be accepted as established. However, more statistical information is required on current tails in the range of about 100 μs to several milliseconds. In this context it would appear fruitful to study the effect of soil resistivity. Present knowledge of the characteristics of intra-cloud discharges is examined in Chapter 6.

Acknowledgements

I am indebted to Professor K. Berger for his constructive criticism of this chapter.

References

Alizade, A. S., Kuliev, D. A. and Kalantarov, V. A. (1968). Investigation of the slope and amplitude of the current in direct lightning strokes to transmission lines. *Elektrichestvo* **7**, 73–77.

Allibone, T. E. and Meek, J. M. (1938). The development of the spark discharge. *Proc. R. Soc.* **166**, 97–126.

Anderson, J. G. and Hagenguth, J. H. (1958). Magnetic fields around a transmission line tower. *Trans. Am. Inst. elect. Engrs* **77**, 1644–1650.

Anderson, R. B. (1971). A comparison between some lightning parameters measured in Switzerland with those in South Africa. C.S.I.R. report ELEK 6, Pretoria, South Africa.

Anderson, R. B. and Jenner, R. D. (1954). A summary of eight years of lightning investigation in Southern Rhodesia. *Trans. S. Afr. Inst. elect. Engrs* **39**, 217–272.

Appleton, E. V. and Chapman, F. W. (1937). On the nature of atmospherics IV. *Proc. R. Soc.* **158**, 1–22.

Baatz, H. (1951). Blitzeinschlag-Messungen in Freileitungen. *Elektrotech. Z. Ausg. A.* **72**, 191–198.

Barlow, J. S., Frey, G. W. and Newman, J. B. (1954). Very low frequency noise power from the lightning discharge. *J. Franklin Inst.* **258**, 187–203.

Barnard, V. (1951). The approximate mean height of the thundercloud charges taking part in a flash to ground. *J. geophys. Res.* **56**, 33–35.

Bell, E. (1940). Lightning investigation on a 220 kV system, Part III. *Trans. Am. Inst. elect. Engrs* **59**, 822–828.

Bell, E. and Price, A. L. (1930). Lightning investigations on the 220 kV system of the Pennsylvania Power and Light Company. *Trans. Am. Inst. elect. Engrs* **50**, 1101–1110.

Bellaschi, P. L. (1935). Lightning currents in field and laboratory. *Trans. Am. Inst. elect. Engrs* **54**, 837–843.

Berger, K. (1929). Die ersten Beobachtungen des Verlaufes von durch Gewitter verursachten Spannungen in Mittelspannungsnetzen mittels des Kathodenstrahl-Oszillographen des S.E.V. *Bull. Schweiz elektrotech. Ver.* **20**, 321–338.

Berger, K. (1943). Gewittermessungen des Jahres 1936 und 1937. *Bull. Schweiz elektrotech. Ver.* **34**, 353–365.

Berger, K. (1955). Messungen und Resultate der Blitzforschung der Jahre 1947 . . . 1954 auf dem Monte San Salvatore. *Bull. Schweiz elektrotech. Ver.* **46**, 405–424.

Berger, K. (1967). Novel observations on lightning discharges: results of research on Mount San Salvatore. *J. Franklin Inst.* **283**, 478–525.

Berger, K. (1972). Methoden und Resultate der Blitzforschung auf dem Monte San Salvatore bei Lugano in den Jahren 1963–1971. *Bull. Schweiz elektrotech. Ver.* **63**, 1403–1422.

Berger, K. and Vogelsanger, E. (1966). Photographische Blitzuntersuchungen der Jahre 1955 . . . 1965 auf dem Monte San Salvatore. *Bull. Schweiz elektrotech. Ver.* **57**, 599–620.

Berger, K., Anderson, R. B. and Kröninger, H. (1975). Parameters of lightning flashes. *Electra*, **40**, 101–119.

Bermbach, (1908). Ist der Blitz eine oszillatorische Entladung? *Elektrotech. Z. Ausg. A.* **29**, 40–41.

Biermanns, J. (1949). "Hochspannung und Hochleistung". Carl Hanser Verlag, München.

Binder, L. (1928). Einige Untersuchungen über den Blitz. *Elektrotech. Z. Ausg. A.* **49**, 503–507.
Brook, M., Kitagawa, N. and Workman, E. J. (1962). Quantitative study of strokes and continuing currents in lightning discharges to ground. *J. geophys. Res.* **67**, 649–659.
Bruce, C. E. R. (1944). The initiation of long electrical discharges. *Proc. R. Soc.* A **183**, 228–242.
Bruce, C. E. R. and Golde, R. H. (1941). The lightning discharge. *J. Instn. elect. Engrs* **88**, Part II, 487–520.
Burgsdorff, W. W. (1941). The effect of soil conductivity on lightning current (in Russian). *Elekt. Sta. Mosk.* Nos 11 and 12, 26–29.
Cianos, N. and Pierce, E. T. (1972). A ground-lightning environment for engineering usage, Stanford Research Institute, Menlo Park, California, Tech. Report 1.
Creighton, E. E. F. (1924). Lightning. *Trans. Am. Inst. elect. Engrs* **43**, 1197–1204.
Davis, R. and Standring, W. G. (1947). Discharge currents associated with kite balloons. *Proc. R. Soc.* A **191**, 304–322.
De Blois, L. A. (1914). Lightning protection for buildings. *Trans. Am. Inst. elect. Engrs* **33**, 519–544.
Dennis, A. S. and Pierce, E. T. (1964). The return stroke of the lightning flash to earth as a source of VLF atmospherics. *Radio Sci.* **7**, 777–794.
Emde, F. (1910). Die Schwingungszahl des Blitzes. *Elektrotech. Z. Ausg. A.* **31**, 675–680.
Fisher, R. J. and Uman, M. A. (1972). Measured electric field rise times for first and subsequent lightning return strokes. *J. geophys. Res.* **77**, 399–406.
Flowers, J. W. (1944). Lightning. *Gen. Elect. Rev.* **47**, 9–14.
Fortescue, C. L. (1930). Direct strokes—not induced surges—chief causes of high-voltage line flashover. *Elect. J.* **27**, 459–462.
Foust, C. M., Maine, B. C. and Lee, C. (1953). Lightning stroke protection at high altitude in Peru. *Trans. Am. Inst. elect. Engrs* Part III. **6**, 383–393.
Garbagnati, E., Giudice, E., Lo Piparo, G. B. and Magagnoli, U. (1974). Survey of the characteristics of lightning stroke currents in Italy—Results obtained in the years from 1970 to 1973. E.N.E.L. Report R5/63–27.
Gary, C., Cimador, A. and Fieux, R. (1975). La foudre: Etude du phénomène. Applications à la protection des lignes de transport. *Revue gen. Élect.* **84**, 25–62.
Golde, R. H. (1946). Lightning currents and potentials on overhead transmission lines. *J. Inst. elect. Engrs* Part II. **93**, 559–569.
Golde, R. H. (1954). Lightning surges on overhead distribution lines caused by indirect and direct lightning strokes. *Trans. Am. Inst. elect. Engrs* Part III. **73**, 437–447.
Golde, R. H. (1967). The lightning conductor. *J. Franklin Inst.* **283**, 451–477.
Golde, R. H. (1973). "Lightning Protection". Edward Arnold, London.
Gosh, J. D. and Khastgir, S. R. (1972). Time-variation of return-stroke current in the cloud-to-ground lightning discharges. *Indian J. pure appl. Phys.* **10**, 556–557.
Griscom, S. B., Caswell, R. W., Graham, R. E., McNutt, H. R., Schlomann, R. H. and Thornton, J. K. (1965). Five-year field investigation of lightning effects on transmission lines. *Trans. Am. Inst. elect. Engrs* Part III. **8**, 257–280.
Gross, I. W. and Lippert, G. D. (1942). Lightning investigation on 132 kV transmission system of the American Gas and Electric Company. *Trans. Am. Inst. elect. Engrs* **61**, 178–184.

Hagenguth, J. H. and Anderson, J. G. (1952). Lightning to the Empire State Building, part III. *Trans. Am. Inst. elect. Engrs* Part III. **71**, 641–649.

Hagenguth, J. H., Rohlfs, A. F. and Degnan, W. J. (1952). Sixty cycle and impulse sparkover of large gap spacings. *Trans. Am. Inst. elect. Engrs* Part III. **71**, 455–460.

Hatakeyama, H. (1958). The distribution of the sudden change of electric field on the earth's surface due to lightning discharge, *in* "Recent Advances in Atmospheric Electricity." (L. G. Smith, Ed.) pp. 289–298, Pergamon Press, London.

Hepburn, F. (1957). Atmospheric wave forms with very-low frequency components below 1 kHz. *J. atmos. terr. Phys.* **10**, 266–287.

Hill, R. D. (1963). Determination of charges in lightning strokes. *J. geophys. Res.* **68**, 1365–1375.

Horváth, T. (1973). Einfluss der Beobachtungsstelle auf die gemessenen Blitzgrössen. International Lightning Conference, Portoroz, Jugoslavia, report No. 1.5.

Humphreys, W. J. (1918). Physics of the air. *J. Franklin Inst.* **186**, 57–75; 211–232; 341–370.

Hyltén-Cavallius, N. and Strömberg, A. (1959). Field measurements of lightning current. *Elteknik*. **1**, 109–113.

Kitagawa, N. (1965). Types of lightning, *in* "Problems of Atmospheric and Space Electricity" (S. C. Coroniti, Ed.) pp. 337–348, Elsevier, Amsterdam.

Kitagawa, N., Brook, M. and Workman, E. (1962). Continuing currents in cloud-to-ground lightning discharges. *J. geophys. Res.* **67**, 637–647.

Kohlrausch, W. (1888). Die Berechnung von Blitzableitern und ein Versuch, die Elektrizitätsmenge der Gewitterladungen zu schätzen. *Elektrotech. Z.* **9**, 123–125.

Komelkov, V. S. (1941). Investigation of the maximum wave-front slope of lightning currents (in Russian). *Elektrichestvo*. **5**, 34–39.

Krehbiel, P., Brook, M. and McCrory, R. A. (1974). Lightning ground stroke charge location from multistation electrostatic field change measurements. Fifth International Conference on Atmosph. Elect. Garmisch-Partenkirchen.

Kröninger, H. (1974). Further analysis on Prof. Berger's San Salvatore lightning current data. C.S.I.R. Special Report ELEK **53**, Pretoria, South Africa.

Lewis, W. W. (1950). "The protection of Transmission Systems against Lightning." John Wiley and Sons, New York.

Lewis, W. W. and Foust, C. M. (1940). Lightning investigation on transmission lines—VII. *Trans. Am. Inst. elect. Engrs* **59**, 227–233.

Lodge, O. J. (1892). "Lightning conductors and lightning guards". Whittaker and Co. London.

Mackerras, D. (1969). Photoelectric techniques for investigating discharge processes in lightning flashes, *in* "Planetary Electrodynamics" (S. C. Coronity and J. Hughes, Eds), Vol. II, pp. 511–524.

Malan, D. J. and Collens, H. (1937). Progressive lightning III—The fine structure of return lightning strokes. *Proc. R. Soc.* A **162**, 175–203.

McCann, G. D. (1944). The measurement of lightning currents in direct strokes. *Trans. Am. Inst. elect. Engrs* **63**, 1157–1164.

McCann, G. D. and Harder, E. L. (1948). Direct strokes and lightning surges on transmission lines. Conf. int. grands. Res. Elect. report No. 322.

McEachron, K. B. (1941). Lightning to the Empire State Building. *Trans. Am. Inst. elect. Engrs* **60**, 885–890; 1377–1380.

McEachron, K. B. and Hagenguth, J. H. (1942). Effect of lightning on thin metal surfaces. *Trans. Am. Inst. elect. Engrs* **61**, 559–564.

McEachron, K. B. and McMorris, W. A. (1936). The lightning stroke: mechanism of discharge. *Gen. Elect. Rev.* **39**, 487–496.
Meese, A. D. and Evans, W. H. (1962). Charge transfer in the lightning stroke as determined by the magnetograph. *J. Franklin Inst.* **273**, 375–382.
Meister, H. (1973). Dynamische Zerstörung einer Erdungsleitung durch einen Blitz. *Bull. Schweiz elektrotech. Ver.* **64**, 1631–1635.
Müller-Hillebrand, D. (1936). Aus der Entwicklung der Überspannungstechnik im letzten Jahrzehnt. *Elektrotechnik und MaschBau.* **54**, 361–366.
Müller-Hillebrand, D. (1962). The magnetic field of the lightning discharge, *in* "Gas discharges and the Electricity Supply Industry", pp. 89–111. Butterworth, London.
Nelson, L. N. (1968). Magnetographic measurements of charge transfer in the lightning flash. *J. geophys. Res.* **73**, 5967–5972.
Neuhaus, H. and Strigel, R. (1935). Der Verlauf von Wanderwellen in elektrischen Maschinen und deren Schutz beim Anschluss an Freileitungen. *Arch. Elektrotech.* **29**, 702–721.
Newman, M. M., Stahmann, J. R. and Robb, J. D. (1967). Experimental study of triggered natural lightning discharges. Federal Aviation Agency, Washington, D.C. report No. DS-67-3.
Norinder, H. (1925). The cathode ray tube as used as high-frequency oscillograph for investigation of transient waves. *Tek. Tidskr. Stockh.* **55**, 152–157.
Norinder, H. (1935). Lightning currents and their variations. *J. Franklin Inst.* **220**, 69–92.
Norinder, H. (1952). Experimental lightning research. *J. Franklin Inst.* **253**, 471–504.
Norinder, H. and Knudsen, E. (1961). Some features of thunderstorm activity. *Ark. Geofys.* **3**, 367–374.
Ollendorf, F. (1932). Der Einfluss des Erdwiderstandes auf den Blitz. *Phys. Z.* **33**, 368–376.
Ouyang, M. (1966). The analysis of magnetic links for the measurement of lightning currents. E.R.A. report No. 5115, Leatherhead, England.
Peek, F. W. (1924). Lightning and other transients on transmission lines. *Trans. Am. Inst. elect. Engrs* **34**, 697–709.
Pierce, E. T. (1955). Electrostatic field changes due to lightning discharges. *Q. Jl R. met. Soc.* **81**, 211–228.
Pierce, E. T. (1956). The influence of individual variations in the field changes due to lightning discharges upon the design and performance of lightning flash counters. *Arch. Met. Geophys. Bioklima.* **9**, 78–86.
Pierce, E. T. (1972). Triggered lightning and some unsuspected lightning hazards. *Nav. Res. Rev.* **25**, 14–28.
Pockels, F. (1897). Über das magnetische Verhalten einiger basaltischer Gesteine. *Annln. Phys. Chem.* **63**, 195–201.
Pockels, F. (1901). Über die bei Blitzentladungen erreichte Stromstärke. *Phys. Z.* **2**, 306–307.
Popolansky, F. (1960). Measurement in lightning currents on EHV lines (in Czech). *Elektrotech. Obz.* **49**, 117–123.
Popolansky, F. (1970). Measurement of lightning currents in Czechoslovakia and the application of obtained parameters in the prediction of lightning outages on EHV transmission lines. Conf. int. grands Res. Elect. report No. 33–03.
Popolansky, F. (1972). Frequency distribution of amplitudes of lightning currents. *Electra.* **22**, 139–146.

Popolansky, F. (1974). The dependance of polarity of lightning current on the frequency distribution measured on objects with varying heights in Czechoslovakia. Conf. int. grands Res. Elect. Study Committee report SC 33–74 (WG 01).
Robertson, L. M., Lewis, W. W. and Foust, C. M. (1942). Lightning investigation at high altitudes in Colorado. *Trans. Am. Inst. Elect. Engrs* **61**, 201–208.
Rump, S. (1926). Frequenz des Blitzes. *Bull. A.S.E.* **17**, 407–427.
Sargent, M. A. (1972). The frequency distribution of current magnitudes of lightning strokes to tall structures. *Trans. Am. Inst. elect. Engrs* **72**, No. 5, 2224–2229.
Schlomann, R. H., Price, W. S., Johnson, J. B. and Anderson, J. G. (1957). 1956 lightning field investigation on the OVEC 345 kV system. *Trans. Am. Inst. elect. Engrs*, Part III **76**, 1447–1459.
Schonland, B. F. J. (1956). The lightning discharge, *in* "Encyclopaedia of Physics", **22**, pp. 576–628. Springer Verlag, Berlin.
Schonland, B. F. J. and Craib, J. (1927). The electric fields of South African thunderstorms. *Proc. R. Soc.* A **114**, 229–242.
Schwab, F. (1965). "Berechnung der Schutzwirkung von Blitzableitern und Türmen." *Bull. Schweiz elektrotech. Ver.* **56**, 678–683.
Shaw, G. E. (1970). A summary of lightning current measurements collected at Mount Lemmon, Arizona. *J. geophys. Res.* **75**, 2159–2164.
Simpson, Sir G. and Scrase, F. J. (1937). The distribution of electricity in thunderclouds. *Proc. R. Soc.* A **161**, 309–352.
Stekolnikov, I. S. (1935). Les idées modernes sur le phénomène de la décharge de la foudre. Conf. int. grands Res. Elect. report No. 318.
Stekolnikov, I. S. (1941). The parameters of the lightning discharge and the calculation of the current wave form. *Elektrichestvo.* **3**, 63–68.
Stekolnikov, I. S. and Lamdon, A. A. (1942). Lightning currents in USSR supply systems during the period 1937–1940 (in Russian). *J. tech. Phys.* **12**, 204–210.
Stekolnikov, I. and Valeev, C. (1937). L'étude de la foudre dans un laboratoire de campagne. Conf. int. grands Res. Elect. report No. 30.
Szpor, S. (1969). Comparison of Polish versus American lightning records. *Trans. Am. Inst. elec. Engrs* **88**, 646–652.
Uman, M. A. (1969). "Lightning". McGraw-Hill Book Co. New York.
Uman, M. A., McLain, D. K., Fischer, R. J. and Krider, E. P. (1973a). Electric field intensity of the lightning return stroke. *J. geophys. Res.* **78**, 3523–3529.
Uman, M. A., McLain, D. K., Fischer, R. J. and Krider, E. P. (1973b). Currents in Florida lightning return strokes. *J. geophys. Res.* **78**, 3530–3537.
Wagner, C. F. (1960). Determination of the wave front of lightning stroke currents from field measurements. *Trans. Am. Inst. elec. Engrs* **79**, 581–589.
Wagner, C. F. (1967). Lightning and transmission lines. *J. Franklin Inst.* **283**, 558–594.
Wagner, C. F. and Hileman, A. R. (1958). The lightning stroke. *Trans. Am. Inst. elec. Engrs* **36**, 229–242.
Wagner, C. F. and Hileman, A. R. (1961). Surge impedance and its application to the lightning stroke. *Trans. Am. Inst. elec. Engrs* **80**, 1011–1022.
Wagner, C. F. and McCann, G. D. (1941). Lightning phenomena. *Electl. Engng. N.Y.* **60**, 374–384; 438–443; 483–500.
Wagner, C. F., McCann, G. D. and Beck, E. (1941). Field investigations on lightning. *Trans. Am. Inst. elect. Engrs* **60**, 1222–1230.
Wang, C. P. (1963). Lightning discharges in the tropics. *J. geophys. Res.* **68**, 1943–1949.
Whitehead, E. R. (1969). Contribution in discussion of "Comparison of Polish versus American lightning records". *Trans. Am. Inst. elec. Engrs* **88**, 652.

Whitehead, E. R., Darveniza, M. and Popolansky, F. (1973). Lightning protection of UHV transmission lines. Conf. int. grands Res. Elect. Working Group 33, Cracow, Poland.

Williams, D. P. and Brook, M. (1963). Magnetic measurements of thunderstorm currents. *J. geophys. Res.* **68**, 3243–3247.

Wilson, C. T. R. (1920). Investigations on lightning discharges and on the electric field of thunderstorms. *Phil. Trans.* A **221**, 73–115.

Wormell, T. W. (1939). The effects of thunderstorms and lightning discharges on the earth's electric field. *Phil. Trans.* **238**, 249–303.

10. Atmospherics and Radio Noise

E. T. PIERCE

Stanford Research Institute, Menlo Park, California, U.S.A.

1. Definitions and Introduction

An "atmospheric" or, more colloquially, a "spheric" is a term that has many meanings. In its purest usage an atmospheric is the transient field (electric or magnetic) generated by a lightning flash, or by any subsidiary feature of the flash. Atmospherics also denotes studies of the transient fields, while spherics is often applied rather specifically to techniques using atmospherics to locate the positions of thunderstorms.

More generally, an atmospheric can also mean any incoming transient that stands out above the inevitable noise background in radio reception. The transient signal may originate naturally but not necessarily in lightning; blizzards, dust storms and corona discharge are among the possible sources. The transients may also be due to man; signals are radiated for example, by cars, by powerlines and by electrical machinery. Indeed, the strongest atmospheric of all is man-made—the electromagnetic pulse created by a nuclear explosion.

When many transients arrive in rapid succession, with their effects often merging into each other, the resulting disturbance is usually termed "radio noise". Radio noise is distinguished according to its origin; thus we have atmospheric (dominantly due to lightning) noise, man-made noise and galactic (cosmic) noise.

In this chapter, we shall be concerned almost entirely with atmospherics and radio noise due to lightning. Transients of other origins will be discussed only passingly. The scientific literature covering atmospherics is formidable; several thousand relevant papers exist. The reference list in this chapter is therefore selective. Well-known results will often be given without any specific citation. Some useful reviews exist; that by Horner (1964), although inevitably outdated, is still the most comprehensive and best balanced. Hagn (1975) is a good guide to recent work.

2. Atmospherics due to Lightning

2.1 General

The characteristics of an atmospheric, A, received at a distance d from the generating lightning flash, are controlled both by the source signal S and by the modifications introduced during propagation P. Generally we may write for a spectral component of the atmospheric at frequency f

$$A(f) = S(f) \times P(d,f). \quad (1)$$

The factors, S and P, are functions of several parameters, and have a marked dependence on f. Three frequency bands can be identified within which both S and P exhibit some uniformity of behaviour. This subdivision is illustrated in Table I. The transition is not, of course, abrupt at the boundaries of the frequency ranges; there is a merging of the behaviour. As regards S, the number of separate impulses per flash increases with increasing frequency to reach a maximum at about 30 MHz and then decrease. Also as frequency increases, propagation is controlled more and more by the higher

Table I

Characteristics of spherics in three frequency ranges

	Approximate characteristics	
Frequency range	Source signals from flash	Propagation
LF and below (<300 kHz)	A few isolated transients. Number increasing with increasing frequency	Channelled in the quasi-waveguide formed by the earth and the lower ionosphere
MF and HF (300 kHz to 30 MHz)	Very numerous impulses	Controlled by reflections from the ionosphere
VHF and above (>30 MHz)	Impulses initially very numerous but becoming much fewer as frequency increases	Signals penetrate the ionosphere. Quasi-line-of-sight propagation

levels of the ionosphere, until for sufficiently large frequencies there is penetration of the ionosphere.

The signals generated by a complete lightning flash can be considered to be produced by many subsidiary dipole radiators of different dimensions. In spherics studies, d is usually much greater than the dimension (l) of the flash; also, f is normally less than $2dc/l^2$ where c is the velocity of light. Under these circumstances the vertical electric field E_z and the horizontal magnetic flux density B_ϕ for an individual radiator are given approximately by

$$E_z \approx \frac{M}{4\pi\varepsilon_0 d^3} + \frac{1}{4\pi\varepsilon_0 cd^2}\frac{\mathrm{d}M}{\mathrm{d}t} + \frac{1}{4\pi\varepsilon_0 c^2 d}\frac{\mathrm{d}^2 M}{\mathrm{d}t^2}, \tag{2a}$$

$$B_\phi \approx \frac{\mu_0}{4\pi d^2}\frac{\mathrm{d}M}{\mathrm{d}t} + \frac{\mu_0}{4\pi cd}\frac{\mathrm{d}^2 M}{\mathrm{d}t^2}. \tag{2b}$$

In Equation (2), ε_0 is the permittivity and μ_0 the permeability of free space, while c is the speed of light. The "charge" moment at any time t is given by $M_t = 2\sum q_z z$ where the summation covers all elementary charges of magnitude q_z at height z. Equation (2) can also be expressed in terms of the "current" moment ($\mathrm{d}M_t/\mathrm{d}t$). When Equation (2) is being applied, to determine variations in E_z and B_ϕ with time, the retarded values of M_t at time $(t-d/c)$ should be used, in order to account for propagation delay. If, as for the classic dipole radiator, M changes with time at a frequency f, the last term in Equation (2), the radiation, electromagnetic, or "far-field" component, will dominate if $d > c/2\pi f$. Thus at 50 km we are already in the far-field regime for frequencies exceeding 1 kHz. In discussing atmospherics we shall almost always be considering far-field signals; the electric and magnetic fields are then directly inter-related and have exactly the same shape for linear channels (Uman *et al.*, 1975).

3. Atmospherics due to Lightning—Source Effects

3.1 General

Figure 1 illustrates some of the main features of the source effects. Changes in E_z at the earth's surface are shown for three frequency ranges. From 1 to 1,000 Hz the signal is dominated by the first term of Equation (2a), the electrostatic near-field component. Slow changes associated with the initial (stepped) leader and the continuing currents accompanying junction and final leaders are indicated. So also are the much more rapid fluctuations due to return strokes and to K recoil streamers; the effects of the latter are exaggerated for ease of recognition (see also Chapter 5.2.1.1).

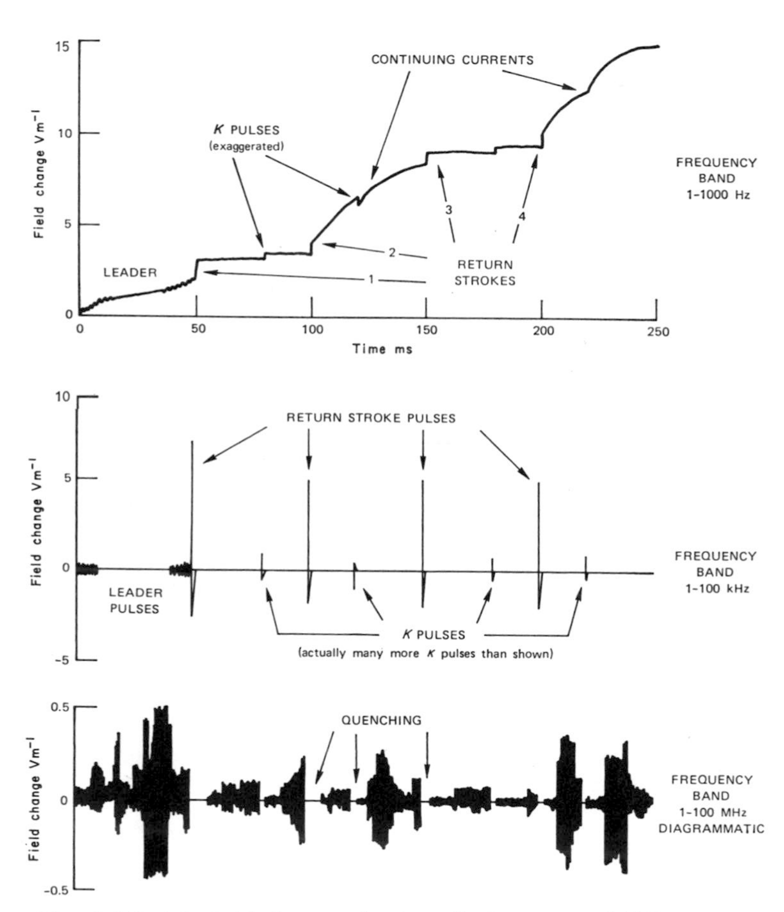

Fig. 1. Electric field changes due to a flash to ground 50 km away.

The second frequency range of Fig. 1 is 1 to 100 kHz; the electromagnetic far-field component now dominates. Strong discrete pulses are generated by return strokes, while isolated signals an order less in magnitude are created by the K processes.

At 1 to 100 MHz the characteristics are entirely different. Most of the record consists of a series of pulses; these occur in too rapid a succession to be resolved on the time scale of Fig. 1. The pulses "quench" markedly after a return stroke and, to a lesser degree, following a K process. The pulse disturbance tends to increase before return strokes and K changes.

As frequency increases above 100 MHz the number of radiated pulses decreases. There is also a gradual change in association from many pulses accompanying leader processes to isolated signals connected with return strokes. Indeed at 11 GHz only the return stroke generates appreciable pulses (Oetzel and Pierce, 1969).

The structure of the noise from an intra-cloud flash follows a pattern rather similar to Fig. 1, but with the notable difference that the effects accompanying return strokes, a feature peculiar to ground discharges, are absent. The VLF/LF signals are now produced dominantly by K streamers (see Chapter 6.5).

The spectral distribution of the source disturbance due to a complete flash is of much practical interest. It is also difficult to define in a completely satisfactory manner. Nor is it easy to reduce the available experimental information for close ($d < 100$ km) ranges to a common basis. A normalization of radiation fields within distance according to d^{-1} is a good approximation since we are usually in the far-field regime and ionospheric effects are slight. However, it is not easy to adjust for different receiver characteristics (Heydt, 1972). Consider a single circuit (in actuality receivers are much more complicated) tuned to a frequency f_0 and with a bandwidth B. Then a sharp incoming impulse will excite an oscillation of frequency f_0 at an amplitude decaying with a time constant proportional to B^{-1}. If the decay is complete before another impulse arrives the normalization of receiver outputs (e_t at time t) is proportional to B. However, if there is a rapid succession of impulses the envelope of the output fluctuates in amplitude and the normalization obeys a $B^{\frac{1}{2}}$ law. The adjustments thus depend both on the temporal structure of the incident noise and on the receiver characteristics.

For an incoming signal the spectral content $S(f_0)$ at f_0 is defined (Horner and Bradley, 1964) by

$$_{t_1}S(f_0)_{t_2} = (0{\cdot}5)\, B^{-\frac{1}{2}} \left(\int_{t_1}^{t_2} e_t^2 \,\mathrm{d}t \right)^{\frac{1}{2}} = \frac{1}{2} \left(\frac{{}_{t_1}I_{t_2}}{B} \right)^{\frac{1}{2}}. \tag{3}$$

The limits t_1 and t_2 of the integral I may be selected to cover a single impulse or the complete sequence of impulses in a discharge of duration T ($t_1 = 0$; $t_2 = T$). The peak amplitude e_p can be immediately obtained from the receiver output. Theory and experiment suggest some empirical relationships when B is of the order of 1 kHz. At VLF, where the impulses are isolated,

I and e_p are both proportional to B, and

$$I/e_p^2 = 0{\cdot}005(250/B), \tag{4a}$$

$$S(f)/e_p = (1/B)(250/800)^{\frac{1}{2}}. \tag{4b}$$

At HF, where the impulses are numerous, I and e_p^2 are proportional to B, and

$$I/e_p^2 = 0{\cdot}015, \tag{5a}$$

$$S(f)/e_p = (3/800B)^{\frac{1}{2}}. \tag{5b}$$

Figure 2 shows experimental information for e_p reduced to a common basis. The representation combines the contributions from the many separate spark processes in the complete flash. The points are from individual measurements; these are too numerous to list in detail but many are referenced in

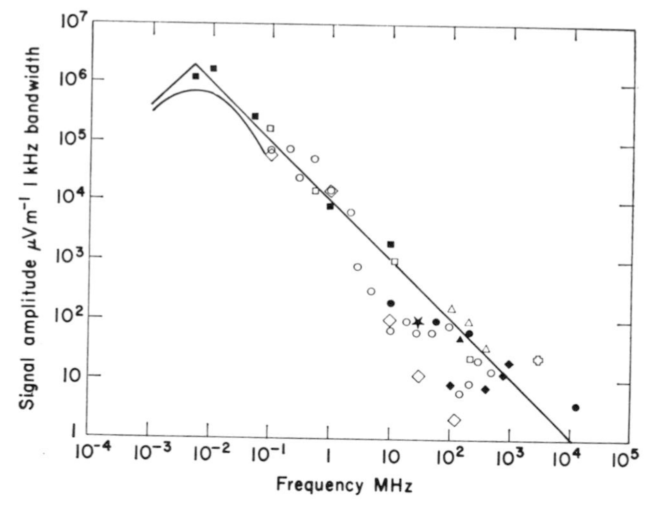

Fig. 2. Peak received amplitude at 10 km for signals radiated by lightning.

Oetzel and Pierce (1969). Over the entire frequency range above 5 kHz an inverse frequency dependence (straight line of Fig. 2) is a good approximation to the amplitude behaviour. The most substantial deviations from this general law are probably at above 10 MHz. However, there is no experimental evidence indicating any particularly strong signals for frequencies near 300 MHz; these are required by Kapitsa's theory of ball lightning (see Chapter 12.4.2). The maximum amplitude is at about 5 kHz; at this frequency the return stroke is the major source of signals. The curve in Fig. 2 is the average amplitude spectrum for a single return stroke pulse (see also Figs 3 and 4).

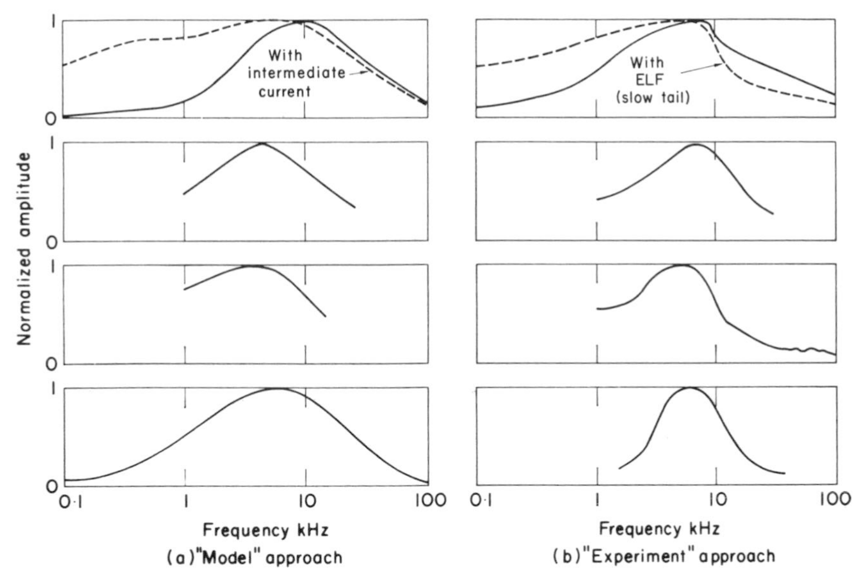

Fig. 3. Amplitude spectrum of source pulse due to return stroke.

3.2 Signals at Frequencies Less Than 300 kHz

3.2.1 The Return Stroke

The return stroke is the dominant source of atmospherics for $f < 300$ kHz. So many papers have been written on these atmospherics that it is even believed, entirely incorrectly, by the partly informed, that the return stroke is the only radiator associated with the lightning flash!

Numerous semi-theoretical derivations of the field radiated during a return stroke exist. If we knew the spatial distributions and temporal histories of the currents energizing the channel, then there would be no difficulty, with modern computing methods, in calculating the field as a function of time. The temporal variation can be transformed into amplitude and phase spectra. Unfortunately, our knowledge of the current-time history of the spatial development of the channel, and especially of the current distribution along the channel, is still not precise. Three types of semi-theoretical approach (see also Chapter 9.4.1.5) are noteworthy.

(a) Bruce and Golde (1941) assumed a current (i_t at time t) instantaneously uniform from ground to the return-stroke tip; this tip ascends at speed v_t.

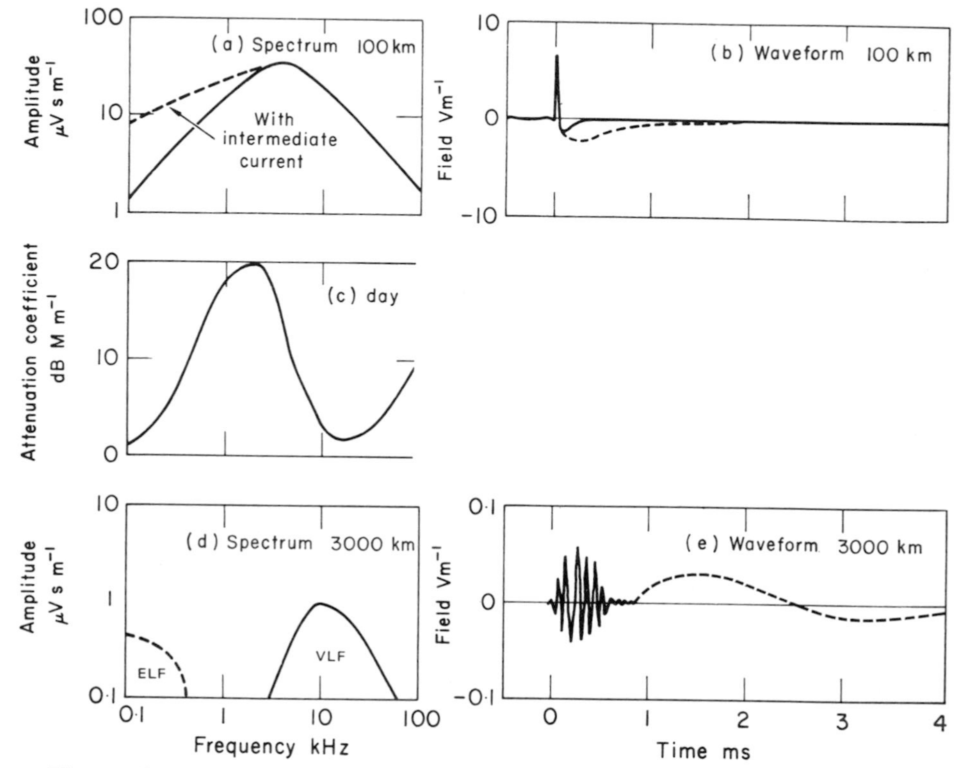

Fig. 4. Illustrating source and propagation factors in the formation of a spheric.

The current moment at t is given, allowing for the image current, by $2i_t \int_0^t v_t\, dt$; the radiated field follows from the last term of Equation (2) (current moment form). Bruce and Golde used direct experimental measurements at the ground end of the channel to express i_t in the well-known double-exponential form

$$i_t = i_0[\exp(-\alpha t) - \exp(-\beta t)]. \tag{6}$$

Boys' camera observations yielded

$$v_t = v_0 \exp(-\gamma t) \tag{7}$$

for the first stroke, with v_t being constant for subsequent strokes. Equation (6) can be extended to include currents succeeding the initial return-stroke phase as

$$i_t = i_0[\exp(-\alpha t) - \exp(-\beta t)] + i_1[\exp(-\delta t) - \exp(-\varepsilon t)]. \tag{8}$$

(b) Dennis and Pierce (1964) recognized that the current cannot be instantaneously uniform along the entire channel; if it were the physical unreality of an immediate transfer of charge from channel base to channel tip is implied. Dennis and Pierce therefore suggested that the current wave travels up the channel with a velocity u. There are the constraints $c \geqslant v_t \geqslant u$. This concept of a travelling wave of current has much in common with regarding the channel as a quasi-transmission line charged by the leader and then discharged by the ascending return stroke. The "transmission line" model has been greatly developed and extended by Uman and his associates (e.g. Uman *et al.*, 1975); see also Chapter 5.5.6.

(c) The two approaches above do not consider any current losses during the ascent of the return stroke. These will be inevitable because of the re-adjustments between the channel and its surrounding sheath of charge deposited by the leader, because of the diversion of currents into branches, and so on.

Very many papers have been written in which the wave-form of a return stroke atmospheric is derived by one of the above approaches. Several of these derivations are grossly incorrect in ignoring the channel development and in thus considering that the wave-form is determined by i_t alone. Other papers are based on variations, in the chosen parameters α, β, etc. that are trivial and have little real justification.

However, substantial variations in the parameter values, and indeed in the selected modelling approach ((a), (b) or (c) above), have surprisingly little effect on the gross characteristics of the return-stroke signal. This is illustrated by Fig. 3. On the left of Fig. 3, the normalized amplitude spectra obtained by four different selected model approaches (Pierce, 1974) are shown; they are generally similar. For comparison, examples of source amplitude spectra deduced by an "experimental" approach are on the right of Fig. 3. The experimental approach uses actual measurements made at known distances; these are extrapolated inwards to a "source distance" thus eliminating the propagation factor of Equation (1).

Modern results suggest the following as round-number values of the parameters in Equation (8) required in modelling:

first strokes:

$$\alpha = 2 \times 10^4 \text{ s}^{-1}, \quad \beta = 2 \times 10^6 \text{ s}^{-1},$$

$$i_0 = 20 \text{ kA},$$

$$v_0 = 10^8 \text{ m s}^{-1}, \quad \gamma = 3 \times 10^4 \text{ s}^{-1},$$

$$i_1 = 2 \text{ kA}, \quad \delta = 10^3 \text{ s}^{-1}, \quad \varepsilon = 10^4 \text{ s}^{-1};$$

subsequent strokes:

$$\alpha = 2 \times 10^4 \text{ s}^{-1}, \quad \beta = 6 \times 10^6 \text{ s}^{-1},$$

$$i_0 = 10 \text{ kA};$$

constant ascent velocity of 10^8 m s^{-1}, other parameters the same.

With these values, calculated radiated fields for $d = 100$ km are a few V m^{-1} in magnitude; this agrees well with experimental results (Lin and Uman, 1973). There is a maximum in the amplitude spectrum near 5 kHz; this maximum is about 50 μV s m^{-1}.

Uman and his associates (1975) have recently made valuable observations of the fields due to close return strokes (see Chapter 9.4.1.6). Their main motivation has been to deduce the variation of i_t with time during the phase of rising current. The current rise-time may be 1 μs or less; the frequency response of the experimental equipment must therefore extend well above 1 MHz. However, the current rise-time although of major importance to the power engineer is of secondary interest in spherics studies and radio engineering. This is because the rise-time has only slight effects on the gross characteristics of the return-stroke pulse. These are determined by the decay of the current surge and the temporal development of the channel; consequently, even for rise-times as different as 0·1 and 10 μs the amplitude spectrum will still peak at a few kHz. The rise-time determines the content of a return-stroke pulse at HF; this, however, is of minor importance in spherics since phases of the discharge other than the return stroke are the stronger radiators at HF (Fig. 1).

The amplitudes of return-stroke atmospherics, whether broadband or narrowband, are distributed log normally when the individual signals are expressed in decibels relative to the median. The standard deviation is about 7 dB. Thus the median signal for $d = 100$ km is 4 V m^{-1}, but 10% of the received pulses will exceed 12 V m^{-1}. Individual storms produce pulses consistently larger or smaller by a factor of two or more than those from an average storm (Horner and Bradley, 1964; Lin and Uman, 1973).

The return-stroke channel below the cloud is predominantly vertical. Consequently the radiated signal is vertically polarized. Malan and Schonland (1951) concluded that the channels within the cloud were also vertical, but others (Pierce, 1955; Brook and Vonnegut, 1960) considered that a quasi-horizontal orientation was more common (see Chapter 6.4.4). Recent deductions from observations of thunder (Teer and Few, 1974) have confirmed that the discharge channel within a cloud is usually horizontal. A horizontal channel generates a horizontally polarized signal, in the usual sense of a transverse electric field, only if the channel is normal to the direction of propagation. Even in this case the signal, because of cancellation by the image current in the ground, is only efficiently radiated for channel altitudes representing a substantial fraction of a wave-length. At 5 kHz the wave-length is 60 km; the constraint is thus important.

3.2.2 K and Leader Processes

A K field-change is produced when a charged leader advancing in the cloud encounters a charge centre of opposite sign (see Chapter 6.5). A discharge then occurs as a streamer recoiling along the channel of the leader. The K effects are superimposed on leaders both for intra-cloud flashes and for discharges to earth. In the latter case, the K process sometimes attains its extreme form, a dart leader, when the recoil is to ground.

Much information on K changes is scattered within the literature. Some of this is conflicting. On the average, however, an individual K process involves a time of 1 ms, a mean current of 1 kA and a streamer length of 1 km. The pulses occur at intervals of 5 to 10 ms. Thus an intra-cloud flash and a discharge to earth will both generate about 40 K pulses; in contrast, only 3 or 4 return strokes occur in a flash to ground. A K signal has about 10% of the amplitude of a return-stroke pulse (Ishikawa, 1961); the maximum in spectral content is near 10 kHz (Wadehra and Tantry, 1967). The statistical distribution is log normal (dB relative to the median) with a standard deviation of 8 dB.

Modelling of K pulses has followed procedures similar to those for the return stroke (Arnold and Pierce, 1964). However, the basic inputs, the variations in i_t and v_t, are quite uncertain. A possible representation for v_t is

$$v_t = v_1 \exp(-kt) + v_2 \exp(-mt), \qquad (9)$$

where $v_1 = 2\times10^7$ m s^{-1}, $v_2 = 10^6$ m s^{-1}, $k = 4\times10^4$ s^{-1} and $m = 2\times10^3$ s^{-1}. The current variation can be written as

$$i_t/i_k = [\exp(-at) - \exp(-bt)] + \tfrac{1}{3}[\exp(-dt) - \exp(-et)], \qquad (10)$$

where $i_k = 15$ kA, $a = 5\times10^4$ s^{-1}, $b = 2\times10^5$ s^{-1}, $d = 5\times10^3$ s^{-1}, and $e = 2\times10^4$ s^{-1}. Analyses using Equation (9) and (10) yield K pulses of about 0·3 V m^{-1} for $d = 100$ km (Pierce, 1974).

The spheric radiated by an individual stepped leader pulse has also been modelled (Arnold and Pierce, 1964). Recent work (Krider and Radda, 1975) suggests that the pulse wave-form used in this analysis was not typical. However, theoretical and experimental results all indicate that the pulse at VLF due to a single step in a leader is only of minor importance, being significantly less than the signals radiated by *K* and, particularly, return-stroke processes.

Stepped leader pulses and to a lesser extent *K* pulses occur not in isolation but in a series. Given the pulse repetition characteristics, the amplitude spectrum for the series of pulses can be deduced from those for a single pulse (Arnold and Pierce, 1964).

3.2.3 ELF Components

Many atmospherics have a substantial content at ELF (3 to 0·3 kHz) and lower frequencies. The ELF signals are particularly evident in spherics that have propagated long distances.

An important source of ELF components, sometimes called "slow tails", is in the intermediate and/or continuing currents that sometimes, but not invariably, follow the initial current surge of the return stroke (see Chapter 9.4.1). An intermediate current lasts for a few milliseconds; physically it represents readjustments between the channel, after the passage of the return stroke, and the sheath of charge deposited around the channel by pre-return-stroke leader processes. Continuing currents have durations of tens of milliseconds; they are due to junction-type leaders probing within the cloud (Chapter 6.4 and 6.5).

Analytical procedures already outlined (Section 3.2.1) for modelling the VLF atmospheric radiated by a return stroke are easily extended to ELF, by adding terms representing the intermediate and continuing currents. Thus Equation (8) is Equation (6) with the addition of an intermediate current term. Figure 3 gives examples of spectra with ELF components, due to intermediate currents, deduced from modelling and from experiments. Continuing currents have magnitudes of the order of a hundred amperes, and this magnitude often fluctuates only slightly. An addition to Equation (8) of 200 A for 50 ms would be appropriate for the continuing current.

Broadband (10 to 1,000 Hz) ELF atmospherics are commonly observed with either positive or negative polarity (Hughes, 1967; Sao and Jindoh, 1974). Return strokes almost always carry negative electricity to ground and should therefore produce positive ELF signals. Sao and Jindoh (1974) therefore ascribe the negative ELF atmospherics to intra-cloud discharges. Jones (1974), however, believes that return strokes are the chief source of all ELF atmospherics for frequencies exceeding 40 Hz.

Continuing currents have similar properties, except for polarity differences, whether the flash is to earth or intra-cloud. Such currents will generate signals principally at frequencies below perhaps 40 Hz. At VLF and ELF (>40 Hz) the return stroke is certainly a very strong source of atmospherics, with the different frequency components being attributable respectively to an initial current surge and to a succeeding intermediate current (Equation 8). This duality of current structure also exists, however, for K changes (Equation 10). Although on the average the currents in K streamers are less than those in return strokes, K processes occur much more frequently and over a wide range in amplitude. The stronger K pulses are therefore comparable in magnitude with the weaker return-stroke atmospherics. Consequently, it seems likely that intra-cloud flashes sometimes generate ELF atmospherics at frequencies considerably exceeding 40 Hz. However, the identification of negative ELF components with intra-cloud discharges is not infallible if the origin is in K streamers; K pulses of either polarity have been observed during individual flashes whether they are intra-cloud or to earth (Pierce, 1974).

3.3 Signals at Frequencies Exceeding 300 kHz

As frequency increases above 300 kHz the number of pulses per flash rises rapidly. Oetzel and Pierce (1969) found that the average rate of pulse emission changed from some 2×10^3 s^{-1} at 3 MHz to reach a maximum of about 10^4 s^{-1} at frequencies between 30 and 100 MHz. With a further increase in frequency there was a sharp drop in pulse emission rate; at 300 MHz, 10^3 s^{-1} was typical. Cloud flashes emitted rather more pulses than did discharges to earth; there were marked differences in behaviour between individual flashes; and occasional discharges produced a maximum pulsing rate approaching 5×10^4 s^{-1}.

Proctor (1973) quoted an average impulse rate at HF/VHF of 5×10^5 s^{-1} for cloud flashes. However, later (Proctor, 1974) he identified two types of cloud flash; one type emitting pulses at 2×10^3 s^{-1} and the other giving pulse rates of 3×10^4 to 5×10^5 s^{-1} (see also Chapter 6.7). Pulse counting is much influenced by receiver characteristics and threshold settings; bearing this in mind, the discrepancies are not serious between the results of Oetzel and Pierce and those of Proctor.

Oetzel and Pierce (1969), using narrowband channels, tuned respectively to frequencies of 10, 30, 90 and 200 MHz, found that some impulses registered only on one or two channels; these could be lower, higher or intermediate channels. Some impulses, however, were recorded on all four channels. Proctor (1974) employed receivers at 30, 250, 600 and 1,430 MHz. When cloud flashes of the low emission type (pulse rate $\approx 2 \times 10^3$ s^{-1}) were monitored, the pulses were consistently observed simultaneously on all four receivers. However, this was not so for the high pulse-rate flashes.

Processes during the initial leader undoubtedly generate pulses at HF and above (Brook and Kitagawa, 1964). This is evident from the temporal associations of Fig. 1. Current surges accompanying the luminous steps are the probable source. Krider and Radda (1975) have shown that the pulse radiated by a step may have a rise time of only a microsecond and a total duration of under 3 μs. Such a pulse will have substantial frequency components within the HF/VHF band.

Pulse sources within the cloud for the low repetition type (2×10^3 s^{-1}) have been investigated by Proctor (1974). By using a time-of-arrival locating technique, he deduced that the sources were linear and about 300 m in length with the strongest emission tending to be from one end of the source. This was so whether the pulses occurred within a cloud flash or during the intervals between the return strokes of a flash to earth. Proctor found the pulse duration to be 1 μs. When, however, pulses were being rapidly emitted ($> 3 \times 10^4$ s^{-1}) Proctor was unable to identify any consistent phenomenology in either source or pulse characteristics.

The HF/VHF pulses from a flash are log-normally distributed in amplitude. The standard deviation (decibels relative to the median) is about 6 dB. Since the dimensions of the radiators involved are a few hundred metres or less, a strongly preferred orientation (vertical or horizontal), and consequently of polarization, is unlikely (except for initial leader steps).

At frequencies near 10 GHz the number of pulses per flash is very low (Oetzel and Pierce, 1969). Typically, isolated pulses occur in close temporal conjunction with return strokes. A plausible source is difficult to postulate. Two possibilities are a very short (≈ 3 cm) channel energized by a strong current, or a travelling current wave with a very sharp ($\approx 10^{-10}$ s) rise time; neither possibility seems likely.

4. Atmospherics due to Lightning—Propagation Effects

4.1 Propagation at Frequencies of less than 300 kHz

Radio propagation for $f < 300$ kHz, and especially below 30 kHz, is channelled within the quasi-waveguide formed by the earth and the lower ionosphere. Two well-known theoretical descriptions of the propagation exist; these are essentially, although not obviously, equivalent (Wait, 1970).

The "ray" or "reflection" theory considers an atmospheric to consist of a series of successive impulses. The signal due to the lightning source first propagates along the surface of the earth as a ground pulse; the second pulse arrives by one ionospheric reflection; the third by two ionospheric reflections and one at the ground; and so on. If the ground and a parallel sharply bounded ionosphere at height h were both perfect reflectors, then the pulses would arrive in a temporal sequence uniquely defined by h and the source distance d.

In the second theoretical approach the flash excites a number of quasi-waveguide "modes" that propagate between the earth and lower ionosphere. If, as above, both boundaries are sharp, parallel and perfectly conducting, then the modes have lower cut-off frequencies defined by $nc/2h$, where n is the mode order and c the velocity of light. Since h is about 70 km by day and 90 km at night, the first-order mode cuts off near 2zkHz.

In fact, the idealization outlined above is never valid (Wait, 1970). Earth curvature is important; the ground and the ionospheric boundaries are not strictly parallel; nor are they perfect conductors. Ground conductivity varies between sea, land or more unusual environments such as permafrost regions and ice caps. The ground boundary is, however, at least well defined, but the ionospheric conductivity changes only gradually with height. At VLF, "reflection" from the ionosphere involves, even at VLF, the form of the conductivity profile over at least 10 km. Furthermore, this profile varies with alterations in ionization production and loss processes; these alterations depend on time of day, season and geographical and geomagnetic location. Finally, at an ionospheric reflection the geomagnetic field can change the polarization state. A vertically polarized incident ray will yield two return components; one will maintain the original vertical polarization but the other will have been converted to a horizontally polarized state. The geomagnetic field influences depend on the size and orientation of the field relative to the propagation path.

Atmospherics may be interpreted by either the ray or the mode approach. In general, if many modes have to be summed (with appropriate phasing) the ray interpretation is preferable; alternatively, if many rays but few modes are involved the mode approach is indicated. Table II outlines some criteria.

Sometimes the decision is clear-cut, mode theory for VLF daytime atmospherics of distant origin; sometimes it is not well marked. In most circumstances, however, the mode theory is preferable.

The vertical electric field $({}_fE_d)_n$, within a narrowband centred at frequency f, and for an atmospheric propagating a distance d in the mode of order n, can be written (Wait, 1970) as

$$({}_fE_d)_n \approx (300/h)(F_n)[P(f)\,\lambda]^{\frac{1}{2}}[r\sin(d/r)]^{-\frac{1}{2}}\exp(-{}_f\alpha_n d). \tag{11}$$

In Equation (11) F_n is the mode excitation factor, $P(f)$ the source power, λ the wave-length, r the radius of the earth and ${}_f\alpha_n$ an attenuation coefficient.

If $P(f)$ is in kilowatts; h, λ, r and d, in kilometres; and ${}_f\alpha_n$ in nepers km^{-1}, then the field is in mV m^{-1}. The attenuation coefficient depends on ground conductivity, profile of ionospheric conductivity and the geomagnetic field.

Table II
The choice between mode and ray theory

	Approach to be preferred	
	Mode	Ray
Distance	$d > 1{,}000$ km	$d < 1{,}000$ km
Frequency	< 30 kHz	30 to 300 kHz
Conditions	Day	Night

Modes can be excited with either the magnetic (TM modes) or the electric (TE modes) field transverse to the direction of propagation. The efficiency (F_n) of the excitation depends on the ground and ionospheric characteristics at the source, and on the orientation of the generating lightning channel relative to the propagation direction. The TE modes are poorly excited principally because the ground acts as an electrical conductor with the effects of the channel image in the ground tending to cancel those of the channel itself (Pappert and Bickel, 1970). Thus, TM modes are usually the more important in spherics studies. Mode conversion, however, can occur during propagation, particularly if there are abrupt changes in the guide characteristics.

Two TM modes ($n = 0$ and $n = 1$) are particularly important. The ELF component propagates in the zero-order mode, and the strongest VLF components ($f < 15$ kHz) in the first-order mode; higher-order modes are significant in the upper part of the VLF range. Attenuation coefficients and phase velocities have been deduced, for the two most important modes, from experimental observations of atmospherics (Challinor, 1967). The results agree well with theoretical calculations.

The variation of phase velocity v_p with frequency is interesting. Experiment and theory show that within the first-order mode v_p is somewhat greater than c the velocity of light, and as frequency decreases from 10 kHz to 3 kHz v_p increases slightly. However, in the zero-order mode as f decreases below 1 kHz so does v_p, and very sharply. Theory therefore predicts a time separation τ between the VLF and ELF components of an atmospheric (Wait, 1970). Approximately,

$$\tau = 0{\cdot}09\,[(d/2h)\,\omega_r^{-\frac{1}{2}} + \sigma^{\frac{1}{2}}]^2. \tag{12}$$

In Equation (12) ω_r is a parameter representing the ionospheric conductivity, and σ accounts for the time separation at the source ($d = 0$). The units of Equation (12) are kilometres and seconds ($\tau, \sigma, \omega_r^{-1}$).

4.2 Propagation at Frequencies Exceeding 300 kHz

In the MF and HF bands (0·3 to 30 MHz) the groundwave signal is very rapidly attenuated, and, except for close thunderstorms, the reception of atmospherics depends entirely on ionospheric reflections. The ray theory familiar in commercial radio is adequate. Many modes of propagation are possible; these modes include "one-hop" (one ionospheric reflection), "two-hop" (two ionospheric reflections and one at the ground), "*M*" (three ionospheric reflections, one internal) and so on. The upper limit on propagating frequencies depends primarily on the angle of incidence at the ionosphere and on the critical frequency f_c of the reflecting ionospheric *F* region. By day f_c may exceed 10 MHz; propagation at frequencies as high as 30 MHz is then possible. However, the lower frequency HF and MF signals are strongly absorbed in daytime when passing through the lower *D* region of the ionosphere. At night the *D* region is much less dense and the absorption is greatly reduced. However, f_c is also less so that the higher frequencies penetrate the ionosphere without reflection. Signals above 30 MHz almost always penetrate the ionosphere, and the modifications (absorption; dispersion) to the signals during this penetration decrease with increasing frequency. Propagation along the earth's surface can only occur, beyond the line-of-sight limit, by the inefficient processes of diffraction and scattering. Thus only local storms produce VHF noise at ground-based receivers. Atmospheric VHF noise is, however, of more significance in satellite communications.

4.3 Summary

It was emphasized in Section 2 that an atmospheric results from both source and propagation influences. The combination is particularly evident for return-stroke atmospherics at VLF/ELF. The source signal is strong and discrete, while the quasi-waveguide propagation is good. Consequently return-stroke atmospherics are easily detected for $d = 3{,}000$ km or more.

Figure 4 illustrates how source and propagation effects combine in a return-stroke atmospheric. Figures 4(a) and 4(b) give the amplitude spectrum and broadband (0·1 to 100 kHz) wave-form for $d = 100$ km; plots are shown with and without an intermediate current (ELF component). Figure 4(c) shows the effective attenuation coefficient by day; the unit is dB Mm^{-1}, not nepers km^{-1} as in Equation (11). Two "windows" of low attenuation exist; at frequencies $<0{\cdot}3$ kHz and centred around $f = 15$ kHz. Within the ELF window the propagation is only in the zero-order mode. For the VLF window the propagation is dominantly, but not entirely, in the first-order mode. The combination through Equation (11) of source (Figs 4a and 4b) and propagation conditions (Fig. 4c) yields the results of Figs 4(d) and 4(e). Note how the sharp impulse of Fig. 4(b) is transformed into a smooth VLF oscillation (Fig. 4e).

Theoretical results for source and propagation influences, and experimental observations of atmospherics, all combine into a harmonious whole for the return-stroke disturbances. There are no major disagreements. The magnitude of the broadband VLF atmospheric decreases by day from a few volts per metre at $d = 100$ km to the order of $0{\cdot}1$ V m^{-1} at $d = 3{,}000$ km. At night, atmospherics of distant origin are twice as large as by day. Propagation moves the maximum in the amplitude spectrum towards higher frequencies (Figs 4a and 4d). When an ELF component is present, the time separation τ between the start of the VLF pulse and the maximum of the ELF signal obeys Equation (12). Typically, the parameter $(2h)^{-1}(\omega_r)^{-\frac{1}{2}}$ is 2×10^{-5} by day and 10^{-5} by night. The source separation $(0{\cdot}09\sigma)$ is some 400 μs; this is consistent with an origin for the ELF component in the intermediate current. Both the mode and the ray theory are successful in explaining the development of atmospherics with distance. Thus the smooth VLF oscillation, characteristic of an atmospheric of distant origin, follows from the dominance of a single mode. Alternatively, it can be regarded as a merging of the successive pulses—themselves smoothed by ionospheric reflections—on the ray picture. The ground pulse, incidentally, is much weaker than the sky pulses for distances exceeding a few hundred kilometres.

Although some "fine-structure" studies could still be rewarding, our general comprehension of return-stroke atmospherics is satisfactory. This is not so for any other kinds of atmospheric. The deficiencies in our knowledge are almost entirely in the "source" area. Propagation, by comparison, is very well understood from the pleasing purity of the mode formulation at ELF/VLF, through the ionospherically reflected ray tracing applicable at HF, to the transionospheric, line-of-sight approach appropriate at VHF and above. At HF the individual flash produces many impulses; the signals can propagate by many ionospherically reflected paths; and many lightning discharges may contribute to the received noise at a particular location. With these complications, it is not easy to deduce source characteristics from observations of HF atmospheric noise structure. At VHF, however, even though there are many source pulses, the source characteristics do not suffer major modifications during the simple line-of-sight propagation. Consequently, the details of the VHF source structure are important in some practical aspects of satellite radio-communication links.

Summarizing our source uncertainties, at ELF and below (<3 kHz) there are disagreements regarding the generation of signals by intra-cloud discharges. Also, present representations of the disturbances radiated by continuing currents in leader processes are inadequately related to the results of measurements close to the discharge. At VLF, existing models of the signals due to K streamers and to stepped leader pulses are based partly in reality but also partly on speculation. However, our knowledge is most vague at higher frequencies. Above about 300 kHz the return-stroke signal falls off as f^{-2} and therefore contributes little to the observed f^{-1} dependence (Fig. 2). This latter dependence results from the combination of various types of spark each radiating over a specific frequency range; the smaller the spark dimension the higher the frequencies generated.

A coherent picture may be emerging. If an advancing leader carrying charge approaches a charged centre of opposite sign, then a discharge, creating luminosity and radiated fields, occurs as a quasi-recoil streamer along the leader. This discharge will be most violent at the tip of the leader. It may range in size from the largest extreme, the dart streamers, through K processes (dimension ≈ 1 km), through the dimensions (≈ 300 m) measured by Proctor (1974), through steps (dimension ≈ 50 m), to the limit of filamentary corona-like streamers. The radiating characteristics will be determined by the magnitude and concentration of the charge centre encountered by the leader, and the distribution of the characteristics may well be continuous rather than discretely separated into distinct types.

5. The Location of Distant Thunderstorms

5.1 Introduction

Spherics methods for locating thunderstorms fall into two categories: techniques that require more than one station, and single-station methods. Similar principles often apply both to the location of close ($d \leqslant 100$ km) lightning (see Chapter 15) and to the fixing of distant storms. We shall here consider only long-range location and distances usually well in excess of a thousand kilometres.

5.2 Multistation Systems

The standard method of locating a lightning flash uses crossed-loop cathode-ray direction finders (CRDF) at two or more stations. Each direction finder consists of two identical vertical loop antennae, respectively orientated north–south and east–west. A vertically polarized incoming atmospheric, originating at an angle ϕ with respect to north, induces signals in the loops proportional, respectively, to $\cos\phi$ and $\sin\phi$. When these signals are applied to the plates of a cathode-ray tube the result is, ideally, a straight line whose angle from the vertical represents the azimuth ϕ of the spheric. Modern electronic digitizing techniques can also be used to indicate ϕ. A whip antenna removes the 180° ambiguity in direction present from the loops alone. A cathode-ray direction finder usually operates within the VLF band.

Intersection of the respective azimuths from at least two stations (typically 500 to 1,000 km apart) gives the flash location. The redundancy provided by more than two stations is helpful in reducing errors. Some time synchronism between stations is necessary to ensure that azimuths are derived for the same atmospheric. However, the incoming data rate of the discrete VLF pulses is rarely large. Inter-station timing with an accuracy of 10 ms is usually adequate, especially if the locations of the stations are judiciously chosen with respect to the thunderstorm centres (Pierce, 1969).

Crossed-loop direction finders experience errors due to horizontally polarized components; these cause the trace on the cathode-ray tube to be elliptical not linear. Horizontal polarization is produced at the source by channel orientation, or introduced during propagation by ionospheric and geomagnetic influences. Because of the propagation effects the latter stages of an atmospheric of distant origin are often almost circularly polarized; this is particularly so at night. Yamashita and Sao (1974a, b) have calculated the polarization errors resulting from propagation and from channel orientation. At 10 kHz both factors introduce errors of a few (0 to 6) degrees; at 5 kHz there is a large error, 10 to 30°, due to the propagation factor; but between 100 and 600 Hz the error due to orientation is zero and that caused by propagation is less than 5°. Apparently the ELF range has advantages for direction finding.

Polarization effects are not the only sources of error (Horner, 1964). If the site is not level, errors of a few degrees can be produced. Site errors are also created by extended conductors such as wire fences or buried cables; re-radiation of signals, induced by horizontal components in the incident atmospheric, is then particularly serious. Finally, when narrowband receivers are used a second impulse arriving before the disturbance due to a first spheric has completely decayed can cause errors.

Another important multistation method is the time-of-arrival (TOA) technique; this avoids many of the errors of the CRDF approach. The TOA technique measures the difference in the times of arrival of the same feature (often the start) of an atmospheric at two stations (Hughes and Gallenberger, 1974). This difference defines a hyperbola upon which the source of the spheric must lie. A TOA system is more logistically complicated than is a CRDF arrangement. In the latter a single station provides a locus (azimuth) on which the flash must lie; two stations give a fix. With the TOA method two stations are necessary for the locus (the hyperbola) and three for a fix which may still contain an ambiguity. The time synchronism constraints for the TOA method are stringent; 10 μs rather than 10 ms (CRDF) is required. Additionally, when fixes are obtained from the intersection of hyperbolas (TOA) rather than straight lines (CRDF) the plotting problems become more complicated.

5.3 Single-station Methods

Single-station techniques combine a crossed-loop direction finder with some measured parameter of the spheric indicative of distance. Many parameters can be used; the possibilities and the practical limitations have been reviewed by Pierce (1956) and Horner (1964).

The magnitude of an individual spheric is only a crude indicator of distance, principally because of the source variability. On the "ray" theory, and if the earth and ionosphere are good reflectors, the successive pulses, making up the atmospheric wave-form, arrive in a temporal sequence (Section 4.1) from which the ionospheric height and d can be determined (Tixier and Rivault, 1974). In actuality few wave-forms—and those usually restricted to short ranges and/or night conditions—can be satisfactorily analysed by the successive reflection mechanism. Identification of the individual pulses is difficult because of modifications at the ionosphere and because of superposition of pulses. Also, h is not constant throughout the wave-form; it changes with pulse order and with the obliquity of pulse incidence on the ionosphere.

Two important techniques are to examine the spectral content of an incoming atmospheric or its dispersive characteristics. The amplitude spectrum of an atmospheric (Fig. 4d) is largely controlled by the frequency-dependent propagational attenuation (Fig. 4c). Thus, for example, the ratio of amplitudes measured at two selected frequencies depends on distance. The choice of frequencies to be monitored needs care. There is an interplay between the most attenuated frequencies (most distance-dependent and therefore apparently desirable for selection) and the size of the signals detectable above the background (Gorodenskii, 1973). The dispersive technique has been particularly applied to the propagation behaviour in the VLF and ELF bands (Equation 12, Fig. 4, Section 5). The most recent developments have been by Sao and Jindoh (1974).

In recent years, sophisticated equipment—now termed the "atmospherics analyser"—has been developed (Harth, 1972) from an original concept by Heydt and Volland (1964). In the equipment, the magnitudes of three parameters dependent on distance are presented, as a function of azimuth, for every incoming atmospheric exceeding certain thresholds. The equipment operates within the VLF band; the three displayed parameters are:

(a) the amplitudes at selected frequencies (Spectral Amplitude, SA);
(b) the ratio of amplitudes at pairs of selected frequencies (Spectral Amplitude Ratios, SAR);
(c) the time differences in arrival of the components at pairs of selected frequencies (Group Delay Time Difference, GDD); this parameter can also be expressed in μs/kHz at the average frequency of the selected pair.

The selected frequencies (1, 2 and 3) usually lie between 5 and 10 kHz; the pairs of frequencies involved (2 and 3) are normally separated by 2 kHz. The atmospherics analyser combines the three approaches of measuring magnitude (broadband or at selected frequencies), amplitude spectrum and dispersion. These approaches, as outlined above, have been separately applied in other investigations.

The GDD parameter is presently preferred for estimating distance. Since the GDD approach (say from 6 to 8 kHz) and that using the VLF/ELF dispersion are basically identical, some comparison is interesting. The time separations, for sources at a few thousand kilometres, are much greater (≈ 2 ms) with the ELF/VLF method than (≈ 200 μs) with the GDD technique. Thus the ELF/VLF technique is intrinsically the less sensitive to source, propagation and instrumental variabilities. The propagation for the GDD can involve more than one mode (Price, 1974a); with the ELF/VLF approach, the propagation at ELF is only in the stable zero-order mode, but that at VLF, as with the GDD, may be by multi-moding. The source for the GDD will usually be a return-stroke pulse of limited variability at the origin, but an entirely different category of pulse, that due to a K process, can cause complications (Pierce, 1974). However, the source variability is much more pronounced in the VLF/ELF case; the magnitudes of the two components have little relationship at the source. Many observations are readily processed on the GDD method to yield an average location for a storm centre. Such processing has not yet been applied to the ELF/VLF approach.

5.4 Summary

All techniques for locating the origins of atmospherics are subject to errors. There are three main categories of errors: those associated with signal variability at the source; those introduced by fluctuations in the propagation characteristics; and those of localized (e.g. topographical influences on direction finding), or instrumental, origin. The relative importance of the categories differs according to the technique involved.

The multistation time-of-arrival (TOA) method is undoubtedly the most accurate way of fixing individual sources. It is also the most costly. Expenses are substantial in installation (elaborate equipment, at least three stations needed), maintenance (accurate timing required) and data processing (complicated geometry). Crossed-loop multistation techniques are much more economical and yield an accuracy adequate for most purposes.

When a single-station is used, with a crossed-loop system to give the direction, the distance is best obtained by either the amplitude-spectrum or the dispersive approach. The latter, in the form of the ELF–VLF separation, gives good results for individual atmospherics with simple equipment; however, many atmospherics cannot be processed by this method. The amplitude-spectrum approach, when only two frequencies are considered, is less satisfactory than the ELF–VLF dispersive technique; however, the potentialities when several frequencies are monitored have not been exhaustively explored. The atmospherics analyser avoids the inaccuracies associated with the fixing of individual atmospherics by statistical processing, intended to group all spherics originating from a specific storm centre. This procedure works well for the intense storms of the tropics where the flashing rate and storm duration are both high. It is less effective for the weak storms of temperate regions that may only produce a few discharges during their entire life; for such storms location of the individual flashes is almost essential.

6. Radio Noise and Special Topics

6.1 Radio Noise–General

Radio noise is studied by engineers rather than physicists. The motivation is practical; noise is often the factor limiting the amount of information carried by a radio communication link.

Figure 5 illustrates the relative contributions of the three main sources of noise. For man-made noise the two lines represent, respectively, an urban environment and a quiet rural location in a developed country. The atmospheric noise curves are for a temperate location with no close thunderstorms. The upper curve is for the hours immediately after local midnight; the lower curve is for the morning daytime hours, a time when propagation is poor and the contributing thunderstorm centres are of low intensity.

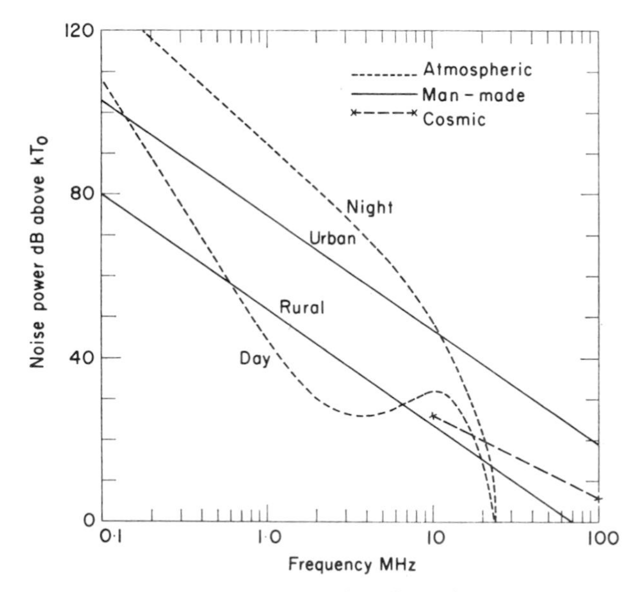

Fig. 5. Sources of radio noise.

There are many statistical descriptors of radio noise. The mean noise power averaged over several minutes is widely used. If the bandwidth, B, is small compared with the centre frequency, then the noise power p_n is proportional to B. Noise power is usually expressed as a noise factor F_a given by

$$F_a = 10 \log \frac{p_n}{kT_0 B}, \tag{13}$$

where k is Boltzmann's constant ($1{\cdot}4 \times 10^{-23}$ J K^{-1}) and T_0 is the reference temperature (288 K). Noise measurements are commonly related to a standard short vertical antenna over a good ground. In these circumstances, E_n, the vertical component of the root-mean-square noise field (dB above 1 μV m^{-1}) is given by

$$E_n = F_a + 20 \log f - 155{\cdot}5 + 10 \log B \tag{14}$$

with f in kHz and B in Hz. The envelope of the noise field $e(t)$, and of any related voltage $v(t)$ within the receiver, will fluctuate with time according to the characteristics of the incoming noise. Two descriptors related to this structure are V_d ($= 20 \log v_{rms}/v_{ave}$) and L_d the difference in decibels between v_{rms} and the antilogarithm of the average logarithm of the envelope voltage.

The most informative descriptor of noise is the amplitude probability distribution (APD). This distribution in its cumulative form shows the fraction of time for which the noise envelope exceeds designated values; it is a good indicator of the impulsiveness of the noise. Values of F_a, V_d and L_d can all be derived from the APD; the reverse process is also possible (Spaulding *et al.*, 1962). Detailed time statistics are useful in many communications and systems applications (Hagn, 1975). These statistics include the pulse-spacing distribution (intervals between end of one pulse and the start of the next), the pulse-duration distribution (times for which pulses exceed designated levels) and the average crossing rate (number of times per second that the envelope voltage crosses selected thresholds) (Nakai *et al.*, 1974).

Systematic measurements of radio noise were made by a global network of stations from 1957 to 1961. These measurements were summarized in a form enabling noise power to be quickly determined at any frequency, for any part of the world at any hour of the day (CCIR, 1964). The measurements are generally consistent with the results deduced by first estimating the strengths of the lightning sources, and then calculating the propagational losses. For example, the typical variation of atmospheric noise power with frequency (Fig. 5) follows from this combination of source and propagation effects. The source strength decreases with increasing frequency (Fig. 2); propagation is better by night than by day, but the ionosphere can rarely support long distance propagation for $f > 25$ MHz.

Two properties of atmospheric noise, polarization and directivity, have been little studied. Most measurements have used vertical antennae. The scanty information available suggests, as would be expected from source and propagation influences, that horizontally polarized components increase in importance with increasing frequency. The azimuthal dependence is important because directional antennae are widely used. Measurements (Posa *et al.*, 1972) show that azimuthal variations are pronounced, are related to the locations of the thunderstorm sources, and can be modelled effectively by combining source and propagation effects.

The literature covering the more sophisticated descriptors such as APDs and time statistics is already massive and still rapidly growing (Hagn, 1975; Aiya *et al.*, 1972). We content ourselves here by discussing some properties of the APD. Figure 6 shows two typical APDs. The coordinate net is of the Rayleigh type, so that for Gaussian noise (Rayleigh envelope distribution) the plot is a straight line of the indicated slope. Log-normal distributions will also plot as straight lines with slopes proportional to the standard deviation of the distribution (Beckmann, 1964). For both curves of Fig. 6 there is a transition from the Rayleigh slope, corresponding to many small incoming pulses of comparable amplitude and distant origin, to the log-normal distribution characteristic of a few high-amplitude pulses of close origin. The transition is more marked at 13 kHz than at 10 MHz. This is entirely compatible with the impulsive character of the source signals at VLF (Fig. 1). Measurements at the same frequency but with different bandwidths show that as bandwidth decreases the log-normal section of the APD is curtailed. This is to be expected since small bandwidths imply long "rings" for each pulse and a consequent merging of the pulse effects, a fundamental property of the Rayleigh concept.

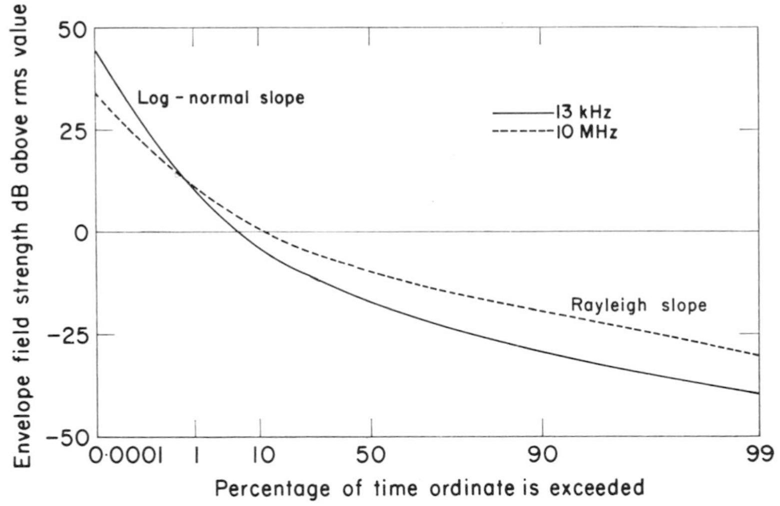

Fig. 6. Amplitude–probability distributions.

6.2 Whistlers

The classic paper of Storey (1953) showed that part of the VLF signal generated by the return stroke often penetrates through the ionosphere into the magnetosphere. Here it can be ducted along a line of force of the geomagnetic field ultimately to return to earth after a second penetration of the ionosphere. The propagation within the magnetosphere is dispersive, so that the sharp impulse originally generated returns to earth as, in terms of an auditory output, a "whistle" of descending pitch with the higher frequencies arriving before the lower.

Whistler studies and those of conventional atmospherics were originally closely associated, but, following the realization that whistlers represented a powerful tool for studying magnetospheric conditions, there was a divergence. Whistler investigations have become increasingly complicated (Helliwell, 1965; Horner, 1970). Both ducted and non-ducted propagation can occur in the magnetosphere; there are variations, associated with magnetospheric dynamics, in the ducted propagation; while the concentrations and properties of the magnetospheric electrons and ions affect both modes of propagation.

6.3 Satellite Measurements

Satellite observations of atmospheric noise have apparent attractions. For example, they offer possibilities for the monitoring of global thunderstorm activity. However, there are many constraints, notably a conflict between resolution and the chance of detecting a lightning signal (Pierce, 1969). A satellite orbiting at 1,000 km, with an antenna system providing a spatial resolution of 5 km, would only observe a thunderstorm cell for about a second. Since the interval between flashes is some twenty seconds, the great majority of storms would not be detected.

Radio signals from lightning, as evidenced by whistlers, can penetrate the ionosphere at VLF. However, a VLF satellite will receive signals from other sources unrelated to sferics. The ionosphere is essentially opaque at LF and MF, but within the HF band (3 to 30 MHz) there is a transition with increasing frequency from almost complete opacity to almost complete transparency. If f_c is the critical frequency, at which a signal, normally incident from below, penetrates the ionosphere, the criterion for penetration at oblique incidence is approximately $f > f_c \sec \phi_0$ where ϕ_0 is the ionospheric angle of incidence. A satellite receiver, operating at frequency f and located immediately above a thunderstorm complex, will accept signals from an area A_f of the complex roughly defined by

$$A_f = \pi h_s^2[(f/f_c)^2 - 1], \tag{15}$$

where h_s is the satellite altitude. In the tropical areas of most storm activity f_c is usually between 4 and 12 MHz.

The anticipated behaviour (Equation 15) at HF stimulated two notable satellite experiments. Horner and Bent (1969) used the Ariel III satellite ($h_s \approx 550$ km) and recorded noise at 5, 10 and 15 MHz, while measurements at 9 MHz from the RAE I satellite ($h_s \approx 6{,}000$ km) are discussed by Herman *et al.* (1973). Unfortunately, the results could rarely be interpreted by considering the noise to be received only from storms immediately below the satellite (Equation 15). The noise was found to propagate, often for considerable distances, between the earth and the ionosphere until it encountered an area where f_c was low and there passed through the ionosphere. Thus the received noise was related to the global morphology of f_c rather than to the positions of the storms. Also, man-made signals at HF, especially those due to powerful transmitters, are often stronger than atmospheric noise (Fig. 5). Since both types of noise could propagate substantial distances before penetrating the ionosphere, it was difficult to distinguish the spheric contribution to the satellite measurements.

At VHF and higher frequencies the noise signals pass through the ionosphere. In so doing, they suffer some absorption, but the major ionospheric influence is dispersive. The area viewed by the satellite is now defined by

$$A_s \approx 2\pi r^2 h_s/(r+h_s), \tag{16}$$

where r is the earth's radius. Chiburis and Jones (1974) have discussed data at three selected frequencies (28, 34, 42 MHz) in the VHF band obtained by the VELA 4B satellite ($h_s \approx 19r$). Major thundery regions, but not individual storms, could be identified from the data, and the identification was facilitated by examining the dispersion (differential-group-time delay) between signals on the selected frequencies.

Table III estimates the peak signal received in a satellite from an individual flash. An ionosphere is assumed with $f_c = 5$ MHz. For comparison purposes, the major non-terrestrial background source, cosmic noise, is also shown. The lightning signals are often stronger than the cosmic noise. However, man-made noise originating from the earth will also supply a substantial background against which the lightning signals must compete.

Summarizing, satellite noise experiments at HF and lower frequencies reveal much more about the properties of the ionosphere and magnetosphere than they do about the characteristics of atmospherics and thunderstorms.

Table III
Field strengths at satellite altitudes
(μV m^{-1} in 1 MHz Bandwidth)

	Altitude—1,000 km Frequency			Altitude—100,000 km Frequency		
Source	10 MHz	30 MHz	100 MHz	10 MHz	30 MHz	100 MHz
Individual lightning flash	2×10^3	5×10^2	10^2	20	5	1
Cosmic noise background	10	10	6	10	10	6

On the other hand, in the VHF band the ionosphere is much less significant, and its influence (notably dispersion) can be advantageously used when the motivation is to study the sources of atmospheric noise. The lower end of the VHF band is particularly suitable for such studies.

Finally, Sparrow and Ney (1971) have detected the optical signals from lightning using photometers in satellites. The practical problems have similarities with those involved in radio detection but the obscuration is now by cloud, and the competing signals are sunlight and moonglow.

6.4 Schumann Resonances

Schumann (1952) predicted that the cavity formed between the earth and the lower ionosphere should resonate at defined frequencies, and that these resonances would be excited by lightning flashes. The resonant frequencies are given approximately by

$$f_m = [m(m+1)]^{\frac{1}{2}} \left(\frac{c}{2\pi r}\right) \left(\frac{Q-1}{Q}\right)^{\frac{1}{2}}, \qquad (17)$$

where m is the order of the resonance, c the velocity of light, r the radius of the earth and Q the damping loss factor. For perfectly conducting boundaries $Q = \infty$ and $f_1 \approx 10{\cdot}6$ Hz; in reality, the finite conductivities of the earth and especially the ionosphere produce losses, and experimental results show that $f_1 \approx 7{\cdot}9$ Hz.

The detailed theory of the Schumann resonances is not simple (Wait, 1970), and its practical application presents many complications. The profile of ionospheric conductivity, including both ionic and electronic contributions, is of vital importance in determining the finer properties of the resonances (Cole, 1965; Ogawa and Murakami, 1973).

The characteristics of a specific resonance, observed at a particular locality, depend principally on the distances to the lightning sources and on ionospheric influences. The latter vary with time of day, season, geography and the morphology of the geomagnetic field; however, the source–receiver distance is usually the more significant factor. The global pattern of resonance behaviour is a function of source–receiver distance characteristic for each individual resonance mode. Thus if several resonances are monitored simultaneously the positions of the sources can in theory be deduced (Polk, 1969). If only a single storm centre were active over the entire world the deduction would be quite precise (Jones, 1969). However, many storms are active simultaneously (Chapter 14). Ogawa and Otsuka (1973) have shown that, in consequence, the resonance measurements are so blurred that their interpretation in terms of storm positions is difficult.

6.5 Special Noise Sources

Whenever there is strong atmospheric electrification small gaseous discharges occur and electromagnetic signals are generated. Thus, although the electrification never relaxes in the massive spark of lightning, the electrified clouds of blizzards, dust-storms and showers all generate noise. Noise is also radiated from a thundercloud before the first flash takes place. However, in all these circumstances the noise signals are slight (Medaliyer and Karmov, 1972).

The so-called precipitation static is more serious. This form of noise is often apparent in a receiving system if the associated antenna is exposed to impinging precipitation particles. Part of the noise is caused directly by the impact of the particles, which are usually charged, on the antenna. Most, however, is due to high atmospheric electric fields associated with the precipitation. An antenna will often, in consequence, go into corona and thus generate noise (see also Chapter 13.2.2.1). Noise can also be picked up by radiations from adjacent objects in corona. Corona-noise interference occurs when an aircraft becomes charged; the dischargers at the wing-tips are designed to minimize this interference (Nanevicz and Tanner, 1964; also Chapter 21.5.5).

The noise radiated by severe thunderstorms differs from that due to average storms. The energy from the severe storm is shifted towards higher frequencies than usual (Lind *et al.*, 1972; Taylor, 1973). Only severe storms spawn tornadoes. However, there is no distinctive "signature" in the severe storm noise characteristics to indicate whether there is an active tornado or not.

Efforts have been made to associate atmospherics phenomenology with large-scale disturbed weather systems. Under the stimulus of the pioneering work of Kimpara (1955) much of this work has been in Japan (Nishino, 1974).

6.6 The Electromagnetic Pulse

A nuclear explosion creates a strong electromagnetic pulse. The characteristics of this pulse are determined by many mechanisms; these have been well reviewed by Price (1974b). The most fundamental mechanism is the emission of γ-radiation by the explosion. The γ-rays create Compton electrons by various interactions, notably that with the ambient atmosphere. The current represented by the movement of the electrons then becomes asymmetric for several reasons; among these are deflection of the electrons by the geomagnetic field, and the non-uniformities represented by the atmospheric density gradient and by the air–earth interface. The asymmetry of the Compton currents results in the radiation of the pulse. Because of the many mechanisms possibly involved, an electromagnetic pulse may have significant frequency components at as low as 1 Hz or as high as 100 MHz.

Johler and Morgenstern (1965) have presented a wave-form for the pulse from a nuclear explosion 45 km away. The wave-form has some similarities with the spheric due to a return stroke; there are also differences. As compared with the usual return-stroke spheric, the polarity, initially negative going, is different; the zero-to-peak amplitude is much greater—60 V m^{-1} as compared with a likely 10 V m^{-1}; and the frequency content tends towards higher frequencies. However, after long-distance propagation and its consequent filtering effect, it is often difficult to distinguish between a nuclear pulse and a natural atmospheric (Price, 1974b).

Acknowledgements

This article was substantially written while I was on leave of absence from the Stanford Research Institute. During this leave I was employed by the U.S. Office of Naval Research in establishing a Scientific Liaison Group at the U.S. Embassy in Tokyo. I much appreciate the actions of the Stanford Research Institute in allowing me to take the leave of absence, and of the Office of Naval Research in encouraging me to write this chapter. I am indebted to my colleague, Mr G. H. Price, for reading and commenting on the manuscript. The logistics of preparation were supported by the Office of Naval Research under Contract N00014–74–C–0134 with the Stanford Research Institute.

References

Aiya, S. V. C., Sastri, A. R. K. and Shivaprasad, A. P. (1972). Atmospheric radio noise measurements in India. *Indian J. Rad. space Phys.* **1**, 1–8.

Arnold, H. R. and Pierce, E. T. (1964). Leader and junction processes in the lightning discharge as a source of VLF atmospherics. *Radio Sci.* **68** D, 771–776.

Beckmann, P. (1964). Amplitude-probability distribution of atmospheric radio noise. *Radio Sci.* **68** D, 723–736.

Brook, M. and Kitagawa, N. (1964). Radiation from the lightning discharges in the frequency range 400 to 1,000 Mc/s. *J. geophys. Res.* **69**, 2431–2434.

Brook, M. and Vonnegut, B. (1960). Visual confirmation of the junction process in lightning discharges. *J. geophys. Res.* **65**, 1302–1303.

Bruce, C. E. R. and Golde, R. H. (1941). The lightning discharge. *J. Instn. elect. Engrs* **88**, 487–520.

C.C.I.R. (1964). World distribution and characteristics of atmospheric radio noise. Report 322, International Telecommunication Union, Geneva.

Challinor, R. A. (1967). The phase velocity and attenuation of audio-frequency electromagnetic waves from simultaneous observations of atmospherics at two spaced stations. *J. atmos. terr. Phys.* **29**, 803–810.

Chiburis, R. and Jones, R. (1974). Severe storm observations from the VELA 4B satellite, *in* "Proceedings of Waldorf Conference on Long-range Geographic Estimation of Lightning Sources", pp. 264–282. NRL Report 7763, Naval Research Laboratory, Washington, D.C.

Cole, R. K. (1965). The Schumann resonances. *Radio Sci. J. Res. N.B.S.* **69 D**, 1345–1349.

Dennis, A. S. and Pierce, E. T. (1964). The return stroke of the lightning flash to earth as a source of VLF atmospherics. *Radio Sci.* **68** D, 777–794.

Gorodenskii, S. N. (1973). Estimation of the accuracy of spectral measurements of the distance to thunderstorms. *Geomagn. Aeron.* **12**, 223–227.

Hagn, G. H. (1975). Radio noise of terrestrial origin, *in* "Review of Radio Science 1972–1974" (S. A. Bowhill, Ed.), pp. 127–133. International Union of Radio Science, Brussels.

Harth, W. (1972). VLF-Atmospherics: Ihre Messung und Ihre Interpretation. *Z. Geophys.* **38**, 815–849.

Helliwell, R. A. (1965). "Whistlers and Related Ionospheric Phenomena." Stanford University Press, Stanford California U.S.A.

Herman, J. R., Caruso, J. A. and Stone, R. G. (1973). Radio astronomy explorer (RAE)—I. Observations of terrestrial radio noise. *Planet Space Sci.* **21**, 443–461.

Heydt, G. (1972). Remarks on the comparability of VLF atmospheric counts in broad- and narrow-band operation. *Met. Runds.* **25**, 20–23.

Heydt, G. and Volland, H. (1964). A new method for locating thunderstorms and counting their lightning discharges from a single observing station". *J. atmos. terr. Phys.* **32**, 609–621.

Horner, F. (1964). Radio noise from thunderstorms, *in* "Advances in Radio Research" (J. A. Saxton, Ed.), Vol. 2, pp. 121–204. Academic Press, London and New York.

Horner, F. (1970). The use of atmospherics for studying the ionosphere. *J. atmos. terr. Phys.* **32**, 609–621.

Horner, F. and Bent, R. B. (1969). Measurement of terrestrial radio noise. *Proc. R. Soc.* **311** A, 527–542.

Horner, F. and Bradley, P. A. (1964). The spectra of atmospherics from near lightning discharges. *J. atmos. terr. Phys.* **26**, 1155–1166.

Hughes, H. G. (1967). A comparison at extremely low frequencies of positive and negative atmospherics. *J. atmos. terr. Phys.* **29**, 1277–1283.

Hughes, H. G. and Gallenberger, R. J. (1974). Propagation of extremely low-frequency (ELF) atmospherics over a mixed day-night path. *J. atmos. terr. Phys.* **36**, 1643–1661.

Ishikawa, H. (1961). Nature of lightning discharges as origins of atmospherics. *Proc. Res. Inst. Atmos. Nagoya Univ.* **8** A, 1–273.

Johler, J. R. and Morgenstern, J. C. (1965). Propagation of the ground wave electromagnetic signal with particular reference to a pulse of nuclear origin. *Proc. Inst. elect. electron. Engrs* **53**, 2043–2053.

Jones, D. L. (1969). The apparent resonance frequencies of the earth-ionosphere cavity when excited by a single dipole source. *J. Geomagn. Geoelect.* **21**, 679–684.

Jones, D. L. (1974). Extremely low frequency (ELF) ionospheric radio propagation studies using natural sources". *Trans. Inst. elect. electron. Engrs* Part I. COM-**22**, 477–484.

Kimpara, A. (1955). Atmospherics in the Far East. *Proc. Res. Inst. Atmos. Nagoya Univ.* **3**, 1–28.

Krider, E. P. and Radda, G. J. (1975). Radiation field wave forms produced by lightning stepped leaders. *J. geophys. Res.* **80**, 2653–2657.

Lin, Y. T. and Uman, M. A. (1973). Electric radiation fields of lightning return strokes in three isolated Florida Thunderstorms. *J. geophys. Res.* **78**, 7911–7915.

Lind, M. A., Hartman, J. S., Takle, E. S. and Stanford, J. L. (1972). Radio noise studies of several severe weather events in Iowa in 1971. *J. atmos. Sci.* **29**, 1220–1223.

Malan, D. J. and Schonland, B. F. J. (1951). The distribution of electricity in thunderclouds. *Proc. R. Soc.* **209** A, 158–177.

Medaliyer, K. K. and Karmov, M. I. (1972). Radio emission of clouds. *Trans. Vysokogorn. geofiz. Inst.* **21**, 92–103.

Nakai, T., Nagatani, M. and Nakano, M. (1974). Statistical parameters on the electric-field-intensity changes at the source of atmospherics. *Proc. Res. Inst. Atmos. Nagoya Univ.* **21**, 1–7.

Nanevicz, J. E. and Tanner, R. L. (1964). Some techniques for the elimination of corona discharge noise in aircraft antennas. *Proc. Inst. elect. electron. Engrs* **52**, 53–64.

Nishino, M. (1974). The occurrence and the movement of the source of atmospherics accompanied with the cold fronts. *Proc. Res. Inst. Atmos. Nagoya Univ.* **21**, 9–17.

Oetzel, G. N. and Pierce, E. T. (1969). The radio emissions from close lightning, *in* "Planetary Electrodynamics" (S. C. Coroniti and J. Hughes, Eds), Vol. 1, pp. 543–571. Gordon and Breach, New York.

Ogawa, T. and Murakami, Y. (1973). Schumann resonance frequencies and the conductivity profile in the atmosphere. *Contr. geophys. Inst. Kyoto Univ.* **13**, 13–20.

Ogawa, T. and Otsuka, S. (1973). Comparison of observed Schumann resonance data with the single dipole source approximation theories. *Contr. geophys. Inst. Kyoto Univ.* **13**, 7–11.

Pappert, R. A. and Bickel, J. E. (1970). Vertical and horizontal VLF fields excited by dipoles of arbitrary orientation and elevation. *Radio Sci.* **5**, 1445–1452.

Pierce, E. T. (1955). The development of lightning discharges. *Q. Jl R. met. Soc.* **81**, 229–240.

Pierce, E. T. (1956). Some techniques for locating thunderstorms from a single observing station, *in* "Vistas in Astronomy" (A. Beer Ed.), Vol. 2, pp. 850–855. Pergamon Press, Oxford.

Pierce, E. T. (1969). Monitoring of global thunderstorm activity, *in* "Planetary Electrodynamics". (S. C. Coroniti and J. Hughes, Eds), Vol. 2, pp. 3–16. Gordon and Breach, New York.

Pierce, E. T. (1974). Source characteristics of atmospherics generated by lightning, *in* "Proceedings of Waldorf Conference on Long-range Geographic Estimation of Lightning Sources", pp. 64–79, NRL Report 7763, Naval Research Laboratory, Washington, D.C.

Polk, C. (1969). Relation of ELF noise and Schumann resonances to thunderstorm activity, *in* "Planetary Electrodynamics" (S. C. Coroniti and J. Hughes Eds), Vol. 2, pp. 55–83. Gordon and Breach, New York.

Posa, L. M., Materuzzi, D. J. and Gerson, N. C. (1972). Azimuthal variation of measured HF noise. *Trans. Inst. elect. electron. Engrs* Part II, EMC-**14**, 21–31.

Price, G. H. (1974a). Multimodal influences on the apparent propagational dispersion of VLF sferics, *in* "Proceedings of Waldorf Conference on Long-range Geographic Estimation of Lightning Sources." p. 211–240, NRL Report 7763, Naval Research Laboratory, Washington, D.C.

Price, G. H. (1974b). The electromagnetic pulse from nuclear detonations. *Rev. Geophys. space Phys.* **12**, 389–400.

Proctor, D. E. (1973). Comments on "A Technique for Accurately Locating Lightning at Close Ranges". *J. appl. Met.* **12**, 1419–1423.

Proctor, D. E. (1974). Sources of Cloud-Flash Sferics. C.S.I.R. Special Report TEL 118, Council for Scientific and Industrial Research, P.O. Box 3718, Johannesburg 2000, South Africa.

Sao, K. and Jindoh, H. (1974). Real time location of atmospherics by single station techniques and preliminary results. *J. atmos. terr. Phys.* **36**, 261–266.

Schumann, W. O. (1952). Über die strahlungslosen Eigenschwingungen einer leitenden Kugel, die von einer Luftschicht und einer Ionosphärenhülle umgeben ist. *Z. Naturf.* **72**, 149–154.

Sparrow, J. G. and Ney, E. P. (1971). Lightning observations by satellite. *Nature* **232**, 540–541.

Spaulding, A. D., Roubique, C. J. and Crichlow, W. Q. (1962). Conversion of the amplitude-probability distribution function for atmospheric radio noise from one bandwidth to another. *J. Res. natn. Bur. Stand.* **66** D, 713–720.

Storey, L. R. O. (1953). An investigation of whistling atmospherics. *Phil. Trans. R. Soc.* **246** A, 113–141.

Taylor, W. L. (1973). Electromagnetic radiation from severe storms in Oklahoma during April 29–30, 1970. *J. geophys. Res.* **78**, 8761–8777.

Teer, T. L. and Few, A. A. (1974). Horizontal lightning. *J. geophys. Res.* **79**, 3436–3441.

Tixier, M. and Rivault, R. (1974). Study of typical night time atmospheric waveforms observed at Poitiers (France), *in* "ELF-VLF Radio Wave Propagation" (J. A. Holter, Ed.), pp. 245–249. Proc. NATO Adv. Study Inst., Spatind.

Uman, M. A., Brantley, R. D., Lin, Y. T., Tiller, J. A., Krider, E. P. and McLain, D. K. (1975). Correlated electric and magnetic fields from lightning return strokes. *J. geophys. Res.* **80**, 373–376.

Wadehra, N. S. and Tantry, B. A. P. (1967). Audio-frequency spectra of *K.* changes in a lightning discharge. *J. Geomagn. Geoelect.* **19**, 257–260.

Wait, J. R. (1970). "Electromagnetic Waves in Stratified Media". Pergamon Press, New York.

Yamashita, M. and Sao, K. (1974a). Some considerations of the polarization error in direction finding of atmospherics—I. Effect of the earth's magnetic field. *J. atmos. terr. Phys.* **36**, 1623–1632.

Yamashita, M. and Sao, K. (1974b). Some considerations of the polarization error in direction finding of atmospherics—II. Effect of the inclined electric dipole. *J. atmos. terr. Phys.* **36**, 1633–1641.

11. Thunder

R. D. HILL

*Montecito, California, U.S.A.**

1. Introduction

Why does thunder precede lightning in our expression "thunder and lightning"? Does this expression date only from the time of Aristotle who taught that, despite appearances, lightning follows thunder?

For over two millennia there was speculation and confusion concerning the generation of thunder and it is only in this century that a reasonable consensus on the origin of thunder has been reached. Yet, even recently, Colgate (1969) was induced to remark in discussion that "I don't think we really understand thunder this well".

As is often the case, advances in understanding thunder have been attained through the use of new techniques, such as the oscillograph, microphone, magnetic tape recorder and digital computer. The aim of this chapter is to review these modern developments and to recognize remaining problems.

In the narrowest sense of the word, thunder is the sound that accompanies lightning. In its wider sense, thunder is associated with all hydrodynamic aspects of the atmosphere surrounding storms, and therefore includes the phenomenon of infra-sound which has too low a frequency to be heard, at least by human beings.

Subjective descriptions have formed the basis of scientific discourse on thunder for hundreds of years. Golde (1945) reviewed published oral descriptions of thunder and used this information to guide a statistical survey of the frequency of close lightning. Malan (1963) gave the following description in modern terminology of the sound from a close thunder and lightning event. When a lightning flash strikes within a hundred yards of an observer, the sounds are: first a click, then a crack like that of a whip, and finally the characteristic continuous rumble of thunder. The click is caused by a discharge

* Formerly General Research Corporation and University of Illinois, U.S.A.

streamer directed upward from the ground towards the downward leader just prior to the commencement of the return stroke. The crack is caused by the intense return stroke in the lightning channel nearest to the observer. The rumbles come from the multiplicity of sound sources distributed along the lightning channel. In cases when lightning strikes several hundreds of yards away from an observer, according to Malan, the first sounds heard are like the tearing of cloth. As Lucretius said:

> "it . . . imitates the tearing sound of sheets of paper—even this kind of noise thou mayst in thunder hear" (Book VI, l. 118).

This tearing sound which lasts an appreciable part of a second is attributed by Malan to many discharge streamers issuing from a large number of individual sharp points spread out over a wide area in the vicinity of the lightning strike (see Chapter 5.5.2). However, according to Brook (private communication), sounds like the tearing of cloth could arise from a straight-line sequence of thunder sources. Such sounds have been simulated by a long string of primacord explosions detonated in rapid sequence.

In a modern review, Remillard (1960) surveyed the extensive historical aspects of thunder and summarized the "accepted" facts. Seven of these facts (out of an original twelve given by Remillard) are given below. However, as pointed out by Remillard, practically every one of these facts was at one time disputed.

(a) Cloud-to-ground lightning flashes in general produce loudest thunder.
(b) Thunder is seldom heard over distances greater than about 10 miles.
(c) The time interval between the perception of lightning and the first sound of thunder can be used to estimate the distance of the lightning stroke.
(d) Atmospheric turbulence reduces the audibility of thunder.
(e) There is often a heavy downpour of rain immediately after a strong clap of thunder.
(f) The intensity of a pattern of thunder in one geographical location appears different from the pattern in another location.
(g) The pitch of thunder deepens as the rumble persists.

2. Experimental Techniques and Observations

2.1 Techniques prior to 1960

The first accurate records of the pressure variations in air caused by thunder appear to have been made by Schmidt (1914) who constructed two novel types of pressure-sensitive device. Since these devices have been superseded by improved detectors, only a brief description will be given of them.

The first device was a resonant cubical box with 60 cm sides, and a 15 cm diameter hole in one side. This hole was almost completely covered with an aluminium diaphragm suspended from outside the box by long threads that allowed a long period disturbance to be detected. Movements of the diaphragm which were caused by air-pressure differences between the inside and outside of the box were recorded by the movement of a pen across an advancing chart. This equipment was resonant to a disturbance of approximately 3 Hz central frequency, with a total width of the resonance also equal to 3 Hz.

The second device used by Schmidt consisted of the smoky plume of a flame from a turpentine lamp burning in a short vertical chimney that was connected to a sound-collecting gramophone horn. The smoke from the plume was allowed to deposit a trace on a moving strip of paper, thereby producing a permanent record of the pressure variations. This device was capable of registering variations having frequencies in the 15 to 200 Hz range.

In attempting to summarize the many events he observed Schmidt concluded that thunder frequencies usually peaked between 15 and 40 Hz, but that there was a smaller peak between 75 and 120 Hz. In spite of this peaking, however, Schmidt pointed to a remarkable feature in the events he observed; namely, that most of the energy radiated in thunder lay in the sub-audible region below 5 Hz, with a peak at 1·85 Hz.

Schmidt's results on the sub-audible content of thunder were confirmed by Arabadzhi (1952) who used techniques modelled after those of Schmidt. Arabadzhi's infra-sonic peak was observed at 0·5 Hz.

2.2 Techniques after 1960

A new phase of thunder experimentation was initiated at the New Mexico Institute of Mining and Technology (NMIMT) in the early 1960's. Brook *et al.* (1962), Williams and Brook (1963) and Latham (1964) reported on thunder-recording devices used largely as aids in sound-ranging on lightning strokes.

Thunder experiments were also initiated in Canada by Bhartendu and Currie (1963). Bhartendu (1964, 1968, 1969) employed three types of pressure-sensitive device, the signals from which were recorded by electrical oscillographs. These devices were: (a) a hotwire microphone located in a tube attached to a Helmholtz resonator, (b) a 12-inch diameter speaker acting as a microphone and attached to one end of a cylindrical collimating tube, (c) a wide-frequency-range crystal microphone. Bhartendu used power-spectrum analysis of the pressure-variation data and in the initial report of thunder spectra he gave maxima at frequencies of 22 to 28, 52 to 56 and 66 to 67 Hz. These results obtained from hotwire and crystal microphones, using sound-level meter output, were subsequently criticized by Brook *et al.* (1968), McCrory and Holmes (1968), and Few (1968) because the output signals were partly rectified and smoothed. The approach of using power-spectrum analysis was nevertheless recognized as an important advance.

In 1967, a report of work on the frequency spectrum of thunder was presented by Few *et al.* (1967), a combined group from Rice University (RU) and NMIMT. According to the NMIMT group, which analysed thunder from 12 cloud-to-ground flashes and 11 intra-cloud flashes, there were no dominant peaks in the acoustic spectrum. They found, instead, that the power rose by two orders of magnitude in the range from 1 to 200 Hz, at which frequency there was a broad maximum, after which the power fell approximately by one order of magnitude in the range from 200 Hz to 20,000 Hz. However, because of an incorrect frequency normalization used in this report, the NMIMT spectrum maximum at 200 Hz was subsequently found to be too high and the proper value should have been below 100 Hz (Brook, private communication). According to Few and Dessler of the RU group, their two records of thunder indicated that the dominant frequency occurred at 200 Hz. This frequency analysis was also subsequently reported to have been incorrect, and the data were revised by Few (1969a) who indicated a maximum in the power spectrum at 40 Hz.

2.3 Most Recent Techniques and Results

2.3.1 Rice University Measurements

The equipment used at RU by Few *et al.* (1967), Few (1968, 1969a), Teer (1972, 1973) and Few and Teer (1974) consists of an array of three commercial (Globe) capacitor microphones located 7 cm above ground-reflecting surface and at three corners of an equilateral triangle with 100 m side. The pressure-sensitive element in a Globe microphone is a circular, parallel-plate capacitor formed by a thin, aluminized mylar film that is stretched close to a perforated metal plate. Behind the plate is a closed space containing air at atmospheric pressure. Fluctuations of air pressure impinging on the mylar film cause changes in the capacitance, these changes in turn causing output voltage variations proportional to the magnitudes of the air pressure fluctuations. High sensitivity of 0·225 V dyn^{-1} cm^2 was obtained and the frequency response was reported as essentially flat over the range from 1 to 100 Hz. The purpose of the array of microphones was to distinguish between desired thunder signals and any background disturbances due to noise and wind. The direction of propagation could be determined from a cross-correlation of signal characteristics in the separate records.

Few (1968, 1969a) performed a detailed cross-correlation and power-spectrum analysis on a single, well-defined thunder event, recorded under good conditions of background wind noise. In reducing the data, the procedure used by Few was as follows: (a) the analog microphone signals on magnetic tape were converted to digital information on magnetic tape for computer analysis, (b) the pressure readings were corrected for any efficiency of microphone-response variation, thus yielding a digitized pressure $p(r, t)$, for a particular microphone position r at time t, (c) the scalar acoustic energy flux $f(r, t)$ was calculated from the square of $p(r, t)$ and appropriate factors of velocity and air density, (d) the Fourier transform $G(\phi, t_0, T)$ of $f(r, t)$, representing the energy flux of frequency ϕ in the finite interval of time T was then calculated, starting at selected initial times t_0. In the case of Few's single event analysed, T was chosen as 8·19 s.

The thunder power-spectra, $G(\phi, t_0, T)$, given in units of 10^{-4} erg cm^{-2} s^{-1} Hz^{-1} as a function of frequencies in Hz, are shown for five different initial t_0 times of the same event in Fig. 1. These curves were interpreted as indicating that the power spectrum of thunder, at least in this one event, was essentially independent of the time delay between lightning flash and observation time. In this event, which was a case of cloud-to-ground lightning, the peak of the thunder power-spectrum occurred at a frequency between 30 and 50 Hz.

2.3.2 New Mexico Institute Measurements

In the NMIMT work, including that reported by Holmes *et al.* (1971) and McCrory (1971), modified capacitor microphones, made commercially by Bruel and Kjaer, were used. Each microphone was mounted at the end of a 1-inch diameter tube which protruded approximately 1 inch beyond the side of a mounting box. The capacitor element of the B–K microphone is similar to that of the Globe instrument used by RU except that the diaphragm is a 5 micron-thick foil of nickel. The microphone was stated to be essentially omni-directional at its location and its frequency response was given as flat between 0·3 Hz (with 3 dB down at 0·1 Hz) and 20,000 Hz.

Initially, in the NMIMT experiments, thunder signals were obtained from a network of four microphones monitored by telemetry and recorded on paper tape. In later work, signals were recorded by high-speed magnetic tape from a single microphone located at a laboratory site. Measurements of electric field changes during microphone recordings were also simultaneously taken with the thunder data. According to the NMIMT group, the correlation of acoustic data and electric field changes enabled intensity and frequency characteristics of thunder to be identified with different types of lightning.

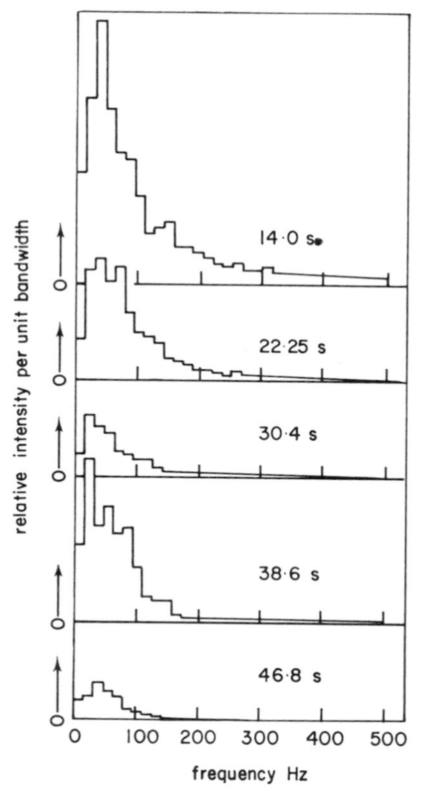

Fig. 1. Power spectra of single thunder event. (After A. A. Few, 1969.) Five histograms of power versus frequency are shown for five time windows. Times on each histogram are starting times of each window measured from zero time of flash associated with thunder event.

The NMIMT records, usually of the order of 10 s duration, were originally digitized electronically at a spacing of 15,000 points s^{-1}. However, it was soon found that no significant power was contained in the thunder spectrum above a frequency of certainly no more than 500 Hz. At a later date, the thunder records were therefore digitized with a spacing of 1,000 points s^{-1}. Each thunder event was divided into contiguous time windows, each window being approximately 1 s duration, and the pressure-variation data in each window were then analysed by a fast Fourier transform algorithm. The complex amplitude components in each frequency bin, for each time window, were squared in order to obtain the power-amplitude spectrum. The widths of the frequency bins were, of necessity, defined by the number of data points taken within a window and for normal practice the width was 5 Hz. However, the power spectral amplitudes were evaluated in terms of the acoustic power per unit frequency bandwidth. Results were displayed in two types of plot.

In the spectral power versus frequency plot, an example of which is shown in Fig. 2(a), the powers for each window in the event were lumped together as a function of frequency in order to obtain an average power over the whole duration of the thunder event. However, as emphasized by the NMIMT group, the thunder spectrum from one window to another is highly variable. Both the RU and NMIMT groups have stressed the point that thunder is an example of non-stationary random data which are not amenable to standard power-spectra-analysis techniques. Further, the power spectrum is subject to large variability with location of the recording; a problem which can only be overcome by spatial averaging of data from many microphones.

In the second type of plot, a power contour is taken as a function of thunder frequency and of time of arrival. This type of display was originally used by Pfeffer and Zarichny (1963) to show the time-varying Fourier components of pressure waves from nuclear explosions. A similar plot of pressure contours was also used later by Nakano and Takeuti (1970) as a display of "the dynamic spectrum of thunder". For the same NMIMT event, as in Fig. 2(a), the associated contour plot is shown in Fig. 2(b). Generally characteristic of the contour plots are one or more long tongues of intense sound indicative of claps and peals. The terms "clap", "peal" and "rumble" are used rather confusingly. Remillard (1960) drew attention to the need to clarify these terms, and Bhartendu (1964) subsequently suggested a set of definitions which are as follows—"clap": a sudden isolated sound that lasts only for a short time-interval of the order of a second; "peal": a sudden sound which repeats itself and which lasts for a long time (not specified); "rumble": a weak murmuring sound which persists for a considerable time. Despite these definitions, the distinction between peal and clap is still not very clear and some workers prefer not to use the terms at all.

The NMIMT findings indicate that there is no one characteristic spectrum or dominant frequency of thunder. Peaks in thunder power spectra were found in the range of frequencies from less than 4 Hz to 125 Hz. Cloud discharges, and some discharges to ground, have power spectrum peaks in the infra-sonic frequencies and other discharges to ground have peaks in the sonic frequency range. The fact that the spectral peak frequency was observed to shift with time after the flash indicated to the NMIMT group that there is probably more than one mechanism to the production of thunder.

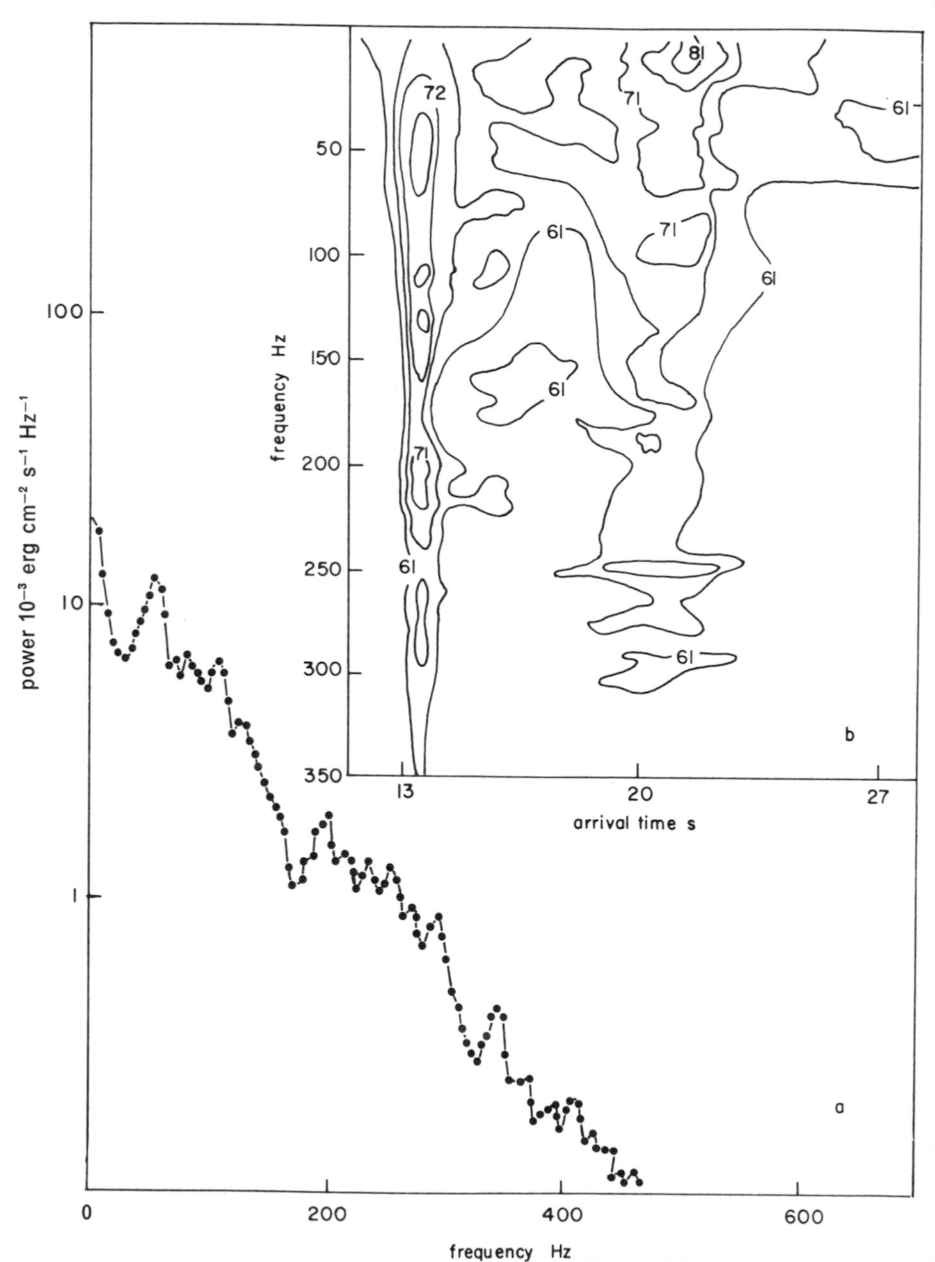

Fig. 2. Spectra of thunder from a flash to ground. (After Holmes *et al.*, 1971.) (a) Power spectrum versus frequency, averaged over 21 windows each of 1 s duration. (b) Power contours versus frequency and arrival time. Initial thunder follows flash by 9 s. Power contours are in dB above the standard level of 10^{-9} erg cm^{-2} s^{-1}. Unlabelled contours are 5 dB above surrounding contour.

3. Theories of Thunder

3.1 General Concepts

The generally accepted theory of the cause of thunder is that an intense lightning current generates a high-pressure shock wave which at large distances degenerates into loud audible sound. However, the original experiments of Schmidt (1914) and recently the extensive NMIMT measurements (described above) have shown that there are also intense sub-audible sounds associated with thunder. Although Holmes *et al.* (1971) have proposed that the sub-audible sounds are produced by a second, entirely different, mechanism from the hot channel mechanism production of thunder, there is still an open question of what is the direct evidence of either mechanism and of the degree to which these two mechanisms contribute to the observed pressure variations of thunder.

3.2 Lightning Channel Theory of Thunder Generation

It was originally shown theoretically by Abramson *et al.* (1947) that a sudden expansion of plasma, accompanied by a shock wave, occurs when spark breakdown and heating takes place in a gas. An analytical method was developed for solving this hydrodynamic problem in the idealized case of an instantaneous energy release along an infinitesimal width line source. This analysis was later extended by Drabkina (1951) to include the case of a more gradual energy deposition into the breakdown channel, and this theory was extended further and applied to the case of lightning by Braginskii (1958). Similar analytical solutions for instantaneous energy releases along infinitesimal width line sources were given by Sakurai (1953) and by Lin (1954). At the same time, Brode (1955) had extended techniques earlier developed by von Neumann and Richtmyer (1950) for carrying out digital computer calculations of high-energy explosions in spherical and cylindrical geometries.

A complete description of lightning-channel growth is complicated by numerous factors such as radiation transport, initial conditions in the channel before the main return current starts, finite size of the initial channel, time distribution of input current, conversion of electrical energy to heat in the channel plasma, losses from the channel, and so forth (Chapter 5.5.5). It is not surprising that a complete theory has not yet been given, but there have been attempts by Troutman (1969), Colgate and McKee (1969), Hill (1971) and Plooster (1971a) to elaborate the channel-growth problem more specifically towards the lightning-channel case.

All treatments so far consider the initial energy to be distributed with cylindrical symmetry, in fact all treatments are only one dimensional, i.e. radial. No attempt has been made to model an actual tortuous lightning channel. For a restricted finite line source, however, all results are in essential agreement in demonstrating that, with the extremely high-energy depositions per unit length in the lightning channel, an intense shock wave of overpressure is generated. An example of the overpressures generated as a function of very short distances from a channel of a typical lightning return stroke is given in Fig. 3.

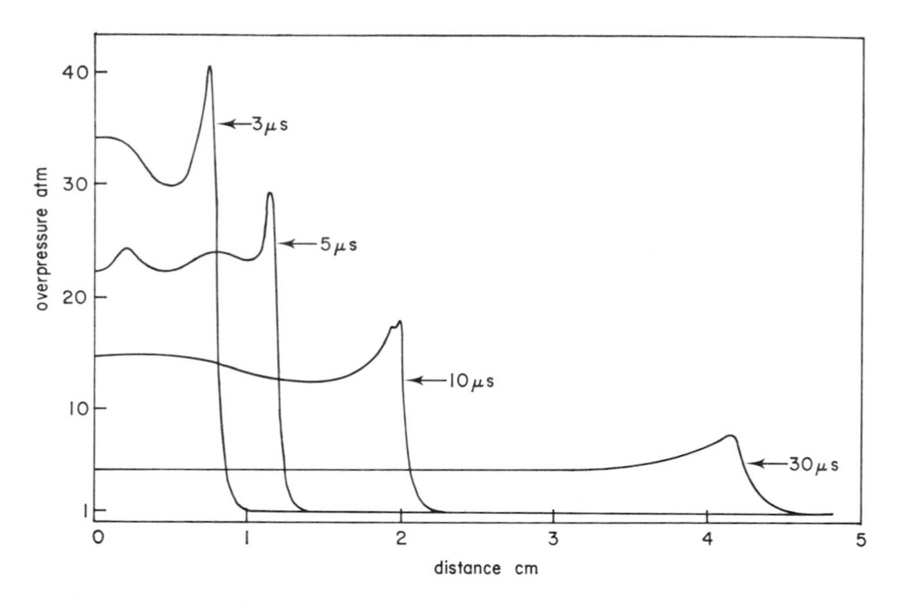

Fig. 3. Decay of shock front with distance from lightning channel. (After R. D. Hill, 1971.) Overpressures in channel and shock front versus distance are given for four times following onset of return stroke. Initial line source radius, 0·6 mm. Lightning current pulse, $I = I_0[(\exp - at) - (\exp - bt)]$, where $I_0 = 30{,}000$ A, $a = 3 \times 10^4$ s^{-1}, $b = 3 \times 10^5$ s^{-1}.

The only known measurements taken close to a lightning channel have been made by Newman *et al.* (1967) in experiments on triggered natural lightning (Chapter 21.3). At a distance of 35 cm from the channel produced by a 10 cm spark across a gap in series with a lightning stroke, they measured a peak overpressure of approximately two atmospheres. Dawson *et al.* (1968) pointed out that this overpressure was consistent with shock-wave theory if the 10 cm spark at a distance of 35 cm was considered to be a "point" source and if the energy dissipated in the lightning channel was typical and equal to 10^{10} erg cm^{-1}.

3.3 Generation of Thunder

3.3.1 Shock-wave Propagation

Although it is clear that the rapidly expanding lightning channel must be responsible for the generation of, at least, a part of thunder, it is not yet clear that the precise forms of observed thunder can be derived from lightning channel sources. There are mainly two reasons for the difficulty of developing a complete description: (a) a detailed analysis, or experimental measurement, of the propagation of the channel disturbance over the whole range of distances from the initial strong shock region near the channel to the remote acoustic thunder region has not yet been made, (b) the effects of either a continuous or a discrete sequence of finite length sources, distributed both in space and time along the length of a lightning channel, have not yet been adequately modelled.

The expansion of a shock wave from a straight-line source as a function of distance from the source was first raised as an unsolved problem by Jones *et al.* (1968). This problem is complicated by the fact that the classical strong-shock theory fails when the pressure ahead of a shock front is no longer negligible compared to the pressure in the shock front itself. Jones *et al.* developed an empirical expression to fit the overpressure decay of a cylindrical shock wave (falling off as $1/R^2$ in the strong-shock region) to the decay in the weak-shock region (where sonic boom and projectile wake shocks have been found to decay as $1/R^{\frac{3}{4}}$, R being the distance of the point of observation from the source).

The use of the Jones *et al.* expression for the calculation of overpressures at kilometre distances from lightning channels has been criticized by Few *et al.* (1970) on the grounds that it does not meet the criterion that a lightning channel is not a straight-line source but instead is extremely tortuous. Few *et al.* pointed out that the Jones *et al.* curves indicated that overpressures between 1 and 10 mb should be observed at a distance of 1 km, if the energy dissipation in the channel was typically between 10^{10} and 10^{11} erg cm^{-1}. They further stated that their own measurements (Few *et al.*, 1967; Few and Garrett, unpublished), as well as those of Bhartendu (1968), indicate that the observed overpressures are generally less than 0·1 mb at 1 km. If a large body of these data could be assembled so that the "typical" energy could be assured, and especially if a profile of overpressures versus distance could be established for thunder, this type of information would seem to be able to settle the issue of the cylindrical symmetry theory versus the tortuous theory.

Although adequate experiments have not yet been carried out on the propagation of lightning-produced shocks, valuable measurements on the decay of shocks from long sparks have been made at Westinghouse Research Laboratories by Uman and co-workers (see Orville *et al.*, 1967; Dawson *et al.*, 1968; Krider *et al.*, 1968; Uman *et al.*, 1968a; Uman *et al.*, 1970).

In the experiments by Uman *et al.* (1970), a piezoelectric microphone was used to record pressures and shapes of shock waves within 2 m of a 4 m length spark, dissipating approximately 2×10^4 J or 2×10^{11} erg. At distances between 3 and 16·5 m from the spark, a Bruel and Kjaer capacitor microphone, similar to that used in the NMIMT thunder experiments, was used to record shocks. Results of the overpressure measurements as a function of distance from the spark are given in Fig. 4.

While the broad features of the spark measurements tend to confirm shock theory, there is a certain difficulty in the interpretation of Fig. 4 which is as yet unexplained. This difficulty concerns the energy expended in the spark. The measured energy, as determined from the integration of the power curve with respect to time, was equal to 50 ($\pm$1) J cm^{-1}. As seen in Fig. 4, the theoretical curve derived from Plooster's numerical computations for 50 J cm^{-1} expended in a cylindrical spark lies well above the measured points. Plooster (1971b) gave this problem considerable attention and he finally attributed the discrepancy to what he felt was an incorrect measurement in the channel energy dissipation. In support of this viewpoint Plooster emphasized that his numerical simulation of the spark channel led to a channel dissipation of only 4 to 5 J cm^{-1} and to temperatures for the channel agreeing with those observed by Orville *et al.* (1967). Using the lower value of the energy dissipation in the channel, Plooster derived the second curve shown in Fig. 4, which appears to agree reasonably well with the observed points.

Uman *et al.* (1970), on the other hand, tended to accept the suggestion offered initially for the lightning case by Few (1969a), namely that since the spark channel was not truly straight, the shock was not characteristic of a cylindrically symmetrical source but rather of a short segment of the channel behaving as a "point" source. The third curve of Fig. 4 was derived from Brode's numerical computations assuming a total source energy of 2,500 J, i.e. a 50 cm length of channel with 50 J cm^{-1} energy dissipation per unit length. This curve is also seen to agree reasonably well with observations.

Also shown in Fig. 4 are points calculated from the expression of Jones *et al.* (1968) for the overpressures in the intermediate and weak shock regions from a straight source of 50 J cm^{-1}. It is clear from the disagreement between

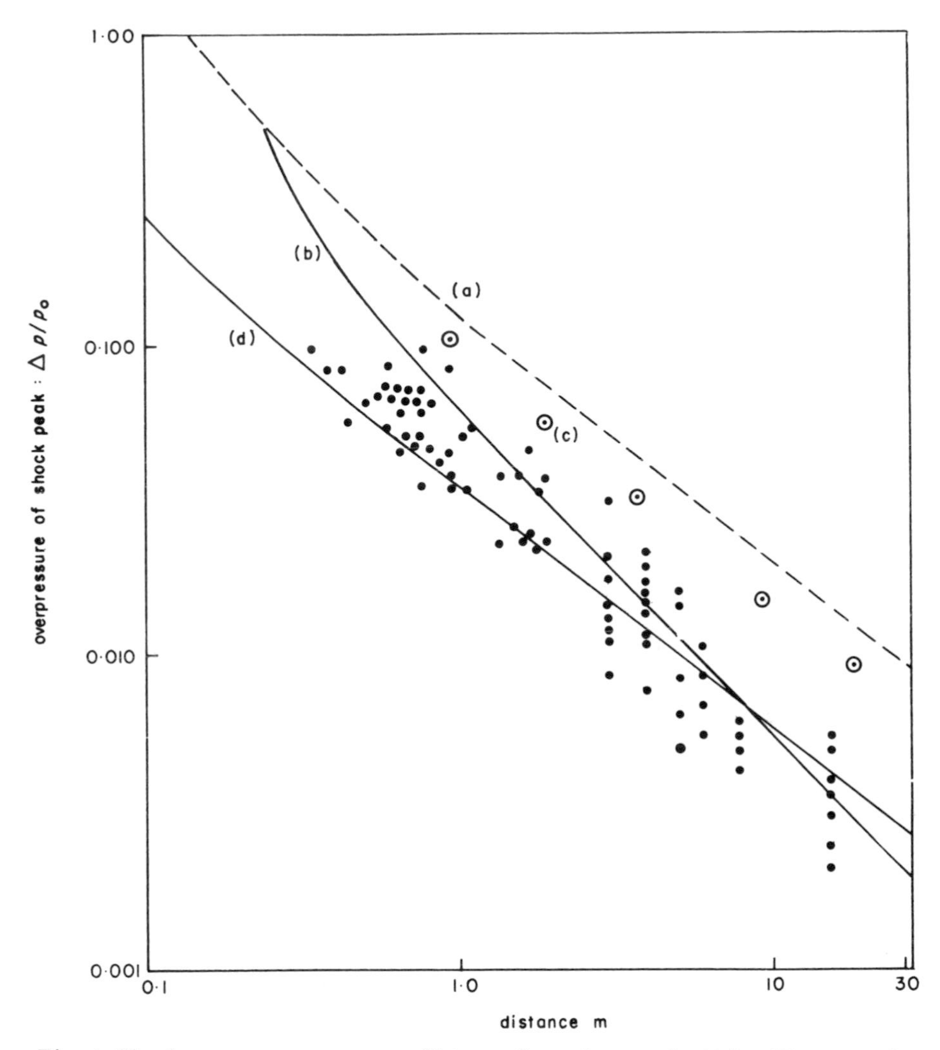

Fig. 4. Shock overpressures versus distance for a 4 m spark. (After Uman *et al.*, 1970.) Comparison of experimental points (●) with theoretical models: see text. Curve (a) 50 J cm^{-1}, cylinder (Plooster); curve (b) 2,500 J cm^{-1}, sphere (Brode); point (c) 50 J cm^{-1}, cylinder (Jones *et al.*); curve (d) 4·2 J cm^{-1}, cylinder (Plooster).

these points and the observations that this aspect of the theory does not provide an explanation of the discrepancy.

3.3.2 Shock-wave Pulse Shape

Few (1969a) extended the model of a thunder source as a sequence of point sources to a proposal for predicting the frequency spectrum of thunder radiated by a tortuous lightning channel. According to Few's proposal, a thunder power spectrum is characteristic of a spherically symmetrical expanding shock wave. In determining this spectrum he assumed that the average segment length of channel which behaves as a "point" is equal to $\frac{4}{3}$ times the characteristic radius, R_0, of a channel, which can be defined by the equation $R_0 = (E/\pi P_0)^{\frac{1}{2}}$, where E is the energy dissipation per unit length of channel and P_0 is the ambient pressure. (This definition is slightly different from the thermodynamic characteristic radius defined by Plooster (1970) as equal to $(4E/b\gamma P_0)^{\frac{1}{2}}$.) Few then proposed that the power spectrum is the same as that derived from the Fourier transform of a Brode (1956) shock-wave pulse from a spherical or point source measured at the most distant radius numerically computed by Brode.

The frequency, f_m of the power spectrum maximum, Few found to be proportional to $C_0(P_0/E)^{\frac{1}{2}}$, where C_0 is the sound velocity. The equation given by Few for f_m was: $f_m = 0{\cdot}63C_0(P_0/E)^{\frac{1}{2}}$, where the proportionality constant was derived from the spectrum of a Brode pulse for a value of E equal to 10^{11} erg cm^{-1}.

The experimental results obtained by the NMIMT group are strong evidence against the Few proposal of a generalized thunder power-spectrum. Not only does there appear to be no dominant frequency of thunder but also there are many peak frequencies pertaining to the observed thunder power-spectra that would require impossible energies in the channel in order to be consistent with Few's proposal. Holmes *et al.* (1971) pointed out, for example, that a peak frequency of 5 Hz (which they observed in a number of spectra) would require a total energy input to the channel of 10^{18} erg, a value which is at least several hundred times larger than the accepted energies (Chapter 9.5.3). Holmes *et al.* (1971) also pointed out that, whereas the average energy in cloud flashes is a factor of three less than in ground flashes, the average peak frequency for cloud flashes is approximately a factor of two higher than for ground flashes, a behaviour which is entirely the opposite to the dependence required by Few's proposal.

Few's proposal also raised a number of basic issues. For example, Remillard (1969) originally questioned Few's proposal that there could exist a dominant frequency which is emitted from a linear lightning channel if the wavelength tended to lengthen with distance from the channel. Although Few (1969b) rebutted this argument with the prevailing evidence that thunder is emitted by a tortuous line source, it did not appear that he justified the assumption that the wave-form of non-shock wave character at a very distant point, say, 5 km (or several thousand R_0 values) from the source would be identical with a strong shock wave at a point very near the source, or approximately 20 m (or 10 R_0 values) from where the Brode pulse form was taken.

Other issues raised by the dominant frequency model are (a) that the choice of $\frac{4}{3}R_0$ for the length of a "point" source is rather arbitrary, and (b) that the 10^{11} erg cm^{-1} is a factor of 10 higher than the typically accepted value of E given by Krider *et al.* (1968). Certainly the length $\frac{4}{3}R_0$ is a feasible value for the scale of tortuosity, but it is well known that the extent of tortuosity varies from hundreds of metres to centimetres. Few's answers to these issues were that these factors entered into the expression for f_m only as the square-root power.

Shapes of shock-wave pulses emitted by the 4 m Westinghouse Laboratory spark were measured at a number of distances from the spark by Uman *et al.* (1970). Although it was observed that the shapes in general resembled fairly well the characteristic shape calculated by Brode, changes in the shape with distance from the spark were not in conformity with theory, e.g. at a distance of less than 2 to 3 m (i.e. $\leqslant 25R_0$) the wave had a single shock front, at 8 m (i.e. $\approx 100R_0$) there were several superimposed shock peaks, and at the most distant point measured, 16·5 m (or $\approx 200R_0$) the shape often reverted to a single shock front again. Uman *et al.* further observed that the shocks were different from either the theoretical spherical or cylindrical shocks in that the underpressure portions were significantly shorter. The differences between theory and experiment remain unexplained.

3.3.3 *Attempts to Simulate Thunder Wave-forms*

In terms of the characteristic radial lengths, the observations of thunder have thus far been made at considerable distances from their sources, and although the minimum distances at which spark recordings were made by Uman *et al.* (1970) were considerably less, it must be noted that even these distances were still at least a factor of two more than the distance at which digital calculations were performed. The reason for only short computer distances is that it has so far been prohibitively expensive to extend computer times in order to obtain large ranges. There are two alternatives which have been followed: (a) an attempt to calculate analytically the propagation laws and (b) an attempt to model experimentally propagation and thunder characteristics.

Wright (1964) and Wright and Medendorp (1967) have studied, both by analytical and experimental methods, the shock-waves generated by a 1 cm spark of 0·01 to 0·1 J energy. These shocks have been analysed at considerable distances, $\geqslant 60R_0$, from the spark and at a number of polar angles θ between 90° and 180° to the axis of the spark. They concluded that the typical *N*-shaped shock wave (which is similar to the Brode shock pulse) only retains its initial shape when the point of observation lies in the equatorial plane around the spark (i.e. at $\theta = 90°$). At angles different from 90°, the over- and under-pressure portions of the *N* shock become rounded. The two portions also begin to separate in time (and space), so much so that at $\theta = 0°$ and 180° the portions have separated to a duration of twice the time-length of the initial *N* pulse. Wright and Medendorp also found that the shock-peak overpressure, p_θ, at an angle θ is a very steep function of θ, namely,

$$p_\theta = p_0[1-(1/\psi)\sin|90° - \theta|],$$

where p_0 is the peak overpressure at $\theta = 90°$ and ψ is equal to CT/λ, where C is the velocity of propagation, T is the total time length of the *N* pulse and λ is the length of the spark. This means that the contributions to the sound intensity at a point from lateral sections of a channel (which are at angles $\theta \neq 90°$) are relatively smaller and more distorted than from the equatorial section. Wright and Medendorp's analysis on short-spark shocks was based on the two assumptions: (a) that a shock pressure can be obtained from a linear superposition of component pressures and (b) that a source is a series of elements which radiate in phase and without interaction. The extent to which these assumptions are valid is unknown, but as far as the Wright–Medendorp analysis was concerned it was found that the linear theory led to an adequate description of the observations.

Uman *et al.* (1970) also attempted to calculate the shock-wave form from the 4 m spark by assuming that each point on a tortuous channel generated a spherical shock wave. However, they stated that they were unsuccessful in deriving any of the experimental wave-shapes observed. They suggested that neglect of non-linear interactions in and near the channel could have been instrumental in causing this lack of correspondence.

A similar attempt was made earlier by Uman *et al.* (1968b) to simulate thunder wave-forms using sound-wave emissions from the most important straight-line sections of typical lightning strokes. They were able to reproduce certain qualitative features of thunder such as (a) low-amplitude rumblings from a tortuous source and (b) intermittent claps or peals which were produced when a section of the channel, or a branch of the main channel, was perpendicular to the line of sight of the observer. This second feature agrees with Wright and Medendorp's analysis, and is also consistent with the widely used statement that a clap or peal comes from that section of a channel which is perpendicular to the line to the observer.

Brook (1969) made reference to unpublished experiments and computer calculations carried out by Moore, Colgate and Brook (private communication) to simulate thunder. Two configurations of point explosions were generated by detonating short lengths of primacord strung by wires from balloons at several thousand feet. In the first configuration, the primacord lengths were arranged in a straight line and exploded from the top. In this case it was found that the overall pattern of sound did not simulate thunder. However, in the second configuration, when the primacord lengths were distributed in a connected, yet tortuous, path simulating the tortuosity of lightning (Hill, 1968), the qualitative features, such as peals and rumbles of thunder, were reproduced. An example of a typical pressure-variation record of thunder is shown in Fig. 5.

In summing up this section on the theoretical understanding of thunder, one can conclude that certain features of the channel have been identified as having a definite influence on the nature of thunder. There is little doubt that tortuosity of the channel and orientation, with respect to the observer's direction, of nearly straight sections of the channel have a marked effect on sounds received by an observer. There appears to be a lack of detailed analytical understanding, however, of the manner in which sounds from different sections of a tortuous channel overlap, especially in the very distant audible-sound regions where most of the existing observations of thunder have been made.

4. Infra-sound

4.1 Observations

An intriguing aspect of the topic of thunder is the elusive reality of infra-sound. Schmidt (1914) believed that he had identified most of the energy of thunder as being in the subsonic region below 5 Hz. Arabadzhi (1952) also concluded that the subsonic peak at 0·5 Hz in his experiments was strongest. However, there was part conflict between Arabadzhi and Schmidt in that the former claimed that the strongest part of the pressure wave was a compression while the latter claimed it was a rarefaction. There also appears to be some question on the part of modern workers about whether the resonant-box type of detectors used by Schmidt and Arabadzhi did not grossly overestimate

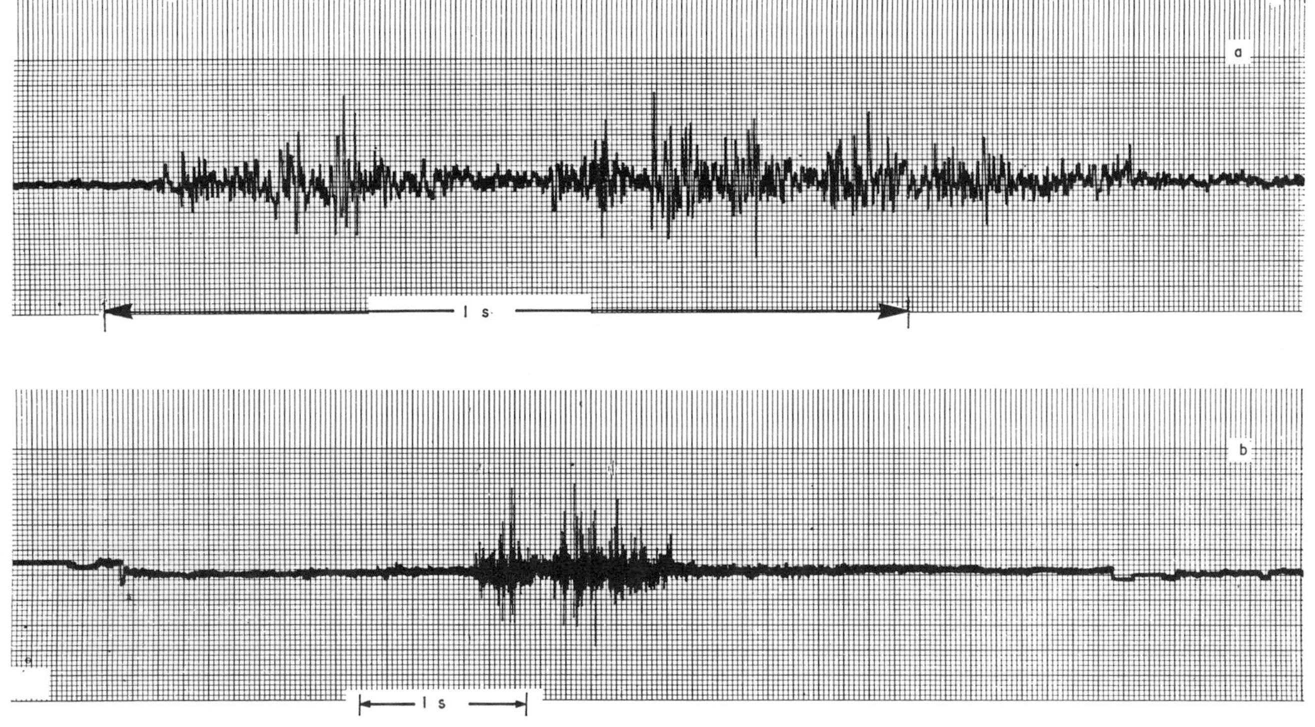

Fig. 5. Microphone recording of a single sharp thunder clap—(a) and (b) are identical records, (a) has expanded time scale. (Courtesy C. R. Holmes and M. Brook, NMIMT Group.) This record clearly shows visual evidence of frequency components between 1 and 300 Hz. A strong modulation of approximately 10 Hz is evident. The normal time window Fourier-analysed by the NMIMT group was 1 s.

the subsonic frequencies of thunder. Later experimental work using geometrically smaller microphone detectors has tended to down-grade the reliability of the earlier results. Few *et al.* (1967) at first found no evidence for a significant infra-sonic component in the thunder power-spectra below 20 Hz, even though they claimed that their equipment was capable of measuring frequencies with high sensitivity to as low as 0·1 Hz. In later papers from NMIMT, although McCrory (1971) expressed surprise that infra-sonic frequencies were detected so strongly in their experiments, Holmes *et al.* definitely established that even peak intensities of infra-sound below 10 Hz exist in thunder power-spectra. This conclusion was strikingly demonstrated by Holmes *et al.* (1971) in the contour plot shown in Fig. 2(b) of a thunder event having two intensity peaks: one, at 13 s after flash, when a peak at a frequency of 61 Hz had a power level of 81 dB, and two, at 21 s after flash, when a second peak at a frequency of 5 Hz had a power level of 83 dB. Holmes *et al.* suggested that the shift in the frequency of the spectral peak was associated with a change in the mechanism of the production of thunder. The first peak at sonic frequencies could be associated with hot-channel production of thunder and the second peak at subsonic frequency they attributed to the electrostatic mechanism proposed originally by Wilson (1920). This theory of infra-sound generation will now be very briefly discussed.

4.2 Theories of Infra-sound Generation

If Wilson (1920) did not refer in his classical paper on thunderstorms to the work of Schmidt on thunder it is probably because he did not investigate the frequency characteristics of the new thunder-generation mechanism which he proposed. Basically Wilson proposed that the sudden contraction of a large volume of a thundercloud would furnish "a by no means negligible contribution to thunder". Wilson pointed out that the hydrodynamic pressure within a portion of a charged thundercloud would be reduced to the extent of approximately 0·3 mm Hg when the electric stress, given by $E^2/8\pi$, where E is typically 100 stat V cm^{-1}, was relieved.

Two analyses of the Wilson mechanism have been made since 1969. The first was made in connection with the leader phase of a lightning discharge by Colgate and McKee (1969). They estimated that a stepped leader would deposit approximately one-tenth of the total charge typically deposited in an average lightning return stroke, and they calculated that this charge would spread out radially from an initial 2·5 m column to an approximately 15 m radius column during the time of advance of the leader. Associated with this spreading of charge, a compression wave in air would be generated by the

Wilson mechanism. According to Colgate and McKee (1969), the dominant frequency of this wave would correspond to the inverse time that the rarefaction, which is associated with the compressive wave but which is propagated in the reverse direction, takes to reflect from the axis of the leader column. For an initial 2·5 m radius cylinder, and for a velocity of approximately sound speed, the dominant frequency would be approximately 130 Hz. The overpressure ratio $\Delta p/p_0$ of the leader pressure wave was estimated to be only approximately 4×10^{-4}. However, it would be expected to precede the intense channel thunder wave by the time that the leader was ahead of the return stroke and therefore might be detectable.

Colgate and McKee further estimated that the sudden release of stress in the cloud supplying charge for the leader would be expected to develop an underpressure wave of approximately 40 dyn cm^{-2} and a frequency of the order of 0·5 Hz. This wave should also precede the normal hot air channel sound.

The second analysis of infra-sound sources has recently been made by Dessler (1973). He has developed two particular models of cloud electrification that produce, on discharge, pressure pulses by the Wilson-type mechanism. The first model is of a flat cloud distribution of 5 km diameter and 0·5 km thickness. For an electric field of between 10^5 and 10^6 V m^{-1} at the cloud surface, and for the discharge of 40 C in a typical lightning flash, the pressure amplitude was calculated to be between 1 and 50 times 10^{-6} atm and the frequency was estimated to be between 0·2 and 2 Hz. These pressure amplitudes are in reasonable agreement with Wilson's. The second model of a long cylindrical cloud gave very similar values to the above.

An interesting aspect of the infra-sound mechanism raised by Dessler was the notion that the propagation of sound could be highly directive. In the case of the horizontal flat layer of cloud, Dessler estimated that a sound wave of 0·5 Hz frequency and 0·6 km wavelength, originating from a disk cloud of 4 km diameter, would be propagated as an almost cylindrical vertical beam, extending both upward and downward. The intensity at the edge of the beam was calculated to fall off very rapidly with angle from the vertical, to the extent of 10 dB in 8°. Similarly restricted directivity was also calculated for a beam propagated perpendicularly to the axis of a cylindrical cloud.

Dessler suggested that such highly directed channels of infra-sound might explain, in part, the lack of observation of infra-sound and the considerable variations of infra-sound that might be expected from one event to another when observed. The fact that infra-sound is so highly directive suggested to Dessler that it could be a significant source of disturbance in regions (e.g. the ionosphere) where the ambient atmospheric pressure is comparable with the magnitude of the infra-sound overpressure. In fact there have been reports of unexplained radio-transmission disturbances in the ionosphere during thunderstorms which might possibly be attributed to long-range infra-sound propagation (see Davies and Jones, 1973).

Holmes *et al.* (1971) pointed out that the mean linear dimension of stored charge released by a lightning stroke was 300 m and, for a sound velocity of 300 m s^{-1}, the expected fundamental frequency excited by a Wilson-type mechanism would be approximately 1 Hz. They suggested that their observations of second infrasonic peaks of less than 10 Hz were consistent with the expected frequencies of a hydro-electrostatic mechanism.

Mention should also be made of the analysis by McGehee (1964) on the influence of thunderstorm space charges on pressure. McGehee made no reference to thunder or to Wilson's paper, but his analysis confirmed and elegantly extended Wilson's estimates of the pressure changes. In the more general context of the influence of thunderstorms on the hydrodynamics of air and the generation of subsonic frequencies, mention should also be made of the observations by Young *et al.* (1968) of infra-sound in the period range from 1 to 30 s. This sound is radiated from regions around thunderstorms and squall lines. They further noted that winds from storms quite often interrupt reception of this infra-sound.

5. Other Topics Associated with Thunder

Two other topics will be mentioned but not discussed. The first is the effect that the properties of the atmospheric medium have on the transmission of thunder. Discussion of this topic lies more appropriately in a chapter on meteorology. Such a chapter would include discussion of such phenomena as refraction of sound, reflection of sound at hard surfaces and also at the interfaces between layers of air, effects of rain and water vapour on sound transmission, reflection and absorption, zones of transmission in the atmosphere, etc. Discussions of many of these effects have been given originally by Remillard (1960) and later by Few (1968).

The second topic concerns the application of thunder to the location of hot lightning channel sources in storm clouds. This application has been developed recently by Teer (1972, 1973) and Few and Teer (1974).

Acknowledgements

Information for this chapter was gathered partly from work performed under contract for the U.S. Office of Naval Research. I wish to thank Mr J. Hughes of the Office of Naval Research for encouraging the writing of this chapter. I also wish to acknowledge Dr Marx Brook's valuable criticism of an initial draft of this material.

References

Abramson, I. S., Gegechkori, N. M., Drabkina, S. I. and Mandel'shtam, S. L. (1947). On the channel of a spark discharge. *Zh. eksp. teor. Fiz.* **17**, 862–867.

Arabadzhi, V. I. (1952). On some characteristics of thunder. *Dokl. Akad. Nauk SSSR* **82**, 377–378.

Bhartendu (1964). Acoustics of Thunder, Ph.D. Thesis, Univ. Saskatchewan.

Bhartendu (1968). A study of atmospheric pressure variations from lightning discharges. *Can. J. Phys.* **46**, 269–281.

Bhartendu (1969). Audio frequency pressure variations from lightning discharges. *J. atmos. terr. Phys.* **31**, 743–747.

Bhartendu and Currie, B. W. (1963). Atmospheric pressure variations from lightning discharges. *Can. J. Phys.* **41**, 1929–1933.

Braginskii, S. I. (1958). Theory of the development of a spark channel. *Soviet Phys. JETP* **34**, 1068–1074.

Brode, H. L. (1955). Numerical solutions of spherical blast waves. *J. appl. Phys.* **26**, 766–775.

Brode, H. L. (1956). The blast wave in air resulting from a high-temperature, high-pressure sphere of air. RAND Research Memo. RM–1825–AEC, Fig. 11, p. 21.

Brook, M. (1969). Discussion on the Few–Dessler paper, *in* "Planetary Electrodynamics". (S. C. Coroniti and J. Hughes, Eds), Vol. I, p. 579. Gordon and Breach, New York.

Brook, M., Kitagawa, N. and Workman, E. J. (1962). Quantitative study of strokes and continuing currents in lightning discharges to ground. *J. geophys. Res.* **67**, 649–659.

Brook, M., Holmes, C. R. and McCrory, R. A. (1968). Acoustic spectra of thunder. *Trans. Am. geophys. Un.* **49**, 688 (Abstract).

Colgate, S. A. (1969). Discussion on the Few–Dessler paper, *in* "Planetary Electrodynamics". (S. C. Coroniti and J. Hughes, Eds), Vol. I, p. 580. Gordon and Breach, New York.

Colgate, S. A. and McKee, C. (1969). Electrostatic sound in clouds and lightning. *J. geophys. Res.* **74**, 5379–5389.

Davies, K. and Jones, J. E. (1973). Acoustic waves in the ionospheric F_2-region produced by severe thunderstorm. *J. atmos. terr. Phys.* **35**, 1737–1744.

Dawson, G. A., Uman, M. A. and Orville, R. E. (1963). Discussion of paper by E. L. Hill and J. D. Robb, "Pressure pulse from a lightning stroke". *J. geophys. Res.* **73**, 6595–6597.

Dawson, G. A., Richards, C. N., Krider, E. P. and Uman, M. A. (1968). Acoustic output of a long spark. *J. geophys. Res.* **73**, 815–816.

Dessler, A. J. (1973). Infrasonic thunder. *J. geophys. Res.* **78**, 1889–1896.

Drabkina, S. I. (1951). On the theory of development of the channel of a spark discharge. *Zh. eksp. teor. Fiz.* **21**, 473–483.

Few, A. A. (1968). Thunder, Ph.D. Thesis, Rice University.

Few, A. A. (1969a). Power spectrum of thunder. *J. geophys. Res.* **74**, 6926–6934.

Few, A. A. (1969b). Reply. *J. geophys. Res.* **74**, 5556.

Few, A. A. and Teer, T. L. (1974). The accuracy of acoustic reconstructions of lightning channels. *J. geophys. Res.* **79**, 5007–5011.

Few, A. A., Dessler, A. J., Latham, D. J. and Brook, M. (1967). A dominant 200-Hertz peak in the acoustic spectrum of thunder. *J. geophys. Res.* **72**, 6149–6154.

Few, A. A., Garrett, H. B., Uman, M. A. and Salanave, L. E. (1970). Comment on letter by W. W. Troutman "Numerical calculation of the pressure pulse from a lightning stroke". Westinghouse Research Lab. Paper 70–9C8–HIVOL–PI (unpublished).

Golde, R. H. (1945). Frequency of occurrence of lightning flashes to earth. *Q. Jl R. met. Soc.* **71**, 89–109.

Hill, R. D. (1968). Analysis of irregular paths of lightning channels. *J. geophys. Res.* **73**, 1897–1906.

Hill, R. D. (1971). Channel heating in return-stroke lightning. *J. geophys. Res.* **76**, 637–645.

Holmes, C. R., Brook, M., Krehbiel, P. and McCrory, R. A. (1971). On the power spectrum and mechanism of thunder. *J. geophys. Res.* **76**, 2106–2115.

Jones, D. L., Goyer, G. G. and Plooster, M. N. (1968). Shock wave from a lightning discharge. *J. geophys. Res.* **73**, 3121–3127.

Krider, E. P., Dawson, G. A. and Uman, M. A. (1968). Peak power and energy dissipation in a single-stroke lightning flash. *J. geophys. Res.* **73**, 3335–3339.

Latham, D. J. (1964). A study of thunder from close lightning discharges, M.S. Thesis, N.M.I.M.T.

Lin, S. C. (1954). Cylindrical shock waves produced by instantaneous energy release. *J. appl. Phys.* **25**, 54–57.

Malan, D. J. (1963). "Physics of Lightning", p. 162. English Universities Press, London.

McCrory, R. A. (1971). Thunder and its Relationship to the Structure of Lightning, Ph.D. Thesis, N.M.I.M.T.

McCrory, R. A. and Holmes, C. R. (1968). Comment on paper by Bhartendu, "A study of atmospheric variations from lightning discharges". *Can. J. Phys.* **46**, 2333–2334.

McGehee, R. M. (1964). The influence of thunderstorm space charges on pressure. *J. geophys. Res.* **69**, 1033–1035.

Nakano, M. and Takeuti, T. (1970). On the spectrum of thunder. *Proc. Res. Inst. Atmos. Nagoya Univ.* **17**, 111–113.

Newman, M. M., Stahmann, J. R. and Robb, J. D. (1967). Experimental study of triggered natural lightning discharges. FAA Report DS–67–3 (unpublished).

Orville, R. E., Uman, M. A. and Sletten, A. M., (1967). Temperature and electron density in long air sparks. *J. appl. Phys.* **38**, 895–896.

Pfeffer, R. and Zarichny, J. (1963). Acoustic gravity wave propagation in an atmosphere with two sound channels. *Geofis. pura appl.* **55**, 175–181.

Plooster, M. N. (1970). Shock waves from line sources. Numerical solutions and experimental measurements. *Phys. Fluids*, **13**, 2665–2675.

Plooster, M. N. (1971a). Numerical model of the return stroke of the lightning channel. *Phys. Fluids* **14**, 2124–2133.

Plooster, M. N. (1971b). Numerical simulation of spark discharges in air. *Phys. Fluids* **14**, 2111–2123.

Remillard, W. J. (1960). The acoustics of thunder. Tech. Memo. No. 44. Acoustics Research Lab., Harvard University, Cambridge (unpublished).

Remillard, W. J. (1969). Comments on paper by A. A. Few *et al.* "A dominant 200-Hertz peak in the acoustic spectrum of thunder". *J. geophys. Res.* **74**, 5555.

Sakurai, A. (1953). On the propagation and structure of the blast wave. *J. phys. Soc. Japan* **8**, 662–669.

Schmidt, W. (1914). Über den Donner. *Met. Z.* **31**, 487–498.

Teer, T. L. (1972). Acoustic Profiling, M.S. Thesis, Rice University.

Teer, T. L. (1973). Lightning Channel Structure, Ph.D. Thesis, Rice University.

Troutman, W. W. (1969). Numerical calculation of the pressure pulse from a lightning stroke. *J. geophys. Res.* **74**, 4595–4596.

Uman, M. A., Orville, R. E., Sletten, A. M. and Krider, E. P. (1968a). Four meter sparks in air. *J. appl. Phys.* **39**, 5162–5168.

Uman, M. A., McLain, D. K., and Myers, F. (1968b). Sound from line sources with application to thunder. Westinghouse Research Lab. Report 68–9E4–HIVOL–R1 (unpublished).

Uman, M. A., Cookson, A. H. and Moreland, J. B. (1970). Shock waves from a four-meter spark. *J. appl. Phys.* **41**, 3148–3155.

Von Neumann, J. and Richtmyer, R. D. (1950). A method for the numerical calculation of hydrodynamic shocks. *J. appl. Phys.* **21**, 232–237.

Williams, D. P. and Brook, M. (1963). Magnetic measurements of thunderstorm currents. *J. geophys. Res.* **68**, 3243–3247.

Wilson, C. T. R. (1920). Investigations on lightning discharges and on the electric fields of thunderstorms. *Phil. Trans. R. Soc.* **221**, 73–115.

Wright, W. M. (1964). Experimental study of acoustical N-waves. *J. acoust. Soc. Am.* **36**, 1032 (Abstract).

Wright, W. M. and Medendorp, N. W. (1967). Acoustic radiation from a finite line source with N-wave excitation. *J. acoust. Soc. Am.* **43**, 966–971.

Young, J. M., Green, G. E. and Bowman, H. S. (1968). Infra sound from thunderstorms, frontal systems and squall lines. *Trans. Am. geophys. Un.* **49**, 688 (Abstract).

12. Ball Lightning

S. SINGER

Athenex Research Associates, Pasadena, California, U.S.A.

1. Properties

Ball lightning, the luminous, sometimes fiery globe seen in thunderstorms, has proved a difficult problem to scientists for two centuries. Although a number of experiments have produced glowing balls which resemble the natural object, none of these has been conclusive in establishing its identity. Eminently reasonable theories have been presented, but an almost equal number of implausible theories has appeared, even in the latest work.

The infrequent and irregular occurrence of ball lightning renders ineffectual the experimental methods of observation which provide so much information on ordinary lightning. Nevertheless, judging by the number of publications, this subject has received as much attention in recent years as any other aspect of lightning.

Despite the extremely limited information available on ball lightning from measurements made during its appearance in nature, its general characteristics are well known. These have been obtained by study of approximately one thousand random observations by chance observers recorded over the past century and a half in the general scientific and meteorological literature.

The glowing spheres are usually associated with ordinary lightning in severe thunderstorms. In contrast to the common flashes of lightning, however, these globes remain visible for an appreciable time while floating freely through the air in extended paths which may take them into houses. Their velocity is relatively moderate, as indicated by witnesses who have escaped being struck by the flying balls by leaping aside (Reimann, 1887).

The globes are approximately spherical in all but 10 to 20% of the cases. The ring shape has also been reported. A blue halo or corona extending from the central mass is observed occasionally, and sometimes sparks or rays are emitted. The balls average 25 cm in diameter. Most are between 10 and 100 cm; extremes from approximately half a centimetre up to several metres have been reported.

The colour of the spheres seems to be comparable to that of a flame or an electrical discharge in the atmosphere. Orange and red are most frequent. Yellow, blue and green are also noted as is an intense white which is especially bright and dazzling. The fireballs usually exist for 1 to 5 s, but some have been visible for a few minutes before disappearing.

Certain characteristic paths are followed by the moving globes. Those first seen at some height in the sky fall directly down. Some of these change direction suddenly near the ground and continue to move horizontally. A meandering path is common in a large number which appear or are seen only near the ground. Several seem to roll along the surface of the earth. A few rise in upward flight. Motion directly against the wind is reported, and rapid rotation is observed in many cases. The luminous balls enter buildings through doors, windows and commonly chimneys, often progressing at a moderate velocity of 1 to 2 m s^{-1}.

Hissing or crackling sounds have been noted during appearances of ball lightning although many are specifically reported to be silent. The sound evidently resembles that of an electrical discharge in the air, but this has seldom been stated by witnesses. The odours reported occasionally as "sulphur", ozone or nitrogen dioxide can also be associated with such a discharge.

Many of the spheres disappear silently without leaving a trace, but a large number explode. Some of the explosions cause no damage despite an ear-shattering noise, but others result in destruction, collapse of primitive buildings, death of domesticated animals and shattering of part of a wharf. Sometimes intense heat is radiated from the non-exploding balls. Others are not noticeably warm according to incidents in which the globes floated close to eye-witnesses.

The properties of ball lightning have been considered in greater detail by de Jans (1910–12), Brand (1923) and Singer (1971). No surveys or collections of reports have been published since that last major review of 1971, but a number of individual observations have been reported. Additional encounters with ball lightning which confirm previously noted characteristics occur continually. Some of the recent reports describe properties of the fireballs observed in previous cases which might well appear so unique as to be doubtful to those unfamiliar with the subject if found in only a single observation.

Additional sightings at sea have been reported (Minter and Bird, 1972). The entrance of ball lightning through the chimney into a home was again observed (Lazarides, 1971). A recent event involved a whitish yellow globe slightly larger than a tennis ball which exploded in a restaurant (Wagner, 1971). A dog was disturbed by the sight of the moving globe, and the explosion drew the attention of people who had not seen the ball itself.

A stationary brilliant white fire-ball reported on a power-line mast during a thunderstorm (Gibbs-Smith, 1971) appears to have been a power arc.

Ball lightning which drifted for 800 m was observed following forked lightning (Galton, 1971). Another brilliant white sphere 20 cm in diameter was formed near the metal supports of an anemometer in a meteorological office after lightning struck the upper tube which extended out through the roof (Bromley, 1970). The ball floated 2 m from an observer who said that it did not radiate heat, and the sphere disappered in 3 to 5 s.

A glowing ball which fell from the clouds in a severe storm drifted a considerable distance, and after remaining stationary for a moment continued its flight until it struck a wharf and exploded (Covington, 1970). The detonation shattered a pile into long splinters. The energy released was estimated at $1{\cdot}3 \times 10^5$ J from the heat required to form water vapour at a pressure equal to the tensile strength of the wooden piling (Zimmerman, 1970). This is a much lower energy than in previous estimates and probably results from the efficient oversimplified process depicted as producing the explosion.

Further encounters with ball lightning on aircraft in flight were reported. A glowing white sphere appeared on the wing-tip of a Boeing 727 and disappeared in 5 s with a soft pop (Felsher, 1970). An Ilyushin 18 was jolted abruptly and the crew was momentarily blinded by a direct collision with a large bright ball (Aulov, 1973).

A bright blue-purple globe 10 cm in diameter with a flame-coloured halo burned an irregular hole of 4×11 cm^2 in the dress of a witness indoors who felt the heat and heard a rattling sound (Stenhoff, 1976). She brushed at the ball as it hit her, and her hand and legs reddened from the contact. Wooding (1976) estimated the energy of the globe at 2 to 3 kJ from these effects, much less than the 10 MJ derived from the ball which heated a cask of water to boiling (Singer, 1971).

The absence of complete data on significant properties in such incidents is common, and only rarely can questions be clarified by access to the witnesses (Lilienfeld, 1970; Davies, 1976). Unpublished incidents which reveal the inexplicable but characteristic behaviour of ball lightning come into the hands of students of the phenomenon. The following two cases illustrating quite diverse behaviour were described by reliable witnesses.

A large blue-white sphere approximately 1·5 m in diameter was observed in an intense thunderstorm by E. H. Sadler, an electrical engineer with war-time experience in explosives, and the three other members of his family.

The stationary globe appeared at the top of the tallest tree in a wood while an unusual ripping or tearing noise was heard. It disappeared with an explosion comparable to the noise from 2 kg of guncotton. Sadler noted that the blinding sphere was unlike ordinary corona or St Elmo's fire with which he was acquainted.

A small bright light as large as a golf ball floated in the door of a kitchen during a thunderstorm where it was observed by Anna Ludwig Cossé. It continued at a leisurely walking speed, bobbing slightly as it moved at a man's height to the back of the room. There it passed over a large stove and then returned, making a horseshoe curve to the middle of the room, and floated out through the door in the space of half a minute without leaving a trace.

2. Theories

Attempts to explain the formation and properties of ball lightning have thus far been inconclusive. Although many theories, several of which invoke reasonable physical principles, have been proposed, none has been decisive. On the other hand, very few have been conclusively rejected. Considering the accessibility of the most reasonable theories to experimental test, the present state of the field is evidently the result of failure to pursue the study of the most promising suggestions.

In view of this situation the profusion of theories on ball lightning from the last two centuries in which the problem has drawn scientific attention demands rather detailed examination. This has already been done to a considerable extent in the three monographs cited previously which reflect the scientific knowledge at the time they were published. Although historic theories with unusual concepts often reappear in modern guise, the residue of evidently reasonable hypotheses in light of present knowledge may be summarized relatively briefly.

The long lifetime of ball lightning compared with that of ordinary lightning has presented a major problem. The length of time which a high-temperature object capable of floating freely in the air can radiate light is several orders of magnitude below the lifetimes reported for ball lightning. For example, a high-energy fire-ball with a diameter of 10 cm would exist for approximately 0·01 s according to estimates from very large fire-balls generated by nuclear bombs and very small ones from focused laser discharges (Kapitsa, 1955; Askar'yan *et al.*, 1967). Two theoretical alternatives have been invoked to overcome this problem: the ordinary decay of such a ball is said to be retarded by special processes, or an external source is proposed to give a continuing supply of energy into the ball.

Suggestions for retarding decay depend on special structures or configurations within the sphere which separate ionic charges and prevent recombination (Hill, 1960; Silberg, 1962). Alternatively, excitation of long-lived radiating species from air molecules has been proposed (Bruce, 1963; Powell and Finkelstein, 1969). Neither of these suggestions for extending the visible life of the fire-balls is adequately supported as yet, by either theory or experiment. On the other hand, there have been a few sometimes informal observations of glowing spheres capable of free motion produced by electric discharges in metal wire (Jones, 1910; Turpain, 1911; Silberg, 1962) and by ignition of low concentrations of fuel gases in air (Barry, 1968), which exhibit the desired long lifetimes. The behaviour of these globes, and in particular their prolonged existence, has not been explained.

The provision of continuing energy from an outside source to maintain the luminous sphere is found in theories which utilize either the direct current of the thunderstorm (Powell and Finkelstein, 1969) or natural high-frequency radio waves produced by a thunderstorm process (Kapitsa, 1955). The power available from a thunderstorm is certainly ample for the maximum estimate derived from the activity of ball lightning. It is not clear from the numerous experiments with d.c. discharges, however, that a single luminous ball with the properties of ball lightning can be generated by such a discharge in air. It has also not been established that sufficient radio-frequency power is present in storms to produce a bright, luminous globe. Recent study of this question is discussed in Section 4.2.

Planté (1875) was an early proponent of the theory that ball lightning is formed in a natural d.c. discharge on the basis of his extensive study of storage batteries and the properties of the discharges they produce under different conditions.

The reports of ball lightning separating from ordinary lightning flashes (Norinder, 1939) gave rise to several theories suggesting how this might occur; for example, by formation of a rapidly rotating vortex at a sharp change in direction of the lightning channel (Meissner, 1930) or by the escape of a high-pressure, ionized jet through the weakened magnetic sheath at a bend in the channel (Bruce, 1963). Thus, the globe would be composed of the same material as ordinary lightning, modified perhaps in the process of formation which also determines the physical shape. Observations of ball lightning rotating rapidly and reports of fire-balls associated with whirlwinds, cyclones and tornadoes encouraged consideration of vortex structures. Continued rotation of the separated sphere in the viscous drag of the atmosphere, for the lifetime associated with ball lightning, presents an additional energy requirement, however, and rapid dissipation would be expected.

Since the earliest concepts of the electrical nature of lightning the presence of electric charges, ions and electrons has been suggested in accounting for properties of ball lightning. Recent knowledge of plasmas, the high-temperature ionic fluids, has been similarly applied. The electric charges have been said to reside on fine droplets or dust particles. Electrically charged ash has been proposed as the cause of volcanic fire-balls (Perret, 1924). Gaseous ions and electrons have also been credited with a role, occasionally in concentrations even greater than those which have thus far been attained in thermonuclear fusion experiments. The energy of ball lightning and the properties of an ionic fluid in spherical form have been studied from the aspect of plasma physics. Unfortunately, the information available from this field has also been unsuccessful in the difficult questions of the physical shape and the lifetime. Recent study of this question is discussed in Section 4.3.

3. Scepticism Toward Ball Lightning

3.1 General Considerations

The history of the study of ball lightning has been marked by contention over the reality of the phenomenon. The unusual properties reported by observers, the failure of methods of systematic study which have succeeded with other forms of lightning and the continued lack of a conclusive theory combined to strengthen doubts that ball lightning exists. From the earliest publications on the subject alternate explanations have been proposed for the observations, including optical illusions or well known natural events, such as meteors or St Elmo's fire (Chapter 2.2). This controversy is similar to the one concerning the origin of meteorites, which was finally resolved in 1803 when a fall of thousands of meteorites occurred in a localized region. After a century during which several noted scientists held a negative opinion on the reality of ball lightning, it appears that in the last decade most meteorologists and perhaps a majority of physical scientists consider the existence of ball lightning well established.

Since the last extensive discussion of the field, however, additional papers supporting the sceptical attitude have been published, including perhaps the most thorough and rational representation of this view by an investigator with long experience in lightning and high-voltage work (Berger, 1973). After thirty years of research at the Mount San Salvatore observatory in Switzerland (Chapter 5), Berger concluded that there was no indication of ball lightning in the thousands of lightning photographs and electric field oscillograms obtained in the studies. The photographs recorded many remarkable lightning tracks, including loops and single bright points; but he stated that these and all published photographs of ball lightning can be otherwise explained. He rejected the theories which have been presented to account for the glowing spheres on grounds that conditions necessary for the proposed processes do not occur, even in an intense lightning discharge, and the properties reported for ball lightning, particularly its long lifetime, cannot reasonably be expected from the proposed models. Berger concluded that the explanation of ball lightning observations would probably be found in physiological study of after-images and the optical effects of short duration, intense lights.

3.2 Photographic Evidence

The most recent edition of the "Encyclopaedia Britannica" contains the statement that no authentic photographs of ball lightning exist (Orville, 1974), in agreement with Berger and duplicating the comment of Brand made fifty years earlier. The assertions by Brand and the "Encyclopaedia Britannica" extend only to the subject of photographs, however, and not to the existence of ball lightning itself.

Concerning the well-known "fireworks" photographs of Jensen (1933), Berger records the story of Norinder, who narrated his investigation of Jensen's work during a visit to Lincoln, Nebraska (U.S.A.). He learned from Jensen's assistant that Roman candles had been fired during a storm to mislead the physicist. Jensen published the pictures, identifying them as ball lightning. Berger cited comparison of Jensen's photographs with those of firework displays in support of this story.

It is significant that Norinder himself presented the photograph of an oval ball lightning by Schneidermann (Fig. 1) as authentic.

A photograph somewhat similar to Jensen's in which sparks are being emitted from the glowing mass was published by Kuhn (1951), Fig. 2. According to Berger, Schwenkhagen, then chairman of the German Committee on Lightning Rods, visited the site of the event and concluded that the storm had caused a short circuit on a power line in the location. The spraying, molten metallic drops which were formed were mistaken for ball lightning. The photographer who also observed the occurrence has not accepted this explanation, repeating his belief that ball lightning was indeed involved. Berger contributed to this discussion his own observation of a very rapid lightning flash which generated a long lasting flame on a 10 kV power-line mast. While flashover or power arcs are liable to occur, the assertion of reasonable alternatives which might have taken place is inconclusive without the support of specific information on the power aspect at the time or distinct traces remaining on the tower from the incident. The generation of a hot metal globe by an ordinary lightning flash has, indeed, been proposed to explain ball lightning. The limited knowledge of such globes is responsible in part for failure to confirm this theory.

These two photographs are evidently authentic ball lightning snapshots supported by direct observation.

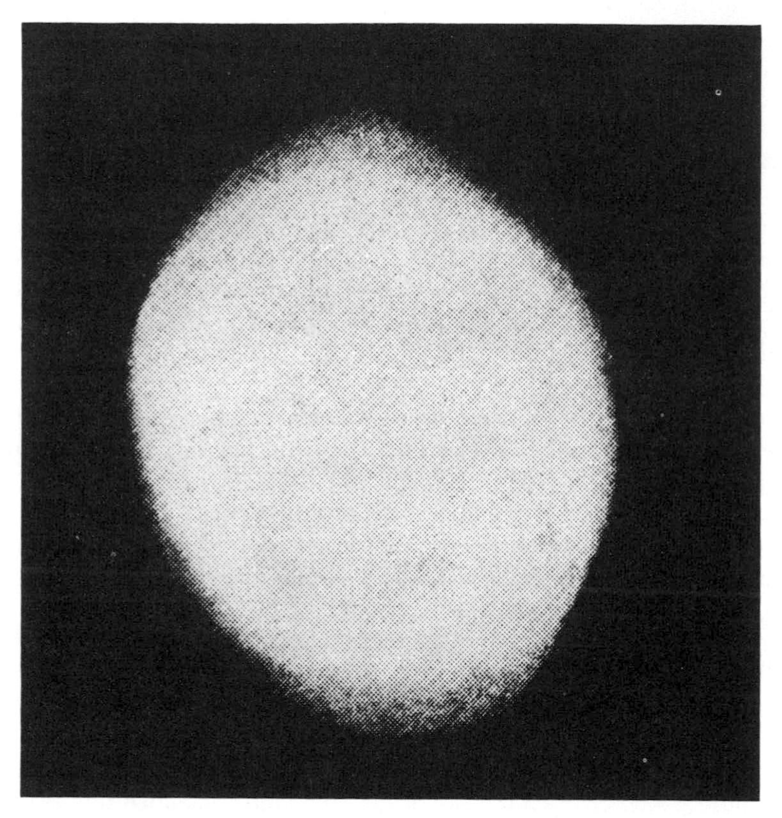

Fig. 1. Schneidermann's ball lightning photograph (Norinder, 1939).

On the other hand, a photograph of a presumed ball lightning path in which the bright track is broken into a regular dashed line (Jennings, 1962) was found on investigation not to show the object of interest at all. Photographs with this characteristic aspect which are not supported by observation were previously considered doubtful (Singer, 1971). Careful analysis of this photograph led to the conclusion that the broken light path was given by a sodium vapour street lamp photographed with an unsteady camera (Davies and Standler, 1972). The investigators conceded that the tremendous explosion reported by the photographer was not explained by this analysis.

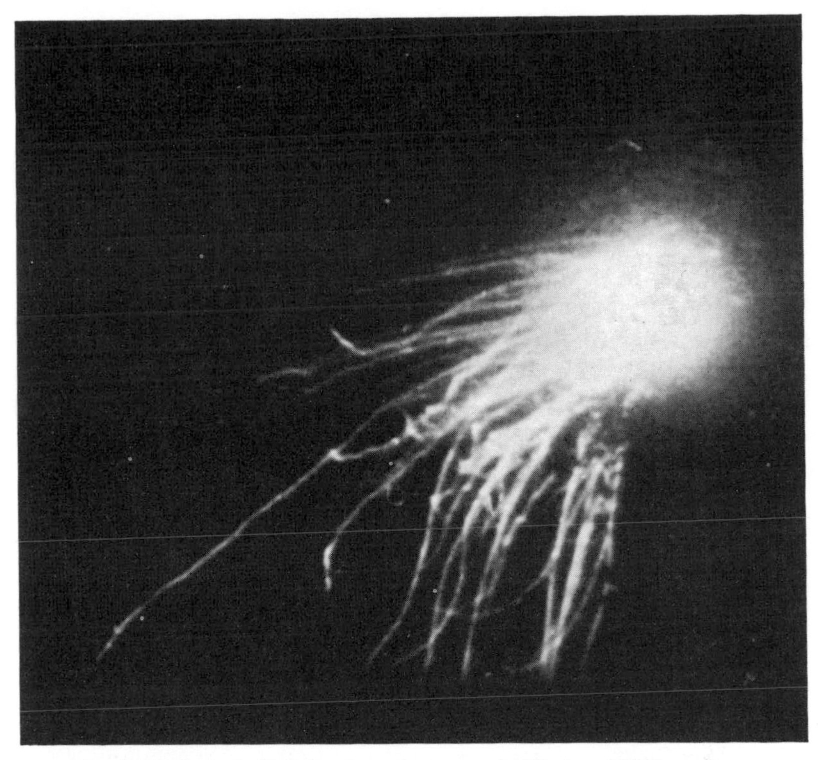

Fig. 2. Kuhn's ball lightning photograph (Kuhn, 1951)

3.3 Optical Illusions

The contention that visual after-images resulting from short duration, intense lights account for reports of ball lightning has been renewed (Argyle, 1971; Berger, 1973; Mortley, 1973). A suggested cause of the after-image is the concentrated, intense light produced by ordinary lightning striking a potentially incandescent material. Intense lights are associated with positive after-images (negative images have also been suggested). Optical effects have been credited with many of the characteristics reported for ball lightning including the extended lifetime ending in sudden disappearance as well as the motion and colour. The reported odours and sounds are ascribed to psychological association with the illusion. The disappearance with a soft pop of the fire-ball on the wing of an airplane (Felsher, 1970) would be in accord with this suggestion, but a psychological effect would seem to be ruled out in the case of the globe which disappeared with a loud bang outside a closed door, causing the occupants of the room who could not see the ball to come out and inquire about the source of the noise (Wagner, 1971). A number of illustrations of this type are found in the literature; these have been selected only from recent reports presented in Section 1.

Initial qualitative work with bright artificial lights is reported to give some effects supporting the theory of optical illusions (Argyle, 1971).

Exposure of human subjects to neutrons has produced star-like flashes and short streaks similar to the effect of cosmic rays on astronauts during lunar flights (Budinger *et al.*, 1971).

As in previous discussion of this question which has extended over a lengthy period, there are several rational replies to the arguments supporting the non-existence of ball lightning; and in the recent literature a number of these have been given. One observer emphasized his certainty that the ball lightning he reported was not an illusion and remarked from his knowledge of ophthalmology that after-images never disappear instantaneously (Galton, 1971). Another eye-witness noted several aspects of the fire-ball he saw close at hand which were inconsistent with its being an after-image (Jennison, 1971). The ball moved from 2 m to within 50 cm of him. He pointed out that ball lightning passes behind objects and disappears from view. It obscures those which it passes in front of. There is a marked change in its angular diameter to the eye on close approach, as is necessary for an object of constant size. These properties are not characteristic of an after-image.

A university lecturer in ophthalmic optics pointed out that after-images fail to account for the constant size during motion, constant colour, complete disappearance without recovery after blinking, as well as the sounds and odours often reported with ball lightning (Charman, 1971, 1972).

Another author has remarked that there are numerous potential sources of after-images at all times, but the particular association of ball lightning with thunderstorms is inconsistent with the role of these optical illusions proposed by some investigators (Davies, 1971). Davies objected further that efforts to classify ball lightning as an illusion represent an undesirable philosophy in science, the rejection of difficult problems in nature by denying the existence of the phenomenon, providing a physiological or psychological explanation, and ignoring physical evidence which contradicts the explanation.

3.4 Ball Lightning Versus other Atmospheric Phenomena

The consideration of ball lightning by the distinguished scientist F. A. Paneth (1956) is of interest in this discussion. Paneth wished to discern whether the increase in number of observations in the records of meteorite falls over a period of 140 years represented a real change in frequency of occurrence.

Certain human factors which change over the span of time and are difficult to assess presumably affect the observations, such as the change in population and conditions of human existence.

If the occurrence of ball lightning in nature were essentially constant, the change in number of ball lightning observations over the same period would afford a measure of such factors with which to compare the record of meteorites. Paneth noted that observations of ball lightning and of meteorites are similar in several respects: the element of surprise, the possibility of danger and the rarity of occurrence. The last factor results in observers who are seldom scientists forcing us "to rely on the narratives of common folk".

Paneth examined and strongly criticized Humphreys' study of 280 personally collected ball lightning reports (1936) with regard to treatment of the material, which was loosely re-edited, and systematic, unfounded misinterpretation. Humphreys dismissed many of the observations as optical illusions or explained them in terms of alternative and often unique natural events.

Table I

Frequencies of meteorites and ball lightning

Years	No. of meteorite observations	No. of ball lightning observations
1800–09	17	1
1810–19	21	0
1820–29	23	3
1830–39	22	3
1840–49	32	4
1850–59	33	7
1860–69	53	13
1870–79	49	8
1880–89	40	47
1890–99	46	50
1900–09	52	36
1910–19	55	30
1920–29	68	—
1930–39	72	—

Paneth then turned his attention to Brand's critical survey of ball lightning observations with which he compiled Table I. He concluded that ball lightning observations provide the desired measure of human factors affecting the meteorite record under study. If meteorite falls were constant over the time span shown, the frequency of meteorite observations should vary in the same manner as observations of ball lightning. A comparison of the meteorite and ball lightning observations in Table I shows that the increase in the interval 1880–1919 over 1800–79 is far less in the case of meteorites. Paneth concluded that there was a definite decrease in the actual number of meteorite falls since the first half of the nineteenth century.

The occurrence of ball lightning has been used to explain a number of the observations referred to as unidentified flying objects. The existence question, as indeed all the information concerning the latter, is in strong popular dispute. Much of the controversy concerns questions quite similar to those which have been debated for almost two centuries on ball lightning, and it appears that both sides are unacquainted with the existence of this precedent. Many of the questions are familiar to those who have considered the problem of ball lightning, as are the results of the debate. The similarities of the two subjects in this respect do not extend to the nature of the observations or the information they provide. A definitive discussion of this issue which concludes that unidentified flying objects do not exist has been presented by Jones (1968).

4. Recent Studies of Ball Lightning Theories

4.1 General Survey

In the five years which have elapsed since the last comprehensive review of this field there have been a number of publications on the major theories of ball lightning, and a few novel explanations have been proposed.

The species, processes and energy which would be associated with recent theories have been examined by Smirnov of the Kurchatov Atomic Energy Institute (1975). He concluded that only chemical reaction can explain the extended lifetime of ball lightning if external sources of energy are excluded. Table II summarizes his results.

4.2 Kapitsa's Radio-frequency Theory

Favourable indications for the radio-frequency theory were obtained in plasmoids generated by a radio-frequency discharge, which remained visible for 0·5 to 1 s in the laboratory apparatus after the radio frequency power was shut off (Powell and Finkelstein, 1969).

Meteorological expeditions in the U.S.S.R. searching for appropriate radio-frequency radiation in nature, as proposed by Kapitsa (1955), continued their study of decimetre radiation emitted from lightning (Kosarev and

Table II

Properties of ball lightning models

Type of ball	Source of internal energy	Type of decay	Typical decay process	Relaxation decay constant ($cm^3\ s^{-1}$)	Particle density for ball decay in 10 s (cm^{-3})	Internal energy for decay of 20 cm ball in 10 s (J)
Plasma, ions and electrons	Recombination of charged particles	Dissociative recombination	$e + O_2^+ \rightarrow O + O$ $e + N_2^+ \rightarrow N + N$ $e + NO^+ \rightarrow N + O$	2×10^{-7} 2×10^{-7} 4×10^{-7}	5×10^5	10^{-8}
Plasma, positive and negative ions	Recombination	Three-body recombination	$O_2^- + O_4^+ + O_2 \rightarrow 4O_2$ $NO^+ + NO_2^- + N_2 \rightarrow NO + NO_2 + N_2$	$1{\cdot}6 \times 10^{-25}\ cm^6\ s^{-1}$ $10^{-25}\ cm^6\ s^{-1}$	3×10^4	10^{-9}
Ions of one type	Electric field of charged particles	Ion separation by action of field	Mobility of large ions greater than $0{\cdot}1\ cm^2\ (V\ s)^{-1}$		10^7	10^{-7}
Excited gas, atoms or molecules in upper electronic levels	Electronically excited atoms	Transition to lower level	$O_2(^1\Delta_g) + O_2 \rightarrow 2O_2$ $O_2(^1\Sigma_g^+) + N_2 \rightarrow O_2 + N_2$ $N_2(^3\Sigma_a^+) + O_2 \rightarrow N_2 + O_2$ $O(^1S) + O_2 \rightarrow O + O_2$ $O(^1D) + O_2 \rightarrow O + O_2$	2×10^{-18} 2×10^{-17} 4×10^{-12} 3×10^{-13} 5×10^{-11}	Lifetime normally $<0{\cdot}1$ s	
Excited gas, vibrationally excited molecules	Vibrationally excited molecules	Energy emitted by vibrational de-excitation	$N_2^* + N_2 \rightarrow 2N_2$ $O_2^* + O_2 \rightarrow 2O_2$ $N_2^* + CO_2 \rightarrow CO_2^* + N_2$	$\sim 10^{-19}$ 10^{-17} 6×10^{-15}	Lifetime normally $<0{\cdot}01$ s	
Chemical	Chemical reactions	Depletion of reacting species	$2O_3 \rightarrow 3O_2 +$ 69·4 kcal mol^{-1} $2NO_2 \rightarrow N_2 + 2O_2 +$ 16 kcal mol^{-1}	4×10^{-20} at 400 K —	$<2 \times 10^{19}$	3×10^4 *a* 6×10^3 *b*

a: Concentration of ozone; *b*: Concentration of NO_2.

Serezhkin, 1974). Contrary to other measurements which generally suggest a smooth, quasi-continuous spectrum in this high-frequency region, the investigators from the Institute of Physics Problems in Moscow reported narrow bands at 0·1 to 0·2 GHz. These bands depart from the regular variation in power with the reciprocal of the square of the frequency υ^{-2} they observed at longer wavelengths. Measurements were made in the Lenin Hills of Moscow which agreed approximately with earlier studies in Pushkin Pass in the Armenian S.S.R. The present work utilized an improved receiver. The microwave radiation is observed 0·1 to 0·4 s after the lightning leader. The longest duration was 0·17 s. Only 10^{-9} to 10^{-8} W of power were measured at a distance of 1 km from the lightning. The authors suggested that the radiation is generated in the lightning channel by transverse plasma resonances. These occur when the magnetic field of the lightning stroke amplifies magnetic Bremsstrahlung and Cerenkov radiation from electrons in the channel. Measurements closer to the discharge facilitated by artificial stimulation of lightning were proposed to aid detection of higher intensities of the narrow-band radiation sufficient to account for ball lightning.

Kapitsa's (1969) work was the basis of his theory of ball lightning and provided information on properties of the discharge with which the theory is concerned. The goal of the investigation, however, was the production of high-temperature plasmas applicable to thermonuclear reactions. Using high-power sources, spherical and filamentary discharges with dimensions of the order of 10 cm were formed by up to 20 kW at pressures of several atmospheres. Rotation of the gas stabilized the discharge. The plasmoids contained a central hot core with an electron temperature of the order of 10^6 K surrounded by a partially ionized cloud with a temperature somewhat below 10^4 K.

A theoretical study of diffusion in high-frequency gas discharges of this type (Meyerovich, 1972) indicated that electron attachment to atoms producing negative ions would be a significant process. In addition, the presence of negative ions leads to more rapid recombination. The plasma temperature was assumed uniform in this study, the difference in temperature between electrons and ions being neglected. The surface resistance calculated by Meyerovich differed greatly from that derived from experimental measurements in Kapitsa's work.

The suggestion that a standing wave of electromagnetic radiation produces ball lightning has been renewed by Jennison (1973).

The structure of ball lightning as a microwave-filled radiation cavity suggested by Dawson and Jones (1969) has been considered further by Trubnikov (1972). The current of repeated strokes in a flash of ordinary lightning is said to flow mainly in the outer portion of the lightning channel.

This plasma envelope reflects sub-millimetre wavelengths and confines them within the channel where they accumulate. Sausage instabilities appear, cutting off a segment of the channel. The stable vertical position of the ball and its horizontal motion result from the momentum of cold air entering the sphere and being ejected upward or to the rear after heating. The internal microwaves maintain the cavity by ionizing external gases as they approach the envelope, producing a thin shield which electrodynamically ejects the ions formed. The momentum of this process increases the effective external pressure and thereby the energy stored in the cavity.

An internal field supporting the ball was ascribed to a rotating electric dipole produced by misalignment of a descending negative lightning leader with rising positive charge (Endean, 1976). The outward electromagnetic pressure requires a peripheral field velocity greater than the speed of light. A ball of 10^7 J can be rationalized using 10% of the energy of a typical lightning discharge and 10% of the leader charge. The instabilities noted in plasma confinement experiments were dismissed on the grounds that the field is within the favourable curvature rather than external and that the rapid rotation of the magnetic boundary field gives high stabilizing shear. This ball is an independent entity transporting its own energy.

Berger (1973) made the apt suggestion that if Kapitsa's theory were correct ball lightning should be seen especially in the neighbourhood of short-wave transmitters, but this has never been reported. Very short waves not in general use are required according to the theory, however, to provide energy by resonance with a relatively small discharge region. The usual transmitters give radio waves of appreciably greater length. Such waves presumably disperse their energy over a relatively large space and thus fail to provide the localized ionization required for a visible plasmoid.

4.3 Plasma Properties

The repeated consideration of ion and electron assemblies in explaining properties of ball lightning indicates that the plasma state may be significant in the problem.

The existence of plasma satellites of the sun was suggested from radio-telescopy at wavelengths of 1·5 and 6 m (Vitkevich, 1962).

Spherical forms were obtained in equilibrium solutions of theoretical toroidal plasmas in a vacuum (Morikawa and Rebhan, 1970). Hydromagnetic analogues of Hill's vortex were noted as in previous plasma theories.

The reported rotation of ball lightning, its association with whirlwinds, and the anticipated heuristic benefit of rotational currents in theory encourage study of vortex models. High-voltage discharges to a plane produced a small tornado-like vortex (Ryan and Vonnegut, 1970). With rotation of the gas by an impeller a stable discharge could be maintained (Ryan and Vonnegut, 1971).

The storage of additional energy in ball lightning in the form of increased ionization was suggested by Hill (1970) in order to exceed the limit imposed on a plasma structure by the virial theorem, which refers to the sum of energy stored in electrical, magnetic and internal (kinetic) form. For ball lightning atmospheric pressure is commonly considered the only force at hand for confinement, and this severely limits the maximum energy. Powell and Finkelstein (1970) pointed out, however, that recombination would occur very rapidly if concentrations were greater than permitted by the temperature equilibrium; and if the temperature were high enough to increase ionization appreciably, the energy would also be rapidly radiated, as noted by Kapitsa (1955).

A theory describing a new strong interaction in assemblies of charged particles has been presented, in part to overcome the difficulty of explaining ball lightning with existing electromagnetic plasma theory (Bergström, 1971). The virial theorem is not applicable in the usual fashion to certain bodies of discrete charges, such as ionic crystals. Bergström commented that the existence of these would be ruled out, as are self-confined plasmoids, if the thermal energy were not much smaller than the Coulomb energy. This theory proposes a strong dielectric-diamagnetic attraction (in place of the magnetic confinement of plasma physics) which overcomes conventional electric charge repulsion with a strong short-range interaction. In previous nuclear theory electric polarization was described, arising from charge separation in a quasi-neutral medium. The interaction can be viewed as an action of Le Chatelier's principle, the compensation by a system for a change so as to reverse or diminish the change; for example, when an electric charge is accelerated by an electric field, an electric field opposite to the original is produced by electromagnetic induction. There is symmetry of the electric and magnetic fields with respect to the attraction. The range of the interaction is small compared to a Coulomb field because of an exponential factor in Yukawa-type equations which are derived with the Maxwell equations. The electromagnetic field is given a quantum structure, which introduces an attractive dielectric potential energy in a charged cloud.

In order to apply to ball lightning, the strong interaction must operate on a greater than microscopic scale. Bergström suggested that the intense currents and fields present in a storm can generate a small region with positive charge by depletion of the electron concentration (utilizing the far greater electron mobility). The strong interaction is produced in response to acceleration of the ions in the dense, positively charged cloud. The energy estimated from this model using a radius of 5 cm and an excess positive charge of $6{\cdot}5 \times 10^{16}$ electron charges is 2×10^7 J. This value representing the total electrostatic energy of the ball is of the same order of magnitude found in some reports of ball lightning.

The globe depicted by this theory suggests that ball lightning may be produced experimentally by intense ionization of air inside a metal tank which depletes the electrons and contains the gases until the excess charge required to effect the strong interaction appears. Bergström compared this process with the reports of ball lightning formation in aeroplanes.

4.4 Nuclear Theories

The occurrence of nuclear reactions in ball lightning has been suggested to account for the large energies reported in a few cases, although the manner in which conditions suitable for such reactions would appear in the atmosphere has been difficult to explain.

A physical process similar to the one considered by Bergström (1971) but using only conventional forces was proposed to reduce the external magnetic field needed for confinement of a thermonuclear plasma (McNally, 1972). The process would presumably be effective in confinement of ball lightning or of a thermonuclear reaction in a magnetic bottle. The nuclear reaction is depicted as ejecting positive ions out from the core, leaving electrons in a central negative zone with an excess charge. The ions from the fusion reaction are retarded by the resulting electric field and turn into the outer ion current layer. In a thermonuclear plasma device an external magnetic field confines this sheath, but the ion current contributes to the field. In this way electrostatic forces aid magnetic confinement. The parameters which would give the desired effect in this configuration were estimated. An electric field of 10^4 V cm^{-1} holds a 1 MeV proton in a circular orbit with a radius of 200 cm. If the field increases linearly with the radius, as in uniform net charge distribution, from the centre to the 200 cm radius, the core potential is 1 MV below that at the centre of the ion current layer. For a smaller configuration the electrostatic field would be larger, but the core potential would be approximately the same. If the ion current were approximately 3×10^{14} ions cm^{-3}, an average excess charge of only 10^7 to 10^8 eV cm^{-3} would give the electrostatic field of 10^4 V cm^{-1} needed to bind 1 MeV protons in a radius of 200 cm. A circulating proton current of 2×10^{13} cm^{-3}, about 7% of the plasma density, would give a 100 kG field.

Isotopic abundance measurements following ball lightning (or in its absence) during thunderstorms or tornadoes were suggested to obtain an indication of the occurrence of nuclear reactions (Altschuler *et al.*, 1970).

The reaction of protons accelerated by storm processes to 1 MeV with common ^{14}N and ^{16}O nuclei in air would produce ^{15}O and ^{17}F, respectively. The positron emission of these radioactive species, according to the theory, accounts for the energy and appearance of ball lightning. The radioactive decay forms ^{15}N and ^{17}O, increasing their natural abundance.

The absence of observations of lethal radiation from ball lightning has been cited against such theories (Dmitriev, 1973). In one incident a bright globe passed 2 m from an observer without producing high radiation exposure of film in his possession, although a scintillation radiometer nearby gave a reading (Dmitriev, 1969). The scintillometer reading was attributed to an incorrect response to radio-wave emission from the fire-ball.

In place of determination of the heavy nitrogen and oxygen isotope abundances, radiation was measured over a period of 12 months (Ashby and Whitehead, 1971). Four cases of high radiation were observed. One occurred at the height of a severe thunderstorm with much rain, but the remaining three took place on two consecutive days without storms. The counting rates on the detectors increased by ten- to fifty-fold for only a few seconds, far less than the half-lives of 124 and 66 s of the radioactive species ^{15}O and ^{17}F.

Hill and Sowby (1970) pointed out that the radiation dose from the energetic nuclear ball lightning on a body 2 m distant would be 175 rad s^{-1} from ^{15}O and 325 rad s^{-1} from ^{17}F. In addition to lethal radiation effects on close observers, solid objects exposed to such fluxes should exhibit thermo-luminescence. A study was made of thermoluminescence in a fragment from the steeple of St George's Church in Leicester, which was sundered in a violent thunderstorm in 1846, possibly by ball lightning (Mills, 1971). There was no increased luminescence observed compared to samples from other stones of the same material obtained in the churchyard. However, an emission of light at 110 °C produced by cobalt-60 irradiation of the same material was undetectable after 3 months. Electron spin resonance also give no indication of radiation effects. Examination of the masonry within a relatively short time after exposure to ball lightning would evidently be needed to provide a significant test with this method.

Another study of thermoluminescence was made with brick samples from a doorway through which ball lightning passed (Fleming and Aitken, 1974). A more recent occurrence was involved in this case, but the lapse in time before the measurements was evidently a few years (1970–74). The average equivalent dose measured carefully in control samples distant from the path of the ball was 13 rad, and the samples directly in the doorway averaged 15·5 rad. The difference from the controls was considered not significant. If the slight increase were attributed nevertheless to radiation from the ball lightning at a distance of 2 m, the radioisotopes proposed would be found in a ball with less than 10^5 J energy.

Ashby and Whitehead (1971) suggested that ball lightning with high energy is formed by very small particles of antimatter, meteoritic dust from the upper atmosphere, rather than by radioisotopes. A particle of antimatter 5 μm in radius with a mass of 5×10^{-10} g would liberate 10^5 J on annihilation. The process might occur by concentration in a thunderstorm of antimatter dust with a negative charge from photoemission of positrons and secondary emission from recoil fragments. The negatively charged dust joins the normal negative current flow to ground of thunderstorms. A barrier against instantaneous reaction of antimatter with ordinary matter is required. To produce ball lightning at high frequency, 0·1 to 1·0 times that of ordinary lightning, 20 to 200 g year^{-1} of antimatter must enter the atmosphere, if the average energy of the globes is 10^5 J and 1% of the incoming antimatter actually reacts to generate the fire balls. This quantity is approximately 10^{-11} of the mass of meteoritic material reaching the earth, and the energy available from it is approximately the same as that from cosmic rays. The radiation pulses in the year-long study mentioned previously would be produced by 10^{-11} g of antimatter 500 m distant from the detector according to this speculative suggestion. The obstacle which the short pulse times present to the theory of radioisotopes produced by nuclear reaction would thus be avoided.

Jennison (1971) objected that the antimatter theory cannot explain the suspension of ball lightning a few feet from the ground, its entrance into aircraft or its size and shape. He concluded that the theory would require a number of larger bodies composed entirely of antimatter in the solar system.

Crawford (1972) proposed that the radiation observed by Ashby and Whitehead (photon fluxes with energy corresponding to positron annihilation) originates in cosmic rays. The observations which were not related to thunderstorm activity (three out of four) are thus more readily understood. If cosmic rays are the source of ball lightning, however, the number of incidents should increase with solar flare activity, a relationship said not to exist (Dmitriev, 1973). Two additional objections have been presented to identification of the radiation pulses reported by Ashby and Whitehead as extensive air showers from cosmic rays (Cecchini *et al.*, 1974). The duration of the radiation indicated by the observed pulses is orders of magnitude (at least 10^3) greater than the slowest conceivable process. And the number of events with the observed energy should be much lower (approximately 2%) than the four reported, considering the distribution of cosmic rays and the geometry of the detector.

4.5 Miscellaneous Characteristics

A presumed photograph of a ball lightning path (Dmitriev, 1970) has been studied photometrically, and the variation in film density has been interpreted in terms of the source of light from the ball (Dmitriev *et al.*, 1972). According to the photograph the brightness of the centre of the path changed irregularly and sharply as the ball moved. The brightness across the track from one boundary to the other increases to a maximum at the centre and decreases rapidly with distance from the centre to the two borders. Dmitriev argued that the brightness should be uniform over the whole diameter if light is radiated only from the surface of the sphere, as expected from the fuel combustion model. If ball lightning is a plasmoid, on the other hand, light is emitted from the inner volume; and the intensity should vary in the manner shown by the photograph.

The existence of a strong magnetic field from ball lightning has been discussed by Blair (1973). Marked disturbance of ships' compasses and the restraint of motion of church bells in the presence of fire-balls support the concept, which would be in accord with proposed rotational and plasma vortex structures. The magnetic field necessary to stop the swinging of bells reported in an incident from 1811 was estimated at 150 G. The calculation was based on the current generated by motion of the bells, which were idealized as metal discs, in the external field given by the ball. The field indicated is the one which causes a force on the bells equal to the force applied in pulling them. An eye-witness objected that metallic objects in his pockets did not exhibit any response when ball lightning passed within 50 cm of him (Jennison, 1973).

The flow of water at high velocity through a nozzle produces a glow discharge which was likened to ball lightning (Koldamasov, 1972). An abrupt expansion was used, resulting in strongly pulsating cavitation. The flow was viewed as producing charge separation in the liquid and ionization at the chamber wall by secondary processes. Recombination of electrons at the wall gave a continuous glow with a colour dependent on the wall dielectric; asbestos cement gave a rose colour; Plexiglas caused a yellow light; and ebonite gave blue.

The spherical structure of ball lightning and its motion, in particular its passage through solid surfaces, have been interpreted as evidence of the ether, the theoretical universal medium for electromagnetic phenomena which fell before the test of the Michelson–Morley experiment (Aspden, 1973).

4.6 Coherent Radiation Theories

The agency of an atmospheric laser in generating ball lightning has been suggested in several theories. In one the laser pulse is described as striking a solid surface to produce a plasma vortex (Wooding, 1972). Laser pulses have been obtained in air at atmospheric pressure by a discharge tube using current pulses comparable to a lightning discharge but of much shorter duration. In view of the large energy of a thundercloud a low efficiency of conversion to coherent radiation could still suffice to provide the energy associated with ball lightning. Plasmas which have been formed by relatively modest laser energies (50 J) were cited in support of the theory.

Stimulated sub-microwave "maser" emission in the atmosphere has been investigated as the source of ball lightning by Professor P. H. Handel, University of Missouri (U.S.A.). Large volumes of the atmosphere (10^2 to 10^6 m^3) with inverted population of energy levels are depicted as active in the provision of maser radiation.

4.7 Formation of Ball Lightning by an Earth Flash

The frequent association of ball lightning with ordinary lightning in observations has provided a basis for numerous theories and experiments concerned with lightning flashes which produce ball lightning or conversion processes by which the common form is transformed into a fire-ball. Particularly significant aspects of the possible relationship are considered in recent publications.

Experimental study of high current discharges reproduced some phenomena of interest in lightning observations (Minin and Baibulatov, 1969). Intensified afterglows of extended duration were given by discharges of up to 1 kA with current supplied for less than 0·5 ms. Bead lightning is associated with multiple lightning strokes. The localized bright spots may last up to 2·5 s, but the afterglow is brightest some time after the strongest first stroke. The suggestion that greater luminosity appears at bends in the lightning channel as a result of radiation from a larger volume of the channel is contradicted by experiments with spark discharges. In the work of Minin and Baibulatov (1969) the afterglow in air was brightest 4 to 7 ms after the discharge current stopped. Air, nitrogen and carbon dioxide gave long afterglows; inert gases did not. At slightly reduced pressure (0·3 atm) sausage instability occurred. In a few experiments a ring formed on the plasma tail, and in one a chain appeared along the whole channel. The changes in the physical form of the discharge were ascribed to surface forces from an electric double layer at the plasma boundary with effects similar to those of surface tension on a liquid jet.

A theoretical model has been presented describing ball lightning as a sphere which is produced from a lightning discharge when water is present to solvate the ions (Stakhanov, 1974). The water surrounds both positive and negative ions and hinders recombination, accounting for the long lifetime as in earlier theories. Stability against hydrodynamic oscillations in the surface of the ball requires a density very close to that of the surrounding air; this is provided by a specific amount of water. Surface tension maintains the plasma structure. The surface tension is estimated at 1 to 10 $erg\,cm^{-2}$ from values for liquid metals and solutions of strong electrolytes in water. Stakhanov (1974) suggested experimental study by injection of water into a high-current plasma-pulse discharge.

Crew (1972) described another process involving water. With sufficient humidity the drop in pressure between the compression at the lightning channel and a compensating reduced pressure surrounding it forms droplets. Radiation pressure from the discharge vigorously repels the droplets. If there is a sonic shock, they shatter. Charge separation results. The negatively charged smaller drops are driven further from the lightning channel, resulting in a negative sheath surrounding the positive core. Long-duration flashes have a marked effect, as the distance of the drops from the channel increases with the square of the time.

An incomplete discharge of ordinary lightning which fails to reach ground may form ball lightning, as Planté (1875) suggested from his early studies of d.c. discharges. Low-frequency oscillations appear in the channel (Kozlov, 1975). Only the lowest advancing front of the discharge where the field is greatest is visible. The Maxwell equations applied to the relaxation of the oscillations using reasonable parameters for the partial discharge gave estimates of the radius (6 cm to 3 m) and lifetime (0·5 s to 9 min) of ball lightning depending on cloud charge. The relaxation oscillations of 3 to 30 Hz agree with the frequency of multiple strokes in lightning (Chapter 5.2.2). The temperature is 300 to 500 K, and 50 kW of power are liberated for motion of the ball.

Brovetto *et al.* (1976) described a globe of positive charge extracted from a ground prominence by the approach of a negative step leader of ordinary lightning. A ball of significant mass forms, as each ion extracted is surrounded by many neutrals. An internal vacuum results if the normal discharge is interrupted. Ion coulomb forces support the sphere against atmospheric pressure. The globe can float freely past other ground objects which have like charge induced by the same initiating field. The upward thrust of the field on the ball, however, may be as much as five times its mass estimated with the ion-neutral clusters. A ball of 10 cm radius and $8{\cdot}36 \times 10^{11}$ electron charges per cm^2 (total charge of $1{\cdot}68 \times 10^{-4}$ C) may have 1,700 J energy from the pressure and the electric energy (the latter being three times the first).

Formation of a rotating electric dipole by a descending step leader misaligned with upward surging positive charge has been discussed on page 423.

Webster and Hallinan (1973) found vortexes in electron sheets when charge density exceeds a threshold. Both charge-sheet and current-sheet instability are possible although experimental evidence has been reported as yet only for the first. A typical mechanism involves charge flow along the sheet wherever it is bent, in response to displacement of the potential minimum. The phenomenon was used in explanation of auroral vortices of large extent, e.g. 2 km; but the laboratory vortices were, of course, much smaller in scale, closer to dimensions associated with ball lightning.

The formation of ball lightning as an electrostatic vortex in a comparable mechanism was based on analogy with the hydrodynamic vortex formed by a liquid drop falling in a stationary liquid (Voitsekhovskii and Voitsekhovskii, 1974). The distribution of charge from a thunderstorm or tornado is compared to the unstable system of a heavy liquid above a lighter one. Excess charge density and the corresponding increased electric field in protuberances favour additional deformation as in the heavy liquid penetrating into the lighter. When the field is sufficient to ionize air, a glow discharge appears. The energy in such a ball is limited. If the electric field decreases gradually as the vortex moves, the ball disappears quietly. If the field suddenly increases, a spark discharge can occur producing the apparent explosion of the ball. Proximity to a conductor can especially provide conditions for the latter process. The major distinction between ball and normal lightning according to this picture is that in the first, regions of air are set in motion by electrostatic forces which discharge gradually during the incident. In ordinary lightning a rapid discharge takes place. The high velocity of ball lightning descending from the clouds compared to its moderate velocity in motion close to earth is in accord with the greater change of potential in a vertical direction in the atmosphere compared to that in the horizontal direction. A velocity of 20 m s^{-1} was estimated for the vortex sphere using the maximum electric field permitted in the atmosphere by the breakdown potential of air, 30 kV cm^{-1}, atmospheric density and the gas dynamic resistance to motion of the vortex, which is equal to the electric force on the same vortex surface.

Another theory invoking fine particles of metal, soot or ash produced by a preceding flash of lightning has been presented (Zaytsev, 1972). The conducting, possibly charged particles rise and form a core which focuses the current flowing from cloud to earth. The displacement current of potential ordinary lightning flowing against the resistance of the core particles heats them, giving a visible glow. The visible region appears as ball lightning to the chance observer. The leader current which follows can heat the core sufficiently to cause a thermal explosion. The currents feeding into the core are largely invisible so that only the bright core appears significant to an observer. Motion of the core up or down results from variation in its charge with the currents and an accompanying change in position in the cloud-to-earth field, as in small discharges moving between charged capacitor plates. The ball disappears when the particles have burned completely. A normal earth flash may occur through the same path.

Quantitative consideration of certain radiation properties of luminous spheres, containing metal or carbon in addition to air, heated by lightning had been carried out previously (Lowke *et al.*, 1969). In the absence of the continuing current invoked by Zaytsev, thermal radiation alone would give very rapid cooling and a marked drop in luminosity in a very short time. The use of external currents to maintain the visible glow presents a difficulty in explaining the existence of ball lightning inside buildings. Chemical reactions such as oxidation of the extraneous substances were suggested as a source of additional energy which might provide visible light of uniform intensity over a longer period of time.

Recent observations have been presented as indicating a direct association of ball lightning with normal lightning. Photographic studies were made of lightning triggered by metal wires carried to a height of 700 m by a rocket (Fieux *et al.*, 1975) (see Chapter 5.4.4.3). Bead lightning was often seen following the more intense flashes of lightning. The discharge channel, which is straight at first, becomes more bent with time, and larger beads appear at the bends. These remain visible after the smaller beads have disappeared. The beads are initially about 40 cm in diameter and are visible up to 0·3 s. Sometimes one or two of the larger globes are especially persistent. They float upward with a velocity of 1 to 2 m s^{-1}. In one experiment, a ball, rising from its origin half-way up an experimental tower 24 m high after a lightning discharge lasting 0·66 s, was not affected when another stroke of very brief duration struck the top of the tower. This evidently rules out the action of external currents or fields upon the existing globe. The vaporization of the triggering wire in the discharge was considered. Vaporization is complete when the arc across the gap from the lower end of the wire, which breaks loose as the rocket reaches its height, to its grounded reel reaches a length of approximately 5 m. During current flow in the discharge the lower few metres in this gap between the end of the rising wire and the reel are not as bright as the higher region where there is metal vapour. This difference disappears early in the glow which continues after the discharge current ceases. The decay in both regions occurs at the same rate. Fire-balls exhibiting fully the properties of ball lightning, such as extended horizontal motion, were not reported in this work.

Photographs of approximately 120,000 lightning discharges obtained in the Prairie Meteorite Network have revealed a few images which apparently show ball lightning separating from the end of an ordinary lightning stroke and descending towards the earth. The number of bead lightning events is an order of magnitude larger. The photographs are obtained by time exposures of the night sky with periodic closure of a shutter to provide timing in the sequence of events. Analysis of the photographs, which is carried out at the University of Wyoming, U.S.A., by D. R. Tompkins and associates, requires careful interpretation to identify the unusual events of interest and distinguish them from more common phenomena.

5. Outlook

The lack of a conclusive explanation for ball lightning continues. Recent theories have advanced toward a more rational description of the phenomenon. Some of the otherwise diverse theories agree in describing the ball as a globe of positive charge. The absence of essential knowledge of even the most simple descriptions which have been proposed suggests direct experimental study which could be decisive considering the current state of the problem. The processes and substances described in the most plausible theories are accessible to such study.

Acknowledgements

I am grateful for the valuable material provided by Professor C. Eaborn, Dr R. H. Golde, Professor V. L. Highland, Dr R. M. Horowitz, Y. Ksander, D. Kuhn, Dr E. Kuhn, Professor A. A. Mills, J. A. Newbauer and E. H. Sadler. Table II is reproduced with permission of B. M. Smirnov, Uspekhi Fizicheskikh Nauk, and the American Institute of Physics, Fig. 2 with permission of D. Kuhn and *Naturwissenschaften*.

References

Altschuler, M. D., House, L. L. and Hildner, E. (1970). Is ball lightning a nuclear phenomenon? *Nature* **228**, 545–547.

Argyle, E. (1971). Ball lightning as an optical illusion. *Nature* **230**, 179–182.

Ashby, D. E. T. F. and Whitehead, C. (1971). Is ball lightning caused by antimatter meteorites? *Nature* **230**, 180–182.

Askar'yan, G. A., Rabinovich, M. S., Savchenko, M. M. and Stepanov, V. K. (1967). The fireball of light breakdown in the laser beam focus. *Zh. eksp. teor. Fiz.* PR **5**, 150–154; *JETP Letters* **5**, 121–124.

Aspden, H. (1973). Ball lightning enigma. *New Scient.* **57**, No. 827, 42.

Aulov, A. (1973). On course—ball lightning. *Izvestia* 28 July, p. 6.
Barry, J. D. (1968). Laboratory ball lightning. *J. atmos. terr. Phys.* **30**, 313–317.
Berger, K. (1973). Ball lightning and lightning research. *Naturwissenschaften* **60**, 485–492.
Bergström, A. (1971). Electromagnetic theory of strong interaction. *Phys. Rev.* D **8**, 4394–4401.
Blair, A. J. F. (1973). Magnetic fields, ball lightning, and campanology. *Nature* **243**, 512–513.
Brand, W. (1923). "Ball Lightning." Henri Grand, Hamburg.
Bromley, K. A. (1970). Ball lightning. *Nature* **226**, 253.
Brovetto, P., Maxia, V. and Bussetti, G. (1976). On the nature of ball lightning. *J. atmos. terr. Phys.* **38**, 921–934.
Bruce, C. E. R. (1963). Ball lightning, stellar rotation, and radio galaxies. *Engineer* **216**, 1047–1048.
Budinger, T. F., Bichsel, H. and Tobias, C. A. (1971). Visual phenomena noted by human subjects on exposure to neutrons of energies less than 25 million electron volts. *Science* **172**, 868–870.
Cecchini, S., Di Cocco, G. and Mandolesi, N. (1974). Positron annihilation in EAS and ball lightning. *Nature* **250**, 637–638.
Charman, W. N. (1971). After-images and ball lightning. *Nature* **230**, 576.
Charman, W. N. (1972). The enigma of ball lightning. *New Scient.* **56**, 632–635.
Covington, A. E. (1970). Ball lightning. *Nature* **226**, 252–253.
Crawford, J. F. (1972). Antimatter and ball lightning. *Nature* **239**, 395.
Crew, E. W. (1972). Ball lightning. *New Scient.* **56**, 764.
Davies, D. W. and Standler, R. B. (1972). Ball lightning. *Nature* **240**, 144.
Davies, P. C. W. (1971). Ball lightning or spots before the eyes? *Nature* **230**, 576–577.
Davies, P. C. W. (1976). Ball lightning. *Nature* **260**, 573.
Dawson, G. A. and Jones, R. C. (1969). Ball lightning as a radiation bubble. *Pure appl. Geophys.* **75**, 247–262.
Dmitriev, M. T. (1969). The stability mechanism of ball lightning. *Zh. tekh. Fiz.* **39**, 387–394; *Soviet Phys. tech. Phys.* **14**, 284–289.
Dmitriev, M. T. (1970). New data on ball lightning. *Wiss. Z. Hochsch. Elektrotech. Ilmenau* **16**, 87–90.
Dmitriev, M. T. (1973). New problems of ball lightning. *Priroda.* **4**, 60–67.
Dmitriev, M. T., Deryugin, V. M. and Kalinkevich, G. H. (1972). The optical emission of ball lightning. *Zh. tekh. Fiz.* **42**, 2187–2189; *Soviet Phys. tech. Phys.* **17**, 1724–1725.
Endean, V. G. (1976). Ball lightning as electromagnetic energy. *Nature* **263**, 753–755. 731–739; *Soviet Phys. Uspekhi* **18**, 636–640 (1976).
Felsher, M. (1970). Ball lightning. *Nature* **227**, 982.
Fieux, R., Gary, C., and Hubert, P. (1975). Artificially triggered lightning above land. *Nature* **257**, 212–214.
Fleming, S. J. and Aitken, M. J. (1974). Radiation dosage associated with ball lightning. *Nature* **252**, 220–221.
Galton, E. M. G. (1971). Ball lightning. *The Times Lond.* No. **58174**, 17 May, p. 13.
Gibbs-Smith, C. H. (1971). On fireballs. *Nature* **232**, 187.
Hill, C. R. and Sowby, F. D. (1970). Radiation from ball lightning. *Nature* **228**, 1007.

Hill, E. L. (1960). Ball lightning as a physical phenomenon. *J. geophys. Res.* **65**, 1947–1952.

Hill, E. L. (1970). Ball lightning. *Am. Scient.* **58**, 479.

Humphreys, W. J. (1936). Ball lightning. *Proc. Am. phil. Soc.* **76**, 613–626.

de Jans, C. (1910–1912). A retrospective glance at the attempts to explain ball lightning. *Ciel Terre* **31**, 499–504; **32**, 155–159, 255–261, 301–307; **33**, 18–26, 143–158.

Jennings, R. C. (1962). Path of a thunderbolt? *New Scient.* **13**, 156.

Jennison, R. C. (1971). Ball lightning and after-images. *Nature* **230**, 576.

Jennison, R. C. (1973). Can ball lightning exist in a vacuum? *Nature* **245**, 95.

Jensen, J. C. (1933). Ball lightning. (i) *Physics* **4**, 372–374; (ii) *Scient. Mon. N.Y.* **37**, 190–192.

Jones, A. T. (1910). A laboratory illustration of ball lightning. *Science* (N.S.) **31**, 144.

Jones, R. V. (1968). The natural philosophy of flying saucers. *Phys. Bull.* **19**, 225–230.

Kapitsa, P. L. (1955). On the nature of ball lightning. *Dokl. Akad. Nauk S.S.S.R.* **101**, 245–248.

Kapitsa, P. L. (1969). Free plasma filament in a high frequency field at high pressure. *Zh. Eksp. teor. Fiz.* **57**, 1801–1866; *Soviet Phys. JETP* **30**, 973–1008 (1970).

Koldamasov, A. (1972). Ball lightning in liquids? *Tekhnika Molod.* **8**, 24–29.

Kosarev, E. L. and Serezhkin, Yu. G. (1974). Narrow-band decimeter radiation from lightning. *Zh. tekh. Fiz.* **44**, 361–371; *Soviet Phys. tech. Phys.* **19**, 229–233.

Kozlov, B. N. (1975). A relaxation principle theory of ball lightning. *Dokl. Akad. Nauk S.S.S.R.* **221**, 802–804; *Soviet Phys. Dokl.* **20**, 261–262.

Kuhn, E. (1951). Ball lightning in a snap-shot? *Naturwissenschaften* **38**, 518–519.

Lazarides, M. (1971). Lightning balls. *Nature* **231**, 194.

Lilienfeld, P. (1970). Ball lightning. *Nature* **226**, 253.

Lowke, J. J., Uman, M. A. and Liebermann, R. W. (1969). Toward a theory of ball lightning. *J. geophys. Res.* **74**, 6887–6898.

McNally Jr, J. R. (1972). Speculations on the configurational properties of a fusioning plasma. *Nucl. Fusion* **12**, 265–266.

Meissner, A. (1930). About ball lightning. *Met. Z.* **47**, 17–20.

Meyerovich, B. E. (1972). Diffusion in a high frequency gas discharge. *Zh. eksp. teor. Fiz.* **63**, 549–566; *Soviet Phys. JETP* **36**, 291–299 (1973).

Mills, A. A. (1971). Ball lightning and thermoluminescence. *Nature* **233**, 131–132.

Minin, V. F. and Baibulatov, F. Kh. (1969). The nature of bead lightning. *Dokl. Akad. Nauk S.S.S.R.* **188**, 795–798; *Soviet Phys. Dokl.* **14**, 979–982 (1970).

Minter Jr, B. and Bird, R. (1972). Lightning—Indian Ocean. *Mar. Obsr* **42**, No. 236, 52.

Morikawa, G. K. and Rebhan, E. (1970). Toroidal hydro-magnetic equilibrium solutions with spherical boundaries. *Physics Fluids* **13**, 497–500.

Mortley, W. S. (1973). Ball lightning. *New Scient.* **57**, 42–43.

Norinder, H. (1939). On the nature of lightning. *K. Vetensk. Samh. Upps. Årsb.* 89–95.

Orville, R. E. (1974). Ball lightning, *in* "Encyclopaedia Britannica," 15th ed., Vol. 10, p. 969d.

Paneth, F. A. (1956). The frequency of meteorite falls throughout the ages. *Vistas Astr.* **2**, 1681–1686.

Perret, F. A. (1924). The Vesuvius eruption of 1906. Carnegie Institution of Washington Publication No. 339, p. 93.

Planté, G. (1875). Research on phenomena produced in liquids by high voltage electric currents. *C. r. hebd. Séanc. Acad. Sci., Paris* **80**, 1133–1137.

Powell, J. R. and Finkelstein, D. (1969). Structure of ball lightning. *Adv. Geophys.* **13**, 141–189.

Powell, J. R. and Finkelstein, D. (1970). Ball lightning. *Am. Scient.* **58**, 262–280, 479.

Reimann, E. (1887). Ball lightning. *Met. Z.* **4**, 164–178.

Ryan, R. T. and Vonnegut, B. (1970). Miniature whirlwinds produced in the laboratory by high-voltage electrical discharges. *Science* **168**, 1349–1351.

Ryan, R. T. and Vonnegut, B. (1971). Formation of a vortex by an elevated electrical heat source. *Nature.—Physical Science* **233**, 142–143.

Silberg, P. A. (1962). Ball lightning and plasmoids. *J. geophys. Res.* **67**, 4941–4942.

Singer, S. (1971). "The Nature of Ball Lightning." Plenum Press, New York; Mir Publishers, Moscow (1973).

Smirnov, B. M. (1975). Analysis of the nature of ball lightning. *Usp. fiz. Nauk* **116**, 731–736; *Soviet Phys. Uspekhi* **18**, 636–640 (1976).

Stakhanov, I. P. (1974). Stability of ball lightning. *Zh. tekh. Fiz.* **44**, 1373–1379; *Soviet Phys. Tech. Phys.* **19**, 861–864 (1975).

Stenhoff, M. (1976). Ball lightning. *Nature* **260**, 596–597.

Trubnikov, B. A. (1972). The nature of ball lightning. *Dokl. Akad. Nauk SSSR* **203**, 1296–1298; *Geophys.-Dokl. Akad. Nauk SSSR* **203**, 13–14.

Turpain, A. (1911). Curious effects of a lightning stroke on a receiving antenna for electric waves. *J. Phys. théor. appl.* V **1**, 372–381.

Vitkevich, V. V. (1962). The possible existence of radiating natural plasma satellites moving around the sun. *Isv. Vÿssh. ucheb. Zared. Radiofizika* **5**, 404–405.

Voitsekhovskii, B. V. and Voitsekhovskii, B. B. (1974). The nature of ball lightning. *Dokl. Akad. Nauk SSSR* **218**, 77–80; *Soviet Phys. Dokl.* **19**, 580–581 (1975).

Wagner, G. A. (1971). Optical and acoustic detection of ball lightning. *Nature* **232**, 187.

Webster, H. F. and Hallinan, T. J. (1973). Instabilities in charge sheets and current sheets and their possible occurrence in the aurora. *Radio Sci.* **8**, 475–482.

Wooding, E. R. (1972). Laser analogue to ball lightning. *Nature* **239**, 394–395.

Wooding, E. R. (1976). Ball lightning in Smethwick. *Nature* **262**, 379–380.

Zaytsev, A. V. (1972). A new theory of ball lightning. *Zh. tekh. Fiz.* **42**, 213–216; *Soviet Phys. tech. Phys.* **17**, 173–175.

Zimmerman, P. D. (1970). Energy content of Covington's lightning ball. *Nature* **228**, 853.

13. Measuring Techniques

R. B. ANDERSON

National Electrical Engineering Research Institute, Council for Scientific and Industrial Research, Pretoria, South Africa

1. Introduction

It is the intention of this chapter to refer to some of the more important modern measuring techniques used by lightning researchers which have given satisfactory results, and to draw attention to possible pitfalls which have been encountered and which require special precautions to avoid introducing large errors.

2. The Electric Field

2.1 Principle of Measurement

The electric field at any given point may generally be defined as a stress existing at that point as a consequence of the force exerted by the electric charge (or charges) situated at a distance or in the vicinity of the point of measurement. To this should be added a further stress which occurs when the charges are dissipating or moving thus resulting in an electric current, from which is derived an electromagnetic field.

An electrostatic field may exist on its own as a purely "static" phenomenon so long as there is no change in the position or quantity of the charges involved. Once there is a change such as may be due to a lightning discharge, it becomes a transient phenomenon and its magnitude varies with time. At this stage it is inseparably linked and related to the corresponding electromagnetic field change which is similarly transient in nature.

The electrostatic field strength may be measured by a device which is not influenced by the presence of an electromagnetic field, such as the rotary field mill or flux meter which is discussed in more detail in Section 2.2.1. The electrostatic field-change, by contrast, may be measured by a static device in the form of a metallic plate mounted flush with the surface of, but well insulated from, the earth, the potential on the plate being proportional to any change in field strength.

The electromagnetic field strength is measured by means of conductors suspended in the field across which potentials are induced as a consequence of changes in the magnitude of the magnetic field strength or due to the movement of magnetic flux through the conductor as a consequence of radiation. The conductors are usually arranged in the form of a loop so that the difference of potentials induced on either side of the loop can be measured; this potential difference is proportional to the magnetic field strength. Such loops can be shielded from the effects of an electrostatic field-change by arranging for a non-ferrous shield around the loop which is broken at one point to avoid the shield forming a short-circuited turn.

A single vertical conductor connected to an *RC* measuring circuit by contrast measures the change in the total electric field strength—that is both that of the electrostatic and the electromagnetic field.

2.2 Electrostatic Field Strength Measurements

2.2.1 Measurement at the Earth's Surface

As stated above, the electrostatic field strength is measured by means of a field mill or flux meter, the principle of which is described in the literature (Malan, 1963; Chalmers, 1967; Israel, 1973).

In earlier models of the field mill the output was in the form of an a.c. signal, the magnitude of which was proportional to the field strength and these signals could be either displayed or recorded as they were, or they could be rectified to give a d.c. output. Additional circuits were necessary to determine the polarity of the field being measured. The instruments were calibrated by fitting a pair of plates or grids to them, having an area well in excess of that of the field-mill rotor plates and at a given distance apart, and applying a d.c. voltage between them.

With the advent of fast switching devices, such as solid-state switches and photo-sensitive transistors, it was possible to arrange the circuitry so as to provide a d.c. signal of the correct polarity. Berger (1972) developed a d.c. field mill (see Chapter 5.2.1.1) similar in principle to that of Smith (1954) which employed four pairs of rotating vanes. An improved version of this type was developed (Anderson, 1976) and the circuit diagram is illustrated in Fig. 1.

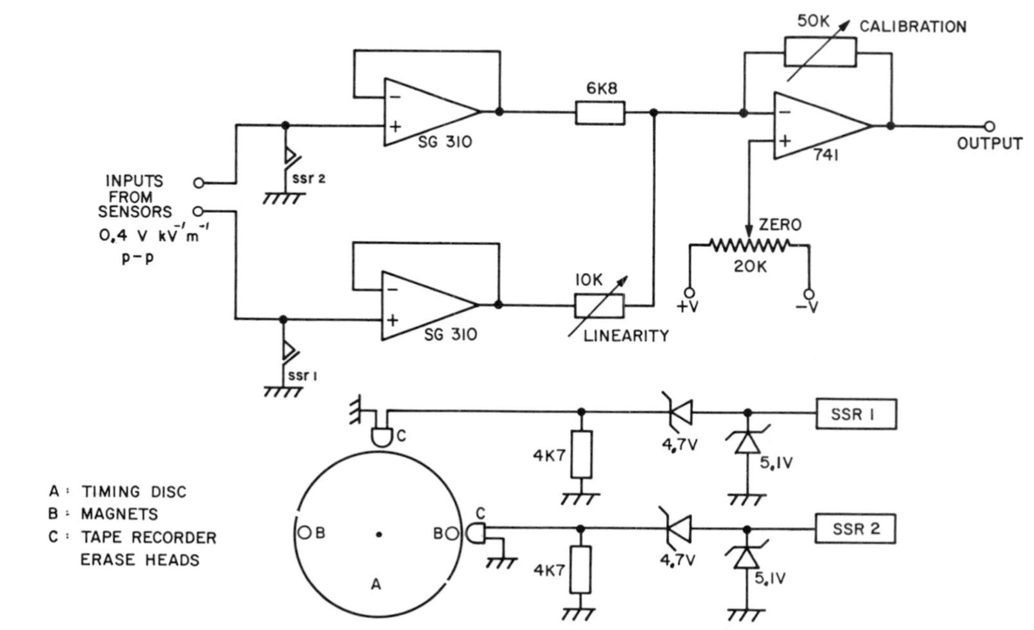

Fig. 1. Circuit diagram of a field mill for electrostatic field-strength measurement.
SSR: Solid State Relay.

This field mill is 150 mm in diameter and 80 mm in length, weighs 1 kg and when used outdoors it is inverted and fitted with a weather shield; the whole is fixed to a swan-neck type of support so that the rotating vanes are about 1·5 m above ground level. In this position the field mill must be recalibrated under natural site conditions during a thunderstorm by comparing its output with that of a laboratory-calibrated unit mounted flush with the ground plane some 10 m away. It was found that the inverted instrument was more sensitive than the ground-mounted instrument by a factor of about 4.5, due no doubt to the field intensification resulting from its mounting configuration. The output of the field mill is recorded on a pen recorder having a 1 s print-out interval, but given a faster recording medium such as an oscilloscope field mills of this type should be able to record transient field changes up to 40 kHz.

Whilst the continuous measurement of the electrostatic field strength at a single point may not provide results capable of any meaningful analytical interpretation, it provides a useful indication as to the general intensity of electrification of thunderstorms compared with one another or with one global situation and another. For example, the results of an analysis of the electrostatic field excursions during 65 thunderstorms in the Pretoria area during the year 1974–75 have been reported (Eriksson, 1975) and the time-interval distribution obtained is shown in Fig. 2. The results may also be processed (Eriksson, 1974) to indicate the starting and finishing times and duration of thunderstorms.

Multi-station measurements promise more exact analytical treatment to determine the charge quantity, distribution and movement in thunderstorms (see Chapter 6.4.3). Mackerras (1974) has also examined the possibility of making the equivalent of multi-station measurements using a low-flying aircraft with field mills mounted on the fuselage, or fast road vehicles, equipped for continuous recording, and using a network of fixed stations to correct for slow field-changes occurring during the traverses. Such measurements result in field-strength profiles being made available on a traverse approach towards and within a thunderstorm which could give at least good qualitative information regarding the location of charges.

So far only ideal flat conditions of the so-called infinitely conducting plane have been considered, but in practice the actual terrain may be far from level, thereby introducing distortion and possible errors. In rolling country where the altitude changes are not large compared with the height of the cloud charges, errors would be small and could be corrected on the basis of the means of measured distribution of field intensity on high ground and on lower ground. Alternatively a reference area may be chosen such as an extensive plateau or a lake, to which all other data may be referred. In mountainous terrain, however, the problem of finding a suitable reference may indeed be difficult, and it would appear that recourse would have to be made to field-plotting techniques such as employed by Berger (1973).

One of the most interesting problems of field-strength measurement relates to the conditions at the top of and surrounding tall structures such as microwave towers or chimneys or the tall towers and conductors of transmission lines.

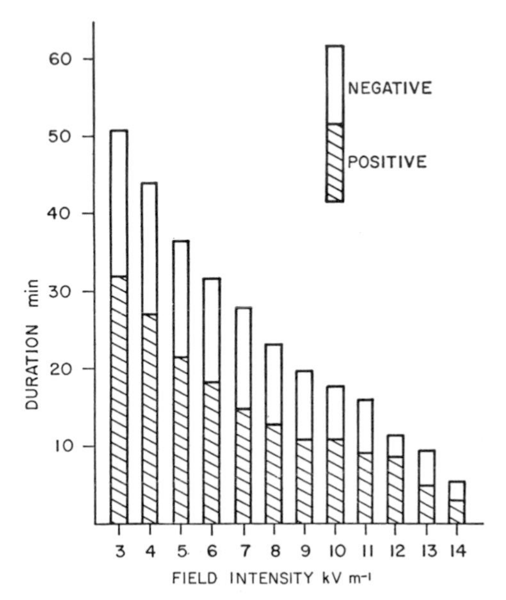

Fig. 2. Distribution of time intervals during which the field strength of the stated values was exceeded (65 thunderstorms 1974–75, South Africa).

When placed in a uniform electrostatic field such as would be encountered with cloud charges situated at a height very much in excess of the height of the structure, charge is raised into the structure and distributed so as to maintain zero potential on all conducting surfaces. This means that the potential at any given point on the structure due to the charge on the structure is equal in magnitude but opposite in polarity to the potential at that point in space in the primary electrostatic field induced by the cloud charges.

If the form of the charge distribution along the structure could be determined, it would be possible to calculate the field strength at the top of the structure, or in the vicinity, say at the earth's surface near the structure, as a ratio to the inducing field and the calculations could thereafter be checked by actual measurement. Alternatively, the errors of measuring the inducing field strength near such structures could be estimated.

If the charge distribution along a vertical structure of length L and radius R is assumed to be linear, i.e. increasing uniformly from zero at the base of the structure to a maximum value at the top—the potential gradient along the structure is found to be uniform for most of the length (Anderson, 1971b), but as the top is approached it reduces rapidly indicating that the charge density in fact should increase more sharply than assumed. It is also surmised that this increase should be greater for lower values of the ratio L/R.

Assuming therefore, as a first approximation, that a portion of the charge is distributed exponentially and that the balance is linear, the amount of this proportion and its exponent may be varied to give the best-fit distribution to that required. Calculations may be facilitated by dividing the structure length into n segments and distributing the segmental charge Δq_r uniformly along each segment, whence it can be shown that if q is the total charge on the structure, then for the rth segment counted from the base of the structure

$$\Delta q_r/q = [n^2(1+K)]^{-1} \times [(2r-1)+A^{r-1}B], \tag{1}$$

where K is the ratio of the total exponentially distributed charge to that of the linearly distributed charge, A is the exponent of the exponentially increasing charge and

$$B = n^2 K(A-1)(A^n-1)^{-1}.$$

The best-fitting values of the variables K and A thus found are shown in Table I for given values of the ration L/R for $n = 100$ points.

Table I

Values of the proportion K and the exponent A of the charge distribution on a vertical structure ($n = 100$).

L/R	K	A
20	0·273	1·99
30	0·205	1·53
100	0·110	1·31
200	0·084	1·26
500	0·054	1·23
1,000	0·038	1·17

$K = \dfrac{\text{total exponential charge}}{\text{total linear charge}}$; A = exponent.

The resulting charge distribution is shown in Fig. 3.

As surmised earlier, Fig. 3 shows that, as the ratio of L/R increases, the proportion of charge distributed exponentially decreases, until at very high values of L/R the charge distribution is almost entirely linear.

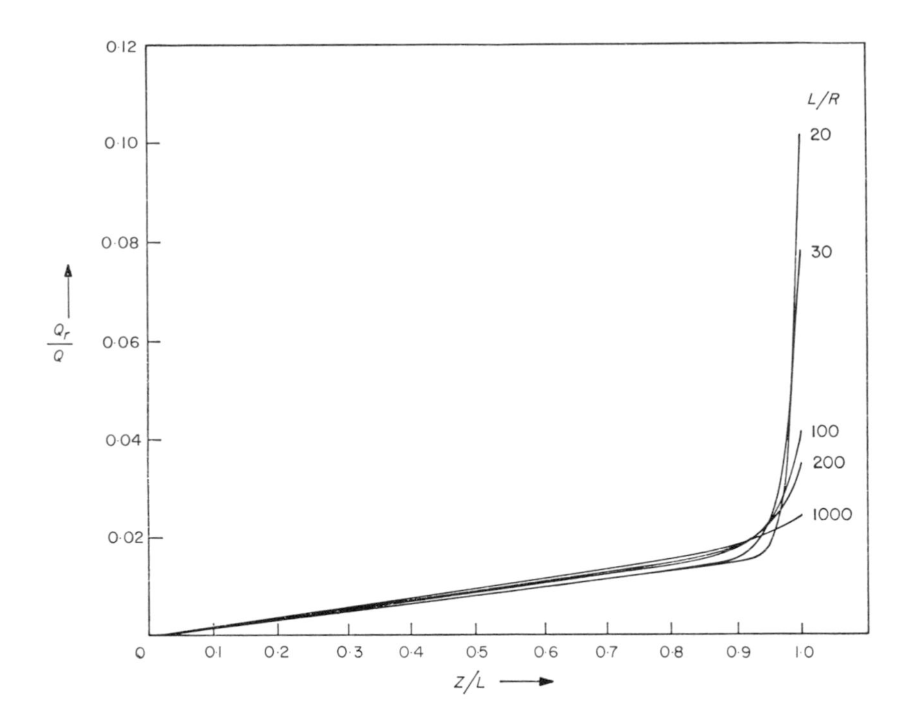

Fig. 3. Variation of charge distribution along a vertical conductor from the base ($Z = 0$) to the top ($Z = L$).

Having determined the approximate charge distribution, the vertical component of the field strength, as a ratio of the inducing field on the centre line of the top of the structure, can be calculated by assuming that the structure forms an open-ended hollow tube with the charge distributed on the annular external surface. Similarly, the horizontal component normal to the surface of the tube can be determined by assuming that the charge is now distributed along the centre line. Such an artifice, whilst not exact, is nevertheless sufficiently accurate to obtain an approximate result for the field intensification. The net result of the vector sum of the two components is that the total field strength E_t at the top of the structure expressed as a ratio to the inducing field E_0 has the calculated values shown in Fig. 4.

It is clear from Fig. 4 that the intensification of the electrostatic field at the top of a conducting structure is not a function of its length (or height) alone but is related to the ratio of its primary dimensions, namely L/R. Tall structures such as microwave towers have high values of L/R especially if tapered to a smaller diameter at the top, and hence the possibility of high field strengths occurring at the top is enhanced. On the other hand, since the phenomenon is not height dependent, short conductors having large L/R ratios are equally liable to high field end-effects. Hence these must be taken into account when measuring field strengths, say, with a field mill mounted on

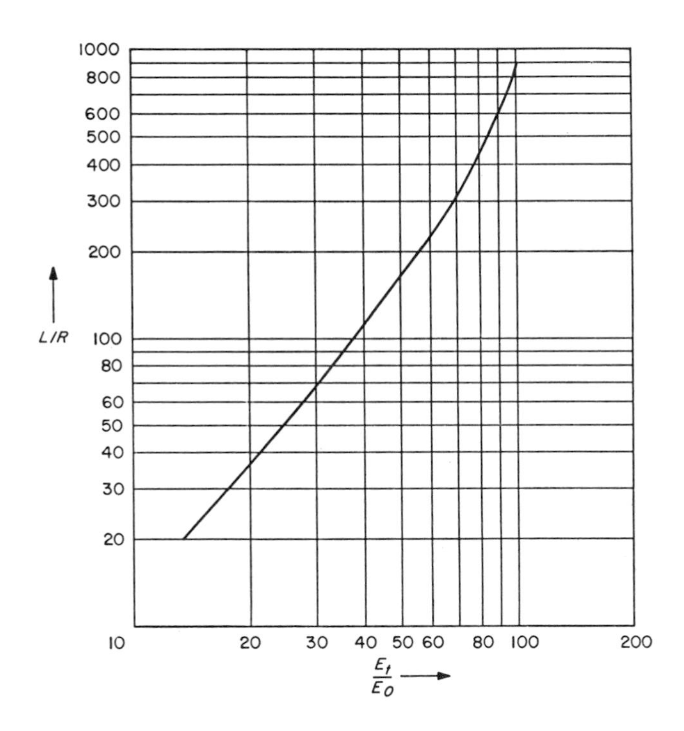

Fig. 4. Ratio of resultant field strength E_t at the top of a structure to the inducing field E_0 for given ratios of structure dimensions L/R.

a short vertical support. Similarly, vertical conductors used for aerials for counters or for field-change measurements could be liable to discharge under high field-strength conditions, and this may be avoided by reducing the value of L/R or by providing stress relief by means of a sphere mounted on the top of the aerial. It should be mentioned that in the event of an aerial being connected through a high impedance to earth—such as the input circuit of an oscilloscope—any discharge which does occur under high electrostatic field conditions is pulsating in nature, and such pulses cause the oscilloscope to trigger continuously at a rate proportional to the intensity of the field. These pulses are mostly due to the rise in potential of the aerial with respect to earth during discharge which then reduces the field strength at the top thereby stopping the discharge momentarily. Alternatively, in the case of a positive electrostatic field, Trichel discharges (see Chapter 4.2) occur from the negative aerial which are also pulsating in character.

Practical structures may not conform to the idealized model assumed. Angles and sharp corners will no doubt be in excess of the calculated value for the structure as a whole. The inception of ionizing fields therefore could occur earlier than anticipated, but once having been established it is highly probable that the resultant corona discharge would tend to modify effectively the physical configuration of the structure top to one tending towards a cylindrical or ellipsoidal shape, in which case the calculated field strength should again be applicable. The corona discharge itself will give rise to space charge above the structure which will tend to accumulate under still air conditions thereby lowering the measured field strength beneath it, since the field produced by the space charge is in opposition to the main inducing field. However, if the space charge is continuously removed by high wind, the more intense the inducing field, the greater will be the amount of corona discharge, at the same time maintaining the field strength at the top of the structure at a stable value.

So far the electrostatic field condition considered assumes that the charges inducing the field are remote—say in the clouds—when uniform field conditions at earth level are valid. In the case of the field induced by a descending lightning leader, however, the field strength near the earth may not be uniform but the field intensification at the top of a structure would conform to a pattern similar to that shown in Fig. 4 so that upward connecting leaders could develop more easily if the structure dimensions were favourable (see Chapter 17.2.2).

The vertical field strength E_x at the earth's horizontal plane for any distance x radially from the structure centre line can also be calculated as a ratio to the inducing field, and this is again found to be dependent upon the L/R ratio of the structure. The resultant ratio of the total field E_x to that of the inducing field E_0 is shown in Fig. 5.

Due to the charge in the structure the electrostatic field is of opposite polarity to that of the inducing field, and hence, close to the structure, it could predominate. This happens when the L/R ratio is low and consequently in such cases there is a point near the base of the structure where the field strength passes through zero. Thus, in order to measure the prevailing electric field strength it is necessary to move away from the structure to a point where

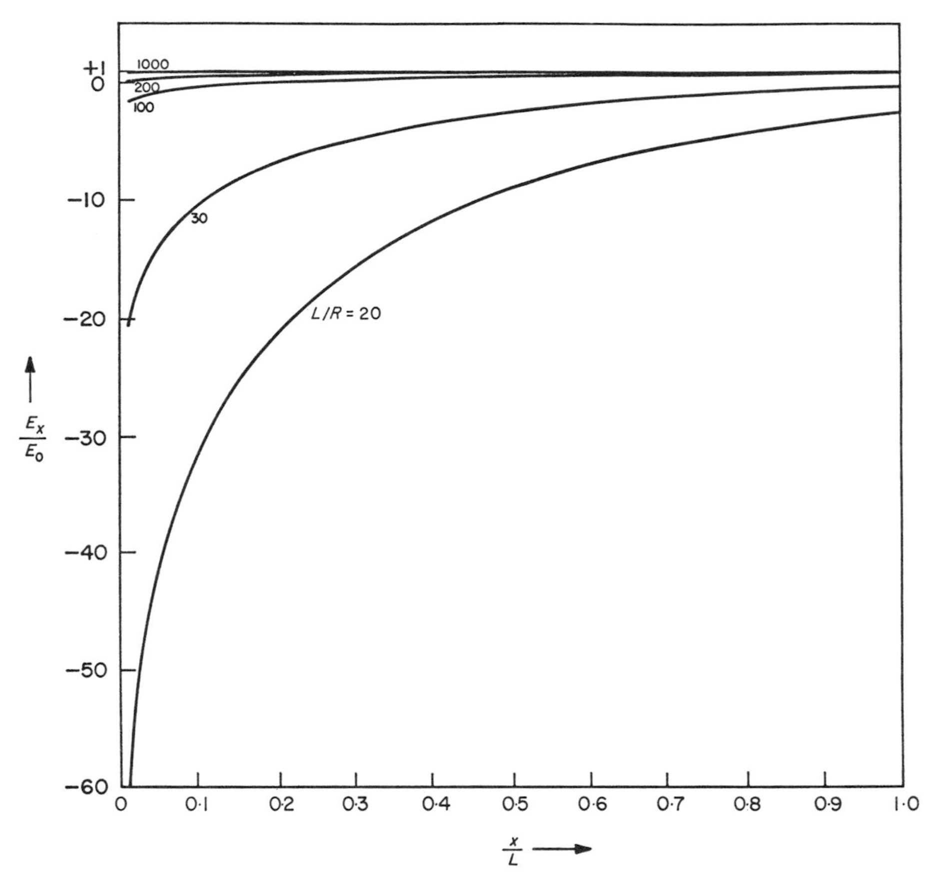

Fig. 5. Ratio of vertical field strength E_x along the horizontal earth plane to that of the inducing field E_0, for given values of L/R.

the error would be, say, less than 5%, or at least where a known correction could be applied. From Fig. 5 the values of x/L, where this condition is fulfilled, can be interpolated and the results are given in Table II.

2.2.2 *Measurement of Field Strength in the Aerospace*

Measurement of the field strength in the space between clouds and the earth, for example by means of aircraft, poses special problems. First of all an aircraft acquires a charge of one polarity on its fuselage as a result of friction or due to the ionic content of the exhaust from its engines. This affects the calibration by increasing the ambient field strength on one of the surfaces, say the upper, according to the polarity of charge, whilst reducing it on the lower surface by an equivalent amount. The employment of two field mills, one on each surface, is therefore necessary, and the mean value of their registrations would closely approximate the true value.

Table II

Approximate distance x from centre line of a structure of height L where the measured field strength E_x is either zero or 5% below that of the inducing field E_0

L/R	x/L ($E_x/E_0 = 0$)	x/L ($E_x/E_0 = 0{\cdot}95$)
20	1·74	5·18
30	1·08	3·51
100	0·16	1·51
200	—	0·90
1,000	—	0·02

In addition, the two field mills need to be calibrated in a known electric field, since the induced charges on the aircraft tend to increase the field strength on both upper and lower surfaces above the true value. Also, there will always be some distortion of the field in the immediate vicinity of the field mills due to their mountings, which can then be corrected. This calibration can be done during natural thunderstorm electric field conditions whilst the aircraft is on, but extremely well insulated from, the ground, the reference field strength being measured at the earth's surface some distance from the aircraft.

Unlike on level ground, the field direction during flight may not be vertical, or the aircraft may not be flying horizontally, and hence the measured field strength with the normal type of field mill could only be approximate. Kasemir (1972) developed a cylindrical field mill capable of measuring the component values of the field strength resolved in three directions at right angles. So if the angle of inclination of the aircraft is known in addition to its flight bearing and position, the resultant maximum field strength and its direction in space can be calculated.

Owing to the finite flying time between two points or along a fixed route through a thunderstorm area, the overall field strength could change due to the charge generation process or as a consequence of a lightning flash. To allow some compensation for these changes ground-mounted field mills should be installed at intervals along the flight route in order to provide a controlled measurement of the overall field-strength changes taking place during the flight. The field measurements on the aircraft could, as a first approximation, be adjusted in proportion to the slow change occurring on the surface of the earth. It is doubtful, however, if the changes due to a lightning flash can be so adjusted, since the complete field configuration will have been altered with such a discharge.

Field measurements using aircraft are usually confined to more specific objectives such as finding the maximum field strength, or the location of charged volumes. Such evidence as exists suggests, however, that the chances of successful missions of this kind are rather remote, primarily because not only are these areas of comparatively small dimensions, but they are also continually changing position as one cell builds up and is dissipated in a matter of 20 to 30 minutes (Held and Carte, 1973) to be replaced by another adjacent cell, and so on (see Chapter 3.3.1).

Alternative methods using radio sondes or rocket-propelled devices could be as effective in achieving such objectives. Unfortunately these introduce other problems; particularly that the field-strength measuring instrumentation has to be powered in flight; also, such devices have to be accurately tracked by radar or by means of radio-location of transmitted signals from the device.

Winn and Moore (1971) devised a rather ingenious rocket-propelled field mill which used the rotational spin of the rocket itself to provide an alternating signal proportional to the field strength. Results obtained by this method are given in Chapter 3.4.5.

As far as I am aware, there have been no experiments of this nature carried out in a cloud at high altitudes on mountains where conditions might approximate those in thunderclouds. It is also conceivable that measurements of the breakdown strength in clouds could be achieved by using a specially devised cell on an aircraft where the application of high voltage to the cell from within the aircraft should not be too difficult.

3. The Electric Field-change

3.1 The Electrostatic Field-change

The electrostatic field-change which occurs during the slow build-up of charge in clouds during the charge-separation process, or when a thunderstorm is moving towards or past an observation station, can be faithfully recorded on a field mill; the field-change associated with a lightning discharge on the other hand, taking place as it does in times of the order of microseconds, could not be recorded accurately by such means and other methods are usually employed.

A simple static instrument which may be termed a field-change meter (Krehbiel *et al.*, 1974) has been made possible as a consequence of the availability of very high-impedance transistor amplifiers to measure the potential of a highly insulated plate with respect to earth and having a decay time of the order of seconds at least.

The principle of the field-change meter is that when a plate of area A and capacitance C is exposed to a vertical field E_0, it acquires a charge $q_0 = \varepsilon_0 E_0 A$ at zero potential (see also Chapter 6.1.2). If the field now changes to a new value E_1, the plate acquires a potential proportional to the field-change $E_c = (E_1 - E_0)$ given by $V = (q_1 - q_0)/C$ and this potential decays exponentially with a time constant of RC, where R is the leakage impedance to earth, usually a resistor of high ohmic value, connected to the plate.

If the time constant RC is large enough, typically 10 s, then even modestly slow recording equipment can be used to measure, with sufficiently good resolution, the field-changes during the relatively slow leader process as well as those during the individual strokes of a multiple flash. The application of this device to a multi-station measurement is described in Chapter 6.4.3. Suffice to say here that it is possible to calculate the charge dissipated during individual strokes of a flash to earth as well as designating the original location of the charge to a known degree or error.

The use of a field-change meter with a suitable high-speed output device, such as a digital transient recorder or an oscilloscope, enables the accurate recording of the electrostatic field change and this in turn provides a useful means of estimating the form of the charge distribution on the leader and the mechanism of the return-stroke discharge process, neither of which can be regarded even today as having been finally settled. For example, I (1971b) have shown that, if it is assumed as a first approximation that the charge lowered by the leader is equal to that dissipated during the return stroke, then the ratio of the electrostatic field-change during the leader to that during the return stroke (when $D \gg H$) is unity if the charge distribution along the leader was uniform, but the ratio increases to two for a linear distribution and even greater for an exponential distribution as assumed by Bruce and Golde (1941).

The time variation of the field-change during the leader process has been calculated by Malan (1963) for the case of a uniform distribution, assuming idealized conditions of a single vertical leader channel without branches. For a linear distribution of charge, on the other hand, the field-change magnitude differs by a factor of two, and the time variation of the field-change varies—for example, the value of D/H, when the net field-change is zero, is changed. For the calculation of the field-change for an assumed exponential distribution of charge on the leader the expanded series for the exponential function may be utilized but a simpler numerical approach is preferred in which small elements of charge are placed at n equally spaced points along the leader, as was done for the charge distribution along a vertical conductor illustrated in Section 2.2.1.

So far in this section only the electrostatic component of the field-change has been discussed, and it may be measured by means of the field mill and field-change meters described, including the electrometer employed by Wilson (1920), since these devices do not respond to the electromagnetic component of the electric field-change. Aerials can be, and have been, used to measure the field-change which occurs both during the leader and return stroke, provided that it is understood that aerials respond to the total electric field, that is including a component of the electromagnetic field-change, and that their frequency response must be adequate. Usually, however, it is not advisable to employ the same circuit for the measurement of the leader field-change as for that of the return stroke because of the very considerable bandwidth needed to accommodate both the slow leader phenomenon lasting several tens of milliseconds and the very fast R component of the return stroke with rise times which are sometimes less than a microsecond. Similarly the field-change which follows the return stroke in the event of continuing current lasting many milliseconds is a comparatively slow process. The same disability applies to the measurement of field-changes due to inter- or intra-cloud flashes having a very slow basic component which may be sustained for as long as a second, with very fast field-changes superimposed on it (see Chapter 6.5).

The equivalent circuit of an aerial connected to a simple RC circuit is shown in Fig. 6. In this circuit C_a is the capacitance of the aerial having an

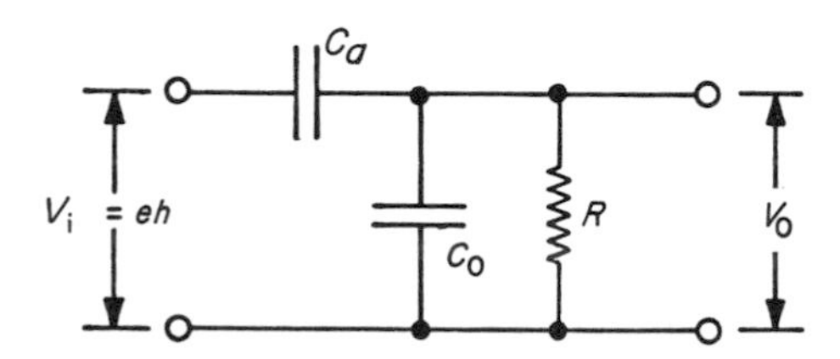

Fig. 6. Equivalent circuit of an aerial of capacitance C_a connected to an RC circuit. C_a = aerial capacitance; C_0 = circuit capacitance.

effective height h and when subjected to a field-change of e V m^{-1}, the equation representing the output voltage V is given by:

$$dV/dt + V/RC = C_a h/C \cdot de/dt, \tag{2}$$

where $C = C_a + C_0$.

Now it is assumed, for example, that the field-change has the exponential form indicated in Equation (3):

$$e = E_c[1 - \exp(-t/T)], \tag{3}$$

where E_c is the overall field-change and T is its exponential time constant. Then putting $T_0 = RC$, the time constant of the measuring circuit, the solution to Equation (2) is given by:

$$V = C_a/C \cdot hE_c[\exp(-t/T_0) - \exp(-t/T)](1 - T/T_0)^{-1}. \tag{4}$$

Solving for the condition of maximum output voltage V_m the total field-change E_c is given by:

$$E_c = C/C_a \cdot V_m/h \cdot \{(1 - T_r)[\exp(aT_r) - \exp(a)]^{-1}\}. \tag{5}$$

Here $T_r = T/T_0 =$ the ratio of time constant of the field-change to that of the measuring circuit and $a = \ln[T_r/(1 - T_r)]$.

The bracket term of equation (5) approaches unity as T_r approaches zero but when $T_r = 0{\cdot}01$ it is already 1·05. This means that the time constant of the measuring circuit must be at least 100 times that of the field-change if the measurement error is not to exceed 5%. Hence, if the rise time to the maximum value of an exponential field-change is, say, five times the time constant, the measurement *RC* circuit must have a time constant of at least 20 times the rise time.

For recording the field-change of the *R* components of the return strokes of a flash, a value of 30 ms has been employed by Malan (1963) and by Müller–Hillebrand (1963), but for recording very slow leaders, cloud flashes or continuing current components, much larger values are needed—of the order of ten to one hundred times his value. Scuka (1965), incidently, gives a circuit for the employment of two separate time constants using the same aerial, which may be useful in this regard.

The aerial capacitance C_a and its effective height h must also be determined, and in the case of a vertical aerial of length L and radius r, mounted with its base just off the ground, $h = L/2$ and the capacitance is:

$$C_a = 2\pi\varepsilon_0\,\varepsilon_r L[\ln(2L/r) - \tfrac{3}{2}]^{-1}. \tag{6}$$

If the aerial is mounted with its base some distance above ground, the effective height is taken as that of the mid-point of the aerial above ground, but the capacitance is also changed to a lower value and is affected by the structure upon which it is mounted if it is conducting.

For other configurations such as inverted L or T aerials, the effective height is taken as that of the top horizontal portion, but the capacitance must take into account both the vertical and horizontal portions and the effect of mounting structures, if conducting (Anderson, 1971b).

3.2 The Electromagnetic Field-change

If in the classical model proposed by Le Jay (1926) the charge q at a point H above ground and at a distance $D \gg H$ from an observer is expressed as a function of time, namely $q(t)$, that is, it is assumed that the charge is dissipated resulting in a current flow of $i(t)$, then the total electric field at the point of observation can be restated as:

$$E = H/2\pi\varepsilon_0 \varepsilon_r \left(\frac{q(t)}{D^3} + \frac{i(t)}{cD^2} + \frac{\mathrm{d}i/\mathrm{d}t}{c^2 D}\right) V/m, \tag{7}$$

where c is the velocity of light.

From the above it may be concluded that if one of the component fields can be measured, or calculated, it should be theoretically possible to derive the other two. However, the complexity of doing this may be illustrated from the expressions derived by the writer (Anderson, 1971b). The model used for the discharge mechanism during a return stroke was that in which the current value and wave-shape $i(t)$ is assumed; the charge is distributed uniformly and the current channel elongates from the earth upwards at a velocity dependent upon the current wave-shape and the charge distribution assumed. For further simplification $D \gg H$, and the three component field-changes are then calculated to be as follows:

$$E_s(t) = -qH/4\pi\varepsilon D^3 \cdot F^2(q, t) \tag{8}$$

where $\varepsilon = \varepsilon_0 \varepsilon_r$ and where $F(q, t) = \int_0^t [i(t)\,\mathrm{d}t/q]$ and which approaches unity as t approaches infinity

$$E_i(t) = -H/2\pi\varepsilon c D^2 \cdot F(q, t) \cdot i(t). \tag{9}$$

$$E_r(t) = -H/2\pi\varepsilon c^2 D \cdot [F(q, t)\,\mathrm{d}i/\mathrm{d}t + i^2(t)/q]. \tag{10}$$

Hence if $E_s(t)$ is measured with a sufficiently good time resolution, it would be possible to derive the time variation corresponding to the current discharged $i(t)$, but absolute values could not be determined without prior knowledge of the values of the other parameters q, H and D. These could be calculated from at least four simultaneous measurements of the total field-change. The procedure for other forms of charge distribution on the leader becomes more complicated, as does the case when D and H are comparable in value, but in all cases a model of the discharge phenomenon is needed before a numerical result is possible.

The total electromagnetic field-change E_m can also be measured by means of a shielded, crossed-loop antenna provided that the bandwidth is sufficiently large. If the distance D is not too great, i.e. of the order of 20 to 30 km, the radiation portion E_r can possibly be neglected compared with the induction component E_i, in which case, apart from the influence of $F(q, t)$ which tends to suppress the magnitude of the field-change at the start, the time variation should approximate that of the current $i(t)$.

Given a model of the current discharge process in the return stroke, it is possible to calculate the induction field in electromagnetic units by conventional means with the same results as above, when $D \gg H$ by converting to electrical units. For an ideal vertical lightning channel the magnetic field is circular and horizontal, and the conversion from magnetic units to electrical units is consistent with the assumption that the field is radiating from the conductor at the velocity of light and at the same time attenuating, and that the electromagnetic field strength at any given point is found by assuming that the magnetic flux cuts a vertical conductor at that point, producing a potential gradient along such a conductor numerically equal to the electric field strength in units of volts per metre.

This mode of propagation explains the effect on a rectangular conducting loop, to the extent that the voltage appearing across the loop is the difference between the voltages induced in the two vertical sides of the loop, which differ both in magnitude and phase (or time difference); this difference is a maximum if the two vertical sides of the loop are aligned with the direction of propagation. Since the separation of the vertical sides of the loop must be small compared with the wave-lengths of the component frequencies of the propagating field, these differences are also very small and usually have to be amplified by using a large number of turns effectively in series as well as by electronic means. Unless the direction of the lightning discharge is also measured relative to the loop, the outputs from two loops mounted at right angles must be summed vectorially for the calculation of the absolute magnetic field strength.

The change in the induction field strength with distance (i.e. neglecting any change of current in the lightning channel) is given by differentiating Equation (9) with respect to D, and if the loop has vertical sides of dimension h and horizontal separation d at an angle of θ to the direction of propagation the voltage difference is:

$$V(t) = H/\pi\varepsilon c D^3 \cdot [F(q, t)\, i(t)] \cdot hd \cos\theta. \tag{11}$$

The product $hd = A$ is the area of the loop, and if n is the number of turns, Equation (11) becomes:

$$V(t) = H/\pi\varepsilon c D^3 \cdot [F(q, t)\, i(t)] \cdot nA \cos\theta. \tag{12}$$

For $H = 4$ km, $D = 20$ km and $nA \cos \theta = 100$

$$V(t) = 6[F(q, t)] \ \mu\text{V A}^{-1}$$

where, as stated previously, $F(q, t)$ varies with time from zero to unity.

The maximum rate of change of current observed by Berger *et al.* (1975) is in the order of 120 kA μs^{-1}, and if the separation of the loop sides was say 1 m corresponding to a time increment of 1/300 μs, the change in current over this increment would be 400 A corresponding to a 2·4 mV change in the above-mentioned example. On the other hand, if this rate of change occurs for a lightning current of 20 kA, corresponding to 120 mV in the example, the error in neglecting the effect of the current change would only be 2%, and this is acceptable, especially since the maximum rate of rise of current was usually observed to be associated with the current peak (see Chapter 9.4.1.1).

Since the measurement of the total electromagnetic field involves the vector sum of the two signals, another, possibly simpler, method would be the subtraction from the total electric field-change, as measured by means of a vertical aerial, of the electrostatic component using a field-change meter of the exposed plate type described previously.

Mention should be made of the question of propagation time and its effect upon measurement and calculation. If $D \gg H$, say when D exceeds five times the height of the channel, radiation from all parts of the channel can be assumed to arrive at the point of observation (or measurement) simultaneously. All the component field-changes will therefore take place at a time D/c later than the actual event. In the case of measurements closer to the stroke path, however, the propagation times differ when radiation occurs from the base of the stroke compared to that of the upper extremity, and is of the same order of magnitude as the wave-front time of the current, and consequently can influence the time variation of the field-change. Calculations can be carried out on a computer using small increments of time as the basis.

For example, if a model of the discharge process is assumed such as that proposed by Bruce and Golde (1941), the channel is divided into small increments of length (including its mirror image), and for each element the current variation with small increments of time at the position of the stroke is then known. The electromagnetic induction field, due to the current flowing in each element, may then be calculated at a distant point, using the same increments of time but delayed by an amount of S/c where S is the actual point-to-point distance of the incremental element of channel length from that distant point. Since for a vertical channel the electromagnetic field is also vertical at the distant point for contributions from any point on the channel, these contributions may be summed arithmetically for each increment of time and may then be referred to time at the point of origin.

Similar calculations can be undertaken for the electrostatic field-change and the electromagnetic radiation field, but these could be derived from the induction field by integration and differentiation respectively, on an incremental basis.

4. Wave-shapes of Lightning Currents

The devices used to record the amplitudes and wave-shapes of lightning currents and the results obtained are discussed in Chapters 5.4.1, 9.3 and 9.4. Many of the most valuable of these results were obtained on Mount San Salvatore in Switzerland. However, I (Anderson, 1971a) have observed that current measurements in which magnetic links on 20 m transmission towers in Rhodesia were used appeared to have values generally exceeding those of Berger (1972), and that other parameters, such as the number of strokes in multiple-stroke flashes and their polarity, differed considerably from values reported elsewhere in more temperate climates. Accordingly, it was decided to make further direct measurements in South Africa using a 60 m tower with insulated stays and base and situated on a small hill, no higher than 100 m above the surrounding terrain, to try to simulate a high transmission tower in normal undulating country. Measurements are carried out at the base of the tower, and the scheme presently used by Eriksson (1974) is illustrated in Fig. 7.

In addition to the current measurements which have averaged about 10 per annum in an area having a lightning ground-flash density of 7 km^{-2} per annum, measurements of corona discharge are made, as well as of electric field strength and field-change as practised by Berger (1973). Furthermore, in order to obtain information on the striking distance of lightning (see Chapter 17.3), the tower is under continuous photographic surveillance from two directions approximately at right angles, and also from a television camera; all this equipment is operated by remote VHF radio switching when overhead thunderstorms are imminent.

Preliminary results obtained by means of the television camera show that a great deal of information regarding the progression of lightning strokes may be obtained with relatively cheap and easily operated equipment. The time resolution available on commercial equipment, whilst not sufficient for detailed work such as the measurement of even the leader-stroke velocity, is at least good enough for the determination of the number of multiple strokes, their time intervals to a moderate accuracy, as well as the total duration, and certainly the general characteristics of the discharge as a whole may be determined.

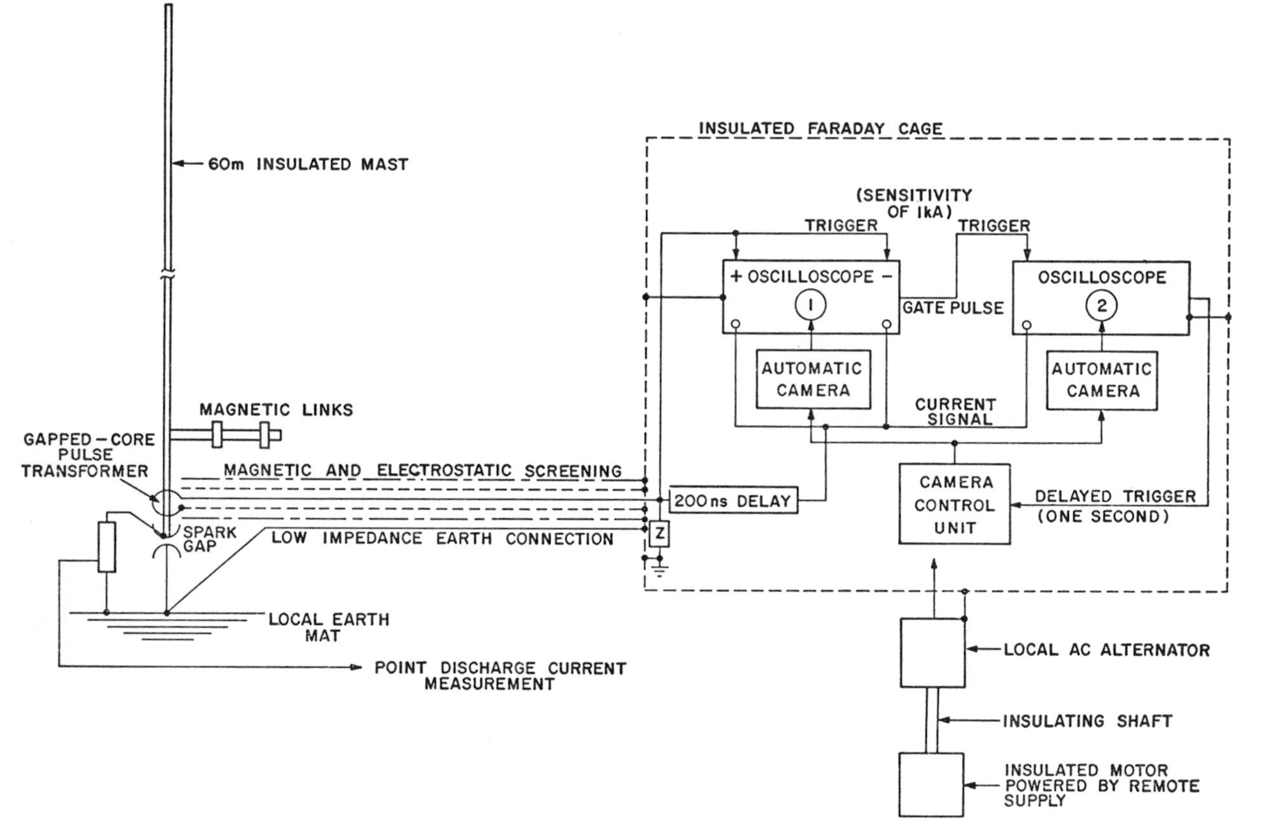

Fig. 7. Instrumentation for the automatic recording of lightning current wave-forms of strokes to a 60 m experimental mast. The gapped-core pulse transformer has a frequency response from 1 Hz to 10 MHz (3 db points), a droop of 0·02% ms^{-1} and a sensitivity (when correctly terminated), of 1·5 V kA^{-1}. Oscilloscope 1 is a dual-beam unit and is used to provide records of the individual stroke wave-forms for both positive and negative currents. The time base normally adopted is 100 μs. Oscilloscope 2 is used in single-beam mode and records the number of strokes and total flash duration. The time-base normally adopted is 1 s.

The medium should be explored to improve its capabilities, particularly in the time domain, but other features, such as luminosity changes, could be investigated as a means of indication of the current flowing in a similar manner to the method employed by Mackerras (1971) using photosensitive diodes.

5. Lightning Flash Density

5.1 Spherics

The subject of spherics from lightning received considerable attention in earlier years from the point of view mainly of interference with radio communication and this culminated in the development by Horner (1960) of a counter which could be used as a measure of the spherics activity arriving at a given point. This counter was designed for a bandwidth of 1 to 50 kHz and a sensitivity level of 3 V m^{-1}, and it uses a vertical aerial. Such an instrument, if calibrated for range, would measure the total lightning activity per unit area at a given location, and was not intended to differentiate between discharges within or between cloud charges, or flashes to earth.

This area of investigation focused attention on the measurement of the spectral frequency distribution of lightning spherics, yielding results such as reported by Hill (1957), see Chapter 10.3.1. Much of this work concentrated on earth flashes which propagated radiation over very long distances in the VLF range, say from 1 to 100 kHz, but little was known regarding the corresponding distribution for inter- or intra-cloud discharges except by the subjective inference that the frequency band should be much lower for the slow discharge and higher, say, in the 100 kHz region for the fast spikes or *K* discharges reported by Kitagawa and Brook (1960).

It was not surprising, therefore, that difficulty has been experienced in designing counters which would discriminate between cloud and earth flashes. One of the first reported efforts in this regard was by Malan (1962) who, from observation, concluded that the ratio of the amplitude of 5 kHz radiation to that at 100 kHz was about 5 : 1 for earth flashes and 1 : 1 for cloud flashes (see Chapter 6.7). Unfortunately, whilst this was obviously true for the general level of radiation, it was not for instantaneous measurements and the counter has accordingly not been further developed.

A similar problem arises in connection with the measurement of the range to lightning flashes by electrical direction-finding (DF) methods (see Chapters 10.5 and 15.5.3), except that other factors such as polarization errors also intrude. These systems were normally designed using crossed-loop antennae operating in the VLF frequency band, but their accuracy was impaired at short ranges (up to 300 km) by changes in polarization errors between night and day as described by Horner (1954), and by the considerable spread in directional errors as a consequence of the large physical extent of the lightning phenomena. Compared with observations using a so-called all-sky camera reported by Kidder (1973), the errors were found to be sometimes in excess of 15° of azimuth for earth flashes, and the direction of cloud flashes could not be reliably measured at all. Recent advances by Krider *et al.* (1976), however, promise to reduce such errors to within a standard deviation of 2°, but the system is still confined to recording earth flashes.

5.2 Lightning Flash Counters

In the development of counters which record a proportion of cloud flashes, the problem of their calibration to find their effective range for both types of flash must be solved, and additionally the ratio of the densities of the respective flashes in any particular area, commonly referred to as the cloud–earth ratio must also be determined. For practical application, the cloud and earth flash densities are required separately, to cover both the question of the probability that aircraft (or other spacecraft) may be struck by lightning, as well as that of damage to objects on the earth.

A lightning flash counter (see also Chapter 15.5.4) was originally proposed by Pierce (1956) and later modified by Golde (1966); it was based upon counting field-changes which exceeded 5 V m^{-1}. This counter was transistorized by Prentice and Mackerras (1969) and finally adopted as standard by CIGRE (Prentice 1972). It was fitted with a six-wire horizontal aerial and had a 3 dB bandwidth of 125 to 2,000 Hz with a maximum response at 500 Hz. Its effective range was determined for Brisbane, using a network of observers, and amounted to 33 km for earth flashes and approximately 20 km for cloud flashes, and this agreed with measurements in South Africa (Anderson *et al.*, 1975) of about 36 and 20 km respectively.

This type of counter has been used extensively in many parts of the world for some 10 to 15 years, but in Europe it is claimed (Frühauf *et al.*, 1966) that the effective range for earth flashes was much less, namely 17 km as against that measured in the two subtropical locations referred to. This question has still to be finally resolved since, quite apart from possible differences in range which may occur in temperate compared with tropical climates, differences in the cloud–earth ratio are also postulated by Pierce (1970); see Chapter 14.4.

Following on the work of Malan, i.e. to differentiate between cloud and earth flashes by using differences in the frequency spectra, the South African group (Anderson *et al.*, 1973) modified the CIGRE counter to a frequency response peaked on 10 kHz with a bandwidth of 2·5 to 40 kHz; at the same time the sensitivity was lowered to provide a threshold level of about 20 V m^{-1} and a vertical aerial of 3 m effective height was used. Since the range of the counter was now much less than that of the CIGRE counter, its response to cloud and earth flashes could be more accurately observed.

Using a network of four radio-switched synchronized direction-finding stations (Anderson *et al.*, 1975) based on all-sky cameras and triggered by the CIGRE counter, it was possible to obtain comparative data regarding the CIGRE 500 Hz counter and what was named the RSA-10 counter, as indicated in Table III, including another version of counter referred to as the RSA 5 C1, which used the identical CIGRE transistorized circuit but with a vertical aerial with capacitance compensation giving a lower sensitivity. The correction factor Y_g is the ratio of earth flashes to total number of flashes registered by a given counter. By identifying the flashes registered by the RSA-10 counter over a period of two years as earth or cloud flashes the values listed in Table III were obtained.

Table III

Comparison of three types of lightning flash counter

Parameter	Unit	CIGRE	RSA 10	RSA 5 C1
Frequency for maximum sensitivity	kHz	0·5	10	0·5
Sensitivity threshold level	V m^{-1}	5	20	35
Effective range to earth flashes R_g	km	36	20	15
Effective range to cloud flashes R_c	km	19	4	11
Correction factor Y_g	—	0·71	0·94	0·57
Apparent cloud–earth ratio N_c/N_g	—	1·5	1·5	1·5

Hence, assuming the total number of flashes registered by any counter is K_t, then the number of earth flashes which occurred is taken to be $K_g = Y_g K_t$ and the number of cloud flashes $K_c = (1 - Y_g) K_t$. Since the effective ranges of the counter to earth and cloud flashes as determined by observation over a long period were R_g and R_c respectively, the respective flash densities are given by $N_g = K_g/\pi R_g^2$ and $N_c = K_c/\pi R_c^2$. The "apparent" ratio of cloud-to-earth flashes is then given by the ratio N_c/N_g.

Whilst the value of N_c/N_g obtained for the RSA-10 counter (1·5) might be regarded as low for a subtropical area such as the Transvaal in South Africa, it is closely dependent on the method of obtaining the cloud flash range, which was limited by the sensitivity of the camera optical system and its triggering level, but which applied equally in the determination of cloud flash range for all counters—hence the use of the term "apparent" in Table III.

In other words, if the camera optical system and triggering level had been more sensitive, many more weak cloud flashes would have been recorded by the cameras which were not registered by the counters. The effective range for cloud flashes for all counters would then have been less than actually measured, and since the correction factor Y_g is not affected, the apparent cloud–earth flash ratio would have been correspondingly greater. The determination of the effective range due to cloud flashes is also hampered by the difficulty of measuring the distance to such flashes which are frequently widely dispersed, sometimes extending for horizontal distances of 50 km or more.

Since all parameters of the short-range RSA-10 counter could be measured with reasonable accuracy, the values obtained for N_c, N_g and N_c/N_g with this counter apply equally to the results of all other types of counter installed in the same measuring station. Since the respective effective ranges of these counters had been determined, the values of the correction factors Y_g could be calculated, and these are shown in Table III. There is an inherent check on the results in that the value of Y_g for any other counter can be calculated independently from the relationship $Y_g = \pi R_g^2 N_g/K_t$. Whence if R_c is known, N_c is determined from the value of K_c and hence N_c/N_g, both of which values should agree with the values obtained for the RSA-10 counter. Any discrepancy found can usually be adjusted by a small variation in the cloud flash range R_c for the counter on the assumption that the accuracy of this particular parameter is not exactly determined.

It may therefore be concluded that the measurement of earth flash density may be carried out using any counter which has been properly calibrated in the area in which it is used. Likewise an approximate value for the cloud flash density may be obtained especially if two counters are used, the one being more responsive to earth flashes and the other to cloud flashes.

The effective ranges should be determined experimentally, but could also be checked by using two identical counters following the method developed by Van Niekerk (1974). This involves installing the counters a distance apart, approximately equal to the supposed effective range, and by means of a telemetering link between them, determining the number of counts registered by the counters simultaneously as a proportion of the total registered. The proportion so obtained over a reasonably long period of observation will be related to the amount of overlap of their effective ranges.

Finally, in view of the growing importance of measuring global differences in the frequency of positive as compared with the more usual negative discharge to earth, the ultimate development of counters should be aimed at differentiating between the two polarities. Preliminary work in South Africa (Anderson *et al.*, 1975) indicates that, whilst there are no technical problems in designing counters which respond either to the one or the other polarity field-change, the field-change itself apparently exhibits dual-polarity characteristics in some flashes. Especially in the case of close discharges, the rapid negative excursion of the field during the approach of a negative leader, for example, could be sufficient to trigger a counter designed for counting positive strokes (i.e. negative field-change) and vice versa (see Chapter 5.2.1.1).

A ratio of seven negative flashes to one positive was measured for at least two years, compared with the about 10 : 1 ratio expected, but during the last year of recording the ratio fell to about 3 : 1, indicating a significant variation. However, as with all counters, the effective range should also be determined for positive flashes, particularly since their discharge characteristics are known to differ (see Chapter 5.3.2), and the question is being pursued.

6. Conclusions

In the literature many differences have been reported regarding the parameters of lightning, and some have been ascribed to differing climatic conditions throughout the world which may well be true. Bitter experience shows that lightning is an extremely variable phenomenon in itself, differing in characteristics such as intensity or polarity from storm to storm and even during a single storm so that, unless each parameter is regarded as having a broad statistical variation, the trends and variability of which must be established over long periods of measurement, the result could be suspect—not of itself, but as representing perhaps too small a sample of the whole phenomenon, often also resulting in unqualified conclusions.

The need exists today, perhaps more so than in previous years, of marshalling all lightning data into comparable geographical or climatological areas, and of making many more measurements than presently exist in order to provide more realistic and statistically significant comparisons. Furthermore, it appears that much more effort is needed to study not only the basic elements individually, but the phenomenon as a whole, relating such factors as lightning intensity with the synoptic situation and the meteorological aspects at the time of observation. It seems that only in this way can a reasonable explanation be found for the reasons for the variability observed, which reasons in themselves may prove to have valuable practical applications.

Acknowledgements

The author is in debt to Dr Golde for his helpful suggestions. Furthermore, thanks are due to the Director of the National Electrical Engineering

Research Institute of the Council for Scientific and Industrial Research of South Africa for permission to disclose information as to its part in the work reported herein, and to his colleagues for their enthusiastic contributions.

References

Anderson, J. W. (1976). An improved field mill for measuring electrostatic field strength. CSIR Special Report ELEK 98, Pretoria, South Africa.

Anderson, R. B. (1971a). A comparison between some lightning parameters measured in Switzerland with those in Southern Africa, a report to CIGRE Study Committee No. 33.01, Working Group on Lightning. CSIR Special Report ELEK 6, Pretoria, South Africa.

Anderson, R. B. (1971b). The lightning discharge, Parts I and II, Ph.D. Thesis, University of Cape Town. CSIR Special Report ELEK 12, Pretoria, South Africa.

Anderson, R. B., Van Niekerk, H. R. and Gertenbach, J. J. (1973). Improved lightning earth flash counter. *Elect. Letters* **9**, 394–395.

Anderson, R. B., Van Niekerk, H. R. and Meal, D. V. (1975). Seventh progress report on the development and testing of lightning flash counters in the Republic of South Africa during 1974/75. CISR Special Report ELEK 71, Pretoria, South Africa.

Berger, K. (1972). Methoden und Resultate der Blitzforschung auf dem Monte San Salvatore bei Lugano in den Jahren 1963–1971. *Bull. Schweiz elektrotech. Ver.* **63**, 1403–1422.

Berger, K. (1973). Oszillographische Messungen des Feldverlaufs in der Nähe des Blitzeinschlages auf dem Monte San Salvatore. *Bull. Schweiz elektrotech. Ver.* **64**, 120–136.

Berger, K., Anderson, R. B. and Kröninger, H. (1975). Parameters of lightning flashes. *Electra* **41**, 23–37.

Bruce, C. E. R. and Golde, R. H. (1941). The lightning discharge. *J. Instn elect. Engrs* **88**, Part II, 487–520.

Chalmers, J. A. (1967). "Atmospheric Electricity", 2nd ed., pp. 138–158. Pergamon Press, Oxford.

Eriksson, A. J. (1974). The measurement of lightning and thunderstorm parameters. CSIR Special Report ELEK 51, Pretoria, South Africa.

Eriksson, A. J. (1975). The measurement of lightning and thunderstorm parameters. Results for the 1974/75 season. CSIR Special Report ELEK 75, Pretoria, South Africa.

Frühauf, G., Amberg, H. and Wurster, W. (1966). Wirkungsweise und Reichweite von Blitzzählern. *Elektrotech. Z. Aus. B.* Suppl. **6**, 1–46.

Golde, R. H. (1966). A lightning flash counter. *Electron. Engr* **38**, 164–166.

Held, G. and Carte, A. E. (1973). Thunderstorms in 1971/72. CSIR Research Report 322.

Hill, E. L. (1957). Electromagnetic radiation from lightning strokes. *J. Franklin Inst.* **263**, 107–119.

Horner, F. (1954). The accuracy of the location of sources of atmospherics by radio direction finding. *Proc. Instn. elect. Engrs* **101**, Part III, 383–390.

Horner, F. (1960). The design and use of instruments for counting local lightning flashes. *Proc. Instn. elect. Engrs* **107** B, 321–330.

Israel, H. (1973). Atmospheric electricity, Part 2, Field, charges, currents. Translation from *Problème kosm. Phys.* **29**, 325–330 (1961). Israel Program for Scientific Translations, Jerusalem.

Kasemir, H. W. (1972). The cylindrical field mill. *Met. Rdsch.* **25**, 33–38.

Kidder, R. E. (1973). The location of lightning flashes at ranges less than 100 km. *J. atmos. terr. Phys.* **35**, 283–290.

Kitagawa, N. and Brook, M. (1960). A comparison of intracloud and cloud to ground lightning discharges. *J. geophys. Res.* **65** (4), 1189–1201.

Krehbiel, P., McCrory, R. and Brook, M. (1974). The determination of lightning charge location from multistation electrostatic field change measurements. Conference on Cloud Physics of the American Meteorological Society, Tucson, Arizona.

Krider, E. P., Noggle, R. C. and Uman, M. A. (1976). A gated, wideband magnetic direction finder for lightning return strokes. *J. appl. Met.* **15**, 301–306.

Le Jay, P. (1926). Les perturbations orageuses du champ électrique et leur propogation à grande distance. *Onde élect.* **5**, 493.

Mackerras, D. (1971). Subtropical lightning. An investigation of the statistic of and occurrence and physical characteristics of lightning in South East Queensland, Ph.D. Thesis, University of Queensland.

Mackerras, D. (1974). Determination of thundercloud charges and related electrical characteristics. CSIR Special Report ELEK 44, Pretoria, South Africa.

Malan, D. J. (1962). Lightning counter for flashes to ground, *in* Conference Proceedings, Gas Discharges and Electricity Supply Industry, pp. 112–122. Butterworth, London.

Malan, D. J. (1963). "Physics of Lightning." English Universities Press, London.

Müller-Hillebrand, D. (1963). Lightning counters I and II. *K. svenska Vetensk.* **4**, 10–11.

Pierce, E. T. (1956). The influence of individual variations in the field change due to lightning discharges upon the design and performance of lightning flash counters. *Archiv Met. Geophys. Bioklim.* Serie A: *Met. Geophys.* **A 9**, 78–86.

Pierce, E. T. (1970). Latitudinal variation of lightning parameters. *J. appl. Met.* **9**, 194.

Prentice, S. A. (1972). CIGRE lightning flash counter. *Electra* **22**, 149–171.

Prentice, S. A. and Mackerras, D. (1969). Recording range of a lightning flash counter. *Proc. Instn. elect. Engrs* **116**, Part II, 294–302.

Scuka, V. (1965). "Measurements of Electric Fields of Thunderstorms." Institute of High Tension Research, Uppsala. Almquist and Wiksell.

Smith, L. G. (1954). An electric field meter with extended frequency range. *Rev. scient. Instrum.* **25**, 510–513.

Van Niekerk, H. R. (1974). Calibration of lightning flash counters. CSIR Special Report ELEK 49, Pretoria, South Africa.

Wilson, C. T. R. (1920). Investigation on lightning discharges and on the electric field of thunderstorms. *Phil. Trans. R. Soc.* **221**, 73–115.

Winn, W. P. and Moore, C. B. (1971). Electric field measurements in thunderclouds using instrumented rockets. *J. geophys. Res.* **76**, No. 21, 5003–5017.

14. Frequency of Lightning Discharges

S. A. PRENTICE

University of Queensland, Brisbane, Australia

1. Introduction

The internationally accepted parameter of lightning incidence is the thunderday (i.e. a calendar day on which thunder is heard at a station), denoted by T. The established method of summarizing thunderday data is the isobront map (lines of equal numbers of annual thunderdays), and from such a map the areas of differing levels of thunderstorm incidence are obvious. The map may be global (e.g. W.M.O., 1956) or regional. Generally speaking, the highest incidence is near the equator and the lowest near the poles. Maps based on thunderdays per month also show the change in seasonal incidence, typically a maximum in the local summer and a minimum in the winter. In the equatorial zone, the variation in incidence is less than elsewhere and there may be two peaks in the annual variation.

The limitations of the thunderday as a basis for comparing lightning activity in different parts of the world are recognized, apart from the difficulties of extraneous acoustic noise preventing the hearing of thunder. "Thunder heard" means that at least one flash has occurred but the same term applies to a series of thunderstorms lasting a whole day with thousands of flashes producing thunder within the range of hearing. Further, this measure does not distinguish between the types of discharge as is necessary in practical applications of lightning data. Thus there has been growing interest in finding an alternative to the thunderday; this is obviously the frequency of occurrence of flashes.

Discharges may be conveniently divided into those which make contact with the ground or an earthed object (earth, or ground, flashes) and those which do not (cloud flashes). The characteristics of each type have been studied over many years and details are given in Chapters 5 and 6. There is, however, little information on the frequency of occurrence of such flashes except in the several regions where thunderstorm research programmes have been set up and appropriate data collected. It has been concluded that in temperate climates the number of ground flashes is about one-third of the total flashes and in tropical climates it is about one-sixth. More detailed information is given in Section 4, the objectives being (a) to summarize data on the frequency of lightning discharges for different regions and climates and (b) to list empirical relationships which will permit estimates of this frequency to be made: for example, several workers have suggested empirical relationships between discharge frequency and T for a particular region. The frequencies observed need to be related to the time incidence and to the surface area over which the flashing is considered to be occurring. This requires that observers define this area and thus give a density value per unit area and unit time.

As discussed in Chapter 10, in communications studies the incidence of the peak activity of any type of flash determines the time at which maximum interference with a received signal occurs; also the distance between the source and the receiver for which such interference affects reception may be thousands of kilometres. In contrast, where personal danger, damage to structures, interference with power system operation, risk of forest fires or risk of damage to electronic equipment near ground level are of concern, only the long-term average of the frequency of occurrence of ground flashes in the immediate vicinity is relevant.

2. Lightning Incidence in Space and Time

2.1 General

To utilize information on the occurrence of lightning flashes, a value which describes the number of events per unit area and time, i.e. the density, is needed. Ideally, to obtain this for each type of flash, an arbitrary area should be agreed on and a record of the occurrence in the area of each type kept for a statistically useful period. These densities will be denoted as N_c, N_g and N_t corresponding respectively to the numbers of cloud flashes, of ground flashes, and of both. Convenient units are $\mathrm{km^{-2}\,s^{-1}}$ or $\mathrm{km^{-2}\,yr^{-1}}$. For regional studies, the area is assumed to be of the order of 1,000 $\mathrm{km^2}$.

It is observed that the flashing rates vary very greatly between regions and, as indicated in Fig. 1, during the life of a thunderstorm. Typically, so does the ratio N_c/N_g. Hence to obtain statistically significant data, observations must be made over a sufficiently long period. As a guide, variation with time in the lightning-caused faults in electricity supply systems in each region indicates the corresponding variation in ground flash density. Thus, in a University of Queensland study, it was found that the probability that a recorded five-year mean fault-rate lies within 50% of the long-term mean was 0·95.

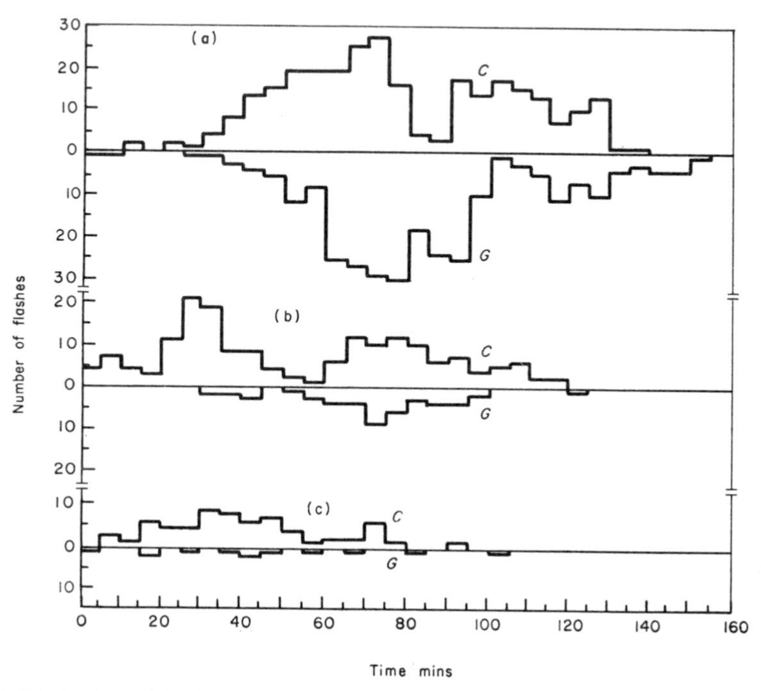

Fig. 1. Variation with time of the number of cloud flashes (C) and ground flashes (G) during temperate climate thunderstorms described as (a) intense; (b) moderate; (c) marginal, over about 1000 km² (adapted from Fuquay and Baugham, 1962).

Reports of extreme values of observed flashing rates during severe thunderstorms are difficult to assess as the circumstances are seldom stated; e.g. Israel (1973) gives 1,000 flashes in 15 minutes and Lane (1966) states that in Pretoria 360 flashes occurred in 3 minutes, and for about an hour 100 flashes per minute were recorded. Shackford (1960) reported a maximum of 1,800 flashes per hour for storms in New England (U.S.A.) within 1 mile of the observer. Presumably a number of cells were active simultaneously and all types of flash were counted. Thomas (1955) reported from the Persian Gulf that on one evening the lightning was so frequent and intense as to appear as one continuous, vivid light.

There appears to be little information on the possible relationship between the types of thunderstorm and the incidence of the different types of flash. From temperate climate studies, Holzer (1953) states that the proportion of ground flashes is higher for frontal than for air mass thunderstorms. He also noted that intra-cloud flashes became relatively more frequent than ground flashes in the latter stages of the life of the cell. Blevins and Marwitz (1968) found that the proportion of ground flashes to total flashes decreased as the total flashing rate increased. Thus the proportion varied from 90% at 0 to 10 flashes per minute to 9% at over 70 per minute. A five-year study in a subtropical climate (Prentice, 1960), in which lightning-caused faults to an electricity distribution system (1,000 km route length) were recorded, showed that frontal and trough thunderstorms together caused about three times the number of faults attributed to air mass thunderstorms (Table I). However, in contrast one air mass thunderstorm produced in 4 hours about half the mean annual number of faults.

Table I

Proportion of thunderstorm types in the Brisbane area, and relationship between thunderstorm type and lightning-caused faults (1954–58)

Data	Thunderstorm type			Totals
	Air mass	Frontal	Trough	
Number of thunderstorms	82	40	90	212
Proportion of each type (%)	39	19	42	100
Number of faults	160	156	194	510[a]
Number of faults per thunderstorm	1·95	3·90	2·15	2·40

[a] Of these, 166 occurred in December 1958.

Global estimates of lightning incidence are few but the following are some examples. Brooks (1925) has given 200 flashes per hour as the most probable mean value for a severe thunderstorm in the temperate zone or for an intermediate thunderstorm in the tropical zone. Later work, reported by Israel (1973), gives an estimate of approximately 60 ground flashes per hour. Brooks (1925) also estimated that about 100 flashes per second occur over the earth's surface. Kolokolov (1971) gave a similar value, 117, noting that 69% were cloud flashes and 31% ground flashes. Loch (1972), from narrow-sector direction-finder studies states that, for the centre of European U.S.S.R., the mean flashing rate (all types) may be taken as $0{\cdot}25\ \mathrm{km^{-2}\,h^{-1}}$. However,

Horner (1964) deduced an average hourly figure of 0·1 discharges km^{-2} during active periods in the tropics, including all types; Horner (1965) gave 10^{-5} km^{-2} s^{-1} as a corresponding estimate for the maximum flashing rate in the main thunderstorm areas. Golde (1966) stated that about 0·15 ground flashes per km^2 and thunderday might be taken as a typical value for temperate regions.

2.2 Methods of Determination of Lightning Incidence

2.2.1 General

The methods described here are applied at ground level or low altitude and are generally applicable only in the vicinity of the thunderstorm. Satellite methods, while seemingly attractive, are at present more appropriate for identifying thunderstorm regions than individual flashes (see Chapter 10.7.3). The first problem is the choice of method of identification and recording of individual flashes.

2.2.2 Visual and Photographic

Experience with visual observations shows that a relatively large number of flashes cannot be classified with confidence. Visibility problems are caused by daylight, precipitation and cloud and building obstructions. A ground flash can often be identified without ambiguity and intercloud flashes can also be classified with confidence when the mainly horizontal discharge channels are clearly visible. Intra-cloud flashes are the commonest classification and yet are the most difficult to identify with certainty as distinct channels are rarely visible. Many appear high enough to be classified as intra-cloud flashes; distant cloud flashes can seldom be distinguished from ground flashes. Broadly, absence of illumination near the horizon supports a decision to classify some of these discharges as intra-cloud. Typically, an observer must guess the identity of many flashes or reject these from his data; if the latter, the apparent ratio N_c/N_g will be lower than the true value.

An estimate of flash density can be made by simple photographic methods but there are important limitations. These are visibility (including lack of contrast), means of determining the distance of the flash from the station and identification of the type of flash. Obviously most studies have required a low ambient light level, and thus only a small proportion of the lightning flashes actually occurring in a given period can be photographed. Studies have been extended to daylight periods by the use of narrow-band interference filters and appropriate film (Salanave and Brook, 1965). A stereoscopic arrangement has been used, e.g. Hagenguth (1947), to enable the distance to be estimated; alternatively radar records and times to thunder, e.g. (Mackerras, 1963), are used. Ground-flash densities derived from photography are included in Table II (p. 470).

2.2.3 *Indirect Methods*

Kreielsheimer and Lodge-Osborn (1971) found that, at frequencies of the order of 10 MHz or more, "gaps" in the noise radiation from lightning follow a different distribution law depending on whether the radiation emanates from a cloud or a ground flash. They claimed that a counter based on this characteristic showed good selectivity to ground flashes.

As discussed in Chapter 13.5.2, several devices have been developed to count the number of flashes occurring within a specified radius. The most widely used is the CIGRE counter (Prentice, 1972). The intention of the originator (Pierce, 1956) was that it should respond selectively to ground flashes; however, a considerable proportion of cloud flashes are counted and hence a correction factor must be applied to the registrations, based on the ratio of N_c/N_g (Section 4). The CCIR counter (Horner, 1960) is designed to respond to both classes. In addition to the gap counter, Gane and Schonland (1948) and Malan (1962) developed counters with separate registers from which the number of each type of flash could be estimated. Many studies of the effects of lightning on power-transmission lines have provided estimates of the number of ground flashes per unit length and time in different regions, e.g. Wagner *et al.* (1942), Golde (1945b), Griscom *et al.* (1965), Popolansky (1970). Some authors (included in Table IV, p. 472) have derived corresponding values of N_g by assuming that the overhead conductors intercept ground flashes which would otherwise strike the ground within a calculated band about the line route. The length of the line and this band width give the "area" used in such derivations. The assumed mechanism of lightning striking an overhead conductor is described in Chapter 22.

Methods based on examination of field-changes require identification of a characteristic feature in recordings of the field-change produced by lightning which will distinguish each type of flash. The absence of impulsive (or sudden) changes in the electric field at a recording station was formerly accepted as evidence that the lightning discharge was a cloud flash rather than a ground flash. This view is oversimplified as sudden changes of field do occur with many cloud flashes (Chapter 6.3) and Mackerras (1968) reported that sudden field-changes do not occur during a small percentage of ground flashes.

Supporting data from luminosity records have also been used to distinguish cloud from ground flashes, e.g. Fuquay and Baughman (1969), Mackerras (1973). With an assumed dipolar charge model, the sign of the field change may be used to identify the discharge as explained in Chapter 6.

2.3 Data on Lightning Incidence

2.3.1 General

The reports of most studies give the number of flashes, classified by type, in one or more local thunderstorms. Peak and average flashing rates are given occasionally but the area of study is seldom defined. It is thought that about one thousand km^2 could be assumed in many cases of visual observations as this is roughly the circular area over which thunder can be heard by an observer at the centre. Where other means of detection are used, the sensitivity of the detector determines the range; in the case of the lightning flash counter the area is determined by the effective ranges of the detecting device to cloud and cloud–ground flashes respectively. Some studies have been restricted to the determination of ground flash density and others to determining the ratio of cloud to cloud–ground flashes in an approximately defined region. As an example of the visual method, Fig. 2 shows the result of several years'

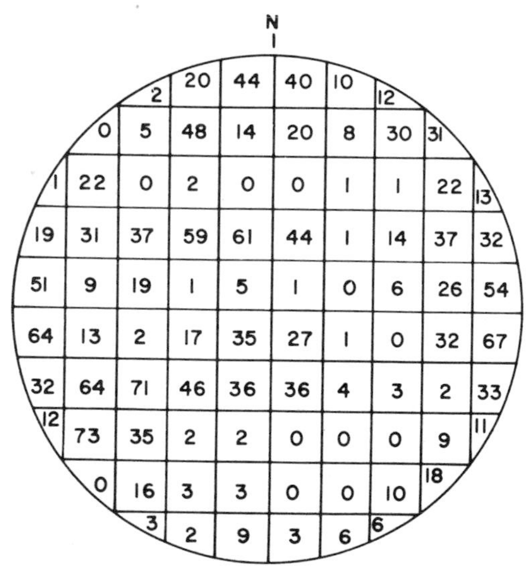

Fig. 2. Areal distribution of ground flashes over a radius of 5 km from observation point (after Kuliev, 1968). Numbers are the observed ground flashes in the areas shown (typically, 1 km^2) from 1960 to 1966.

observation of ground flashes over a radius of 5 km in a mountain area (1,510 m) in the U.S.S.R. (Kuliev, 1968).

2.3.2 *Frequency of Occurrence of Flashes*

From the engineering-design viewpoint, the frequency of occurrence of ground flashes is of the greatest importance in determining lightning risk to structures or equipment above or below the ground. Obviously the same applies to living beings in unprotected situations. This aspect is complicated by the area over which ground currents resulting from the flash may create hazardous conditions (Chapters 16.4.5 and 20.8).

Data from optical and photographic observations in various countries are given in Table II and from recordings, including lightning flash counters, in Table III. In these tables there are large differences in estimates even for the same country. Some differences are likely to be due to errors of method and to short observing periods. Table IV gives data from power-transmission-line performance studies which are typically carried out for several years and hence illustrate well the annual variation in activity. An example is given in Fig. 5 (p. 477). Some of the data are derived from theory and field experience;

Table II

Frequency of occurrence of ground flashes (visual/optical methods)

Country	Regional annual thunder-days	Observations: No. of years	Observations: Radius (km)	Ground flash density ($km^{-2}\,yr^{-1}$)	Reference
Australia	30	10	20	1·2	Mackerras (1976)
East Germany	21	11	—	0·8 (min)	Fritsch (1943)
South Africa	55	—	—	7·7	Schonland (1964)
Switzerland	51	10	2·5	4 (approx.)	Berger (1967)
U.K.	12	17	0·03	2·3	Golde (1945a)
U.K.	—	9	2	0·1	Ashmore (1945)
U.S.A.	33	2	4·8	1·4	Hagenguth (1947)[a]
U.S.A.	11–40	9–19	30	0·4–2·3	Fuquay (private communication)
U.S.S.R.	20	7	5·0	2·8	Kuliev (1968)
West Germany	19	20	—	0·7 (min)	Walter (1933)

[a] With data from Golde (1961).

Table III

Frequency of occurrence of ground flashes (recordings)

Country	Regional annual thunderdays	Method	Observations		Ground-flash density (km^{-2} yr^{-1})	Reference
			No. of years	Radius (km)		
Australia	5–107	C	3–5	30	0·2–3·9	Prentice (1974)
Canada	26	C	5	19	1 (min)	Ellis and Linck (1958)
Central Africa	79	C	6	30	3·4	Jenner (private communication)
Finland	17	C	14	17[a]	1·5	Laitinen (private communication)
Norway	8	C	4	12·5	0·4	Müller-Hillebrand *et al.* (1965)
Singapore	171	C	—	30	12	Horner (private communication)
South Africa	73	C	7	36	6·5	Anderson *et al.* (1975)
Sweden	13	C	5	12·5	0·9	Müller-Hillebrand *et al.* (1965)
Thailand	100	C	—	—	9 (approx.)	Pierce (1968a)
U.K.	14	PD	3	4	1·8	Whipple and Scrase (1936)[b]
U.K.	17	FC	11	4	0·6	Wormell (1939)[b]
U.K.	16	C	—	30	1·0	Horner (private communication)
U.K.	16	C	4	30	0·5	Stringfellow (1974)
U.S.A.	30[c]	FC	2	16	3·3	Trueblood and Sunde (1949)
U.S.A.	50[c]	FC	2	16	7·2	Trueblood and Sunde (1949)
West Germany	15–25	C	3	17·5	1–4·7	VDE (1968)
West Germany	23–35	C	3	17·5	3–5·5	VDE (1968)

C = Counter; PD = Point Discharge; FC = Field Change.

[a] Assumed in reference.

[b] With data from Golde (1945a).

[c] From Golde (1973).

others are adopted values for which the justification appears to be that there is agreement between prediction and performance. Some values are stated to have originated in another country but are considered satisfactory by the reference author. A comparison was made by Prentice (1974) of ground flash density values derived from lightning flash counters and those used in transmission-line protection studies, in Australia. Typically the counters gave less than half the density values used in the transmission-line studies.

Table IV

Frequency of occurrence of ground flashes (power transmission-line studies)

Country	Regional annual thunderdays	No. of years	Ground-flash density (km^{-2} yr^{-1})	Reference
Czechoslovakia	30	6	6·3	Popolansky (1960)[a]
Japan	30	—	4	Owa (1964)
Netherlands	30	—	4–6	Provoost (1970)
Sweden	10	—	1·3	Lundholm (1957)
U.K.	—	10	0·5 (min)	Forrest (1945)
U.K.	—	5	2	Davis (1963)
U.S.A.	39–62	7–8	7·4–12·1	Golde (1945b)
U.S.A.	30	5	6·2	Griscom *et al.* (1965)
U.S.A.	30	—	5·3	Young *et al.* (1963)
West Germany	20–24	4–8	2·3–6·2	Baatz (1951)[a]

[a] With data from Golde (1961).

Regional short-period flashing rates are shown in Fig. 3 as a cumulative frequency distribution from which the extreme and median values during regional thunderstorms can be estimated. One example is from lightning flash counter recordings of 29 thunderstorms during 1969–70 at a subtropical climate station in South-east Queensland. The area studied was about 3,000 km^2. As a number of thunderstorm cells can be assumed to be active simultaneously the flashing rate will be higher than for a single thunderstorm cell. Another example is from observations over a comparable radius at a high-level, temperate-climate station in the northern Rocky Mountains (Fuquay, 1967). These give the distributions of flashes per hour during 31 thunderstorms, with and without hail. More than twice the number of flashes per hour were recorded in the former case.

Data from curve 1 of Fig. 3, combined with duration data, enable the effect of repetition frequency on the accuracy of lightning flash counters to be studied.

Other data on total flashing rates, classified by climate, are shown in Table V. Interpretation is difficult since it is often not clear how many thunderstorm cells may be contributing to the observed or recorded maximum rate. Where the data are for a single cell, these are identified in the table.

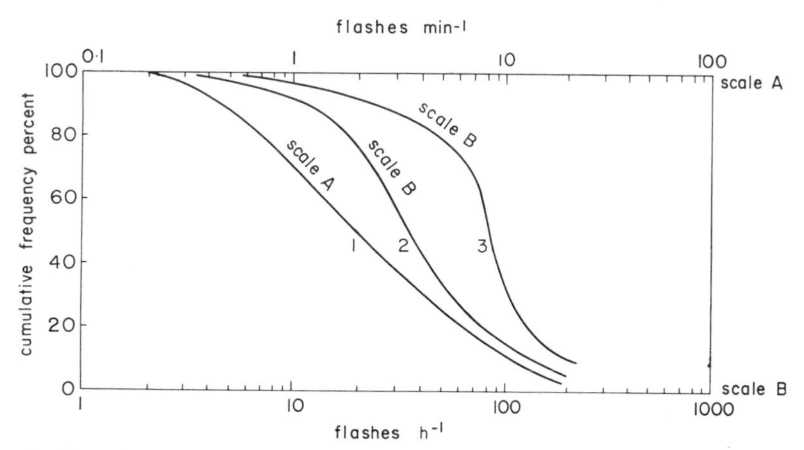

Fig. 3. Cumulative frequency distribution of flashing rates. 1. Flashes min^{-1} (averaged over 5 min), subtropical climate station (29 thunderstorms). 2. Flashes h^{-1}, temperate climate station (17 thunderstorms, without hail) (after Fuquay, 1967). 3. Flashes h^{-1}, temperate climate station (14 thunderstorms with hail) (after Fuquay, 1967).

2.3.3 *Variation of Flashing with Time*

The distribution of lightning flashes with time is an index of the electrical activity of a thunderstorm. It can be shown either as a pattern of incidence of flashing during an individual thunderstorm or as a general pattern over longer periods. Figure 1 (p. 465) shows the observed variation in flashing rate for each type of flash for three classes of thunderstorm in a temperate climate study; Fig. 4 shows the variation during a subtropical thunderstorm, as recorded by a lightning flash counter (see also Chapter 15.3).

The *duration* of the lightning activity is given by such records. It is convenient to identify the commencement and conclusion of such activity by an arbitrary criterion, e.g. in a tropical area, the occurrence of two flashes in 5 minutes (Prentice and Robson, 1968).

The *diurnal variation* in flashing observed from local thunderstorms is similar for land stations, with the maximum flashing rate occurring in the late afternoon and evening. The minimum is some 10 hours earlier. Maxwell *et al.*

Table V
Short period flashing rates (cloud and ground flashes)

Zone	Rate min⁻¹ [a] Average	Maximum	Reference
Temperate		70	Blevins and Marwitz (1968)
	2		Brook and Kitagawa (1960)
		23[b]	Fuquay and Baugham (1969)
	1[c]		Horner (1964)
	3		Norinder and Knudsen (1961)
		47[d]	Russell (1923)
Subtropical	1·6[e]		Eriksson (1974)
		120[f]	Mackerras (1963)
	2[e]		Prentice (refer. Fig. 3)
Tropical	3		Aiya and Sonde (1963)
		8[g]	Aiya (1968)
Global	3[c]	10[c]	Cianos and Pierce (1972)
		50[c]	Dennis (1964)
	5[h]		Sparrow and Ney (1971)

[a] Unless otherwise stated.
[b] Per 5 min.
[c] Single cell.
[d] Lightning flash recorder, 6,924 flashes in 6 hours.
[e] Median value, lightning flash counter records.
[f] Isolated thundercloud.
[g] Median value during peak activity.
[h] Thunderstorm complex, satellite observations. $\lambda = \pm 30°$.

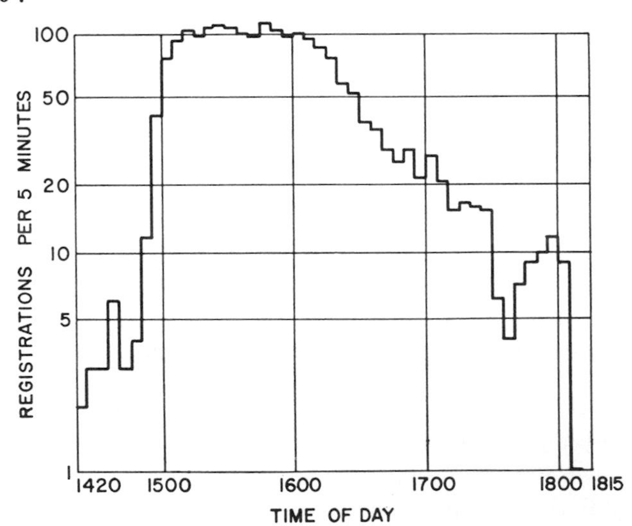

Fig. 4. Variation with time in flashing rate during subtropical thunderstorm.

(1970) compared the mountain and prairie States of U.S.A. with other regions and found that the former have an earlier maximum of activity. They noted that this is consistent with the greater predominance of heat and orographic storms. From observations during tropical thunderstorms, Aiya (1968) noted that peak activity can occur at any hour from 14.00 to 04.00 hours (local time) and that there is a cyclic component in the flashing-rate characteristic arising from the growth and decay of individual cells. From a ten-year study of subtropical thunderstorms in Brisbane, Mackerras (1976) concluded that the maximum activity occurs between 19.00 and 20.00 hours. In the winter months it is earlier and in the summer months later.

The *annual variation* in lightning activity is a local characteristic, dependent on climate. Thus in temperate climates the maximum occurs in mid-summer with occasional activity in the winter; in subtropical climates there is a peak in mid-summer and negligible activity in winter. An example of the latter is given in Table VI (Mackerras, 1976). (The distribution of thunderstorm types is given in Table I, p. 466.)

In tropical climates, a variety of patterns have been reported. Darwin, Australia (lat. 12·5°S) (Prentice and Robson, 1968), has a very similar pattern to that of Brisbane (lat. 27·5°S), Table VI. Aiya (1968) showed that for localities in India below latitude 23°N, thunderstorm activity is at a maximum in the September to November period with a minor peak during March to May. In Malaysia, the most frequent flashing is in the transitional periods between monsoons, i.e. April–May and October–November.

Large *year-to-year variation* in the incidence of lightning is widely reported. Hence estimates of lightning risk should be based on long-period averages. As an example of the studies referred to in Section 2.2.3, the variation over ten years in the occurrence of ground flashes, as indicated by the number of flashes (total 164) to an overhead transmission line, is given in Fig. 5.

The frequency of occurrence of flashes to objects above ground level is discussed in Section 5.

2.4 Discussion

Local information on lightning incidence in space and time is essential if regional activity is to be adequately described; this must await the results of far more extensive field studies. A variety of methods of determination—subjective or objective—are available but a major problem is that the period of observation must be adequate to make the results meaningful statistically. There are obviously very large differences in estimates of flash density for areas with a similar thunderday level. Hence extension to other areas, using an assumed relationship with annual thunderdays, requires caution. In this

Table VI

Annual variation in the occurrence of lightning and rainfall, Brisbane; visual observations (mean values for all recorded thunderstorms, 1958–68)

	July	Aug.	Sep.	Oct.	Nov.	Dec.	Jan.	Feb.	March	April	May	June
Thunderdays per month	0·3	0·4	0·6	2·4	4·4	5·1	2·7	3·2	1·7	0·6	0·3	0·2
Thunderstorms per thunderday	1	1	1	1·12	1·05	1·08	1·15	1·1	1·06	1	1	1
Flashes per thunderstorm	5	36	73	250	267	300	233	120	70	47	71	27
Flashes per month	2	15	44	675	1,235	1,645	725	418	126	28	21	5
Rainfall (mm) per month	49	30	45	77	92	136	142	183	147	77	57	56

connection, it should be noted that the assumptions about the attractive range of an overhead line conductor, referred to in Section 2.2.3, are still being examined (see Chapter 22.3).

However, a direct comparison may be made between the number of lightning-caused faults, N, in a transmission system and the corresponding value of T. Using an exceptionally large amount of data on such faults in the British electricity supply system, Golde (1966a) examined the relationships between the fault frequency per 100 mile-year and the annual thunderdays T. He concluded that N increased more rapidly than T; as an example, for the 33 kV system, $N = 0{\cdot}332T^{1{\cdot}285}$. As N is presumably closely related to the ground-flash density, it is reasonable to assume that this also increases more rapidly than T, in this region. Where estimates of N_g are available, e.g. derived from lightning flash counters, a comparison may be made between N, N_g and T. Prentice (1975) showed that a far closer relationship existed between the fault rate and N_g than between the fault rate and T.

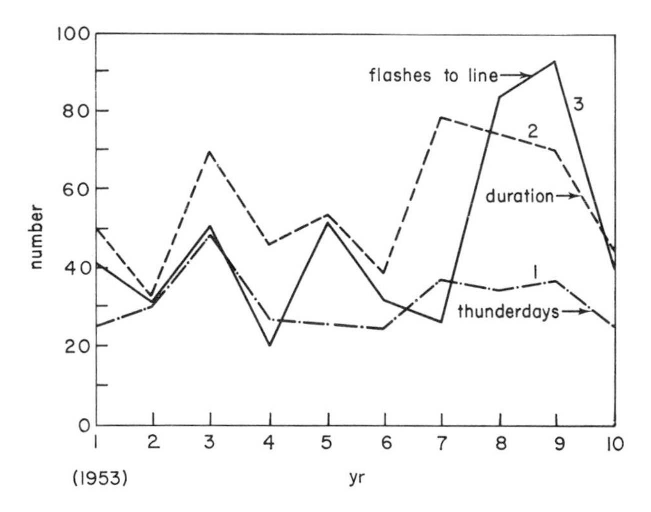

Fig. 5. Variation in thunderstorm activity and number of flashes to a 110 kV transmission line (after Popolansky, 1970). 1. Number of thunderdays. 2. Duration of thunderstorm in hours. 3. Number of flashes to line per 100 km (annual values).

Data on storm duration are of statistical value when planning activities which would be interrupted or otherwise affected by a thunderstorm (see Chapter 15). Also there is a growing preference in some countries for the duration of a thunderstorm in hours over the thunderday as a measure of lightning activity. Figure 5 illustrates the fairly close correlation between lightning incidence and the thunderstorm duration. Thus wider adoption of the latter as a measure of activity may be appropriate until ground-flash densities are generally available.

3. Relationships between Lightning Incidence and Other Parameters

3.1 General

As stated in Section 1, relationships have been proposed between data in a particular region with a view to extending their use to other regions. As an example, Golde (1945b) concluded that a linear relationship existed between ground-flash density N_g and annual thunderdays T. Brooks (1950) considered that this linear relationship applied to both temperate and tropical climates, the latter with a smaller factor of proportionality. Subsequently, from studies with lightning flash counters, the factors of proportionality and the assumption of linearity have been questioned, e.g. Müller-Hillebrand *et al.* (1965). Relationships such as $N_g = aT^b$ derived from temperate climate data suggest surprisingly high ground-flash densities in areas of very high values of T. If true, it seems likely that the consequence, i.e. an exceptionally high fatality rate, would be mentioned in the literature.

In seeking an improved parameter for activity, other studies have examined different correlations and these are discussed in Sections 3.2 and 3.3.

3.2 Lightning Incidence and Thunderstorm Statistics

3.2.1 Comparison of Flash Density and Thunderdays

Possible relationships between actual lightning activity and the accepted measure of relative lightning incidence—the thunderday—have been examined by many workers. The usual comparisons are between flash density and the number of thunderdays on a monthly or annual basis. Empirical relationships between N_g and T and between N_t and T are given in Table VII. No relationship can be singled out as the most reliable. To indicate the variation in estimates, the values of N_g for $T = 10$ range from 0·2 to 1·9; the corresponding values for $T = 30$ are from 1·7 to 5·7.

In addition to the relationships given in Table VII, Pierce (1968b) has suggested the expression $N_{tm} = (aT_m + a^2 T_m^4)^{\frac{1}{2}}$, where $a = 3 \times 10^{-2}$, and Maxwell *et al.* (1970) the expression $N_{tm} = 0{\cdot}06 T_m^{1{\cdot}5}$, where m denotes month.

Table VII

Empirical relationships between lightning flash density and annual thunderdays

Country	Ground–flash density ($km^{-2}\ yr^{-1}$)	Reference
India	$0{\cdot}1T$	Aiya (1968)
Rhodesia	$0{\cdot}14T$	Anderson and Jenner (1954)
Sweden	$0{\cdot}004T^2$ (approx.)	Müller-Hillebrand (1964)
U.K.	aT^b	Stringfellow (1974) [$a = 2{\cdot}6 \pm 0{\cdot}2) \times 10^{-3}$; $b = 1{\cdot}9 \pm 0{\cdot}1$]
U.S.A. (North)	$0{\cdot}11T$	Horn and Ramsey (1951)
U.S.A. (South)	$0{\cdot}17T$	Horn and Ramsey (1951)
U.S.A.	$0{\cdot}1T$	Anderson *et al.* (1968)
U.S.A.	$0{\cdot}15T$	Brown and Whitehead (1969)
U.S.S.R.	$0{\cdot}036T^{1{\cdot}3}$	Kolokolov and Pavlova (1972)
	$0{\cdot}1T^{1{\cdot}3}$*	Kolokolov and Pavlova (1972)
World (temperate climate)	$0{\cdot}19T$	Brooks (1950)
World (temperate climate)	$0{\cdot}15T$	Golde (1966)
World (tropical climate)	$0{\cdot}13T$	Brooks (1950)
World	$0{\cdot}25T$*	Pierce (1966)

* Total flash density.

Experimental work in Nevada by Buset and Price (1975) supports these expressions. A ratio of cloud–ground flashes was assumed. Cianos and Pierce (1972) provided a general guide to the correspondence between thunderdays and total flash density, shown in Table VIII. In their study, values of N_t for some U.S.A. stations were estimated from values of N_{tm} and the expression $N_t = 0{\cdot}02T^{1{\cdot}7}$ is evidently a reasonable fit for the relationship between N_t and T. Total densities are intended to be converted to ground-flash densities by use of the "latitude" formula (Section 4.4).

Table VIII

Relationship between thunderdays and flash density

	Monthly					Annual				
No. of thunderdays	2	5	10	15	20	10	25	50	80	100
Total flash density (km^{-2})	0·2	1	3	6	10	1	4	10	30	50

3.2.2 Comparison of Duration of Thunderstorms and Annual Thunderdays

Several studies comparing the number of lightning-caused faults on transmission lines with T and with the duration of thunderstorms have been made, including that referred to in Section 2.3.3. These have confirmed that the relationship between the number of faults and the duration is the closer. Hence, prediction studies based on thunderstorm-hours are thought to be more reliable than those based on the thunderday. Relationships between these measures have been developed by Popolansky and Laitinen (1972). Kolokolov and Pavlova (1972), from a wide coverage of stations in U.S.S.R. over nine years, obtained for the annual thunderstorm hours, $T_h = 0{\cdot}76T^{1{\cdot}3}$. Combining this with the expression $N_t = 0{\cdot}1T^{1{\cdot}3}$ in Table VII gives a linear relationship between N_t and T_h for these stations.

3.2.3 Comparison of Lightning-flash Counter Registrations and Annual Thunderdays

Data from lightning-flash counter studies are generally given as the number of annual registrations (K) and the corresponding value of T, averaged over a stated period. From data from networks in several countries where daily records were kept, Popolansky (private communication) derived the expression $K = 12{\cdot}74T^{1{\cdot}707}$. Australian data were analysed by Prentice (1974) who obtained a "best-fit" curve, $K = 0{\cdot}023T^2 + 1{\cdot}28T + 23{\cdot}2$ (K in hundreds). Conversion of lightning-flash counter registrations to ground-flash density requires a knowledge of the performance of the counter, i.e. response range and selectivity to ground flashes. Müller-Hillebrand (1964) determined the geographical distribution of ground flashes for southern Sweden for 1958–63, as shown in Fig. 6. Variations in lightning intensity are more appropriately described in this way than by maps of thunderdays.

Prentice (1974) made similar determinations for stations in Australia and Fig. 7 shows the scatter of data for a range of annual thunderdays of approximately 10 to 100. In view of the low density of stations and the great differences in climate, no general relationship between N_g and T is suggested.

3.3 Lightning Incidence and Geographic Latitude

Any relationships between lightning activity and geographic latitude must be consistent with the observed low incidence of thunderstorms at the poles and high incidence in the equatorial belt as shown by the distribution of T

(Brooks, 1925). Brooks noted that the proportion of cloud flashes increases as latitude decreases. As shown in Section 3.2.3, the number of flashes per thunderday tends to increase with T; thus the daily flashing rate increases as latitude decreases. However, latitude can only be a rough indicator of

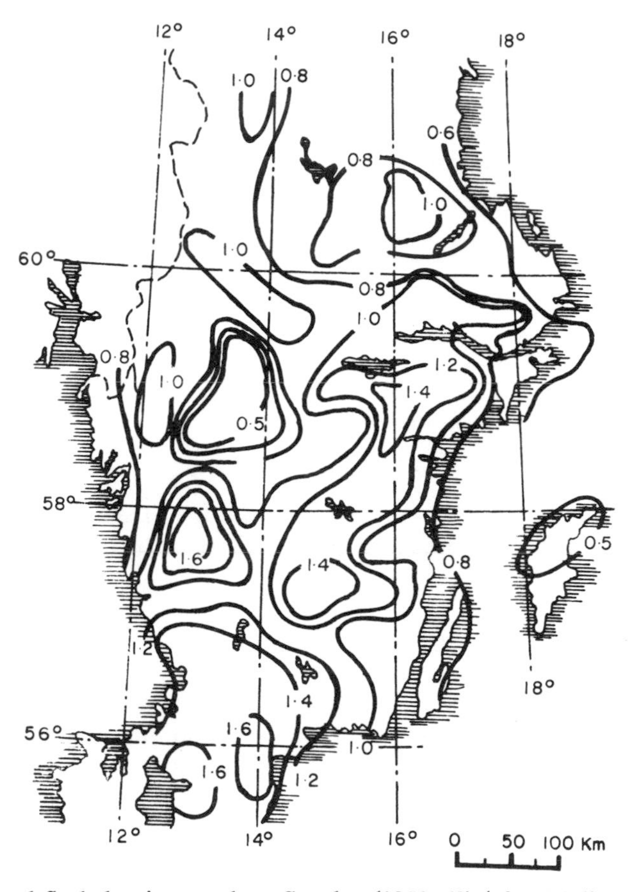

Fig. 6. Ground-flash density, southern Sweden (1958–63) (after Müller-Hillebrand, 1964). Lines on the map join locations having the same average ground-flash density (km^{-2} yr^{-1}).

activity since there are great climatic differences between areas of similar latitude. Also, Popolansky (private communication) has noted from lightning-flash counter studies that for latitudes less than 50°, there is a much higher annual number of flashes in coastal regions than in non-coastal regions.

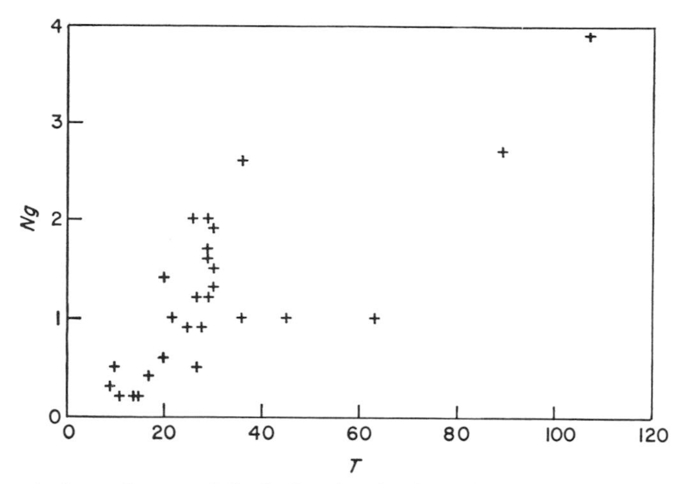

Fig. 7. Variation of ground-flash density (*Ng*) and average number of annual thunderdays (*T*) (Australian stations).

Pierce *et al.* (1962) proposed the relationship

$$N_g = (0{\cdot}1 + 0{\cdot}35 \sin \lambda)\,(0{\cdot}40 \pm 0{\cdot}20)T$$

for the ground-flash density in a region of latitude λ. Later studies, with the parameters λ and T considered separately, are reported in Sections 3.2 and 4.4.

3.4 Lightning Incidence and Precipitation

Thunderstorm precipitation is referred to here because of the observed relationship between frequency of discharges and precipitation (Chapter 3.3.2). Two examples relevant to weather modification studies are given.

From studies in Arizona, U.S.A., Battan (1965) showed that rainfall increases with increase in frequency of cloud–ground lightning from convective clouds. Several years of observations, over about 1,000 km², showed that under "heavy rain" conditions (exceeding 2·5 mm per day) about 56 times more cloud–ground flashes occurred than under "light rain" conditions (less than 0·25 mm per day). Blevins and Marwitz (1968), investigating the incidence of lightning flashes during hailstorms, found that storms having high flashing rates were the most likely to produce hail, and this is consistent with the distributions shown in Fig. 3 (p. 473). Further, they noted a limiting rate of 70 flashes min^{-1} beyond which no hail was reported. It was also noted that large hailstone diameters were associated with low flashing rates and vice versa.

3.5 Discussion

The foregoing shows a wide scatter in data where some reasonable agreement might be expected; there are also conflicts in viewpoints. Hence refinements such as allowance for differences in climate, types of thunderstorm and altitude are not possible with the present limitations in knowledge. Some differences in the relationships shown in Table VII (p. 479) are attributable to the assumed range of magnitudes of lightning current. In some cases, flashes having low magnitude currents (which are least likely to cause damage to structures and engineering systems) may have been excluded. Because the importance of these differences depends on the intended application of the information, the desired accuracy of data should be predetermined.

Probably the relationship of greatest economic importance in protection studies is that between N_g and T, but the commonest assumption, that of linearity, should not be extended beyond the range of values applying to the relevant studies, nor should it be extended to other climates without local supporting evidence. A further difficulty is that the above relationship is statistical only. It is commonly observed in transmission-line studies that a small proportion of thunderstorms produce a large proportion of the ground flashes in the region (Golde, 1966a); in forestry studies, reported by Fuquay (1962), 5% of thunderstorms produced one-third of the ground flashes.

4. Ratio of Cloud Flashes to Ground Flashes

4.1 General

The ratio of cloud flashes to ground flashes, N_c/N_g, is frequently referred to in the literature of thunderstorm phenomena but a search for values which can be used with confidence shows that relatively few stations have made systematic observations and even fewer for more than short periods. Most data on this ratio have been obtained incidentally to general studies of thunderstorms either visually or by analysis of electric field-changes. Other data have come from automatic counting devices triggered by lightning flashes. Specific studies of N_c/N_g have been made as part of research programmes on lightning-caused forest fires, using visual and field-change observations. The very large differences in the values of N_c/N_g between thunderstorms reported in some localities and the large difference during the progress of a single thunderstorm add to the difficulty of obtaining a representative value for this ratio. Differences between different types of thunderstorm are indicated in Section 2.1.

4.2 Fundamental Considerations

It appears that prediction of values of N_c/N_g from fundamental considerations is not yet possible. Various authors have commented on reasons for variation of N_c/N_g with physical conditions such as cloud-base height, dew point, etc. Topography and distance from the sea coast are also factors which could be significant although there is no direct evidence in the literature. Chalmers (1941) stated that the higher the dew point of the air, the more probable is a lightning flash within a cloud in comparison with one to earth, the ratio being roughly proportional to the dew point in °C. Pierce (1970) in discussing variations with latitude in the proportion of lightning flashes which occur to ground, p, and the number of return strokes in a ground flash, n, explained the variations of p and n with λ. He assumed a thundercloud with a net positive charge at a height H, and negative charge centres at a mean height h and separated by an average distance d. The smaller $H-h$ is, relative to h, the greater is the probability of an intra-cloud flash rather than a ground flash; also the chances of the latter involving more than one negative charge centre, that is, of having multiple return strokes, increase according to the relative magnitudes of h and d. Thus the variations of p and n with λ both suggest that h, and therefore also probably that the cloud-base height, is greater in the tropics than in temperate climates.

Brook and Kitagawa (1960) discussed a possible theoretical relationship between the number of intra-cloud flashes and n based on the disposition of regions of charges, following successive cloud flashes, and the subsequent discharge of these regions by the component strokes of the ground flash. Observational evidence is offered in support. They also noted that the number of cloud flashes and ground flashes is not independent; an increase in the former is accompanied by a decrease in the latter.

4.3 Estimates of Ratio

Prentice and Mackerras (1977) gave data on N_c/N_g from many sources. Apart from the expected wide differences in values of N_c/N_g between climatic zones, there are also large differences within any one zone. Some estimates are given in Table IX, as a guide (see also Chapter 13.5.2).

4.4 Empirical Relationships

Several authors have collated observations and developed relationships between N_c/N_g and other variables. Values of N_c/N_g and latitude, and N_c/N_g and annual thunderdays, for particular regions have been used to derive empirical relationships which are a useful guide for areas where no such data are available.

Table IX

Ratio of cloud to cloud–ground flashes

Zone	Ratio N_c/N_g	Reference
Temperate	1·5	Pierce (1955)
Subtropical	3	Mackerras (1976)
Subtropical	4	Horner (1965)
Tropical	9	Aiya and Sonde (1963)
Tropical	6	Horner (1965)
Arid	4	Viemeister (1972)

Using data from 14 stations, Popolansky (private communication) deduced the empirical relationship $O_c/O_g = 0{\cdot}5T^{\frac{1}{2}}$, where O_c and O_g are the *observed* numbers of cloud and ground flashes, respectively. This notation was intended to distinguish between the number of flashes observed during a series of thunderstorms and the actual value of N_c/N_g. Pierce (1968a) gave a relationship with latitude $p = 0{\cdot}1 + 0{\cdot}25 \sin \lambda$. Here p is assumed to equal $N_g/(N_c + N_g)$. With further data, Pierce (1970) modified the relationship to

$$p = 0{\cdot}1\,[1 + (\lambda/30)^2].$$

He also pointed out that, since p is influenced by the type of terrain and type of storm, even with large samples the average value of p may still differ for two adjacent localities with the same λ.

Maxwell *et al.* (1970) in studies of VLF atmospheric noise developed the following expression in terms of latitude and monthly thunderdays

$$p = 0{\cdot}05 + (\sin \lambda + 0{\cdot}05)/(T_m + 3)^{\frac{1}{2}}.$$

Using data from 13 countries, Prentice and Mackerras (1977) developed empirical relations for N_c/N_g versus T and N_c/N_g versus λ. Thus

$$N_c/N_g = 1{\cdot}0 + 0{\cdot}063T \quad (10 \leqslant T \leqslant 84)$$

and

$$N_c/N_g = 4{\cdot}11 + 2{\cdot}11 \cos 3\lambda \quad (0 \leqslant \lambda \leqslant 60°).$$

The range of values over which each equation is intended to apply is shown in brackets. Where local data on N_c/N_g are unavailable, an estimate may be made by averaging the two values corresponding to the values of T and λ respectively.

4.5 Discussion

While an attempt has been made to indicate likely values of N_c/N_g in different regions, it will be evident that data are available from relatively few stations and that the short periods of observation in many cases makes it unwise to assume they are statistically adequate. Additional data on this ratio would assist with the interpretation of data from lightning-flash counting devices whether for electric power systems or communications applications, and with weather modification studies where a possible effect is to influence lightning or thunderstorm occurrence. This must await a programme of observations in representative regions over an adequate period, and the development of appropriate procedures to ensure that results are comparable.

5. Effect of Height of Object and Terrain on Incidence of Ground Flashes

Many factors contribute to the non-uniform areal distribution of ground flashes in a particular region. Some are obvious such as the presence of tall earthed objects; other factors such as topographic features are less evident, e.g. the effect of plains or the channelling of vertical air streams in valleys.

Recognition of the practical importance of height of object and altitude is given in the British Lightning Protection Code (British Standards Institution, 1965). This provides for the assessment of lightning risk by assigning an index figure to particular situations. The greater the risk, the higher the figure. Thus "height of structure above ground" is a component of the risk formula, from which the index figure is obtained, a note being added that structures above 53 m require protection in all cases. Similarly "type of country" is taken into account, varying from flat (at any level) to mountain country above 1,000 m. Thus the height of an earthed object, relative to comparable flat areas in the same region, determines the frequency with which it is struck. For example, there is extensive literature about flashes to, or close to, transmission-line conductors. The occurrences may be detected by instrumentation arranged at intervals along the line or deduced from interruptions to electricity supply. These may vary greatly for different line voltages and constructions in the same region, but a general rule adopted in U.S.A. is that average-height towers (about 30 m) are subject to approximately $2T$ flashes annually per 100 km of route length (see also Chapter 22.6). From European studies, Popolansky (1970) determined that this number is $0{\cdot}06hT$ where h (m) is the average height of the conductor. Both assume linearity between the numbers of flashes and thunderdays and the expressions are in substantial agreement when $h = 30$. The frequency of lightning strokes to isolated structures has been studied by Müller-Hillebrand (1960), Szpor *et al.* (1964) and Popolansky (1964) using magnetic links as a measuring device. The frequency of interception of ground flashes in a particular locality increases as the square of the height of the object, supporting the conclusion of Golde (1961).

Prediction of the probable number of strikes to a tall fixed object requires an assumption about the range of attraction to ground flashes as well as local knowledge of ground-flash density or use of an accepted relationship between this and the annual thunderdays. An example based on the latter and using temperate climate experience is given in Fig. 8, adapted from Müller-Hillebrand (1960). If a more detailed treatment is needed, use may be made of data on monthly and diurnal variation in activity, e.g. Cianos and Pierce (1972).

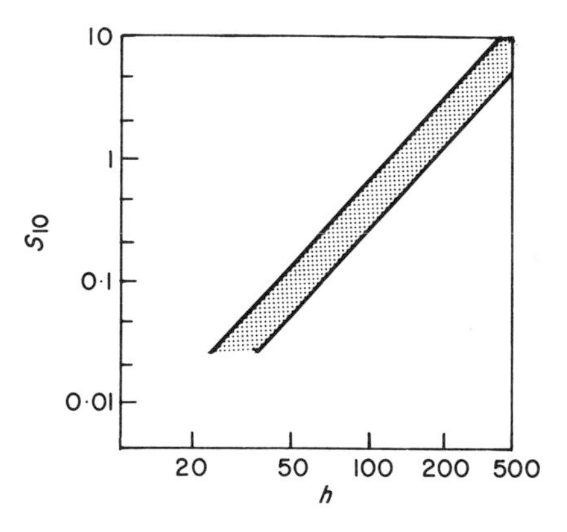

Fig. 8. Frequency of lightning flashes to tall structures (adapted from Müller-Hillebrand, 1960). S_{10}: lightning flashes to structure per year per 10 annual thunderdays. h: height of structure (m).

In addition to the accepted effects of topographical or meteorological influences, lightning nests (small areas of the earth's surface which appear to be struck preferentially) have been reported in some countries and the explanation offered is that there are corresponding local inhomogeneities in the geological formation. This is summed up by Goodlet (1936) thus:

(a) the configuration and nature of the ground favour the formation of a thunderstorm in certain localities;

(b) the lightning flash is guided very considerably by the distribution of space charge of opposite sign below the cloud, produced by point discharge, e.g. from a mast;
(c) the lightning flash tends to fall at places where there is a discontinuity in the soil, e.g. geological faults, outcrops, river banks, underground springs and buried pipes.

Studies in Austria by Fritsch (1963) support (c) and suggest that there are lightning nests in districts of greater geological age. However, Golde (1973) states that observational data have, without exception, been found to be based on wrong conclusions or misapplication of statistical laws. As a further comment on topography, Pierce (1968a) notes that ground flashes are more common in mountainous regions than over flat land. In direct contrast, Kolokolov (1972), from U.S.S.R. studies of convective clouds, states that the ground-flash density (per thunderday) in mountain areas is much less than for plains while the ratio N_c/N_g often increases sharply in the former. These comments are not related to the change in the proportion of ground flashes with geographic latitude, discussed in Section 4.

Berger (1967), in a ten-year study in a mountainous area in southern Switzerland, concludes that ground flashes are distributed at random. Some downward flashes occur in deep valleys between mountains and it appears that the path is guided by the very irregular and variable space-charge distribution. This is influenced by topography and wind direction but probably not by soil conductivity. In the same study, Berger reports that upward strokes occur only on mountain peaks with metal structures (see Chapter 5.3.2). However, Kuliev (1968) considers that the uneven distribution in Fig. 1, while due mainly to meteorological and topographical factors, may be influenced by geophysical factors where strikes occurred far below the observation point. As an indication of the general effect of altitude on the incidence of lightning-caused forest fires in the Rocky Mountains area in U.S.A., Viemeister (1972) states that such fires are believed to be about twice as prevalent between 2,000 to 2,300 m as between 700 to 1,000 m. However, if a mountain peak protrudes into the charge centre of a cloud, there may be fewer ground flashes as the charge can be dissipated by point discharge.

Golde (private communication) has commented on the movement of thunderstorms and the frequency of ground flashes in England where the predominant direction of movement of frontal thunderstorms is from south-west to north-east. In valleys running roughly in the same direction, transmission lines at the bottom of the valley were struck more frequently than lines running along the tops of the hills. Where valleys are at right angles to the direction of cloud movement, lines on the tops of the hills received more flashes than those in the valleys.

The foregoing includes examples of conflicting viewpoints and observational data. To resolve such conflicts, large-scale studies, in which meteorological, topographical and geophysical conditions are adequately described and compared, are needed.

6. Deficiencies in Knowledge

Studies of the incidence and the frequency of occurrence of lightning discharges are an essential part of general thunderstorm research since the mechanisms of charging and discharging are directly related to these. Further, the discharge frequency is a normal parameter for studies of thunderstorm precipitation. The statistical distributions of the many electrical parameters of the lightning discharge, other than the discharge frequency, are given in other chapters. No indication is given here of the relative importance of such parameters in practical applications or of the basis for the selection of appropriate values. Clearly the discharge frequency is a major factor, whether one is considering the simplest form of lightning protection of a small building or the protection of major structures, equipment or systems. This frequency also determines practice in operating and maintaining a variety of services affected by lightning, as examined in several other chapters.

In recent years there have been many advances in our knowledge of thunderstorm discharge phenomena and an increasing number of countries have programmes aimed at providing more detailed information on the frequency of occurrence of lightning. Most data have been obtained in temperate regions, with some from a relatively few stations in subtropical and tropical climates. Conversion of data to flash density is still tentative. The spread of observed values of lightning frequency is surprisingly great even within the same region. One reason is that many published data are the result of observations of relatively few thunderstorms and it is very seldom indeed that long-term values, necessary for practical applications, can be inferred. From considerations of economic and scientific importance, there is the need for widespread recording of this information, over a sufficient period to give the desired accuracy.

Typically, meteorological services do not provide for identification of the type of discharge occurring in thunderstorms, but are chiefly concerned with radar observations of their location for safety reasons, and with estimating precipitation rates. Nevertheless, meteorological organizations in some countries are responsible for lightning-flash counter networks and data from these will build up more realistic maps of lightning intensity than the existing thunderday maps. The chief deficiency is that lightning-flash counters are not sufficiently selective in responding to only one type of flash and, further, their effective range is not established at other than a few research centres. As counters may also be used to record thunderstorm duration, there is the possibility of substitution of the thunderstorm-hour for the thunderday as an interim step in defining the level of lightning activity better.

Because there are still many countries with no reported experience of estimating the frequency of lightning discharges, there is an understandable tendency to use the empirical relationships between parameters established elsewhere. The limitations of this practice are self-evident; corresponding action about precipitation would be regarded as providing almost meaningless data. For example, although a large proportion of the world's population lives in tropical climates, data on flash density for these regions are extremely sparse. Hence there is no adequate basis for the practice of extrapolating from temperate climates to high thunderday regions data stated to connect ground-flash density with annual thunderdays, even linearly. Overconservative and hence uneconomic engineering designs are likely to result from lack of local ground-flash density data.

Clearly, in these circumstances, analyses of records of lightning performance of protected structures and systems should be carried out locally and, if necessary, studies of the lightning environment initiated.

References

Aiya, S. V. C. (1968). Lightning and power systems. *Electrotechnology, Bangalore* **12,** 1–12.

Aiya, S. V. C. and Sonde, B. C. (1963). Spring thunderstorms over Bangalore. *Proc. Inst. elect. electron. Engrs* **51,** 1493–1501.

Anderson, J. G., Fisher, F. A. and Magnusson, E. E. (1968). Calculation of lightning performance of E.H.V. lines, *in* "EHV Transmission Line Reference Book", p. 285. Edison Electric Inst., New York.

Anderson, R. B. and Jenner, R. D. (1954). A summary of eight years of lightning investigation in Southern Rhodesia. *Trans. S. Afr. Inst. elect. Engrs* **45,** 215–294.

Anderson, R. B., van Niekerk, H. R. and Meal, D. V. (1975). Seventh progress report on the development and testing of lightning flash counters in South Africa. Special CSIR Report, ELEK 71, Pretoria, South Africa.

Ashmore, S. E. (1945). Contribution to discussion. *Q. Jl R. met. Soc.* **71,** 89–109.

Baatz, H. (1951). Blitzeinschlag-Messungen in Freileitungen. *Elektrotech. Z. Aus.* A **72,** 191–198.

Battan, L. J. (1965). Some factors governing precipitation and lightning from convective clouds. *J. atmos. Sci.* **22,** 79–84.

Berger, K. (1967). Novel observations on lightning discharges: results of research on Mount San Salvatore. *J. Franklin Inst.* **283,** 478–525.

Blevins, L. L. and Marwitz, J. D. (1968). Visual observations of lightning in some Great Plains hailstorms. *Weather* **23**, 192–194.
British Standards Institution (1965). The protection of structures against lightning. C.P. 326.
Brook, M. and Kitagawa, N. (1960). Some aspects of lightning activity and related meteorological conditions. *J. geophys. Res.* **65**, 1203–1210.
Brooks, C. E. P. (1925). The distribution of thunderstorms over the globe. *Met. Office Geophys. Mem.* **3** (24), 147–164.
Brooks, C. E. P. (1950). "Climate in Everyday Life." Ernest Benn, London.
Brown, G. W. and Whitehead, E. R. (1969). Field and analytical studies of transmission line shielding. *Trans Inst. elect. electron. Engrs* Part III, **88**, 617–626.
Buset, K. and Price K. W. (1975). Lightning flash densities and calculation of strike probabilities to certain vulnerable installations at the Nevada test site. Lightning and Static Electricity Conference, Culham, England.
Chalmers, J. A. (1941). Cloud and earth lightning flashes. *Phil. Mag.* **32**, 77–83.
Cianos, N. and Pierce, E. T. (1972). A ground-lightning environment for engineering usage. Stanford Research Institute, Project 1834.
Davis, R. (1963). Lightning flashovers on the British grid. *Proc. Instn elect. Engrs* **110**, 969–974.
Dennis, A. S. (1964). Final report of Stanford Research Institute, Project 4877.
Ellis, H. M. and Linck, H. (1958). A lightning stroke component counter. *Conf. int. grand Res. Elect.* Report No. 308, 1–11.
Eriksson, A. J. (1974). The measurement of lightning and thunderstorm parameters. CSIR Special Report, ELEK 51.
Forrest, J. S. (1945). Contribution to discussion. *Q. Jl R. met. Soc.* **71**, 89–109.
Fritsch, V. (1943). Mitteilung über die im Blitzversuchsfeld Absroth im Jahre 1941 durchgeführten Arbeiten. *Beitr. Geophys.* **59**, 306–330.
Fritsch, V. (1963). Contribution to discussion. "Problems of Atmospheric and Space Electricity" (S. G. Coroniti, Ed.), p. 435. Elsevier, New York.
Fuquay, D. M. (1962). Project Skyfire. Progress Report to National Science Foundation U.S.A. for 1958–60. Research Paper 71.
Fuquay, D. M. (1967). "Weather Modification and Forest Fires. Ground Level Climatology", pp. 309–325. American Association for the Advancement of Science.
Fuquay, D. M. and Baughman, R. G. (1962). Project Skyfire–Lightning Research. Final report to National Science Foundation, U.S.A. for 1960–61.
Fuquay, D. M. and Baughman, R. G. (1969). Project Skyfire–Lightning Research. Final report to National Science Foundation, U.S.A., for 1965–67.
Gane, P. G. and Schonland, B. F. J. (1948). The ceraunometer. *Weather* **3**, 174–178.
Golde, R. H. (1945a). The frequency of occurrence of lightning flashes to earth. *Q. Jl R. met. Soc.* **71**, 89–109.
Golde, R. H. (1945b). The frequency of occurrence and the distribution of lightning flashes to transmission lines. *Trans Am. Inst. elect. Engrs* **64**, 902–910.
Golde, R. H. (1961). Theoretical aspects of the protection afforded by a lightning conductor. Elect. Res. Ass. Report S/T113, Leatherhead, Surrey.
Golde, R. H. (1966a). Lightning performance of British high-voltage distribution systems. *Proc. Instn elect. Engrs* **113**, 601–610.
Golde, R. H. (1966b) A lightning flash counter. *Electron. Engng* **38**, 164–166.
Golde, R. H. (1973). "Lightning Protection." Edward Arnold, London.

Goodlet, B. L. (1936). Lightning. *J. Instn elect. Engrs* **81**, 1–26.
Griscom, S. B., Caswell, R. W., Graham, R. E., McNutt, H. R., Schlomann, R. H. and Thorton, J. K. (1965). Five-year field investigation of lightning effects on transmission lines. *Trans. Inst. elect. electron. Engrs* **84**, 257–280.
Hagenguth, J. H. (1947). Photographic study of lightning. *Trans. Am. Inst. elect. Engrs* **66**, 577–585.
Holzer, R. E. (1953). Simultaneous measurement of Sferics signals and thunderstorm activity, *in* "Thunderstorm Electricity" (H. R. Byers, Ed.), pp. 267–275. University of Chicago Press.
Horn, F. W. and Ramsey, R. B. (1951). Cable sheath problems and design. *Elect. Engng* **70**, 1070–1075.
Horner, F. (1960). The design and use of instruments for counting local lightning flashes. *Proc. Instn elect. Engrs* **107**, 321–330.
Horner, F. (1964). Radio noise from thunderstorms, *in* "Advances in Radio Research" (J. A. Saxton, Ed.), Vol. 2, pp. 121–204. Academic Press, New York and London.
Horner, F. (1965). Radio noise in space originating in natural terrestrial sources. *Planet. Space Sci.* **13**, 1137–1150.
Israel, H. (1973). "Atmospheric Electricity", Vol. 2. Scientific Publications, Jerusalem.
Kolokolov, V. P. (1971). Distribution characteristics of thunderstorm activity over the globe. Conference Paper, International Conference on Atmospheric Electricity, Moscow.
Kolokolov, V. P. (1972). Relation of electrical activity of a convective cloud to its vertical growth. *Trudy* **277**, Scientific Translation, Jerusalem. (1974), pp. 23–32.
Kolokolov, V. P. and Pavlova, G. P. (1972). Relations between some thunderstorm parameters. *Trudy* 277, Scientific Translation, Jerusalem (1974), pp. 33–35.
Kreielsheimer, K. S. and Lodge-Osborn, D. (1971). New development in lightning-counter design. *Proc. Instn elect. Engrs* **118**, 79–87.
Kuliev, D. A. (1968). The frequency of lightning strikes to earth. *Elekt. Sta. Mosk.* **6**, 61–63.
Lane, F. W. (1966). "The Elements Rage." David and Charles, Newton Abbot, England.
Loch, B. F. (1972). Lightning discharge rates in regions of atmospherics. *Trudy* **277**, Scientific Translation, Jerusalem. (1974), pp. 47–52.
Lundholm, R. (1957). Induced overvoltage surges on transmission lines and their bearing on the lightning performance of medium voltage networks. *Chalmers tek. Högsh. Handl.* **188**, 5–116.
Mackerras, D. (1963). Thunderstorm observations related to lightning flash counter performance. Internal Report UQ/ERB/4, University of Queensland.
Mackerras, D. (1968). A comparison of discharge processes in cloud and ground lightning flashes. *J. geophys. Res.* **73**, 1175–1183.
Mackerras, D. (1973). Photoelectric observations of the light emitted by lightning flashes. *J. atmos. terr. Phys.* **35**, 521–535.
Mackerras, D. (1976). Lightning occurrence in a subtropical area, *in* "Proceedings of the Garmisch Conference" (H. Dolezalek and R. Reiter, Eds). Dietrich Steinkopff Verlag, Darmstadt.
Malan, D. J. (1962). Lightning counter for flashes to ground, *in* "Gas Discharges and the Electricity Supply Industry", pp. 112–122. Butterworth, London.

Maxwell, E. L., Stone, D. L., Croghan, R. D., Ball, L. and Watt, A. D. (1970). Development of a VLF atmospheric noise prediction model. Research Report No. 70–1H2–VLFNO–R1, Westinghouse Georesearch Laboratory, Boulder, Colorado.

Müller-Hillebrand, D. (1960). On the frequency of flashes to high objects. *Tellus* **12** 444–449.

Müller-Hillebrand, D. (1964). Experiments with lightning ground-flash counters. *Elteknik* **4**, 59–68.

Müller-Hillebrand, D., Johansen, O. and Saraoja, E. K. (1965). Lightning-counter measurements in Scandinavia. *Proc. Instn elect. Engrs* **112**, 203–210.

Norinder, H. and Knudsen, E. (1961). Some features of thunderstorm activity. *Ark. Geofys.* **3**, 367–374.

Owa, G. (1964). Study of induced lightning surges and their frequency of occurrence. *Electl Engng Japan* **84**, No. 12, 44–55.

Pierce, E. T. (1955). Electrostatic field-changes due to lightning discharges. *Q. Jl R. met. Soc.* **81**, 211–228.

Pierce, E. T. (1956). The influence of individual variations in the field changes due to lightning discharges upon the performance of lightning flash counters. *Arch. Met. Geophys. Bioklim.* A **9**, 78–86.

Pierce, E. T. (1968a). The counting of lightning flashes. Stanford Res. Inst. Calif. Report 49.

Pierce, E. T. (1968b). A relationship between thunderstorm days and lightning flash density. *Trans. Am. Geophys. Un.* **49**, 686.

Pierce, E. T. (1968c). Lightning, *in* "Encyclopaedia Britannica".

Pierce, E. T. (1970). Latitudinal variation of lightning parameters. *J. appl. Met.* **9**, 194–195.

Pierce, E. T., Arnold, H. R. and Dennis, A. S. (1962). Very-low-frequency atmospherics due to lightning flashes. Final report, SRI Project 3738, Stanford Research Institute.

Popolansky, F. (1960). Measurement of lightning currents on high-voltage lines. *Elektrotech. Obz.* **49**, 117–123.

Popolansky, F. (1964). Study of lightning strokes to high objects in Czechoslovakia. *Elektrotech. Obz.* **52**, 242–246. (English translation, ERA Trans/IB 2291, 1965.)

Popolansky, F. (1970). Measurement of lightning currents in Czechoslovakia. Conf. int. grand Res. Elect. Report No. 33–03, 1–12.

Popolansky, F. and Laitinen, L. (1972). Thunderstorm days, thunderstorm duration and the number of lightning flashes in Czechoslovakia and in Finland. *Studia geophys. geod.* **16**, 103–106.

Prentice, S. A. (1960). Thunderstorms in the Brisbane area. *J. Instn Engrs Aust.* **32**, 33–45.

Prentice, S. A. (1972). The CIGRE Lightning Flash Counter. *Electra* **22**, 149–171.

Prentice, S. A. (1974). The CIGRE Lightning Flash Counter, Australian experience. Conf. int. grand Res. Elect. Report No. 33–04, 1–13.

Prentice, S. A. (1975). Comparison of lightning fault rates and lightning intensity—33 kV transmission system. *Trans. Instn Engrs Aust.* EE **11**, 1–5.

Prentice, S. A. and Mackerras, D. (1977). The ratio of cloud to cloud-ground flashes. *J. appl. Met.* (awaiting publication).

Prentice, S. A. and Robson, M. W. (1968). Lightning intensity studies in the Darwin area. *Trans. Instn Engrs Aust.* EE **4**, 217–226.

Provoost, P. G. (1970). Contribution to discussion. Conf. int. grand Res. Elect., Group 33, II, 5–6.

Russell, S. (1923). The great storm in London. *Met. Mag. Lond.* **58**, 152–153.

Salanave L. E. and Brook, M. (1965). Lightning photography and counting in daylight, using H_{α} emission. *J. geophys. Res.* **70**, 1283–1289.

Schonland, B. F. J. (1964). "The Flight of Thunderbolts." Clarendon Press, Oxford.

Shackford, C. E. (1960). Radar indications of a precipitation–lightning relationship in New England thunderstorms. *J. Met.* **17**, 15–19.

Sparrow, J. G. and Ney, E. P. (1971). Lightning observations by satellite. *Nature* **232**, 540–541.

Stringfellow, M. F. (1974). Lightning incidence in the United Kingdom. Instn elect. Engrs Conf. Pub. No. 108, "Lightning and the Distribution System", pp. 30–40.

Szpor, S., Wasilenko, E., Samula, J., Dyckowski, E., Suchocki, J. and Zaborowski, B. (1964). Results of lightning stroke registrations in Poland. Conf. int. grand Res Elect. Report No. 319.

Thomas, A. C. (1955). Thunderstorm at Sharjab on 14 Nov. 1954. *Met. Mag.* **84**, 355.

Trueblood, H. M. and Sunde, E. D. (1949). Lightning current observations in buried cable. *Bell. System tech. J.* **28**, 278–302.

Verband Deutscher Elektrotechniker (1968). "Blitzschutz und Allgemeine Blitzschutz-Bestimmungen", 8th ed. Verb. Elekt. DH. Verlag, Berlin.

Viemeister, P. E. (1972). "The Lightning Book." M.I.T. Press, Massachusetts, U.S.A.

Wagner, C. F., McCann, G. D. and Lear, C. M. (1942). Shielding of substations. *Trans. Am. Inst. elect. Engrs* **61**, 96–100.

Walter, B. (1933). Über Blitzschutz durch "Fernblitzableiter". *Z. tech. Phys.* **14**, 118–128.

Whipple, F. J. W. and Scrase, F. J. (1936). Point discharge in the electric field of the earth. *Met. Office geophys. Mem.* **68**, 3–20.

World Meteorological Organization (1956). "World Distribution of Thunderstorm Days." Geneva. WMO No. 21.

Wormell, T. W. (1939). The effects of thunderstorms and lightning discharges on the earth's electric field. *Phil. Trans. R. Soc.* A **238**, 249–303.

Young, F. S., Clayton, J. M. and Hileman, A. R. (1963). Shielding of transmission lines. *Trans. Inst. elect. electron. Engrs*, Part III, Special Supplement, 132–154.

Author Index

Numbers in *italic* type indicate pages where References are listed at the end of each Chapter.

L

M

N

T

Subject Index

F

M

W